2nd International Conference on Science and Sustainable Development (ICSSD 2018)

"The Role of Science in Novel Research and Advances in Technology"

IOP Conference Series: Earth and Environmental Science
Volume 173

Ota, Nigeria
14 - 16 May 2018

Editors:

Theophilus Aanuoluwa Adagunodo
Mojisola Rachael Usikalu

ISBN: 978-1-5108-6782-6
ISSN: 1755-1307

Printed from e-media with permission by:

Curran Associates, Inc.
57 Morehouse Lane
Red Hook, NY 12571

Some format issues inherent in the e-media version may also appear in this print version.

Printed by Curran Associates, Inc. (2018)

For permission requests, please contact the Institute of Physics
at the address below.

Institute of Physics
Dirac House, Temple Back
Bristol BS1 6BE UK

Phone: 44 1 17 929 7481
Fax: 44 1 17 920 0979

techtracking@iop.org

Additional copies of this publication are available from:

Curran Associates, Inc.
57 Morehouse Lane
Red Hook, NY 12571 USA
Phone: 845-758-0400
Fax: 845-758-2633
Email: curran@proceedings.com
Web: www.proceedings.com

Table of contents

Volume 173

2nd International Conference on Science and Sustainable Development (ICSSD 2018) "The Role of Science in Novel Research and Advances in Technology"

14–16 May 2018, Center for Research, Innovation and Discovery, Covenant University, Nigeria

Accepted papers received: 25 June 2018
Published online: 17 July 2018

Preface

Papers

Applied Sciences

Environmental Sciences

Geosciences

Renewable Energy

IOP Conf. Series: Earth and Environmental Science **173** (2018) 011001 doi:10.1088/1755-1315/173/1/011001

ICSSD 2018 Preface

The second series of the 2018 International Conference on Science and Sustainable Development (ICSSD 2018), with the theme "The Role of Science in Novel Research and Advances in Technology" was held from 14th to 16th May, 2018, in Center for Research, Innovation and Discovery, Covenant University (CUCRID), Nigeria. The conference was hosted by the Department of Physics, Covenant University, Ota, Ogun State, Nigeria and supported by CUCRID, Covenant University, Petrodata Management Services Limited and some individuals. ICSSD 2018 had series of robust sessions including two (2) keynote speakers (Prof. Paulinus Ekene Ugwuoke and Dr. NANA Ama Browne Klutse), two (2) plenary speakers (Prof. Marvel L. Akinyemi and Prof. Solomon Oranusi), and nine (9) parallel sessions.

Every Innovative society of knowledge emanates from excellent and relevant scientific research results. Science and technology are fundamentally the key drivers to economic advances, improvements in health systems, education and infrastructure. Empirical research results have shown that innovation leads to new and improved products and services, higher productivity, lower prices and new advances in scientific knowledge. Innovation geared towards sustainable development has the potential to lift economic growth, create green jobs, boost social development, while at the same time contributing to environmental protection and conservation. The development of solutions to these key global challenges, and the transition towards 'green societies' will require the mobilization of a wave of creativity and intensive research focused on need.

The Department of Physics is happy to report success in the number of papers submitted and presented in ICSSD 2018, which are greater than the one we recorded in ICSSD 2017 (https://www.rgnpublications.com/journals/index.php/jims/issue/view/62). The full papers were blinded peer reviewed by number of experts and revised by the authors prior to final acceptance. The current issue of *IOP Conference Series: Earth and Environmental Science (EES)* records the proceedings of ICSSD 2018, with six (6) specialized topics. The topics are Applied Sciences, Communication/Atmospheric Physics, Environmental Sciences, Geosciences, Radiation and Health/Medical Physics, and Renewable Energy, which will be useful for government and policy makers.

We are glad to inform you that the next series of ICSSD will be hosted by the same body at the same place. More information will be available at the conference website (http://icssd2018.covenantuniversity.edu.ng/) and communicated to your corresponding e-mails.

Editors:

Theophilus Aanuoluwa Adagunodo, Ph.D
Department of Physics
Covenant University, Nigeria
theophilus.adagunodo@covenantuniversity.edu.ng

Mojisola Rachael Usikalu, Ph.D
Department of Physics
Covenant University, Nigeria
moji.usikalu@covenantuniversity.edu.ng

IOP Conf. Series: Earth and Environmental Science **173** (2018) 011002 doi:10.1088/1755-1315/173/1/011002

Peer review statement

All papers published in this volume of *IOP Conference Series: Earth and Environmental Science* have been peer reviewed through processes administered by the proceedings Editors. Reviews were conducted by expert referees to the professional and scientific standards expected of a proceedings journal published by IOP Publishing.

A Numerical Simulation and Modeling of Poisson Equation for Solar Cell in 2 Dimensions

Akinpelu A[1], Akinojo O.A[2]Usikalu M.R[1], Onumejor C.A[1] and Arijaje T.E[1]

[1]Department of Physics, College of Science & Technology, Covenant University, Ota, Ogun State, Nigeria.

[2]Department of Physics, Faculty of Science, University of Ibadan, Ibadan, Nigeria.

akinwumi.akinpelu@covenantuniversity.edu.ng

Abstract. Solar energy is one of the primary sources of energy replacing fossil fuels due to its abundance. Its versatility and environmental friendliness has made it one of the most promising renewable sources of energy. Solar cells convert solar energy into Electrical Energy. The effort to improve the efficiency of these cells and the reduction of their costs has been a major concern for a long time. Modeling of various structures of solar cells provides an insight into the physics involved in its operation and better understanding of the ways to improve their efficiency. This work modeled Poisson Equation in 2D for an abrupt and linearly graded charge densities system with arbitrary points in space. Linear approximation and differentials, finite difference method, boundary conditions and MATLAB were used to obtain the solution. This is the first step in developing a general purpose semiconductor device simulator that is functional and modular in nature. It was observed that highest electric potential was obtained where the point charge was placed for linearly graded and doping type changed over a small distance compared to the extent of the depletion region for abrupt p-n junction. By solving Poisson equation, voltage, electric field, electric charge density and density of free carriers inside the solar cell can be known.

1. INTRODUCTION

It is no news that there is a looming energy crisis due to the fact that the world reserve of fossil fuels is limited. But renewable energies can potentially replace the energy from fossil fuel and mitigate the crisis. The sun illuminates our planet and supplies it by more than 800 times the global current energy consumption [1]. Light energy from the sun which is a renewable energy can be harnessed by the use of photovoltaic cells also called solar cells to give electricity but with varying degrees of efficiencies. There is a need to optimize the efficiencies of these solar cells prior to fabrication and mass production [2]. This is where device simulation and modeling come in [2]. Solar cell can actually be described by Semiconductor equations [3]. These equations describe the behavior of charge carriers under the influence of an electric field or light in the case of solar cells. One of these equations is the Poisson equation, as this equation relates the voltage gradient of space charge density [4]. Also to provide a tool that can supplement solar cell modeling, semiconductor physics and to construct basic solar cell device equation using MATLAB, Poisson equation needs to be solved numerically, with arbitrary point in space for flexibility purpose. The Poisson equation is a very powerful tool for modeling the behavior of electrostatic systems, but unfortunately can only be solved analytically for very simplified models [5]. As a result of this, numerical simulation must be carried out in order to model the behavior of complex geometries with practical value. Although there are several competing algorithms for achieving this goal, but Finite Difference Method (FDM) is used because of its simplicity [6]. A computational numerical simulation and modeling of Poisson equation for graded charge density in 1D was done by [7] and we will compare it with our results. In this paper, Poisson equation, which describe the potential at each grid point inside the device [8] was modeled in 2D for an abrupt charge density and linearly graded density systems with arbitrary points in space.

2. METHODOLOGY

A semiconductor solar cell is based on a simple p-n junction. The five equations that describe the behaviour of charge carriers in semiconductors under the influence of an electric field and/or light, both of which cause deviations from thermal equilibrium situations [9] can be expressed in equations 1 to 4.

$$\frac{\partial V(x,y)^2}{\partial x^2}+\frac{\partial V(x,y)^2}{\partial y^2}=-\frac{\rho}{\varepsilon}[-p+n-\Sigma p_T+n_T-N_D^+ +N_A^-] \quad (1)$$

$V(x,y)$ is the electrostatic potential, ε is the permittivity of free space, p is the concentration of holes, n is concentration of electrons, ρ is charge density, p_T and n_T are total concentration of holes

and electrons respectively and $N_D^+ andN_A^-$ are donor and electron concentrations

$$J_n = /e/\mu_e n\mathrm{E} + eD_n\nabla_n \quad (2)$$

$$J_p = /e/\mu_p n\mathrm{E} + eD_p\nabla_p \quad (3)$$

μ_e and μ_p are electron and hole mobility, $D_n$and D_p are electron and hole diffusion coefficients. E is electric field, J_n and J_p are electron and hole current density respectively

$$\frac{1}{q}\left(\frac{\partial^2 J_n}{\partial x^2}+\frac{\partial^2 J_n}{\partial y^2}\right)=G_n-R_n \quad (4)$$

$$-\frac{1}{q}\left(\frac{\partial^2 J_p}{\partial x^2}+\frac{\partial^2 J_p}{\partial y^2}\right)=G_p-R_p \quad (5)$$

Where G_n *and* R_n are position dependent volume recombinations and photo generation rates respectively for electrons and G_p*and* R_p for holes, q is electronic charge

But, in this work, Poisson equation was modeled in two dimensions. This is the foundation, in which continuity equations are solved and thereby been able to solve both transport and continuity equations which one can use the combination of these solutions to calculate the solar cell efficiency [9, 10].

2.1. Discretisation of 2D Poisson Equation

Since it is this equation that tells us about electric potential whose unit is joules per coulombs. And Joules is a measure of energy. Computational simulation will be utilized in order to model the behavior of complex geometry with practical values.

$$\nabla^2 V(x,y) = -\frac{\rho(x)}{\varepsilon} \tag{6}$$

Defining a mesh grid of spatial points at which the voltage function will be sampled, if H_x = Vertical Distance, H_y = Horizontal distance and the point that lie on the mesh may be defined by x = iH_x, y= iH_y

$$V(i,j) = V(x,y)$$

$$\rho(i,j) = \rho(x,y)$$

$$\frac{\partial^2 V(x,y)}{\partial x^2} + \frac{\partial^2 V(x,y)}{\partial y^2} = -\frac{\rho}{\varepsilon} \tag{7}$$

Then, applying three point approximation for the second derivative to both part of equation above

$$\frac{\partial^2 V(x,y)}{\partial x^2} = \frac{V_{(x+h,y)} - 2V_{(x,y)} + V_{(x-h,y)}}{h_x^2} \tag{8}$$

$$\frac{\partial^2 V(x,y)}{\partial y^2} = \frac{V_{(x,y+h)} - 2V_{(x,y)} + V_{(x,y-h)}}{h_y^2} \tag{9}$$

Combining the two equations give

$$\frac{\partial^2 V(x,y)}{\partial x^2} + \frac{\partial^2 V(x,y)}{\partial y^2} = \frac{V_{(x+h,y)} - 2V_{(x,y)} + V_{(x-h,y)}}{h_x^2} + \frac{V_{(x,y+h)} - 2V_{(x,y)} + V_{(x,y-h)}}{h_y^2} \tag{10}$$

$$V_{k+1} - 2(1+\lambda)V_k + V_{k-1} + \lambda V_{k+N_x} + \lambda V_{k-N_x} = V_k \tag{11}$$

Where k = (j-1)Nx, $\lambda = \frac{h_x^2}{h_y^2}$

The expression above means V (k) depends only on $\rho(k)$ and voltage at the few nearest neighbors. Expressing the linear relationship between voltage samples into matrix A gives AV=B [2, 10-13], B contains all the information about any charge densities and boundary conditions.

2.2 Boundary Conditions

In order to arrive at a unique solution, some boundary conditions were considered;

I. Newman Boundary condition was applied here. At the sides of the solar cell, $V_{K+1} = V_K$, $V_{K-1} = V_K$, $\frac{\partial v}{\partial x} = 0$ and
II. At the top of the cell, the upper potential would be 2.0 while the lower potential should be zero. (Dirichlet boundary condition)

IOP Conf. Series: Earth and Environmental Science **173** (2018) 012001 doi:10.1088/1755-1315/173/1/012001

We assumed there are 16 points in space, i.e k=16, using eqn 11, the values of electric potential at each point in the cell is written in equation 12a – 12p

$$V_1=V_2-(1+2\lambda)V_1+\lambda V_3=-\frac{\rho}{\varepsilon} \tag{12a}$$

$$V_2=V_3-(1+2\lambda)V_2+\lambda V_4=-\frac{\rho}{\varepsilon} \tag{12b}$$

$$V_3=V_4-2(1+\lambda)V_3+\lambda V_7=-\frac{\rho}{\varepsilon} \tag{12c}$$

$$V_4=-(1+2\lambda)V_4+V_3+\lambda V_8=-\frac{\rho}{\varepsilon} \tag{12d}$$

$$V_5=V_6-(1+2\lambda)V_5+\lambda V_9+\lambda V_1=-\frac{\rho}{\varepsilon} \tag{12e}$$

$$V_6=V7-2(1+\lambda)V_6+V_5+\lambda V_{10}+\lambda V_2=-\frac{\rho}{\varepsilon} \tag{12f}$$

$$V_7=V_8-2(1-\lambda)V_7+V_6+\lambda V_{11}+\lambda V_3=-\frac{\rho}{\varepsilon} \tag{12g}$$

$$V_8=-2(1+\lambda)V_8+V_7+\lambda V_{13}+\lambda V_4=-\frac{\rho}{\varepsilon} \tag{12h}$$

$$V_9=V_{10}-(1+\lambda)V_9+\lambda V_{13}+\lambda V_5=-\frac{\rho}{\varepsilon} \tag{12i}$$

$$V_{10}=V_{10}-2(1+\lambda)V_{10}+V_8+\lambda V_{14}+\lambda V_6=-\frac{\rho}{\varepsilon} \tag{12j}$$

$$V_{11}=V_{10}-2(1-\lambda)V_{11}+V_{12}+\lambda V_{15}+\lambda V_7=-\frac{\rho}{\varepsilon} \tag{12k}$$

$$V_{12}=-(1+2\lambda)V_{12}+V_{11}+\lambda V_{16}+\lambda V_8=-\frac{\rho}{\varepsilon} \tag{12l}$$

$$V_{13}=V_{14}-(1+2\lambda)V_{13}+\lambda V_9+2\lambda=-\frac{\rho}{\varepsilon} \tag{12m}$$

$$V_{14}=V_{15}-2(1+\lambda)V_{14}+V_{13}+\lambda V_{10}+2\lambda=-\frac{\rho}{\varepsilon} \tag{12n}$$

$$V_{15}=V_{16}-2(1+\lambda)V_{15}+V_{14}+\lambda V_{11}+2\lambda=-\frac{\rho}{\varepsilon} \tag{12o}$$

$$V_{16}=-(1+2\lambda)V_{16}+V_{15}+\lambda V_{12}+2\lambda=-\frac{\rho}{\varepsilon} \tag{12p}$$

A MATLAB program was developed to solve this discretized Poisson equation with the above boundary conditions when Finite Difference Method was used. This program calculates the potentials in the problem domain.

3. RESULTS AND DISCUSSION

The results gotten from the simulation with MATLAB are plotted as in Fig. 1 and Fig. 2.

Figure 1 shows the abrupt p-n junction plot, and also the potential at each grid point across the depletion region of the solar cell. The maximum potential is at the edge of depletion region (the yellow part). This potential decreases smoothly reaching zero at midpoint of the region, and change polarity reaches the maximum at the other edge of depletion region. This is the barrier voltage in which the carrier must overcome the barrier in order to cross to the other side. In short, the doping type changes over a small distance compared to the extent of the depletion region, we assumed positive charge is in N region and negative charge in P region.

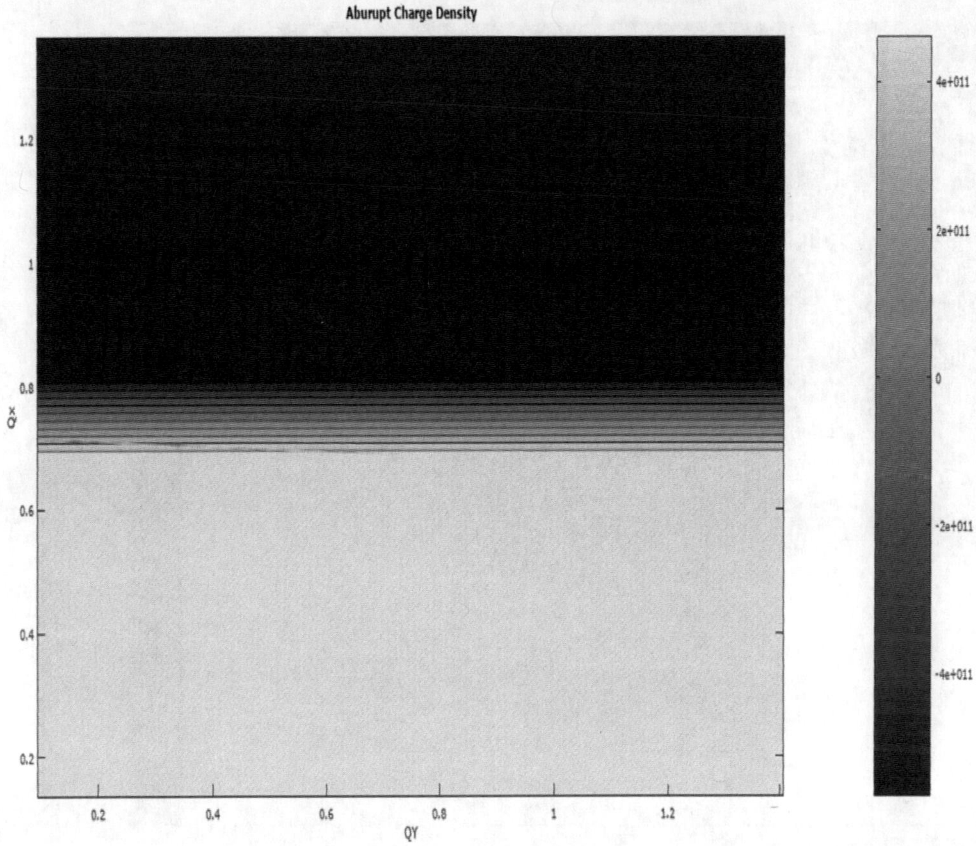

Fig. 1. The abrupt p-n junction plot

Figure 2 is a linearly graded p-n junction which shows the potential curve at each grid point across the depletion region of the solar cell with unequal concentration (a linearly graded). Linearly graded junction has a doping profile, which depends linearly on the distance from the interface.

$$N_d - N_a = ax$$

N_d is density of donors, N_a is density of acceptors.

The depletion layer depends on in built potential this in turn depends on the deletion layer. The maximum potential is at the centre where the point charge was placed and decreases smoothly as it goes farther to the edges of the region. It was also noticed that the system was not affected when the number of grid points were increased, but rather the better it was and the higher the value of electric potential.

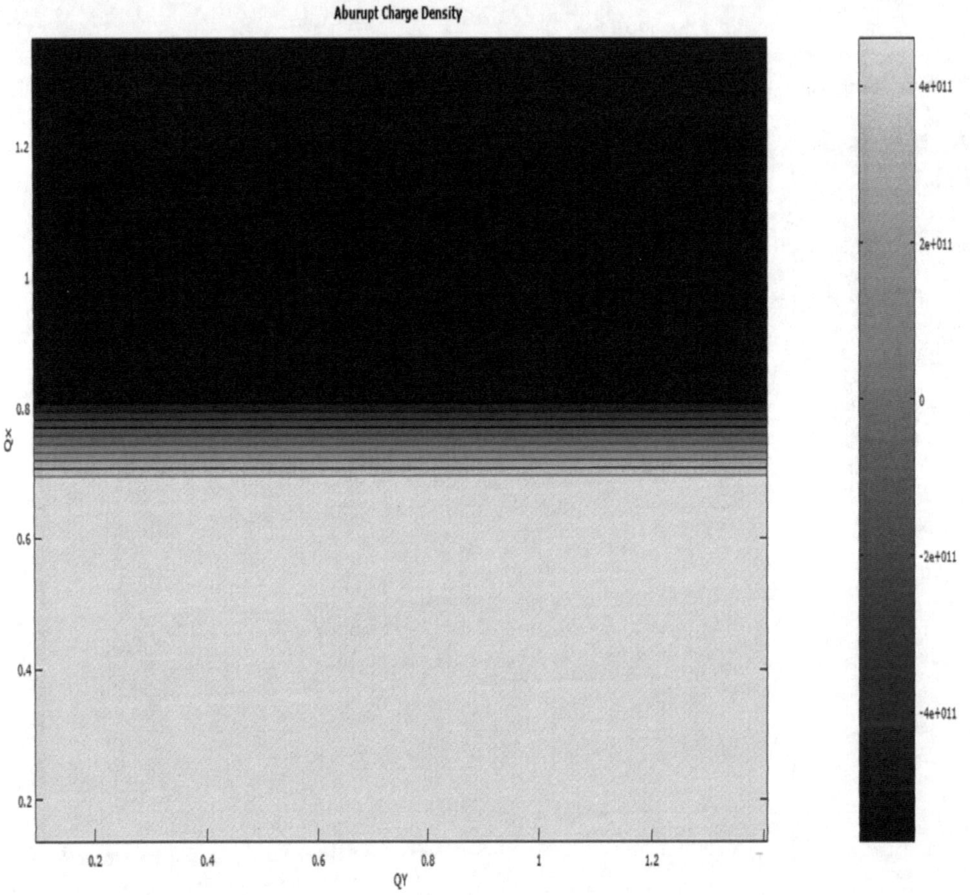

Fig. 2. Linearly Graded Charge Density plot

4.0 CONCLUSION

One of the governing equations of solar cells has been solved numerically. This work has modeled Poisson Equation in 2D for an abrupt and linearly graded charge densities system with arbitrary point in space. A general purpose semiconductor device simulator that is functional and modular in nature has also been developed. Some characteristics can been found, such as Electric potential, space charge density, carrier density and electric field, for the two doping densities. And also a tool that can supplement solar cell modeling, semi conductor physics and construction of basic solar cell device equation using MATLAB tool has been done successfully. However, the study can be extended to 3D to check whether the results will come faster and more accurate.

ACKNOWLEDGMENT
The authors thank Covenant University for conference support

REFERENCES

[1] A. Bahrami and S. Soleimaninezhad, "Photovoltaic cells technology: principles and recent developments," Op. & Quant. Elect. 2013, **(45):** 161-197
[2] C. M. Snowden, "Introduction to semiconductor device modelling," 1985, **(48):** 228-229
[3] H.Ryssel . "Simulations for the design of semiconductor devices," 2014, **(4):** 3-4
[4] A. Khelaf and I. Hamammu, "Computer simulation on solving Poisson's equation for the silicon solar cell" 2013, 1-3
[5] R. J. Nagel, nageljr@ieee.org Department of Electrical and Computer Engineering University of Utah, 2011, lesson note
[6] M. N. Sakidu, "Numerical techniques in electromagnetics" 2nd ed. Boca Raton, Fl: CRCPress, 2001
[7] O. K Akin-Ojo "Numerical analysis and modelling of a solar cell in 1d' 2003 M.Sc. Unpublished
[8] M. Kurata, "Numerical analysis for semiconductor device modelling devices," Heath and Company, 1982
[9] S. M. Sze "Physics of semiconductor devices" John Wiley and sons, New York, 1969. 91-99
[10] C. Kittel, "Introduction to solid state physics" 5th Edition, John Wiley and Sons Inc., 1996
[11] P. B. Zhou, "Numerical analysis of electromagnetic fields. Springer Verlag, 1993.
[12] J. Harold Hovel, "Solar cells for terrestrial applications," 1977, **(19):** 605–615
[13] S. Yıldırım, "Exact and numerical solutions of Poisson equation for electrostatic Potential Problems," 2008, **(8):** 2- 4

Planning materials supply for construction works

Faith Osawaru, Lekan Amusan, Oladipupo Awotinde, Cinwonsoko Akanya, Olatomide Asiyanbola, Faith Akinbo

Department of Building Technology, Covenant University, Ota, Ogun State, Nigeria

faithy012@yahoo.com; lekan.amusan@covenantuniversity.edu.ng; ladiawotinde@yahoo.com; cinwon_akanya@yahoo.com; olatomidefnc@gmail.com; tomiakinbo03@gmail.com

Abstract. This study focused on planning materials supply for construction works. In achieving the objectives, various construction firms and their sites within the area studied was sampled through structured questionnaires. The paper first presented a theoretical background of the study based on the relevance of material planning. A total number of 80 sample size was used for the study; Random sampling technique. Data collected were analyzed using percentage distribution and mean supply index shown in a table to obtain a comprehensive and accurate analysis in the descriptive statistics as applicable. From the results of the analysis of data, conclusion were made and it is recommended that the planning of building materials should be effectively done to avoid delays in construction works caused by untimely ordering and delivery of materials to sites.
Keywords: Planning, Supply, Building Material, Construction Work.

1. Introduction

Planning is a management approach that involves developing a project plan to achieve specified objectives. Planning material supply is very important at the formulation stage of a project. To determine what type of material is required for a particular stage of work, what activity the materials will be required for, how the material will be ordered, stored and utilized, when the materials should be available to site for construction work and who should be responsible for the supply of those materials. This is done so that the work in which the material is needed for can be efficiently delivered at the appropriate time. Building materials are materials used for construction purposes. However, in construction no two projects are similar. The type and quality of construction materials differ from project to project. Although some basic materials like aggregates, wood, cement, steel, water and blocks are common items, others vary with the type of project. For example, most of the finishing materials used in housing projects are not required in industrial project. Materials are planned according to time and usage to avoid delay in transporting and delivery, to avoid delay of execution of work, to prevent improper storage could lead to wastage of materials on site. Materials are sourced according to the contract specifications and construction requirements. Material planning is very important at the feasibility stage of construction, which is significant in materials management, and it has a strong relationship in project planning and control set-up. Both at which this study is to develop a plan for procurement and stocking of construction materials. This is on the direction to determine the type of materials to be used so as to provide the right quantity and quality of materials to site as well as the right prices from the right sources at the right time for work. Planning for materials is a management process that involves materialsidentification, quantities estimation, defining

specifications,forecasting requirements, locating sources for procurement, getting samples of materials approved, designing materials inventory, developing procurement plans and monitoring flow of materials from inception till the construction works are completed (Chitkara 2008).
In the same vein, Olubodun and Kanga (1997) submitted that materials are the most central construction input around which the efficiency and inefficiency of labor, plant and equipment in construction sites revolve. These materials do not just appear on the construction site to achieve the delivery of the construction work but, there is need to plan for materials to be used. This research therefore aims at the importance of planning for materials and to devise effective method of planning for material sup-ply for construction works. The objectives are to determine the various methods of planning for materials, determine the factors affecting the supply of materials to sites and identify the best method of materials supply planning and management. This study is significant because it will help construction and/or project managers to plan and manage materials required for construction work from inception to completion. Also, it will provide insight into criteria that determine effective planning for material supply on sites with a view to guiding against delay in supply and delivery.

2. Literature Review

Planning is said to be a task without end during construction. According to Ogbadu (2003) planning for materials involves budget preparation (what is required and by what means to fulfill the objective of the organization), product research and improvement, value engineering, analysis and standardization of specification. It is pre-pared after careful and extensive research. For a comprehensive material plan, the client and contractor have to: clearly define the materials that will be required for the project in writing, identify all the main materials that are needed at different stages of work, decide budgetary requirement, determine who the suppliers of each material are and how will the materials be made available to sites, what will be the likely duration for the supply of the materials and time for delivery so as to avoid the extension time of project work completion. Management of materials includes material procurement, materials forecasting, recording, plans and monitoring flow of materials from inception to completion. (Chitkara, 2008). This section reviews how to plan for construction materials to avoid delay in delivery of project work. Planning materials for construction works is the key function of material management which helps to develop a plan for procurement and stocking of construction materials so as to provide at site, materials of right quality, in right quantity, at right prices, from the right source and at the right time. Chitkara (2008) said efficient materials management in project environment calls for integrated approach covering numerous functions such as materials purchasing, inventory control, store-keeping and warehousing, materials transportation and handling at site, materials codification and standardization, and the disposal of surpluses. Material management can be done in different ways. Below is a list of various methods of material management.

2.1 Scheduling Method:

Adeweso and Windapo (2008) defined this as the encoding technique used to note the data for material requirements and deliveries. This is the lack of modern planning techniques used in project management by local contractors. In the planning stage, the contractor has to check the availability of materials, quantities, the quality of construction material necessary, the distribution of materials to various parts of the project, purchase of materials, supply and delivery process of materials, specification, the size and dimension of the required materials and communication methods.

2.2 Material Variance Analysis Method

According to Adeweso and Windapo (2008), it is the comparison of definite material usage against the approximation. It is also revenue to track and examine the variability on material cost, quality, size and expenditure.

2.3 *Material recording method*

Material recording comprises of note taking and the use of material record on site. The records include the supplier, receiver, loading and offloading, transit and handling of materials. It is commended that arrangements be made for materials to arrive on time. Once materials are delivered to site, it should be checked for damage, quality, size and requirement (Adeweso and Windapo, 2008).

2.4 *Methods of Planning for Materials*

Planning for materials can be done using different methods. Below are some of the methods of planning for materials for construction works.

i. Demand planning: Prasad (2006) defined this method as a simple technique used. This is based on the historic data of materials. The contractor should derive an approximate consumption of materials for a period of construction and then calculates the quantity of materials needed to be maintained for the work.
ii. Material inventory: Chitkara (2008) defined inventory as the cost of materials in stock at a given time. This stock of materials is held to act as a cushion between supply and demand. It is therefore necessary to hold a planned stock of construction materials at the project site to ensure a timely supply of the expected quantity of materials at the appropriate time for smooth execution of planned construction activities with least inventory.
iii. Material budget planning: Materials budget can be prepared once the requirements are worked out for the activity as follows:

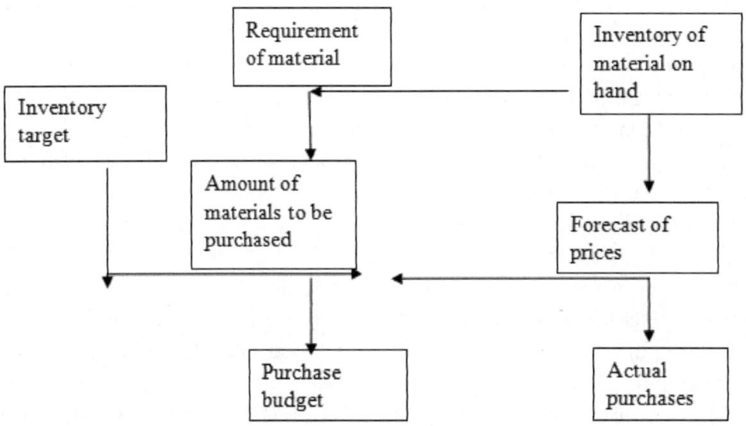

Fig. 1. Material budget planning flow

3. Research Methods

The study used a literature survey and a survey research design to devise effective method of planning for material supply for construction works. Related literature was reviewed to help gather information on the methods used in planning for materials to construction sites. The population of the study was construction professionals in Lagos State, Nigeria. Questionnaire was designed to elicit data from the respondents. A random sampling technique was used to select a sample size of 80 professionals from the frame. Questionnaire was generated for professionals to obtain preliminary information regarding the methods used in planning for materials and how the materials are procured when executing a project from inception to completion. The questionnaire consists of questions that bother on the objectives of the study. The methods of data analysis used in this study were frequency and percentage distribution.

4. Results

Table 1. Results of professional qualification of respondents

Variable Name	No of Respondents	Percentage (%)
MNIOB	18	22.5
ANIA/ARCON	14	17.5
ANIQS	12	15
ANIESV	14	17.5
Others	22	27.5
Total	**80**	**100%**

Table 1 reveals the respondent's level of professions respectively. Eighteen respondents are Builders, 14 are Architect, Twelve are Quantity Surveyors, Twenty-two are Civil Engineers and fourteen are Estate Surveyor.

Table 2. Percentage of years of working experience of the respondents

Variable Name	No of Respondents	Percentage (%)
2-5 years	21	26.25
7-10 years	24	30
11-15 years	16	20
16-20 years	11	13.75
Above 20 years	8	10
Total	**80**	**100%**

The above table shows that twenty-two have 1 to 5 years construction experience, twenty-three have 6 to 10 years, sixteen have 11 to 15 years, eleven have 16 to 20 years and eight has over 20 years construction experience.

Table 3. Results on placing of order for materials

Variable Name	No of Respondents	Percentage (%)
1 week	33	41
2 weeks	31	39
4 weeks	13	16
Others	3	4
Total	**80**	**100%**

Table 3 result identifies placing of materials order; where, 41% of construction sites chose 1 week, 39% of the construction sites chose 2 weeks, 16% of the construction sites chose 4 weeks, while 4% of the construction sites chose "others" which ordering is being done for 2 days and 4days before its requirement date, depending on when the materials are needed to be used.

Table 4. Result on preferred market for procuring of materials to site

Variable Name	No of Respondents	Percentage (%)
Local market	12	15
Foreign/Importation	10	12.5
Through suppliers	36	45
Materials merchants	12	15
Directly from manufacturer	10	12.5
Total	**80**	**100%**

The above Table shows the response on the preferred market for procurement of materials, 15% of construction sites prefer of materials from the local market, 12.5% prefer importation of materials, 45% of construction sites prefer their materials from suppliers, 15% prefer their materials from materials merchants, while 12.5% of construction sites prefer materials directly from the manufacturers.

Table 5. Percentage composition of the methods used in planning for materials

Variable Name	No of Respondents	Percentage (%)
Demand planning/re-order point planning	8	10
Material requirement planning	39	49
Budgeting for materials	16	20
Production planning	8	10
Materials inventory planning	9	11
Total	**80**	**100%**

The table above shows the result of the methods used for planning for materials, 10% of construction sites use demand planning for their materials, 49% of construction sites use materials requirement planning, 20% use budgeting planning for their construction materials on site, 10% chose production planning, while 11% of construction sites use materials inventory planning for their materials. Most construction sites in Nigeria practice the method of materials requirement planning when it comes to the aspect of planning for their materials on site. This is a good practice method to be adopted (Joshua et al. 2018).

Table 6. Percentage of material management methods used on site

Variable Name	No of Respondents	Percentage (%)
Scheduling method	46	57.5
Material variance analysis	0	0
Material recording method	34	42.5
Others	0	0
Total	**80**	**100%**

Table 6 shows the method used for material management, 57.5% of the construction sites use scheduling method for managing their materials on site, 42.5% of the construction sites use material recording method, while material variance analysis method of material management is not used by any construction site which is 0% and "others" show 0%. It was observed that most construction sites in Lagos State and Ota use two methods in the management of their materials on site (Amusan et al. 2017).

Table 7. Result on impact of delays to contract on site

Variable Name	No of Respondents	Percentage (%)
Contract duration is extended	17	21.3
Disputes with client arises	12	15
Extra expenses may be incurred	20	25
Progress of work is slowed	20	25
financial loss may be incurred	11	13.7
Total	**80**	**100%**

On the impact of delays to the contract been experienced on site, 21.3% of the construction sites indicated that contract duration is extended due to material delivery, 15% of the construction sites indicated that disputes with client arises due to material delivery, 25% of the construction sites indicated that extra expenses may be incurred due to material delivery, 25% of the construction sites also indicated that the progress of their work is slowed down due to material delivery and 13.7% of the construction sites indicated that financial loss may be incurred. This is an indication that most construction sites have impact of delays due to material delivery which is caused by the material suppliers (Amusan et al. 2018).

5. Findings

The findings obtained from this study can be said to be as follows:

i. The material requirement method used in planning for materials is averagely practiced but needs to be improved to reduce frequent occurrence of delays.
ii. 75% of construction sites encounter delays due to untimely delivery of some construction materials. This result is found in the research analysis.
iii. This indicates that suppliers are mostly responsible for the procurement and delivery of materials to construction sites.
iv. Most construction site studied; practice the Schedule receipt (placing order on the supplier) in the ordering stage of materials.
v. 80% of construction sites place order for materials between one and two weeks of the requirement date as shown in the result analysis. This is a wrong practice because late ordering for materials may lead to delays by the suppliers.
vi. In the management of materials, scheduling method is mostly used. Where, approximately 58% of the construction sites use scheduling and 43% use the material recording method.

6. Conclusion and Recommendation

Due to the findings of this study, there are some facts gotten that should be of improvement to construction industry on planning for the supply of building materials. These include:

i. The prompt ordering of materials should be done to ensure that suppliers de-liver the materials when required for construction work.
ii. Delivery of materials can be improved by sourcing them in the location where it can be made available to reduce impacts of delay in transportation/ traffic congestion.
iii. It is also recommended that Material Requirement Planning should be adopted because, it is the appropriate method of planning for materials which information includes: when to place order, how much to order, who to order from and when the items need to be on hand.

References

[1] Adeweso, O. S., and A. O. Windapo. "Relationship between material management approach and scheduled project completion time." In Construction and Building Research of the Royal Institution of Chartered Surveyors Conference. 2008.

[2] AmusanLekan.,Jenom, A.,Opeyemi, J.,Ayodeji, O.,Ignatius, O. "Innovative green approach informatics to managing thermal sources and its effect in residential buildings in tropical climate." International Journal of Civil Engineering and Technology, 2017

[3] AmusanLekan, M, OlonijuLeke, I.,Mariam, A.Adebisi, M.,Farayola Hezekiah,Osawaru, F.(2018) Adopting information and communication technology in construction industry International Journal of Mechanical Engineering and Technology, 2018.

[4] Chitkara, K. Construction Project Management. 2nd Edition. Great Britain: McGraw Hill Publisher, 2008

IOP Conf. Series: Earth and Environmental Science **173** (2018) 012002 doi:10.1088/1755-1315/173/1/012002

[5] Joshua, Opeyemi, Kolapo O. Olusola, Kehinde D. Oyeyemi, Ayodeji O. Ogunde, Lekan M. Amusan, David O. Nduka, and Joyce Abuka-Joshua. "Data of the properties of rebar steel brands in Lagos, Nigerian market used in reinforced concrete applications." *Data in Brief* (2018).

[6] Ogbadu, E.E. Introduction to Business, Precious Treasures Ltd. 2003.

[7] Olubodun, F. and Kwanga J. "A Strategy for Effective Control on House Building Sites," A paper presented in the construction and building research conference, 1997.

AREA TARGETING OF HEAT EXCHANGER NETWORK (HEN) USING A MODIFIED PINCH TECHNIQUE

Onyemachi Jerry Okechukwu[1] Oluwatosin Sarafa Azeez[*2] [1] Nile University of Nigeria Abuja
[2] Federal University of Technology Minna [1]shiek2k2@yahoo.com [2] tosin.azeez@futminna.edu.ng
(* = corresponding author)

ABSTRACT

Aspen Hysys version 8.6 which is a modified pinch tool was used to minimize the area of two heat exchanger networks problem and the results obtained were compared with those of other methods used by researchers that have solved the same problems. Stream data adopted from previous research work comprising the inlet and outlet temperatures, heat capacity flow rate, and specific heat capacity were imputed and solved for minimum area using the principle of modified pinch technology embedded in the Aspen Hysys software.The minimum approach temperature, composite curve, grand composite curve and the grid diagram representation are all series of steps employed in the methodology to achieve minimal area targeting using the Aspen Hysys. The problems solved in this research compared well with those of other researchers and even obtained a smaller exchanger area when compared with the non-linear programming technique (NLP). For instance, an area of 22.17m^{2}was obtained in the first example as against 29.84m^2 obtained by the non-linear programming techniques, resulting in a percentage difference of 25.7%. Also in the second example solved in this research, an area of 106.4m^2 was obtained against 188.9m^{2}obtained with the NLP technique which gives a percentage difference of 43.67%.The overall assessment of the results on area targeting on these problems using Aspen Hysys points to the fact that the traditional pinch technique can still obtain results that are as good as those obtained using mathematical programming techniques in heat exchanger networks (HENs).

Keywords: heat exchanger, area, Aspen, Temperature, stream

1.0INTRODUCTION

Process Integration has been used as a holistic approach to process design and optimization, which exploits the interactions between different units in order to employ resources effectively and minimize costs and other variables [1].With the increasing demand for energy in the chemical industries and its apparent cost, the heat exchanger plays a role in bridging the gap between energy demand and cost. The heat exchanger which is basically equipment used in heating up or cooling down streams with the energy exchanged with other streams [2], while the grouping of such interconnected heat exchangers in a plant of is regarded as the heat exchanger network [3, 9]. Pinch technology and Mathematical Programming techniques have both been used for the optimization of heat exchanger networks synthesis (HENS) [3, 4, 5]. The heat exchanger network as a means for energy integration is due to its ability to minimize the utilities demand of the plant and as well as other variables like the cost and area.

IOP Conf. Series: Earth and Environmental Science **173** (2018) 012003 doi:10.1088/1755-1315/173/1/012003

Several techniques in Mathematical Programming have been used to optimize utilities, exchanger area and cost in HENS [3, 9, 12, and 13]. Pinch analysis which uses thermodynamics concept and heuristics as can be seen in [6, 10]resulted in energy integration and minimal utility demand. The principle of pinch technology has been incorporated in software like the Aspen Hysys which is used in process and energy integration. Several variables define a heat exchanger network, top among which are capital cost, annual cost and area of the network. The area of a heat exchanger network is the total space occupied by all the equipment that makes up the network.

1.1 PINCH TECHNOLOGY

The term pinch technology was introduced to represent a new set of thermodynamically based methods that guaranteed minimum energy level in the design of heat exchanger networks. Pinch technology has over the years emerged as the principal technique in process design and energy conservation [10, 11, 15]. Pinch technology software such as Aspen Pinch hasproven to be a valuable resource in the pinch analysis of complex industrial processes [9].In the laws of thermodynamics, the direction of energy flow must always be from the hot stream to cold stream, hence temperature cross over is prohibited. The cooling and heating of streams in a heat exchanger is governed by the minimum approach temperature ($\Delta Tmin$) that exists between the heat exchanger. This minimum approach temperature is the minimum temperature difference between the streams profile in the exchanger [5]. The temperature level at which the minimum temperature approach is observed is called the pinch point and it defines the minimum driving force obtainable in a particular heat exchanger.Integration of heat exchanger network where the area of the network was optimized was extensively researched by [13, 21] with relevant problem solving examples. The non-linear programming technique was used by [13] and was used to solve some problems to obtain the area target. The area targeting non-linear programming technique was used by [13] and the results were compared with those of previous researchers.

1.2 ASPEN HYSYS SOFTWARE

Aspen HYSYS is interpreted to be a comprehensive process modeling tool used by the world's leading oil and gas producers, refineries, and engineering companies for process simulation and process optimization in design and operations [9]. It is user friendly application software that works with the principle of pinch analysis that basically requires the input of data and relatively easy operational skills by users to achieve the desired results. Aspen Hysys can also be used for steady state and dynamic process modeling and simulation like plant design, process design, plant optimization and process optimization. This work uses the Aspen Hysys software to minimize the area of the heat exchanger network, and the result obtained was then compared to that obtained by other researchers that have used different techniques on same problem.

2.0. MATERIALS AND METHODS

The materials used include, the Aspen Hysys software version 8.6, Microsoft Office Visio, process flow diagram and data of three selected heat exchanger problems adopted from [13].

2.1 METHODOLOGY

The Aspen Hysys version 8.6 was used to optimize the area of the heat exchanger synthesis problem discussed in this work. Process data were collected from literature, as in this case heat exchanger analysis problem was adopted from [13]. The process data were extracted properly from the process flow diagram using the required technique for data extraction to form a stream table (heat exchanger data), and consequently the construction of the composite curve, grand composite curve, estimation of ΔTmin, grid diagram representation, and finally evaluating and analyzing the targeted result obtained with those reported by previous researchers on existing plant condition. All thiswere achieved using the Aspen Hysys version 8. The methodology adopted can be summarized as follows: Collection of data, formulation of stream tables, data extraction, construction of composite curves, grand composite curves. It should be noted that some mathematically based materials are embedded in Hysys.

2.2DATA EXTRACTION

This involves the extraction of relevant stream and cost data from the process flow diagram as presented in [13]. The data extracted include the inlet and outlet temperatures, the heat capacities flow rate, heat transfer coefficients, and the specific heat capacities of all streams in the process including that of the utility streams. The general equation for quantification of heat available in a stream is stated in Equation 1.

$Q = Cp \times F \times \Delta T$..............Equation 1
F= mass flow rate of the stream (Kg/s)
ΔT= temperature difference (^{0}C)
Q = heat duty (Kw)
Cp = specific heat capacity of the stream

2.3DETERMINATION (OPTIMIZATION) OF ΔTmin

ΔTmin which is the minimum allowable temperature difference between two streams exiting a heat exchanger [14 – 17] is very important in minimization of utility usage. This is automatically determined by the Aspen Hysys software when the streams and utility load have been imputed. The ΔTmin value for petroleum related heat exchanger problems falls in the range of 20^{0}C to 40^{0}C as observed by various researchers. The value of ΔTmin is important in the determination of heat exchanger area. This can be optimized in the plot of composite curves to achieve minimum energy usage [5, 19, and 21]. The use of composite curves for mass exchange can be seen in [18, 20].

IOP Conf. Series: Earth and Environmental Science **173** (2018) 012003 doi:10.1088/1755-1315/173/1/012003

2.4 GRAND COMPOSITE CURVE

This is also known as the utility composite curve, it is a curve of the temperature and enthalpy profile of all streams in the network with the inclusion of the hot and cold utility provision. The grand composite curve also indicates the type and amount of utility to be used to satisfy the network.

2.5 NETWORK REPRESENTATION ON A GRID DIAGRAM

The grid diagram is a representation of all streams present in the network. In the Aspen Hysys used for this research, the hot streams are drawn at the top with horizontal lines running from left to right while the cold streams are drawn by horizontal line running from right to left at the bottom of the diagram. Each line of hot and cold stream has its respective supply and target temperatures at the end of the line, as well as its heat capacity value. Heat exchangers are represented on the diagram by vertical lines running through the hot and cold streams [3, 5, and 9]. Heat exchangers that violate the pinch rule are also identified by a dash line at the top of the heat exchanger.The streams are match based on their various specific heat capacities value (heat content), hot streams with low specific heat capacity value are matched with cold streams with high specific heat capacity value. In situation where process to process heat transfer based on their specific heat capacities is inadequate to satisfy all the streams, utilities (hot or cold) are used to satisfy the remaining streams.

3.0RESULTS AND DISCUSSION

Two heat exchanger analysis problems were solved in this work, which focused on the minimization of heat exchanger area (area targeting).These two heat exchanger analysis problems were adopted from [13]work on heat exchangers network and solved for minimal area using Aspen Hysys.

3.1 EXAMPLE ONE

This problem is an area targeting problem obtained from the [13] heat exchanger network analysis work. It involved five streams with different heat transfer coefficients of which one among the streams is a cold stream that required heating and the other four are hot streams that required cooling. The film heat transfer coefficient of all streams in the problem, range over two orders of magnitude, which is common with streams containing boiling or condensing streams. The value of ΔTmin chosen for this problem is that which favors utility targeting (ΔTmin =10K), hence the heating and cooling requirements for the set of data is zero. The stream data for this example is shown in Table 1.

Table1 - Stream Data for Example 1 [13]

Stream	Supply temperature (K)	Target temperature (K)	Heat capacity Flow rate (kWK^{-1})	Film heat transfer Coefficient ($kWM^{-2}K^{-1}$)
H1	443	293	0.5	2.0000
H2	416	393	2.0	0.2857
H3	438	408	0.5	0.0645
H4	448	423	1.0	0.0408
C1	273	434	1.0	2.0000
Steam	None			
Water	None			

ΔT_m = 10K (for utility targeting).

Aspen Hysys was used in solving for the area target and the result obtained was then compared with [13] who solved same problem using Non Linear Programming area targeting method and also the result was compared with [21]. The comparison of results is presented in Table 2. [21] and [13] reported same area targets of 29.84m^2 for the problem statement compared to 22.17m^2 obtained with Hysys as can be seen in Table 2, this show a difference of 25.7%. The solution obtained in this research has 2 split streams compared with 4 and 7 splits streams of [13] and [21] respectively which make the network of this solution simpler and less costly in piping.

Table 2 Comparison of areas obtained by using different techniques

	METHOD	Stream Split	AREA (m^2)	% DIFFERENCE TO HYSYS
COLBERG AND MORARI	NON-LINEAR PROGRAMMING (NLP)	4	29.84	34.6
NISHIMURA	Area Targeting Method	7	29.84	34.6
HYSYS	This research	2	22.17	0

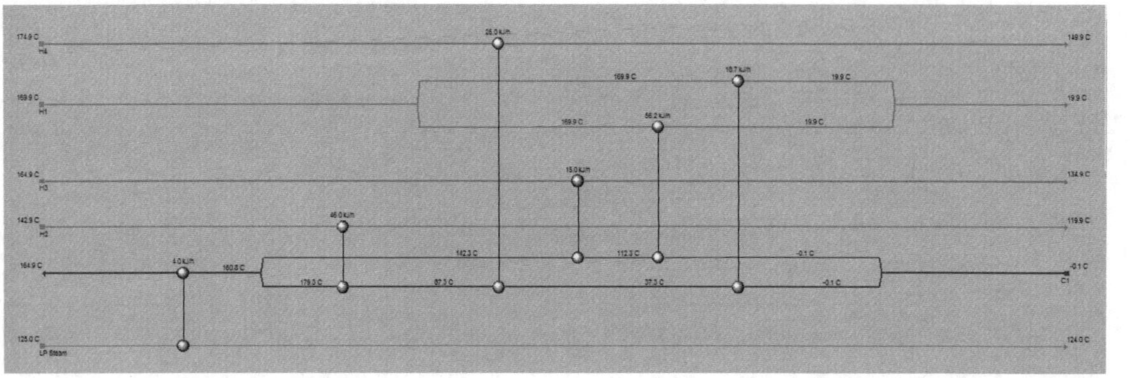

Figure 1 - HEN Structure Generated by Hysys for Area target in Example 1

Aspen Hysys was able to generate a smaller area target with a simpler grid diagram and relatively cheaper operating and capital cost.

3.2 EXAMPLE TWO

The stream data for example two is presented in Table 3 and is adapted from [13]. This set of data has heating and cooling targets of 244.2kW and 172.6Kw respectively. In this problem, minimum area is being targeted for three scenarios

a. When no limit is placed upon the stream matches.
b. For the best selection of matches corresponding to the unit target.
c. For the best selection of matches containing one match more than the unit target.

Table 3 - Stream Data for Example 2 [13]

Stream	Supply Temperature (K)	Target Temperature (K)	Heat capacity Flow rate ($kW\ K^{-1}$)	Film heat transfer coefficient ($kW\ m^{-2}\ K^{-1}$)
H1	626	586	9.802	1.25
H2	620	519	2.931	0.05
H3	528	353	6.161	3.20
C1	497	613	7.179	0.65
C2	389	576	0.641	0.25
C3	326	386	7.627	0.33
C4	313	566	1.690	3.20
Steam	650	650		3.50
Water	293	308		3.50

The area target generated by [13], for the three case scenarios are shown in Table 4. Also it should be noted that the film heat transfer coefficients range over about two orders of magnitude. This problem was solved for area with no streams restriction by [13] using area targeting NLP which decomposes the problem at the pinch.

Table 4: Result of Area target generated by [13]for five different Scenarios

	Limit on Number of matches		
	Above pinch	Below pinch	Area (m^2)
Area targeting NLP	-	-	173.6
	7	5	176.1
	6	5	188.8
Composite curve based target	-	-	227.0
Synthesized HEN	7	5	183.9

The area of the HEN design by [13] for the "no stream restriction" problem which was decomposed at the pinch is 188.9m^2 using non-linear programming method, which is 7.3% greater than the area target generated for seven matches above and five matches below.

Aspen Hysys software which works with the principle of pinch technology and some component of mathematical programmingwas used in this work to solve for area of the network with no stream restriction and the result obtained was compared well with that of [13].
A total area of 106.4m^2 was obtained byHysys when no limit is placed on streams matching and the grid diagram generated by Hysys is shown in Figure 6 which represents a difference of 43.67% in area compared to the result obtained by [13].

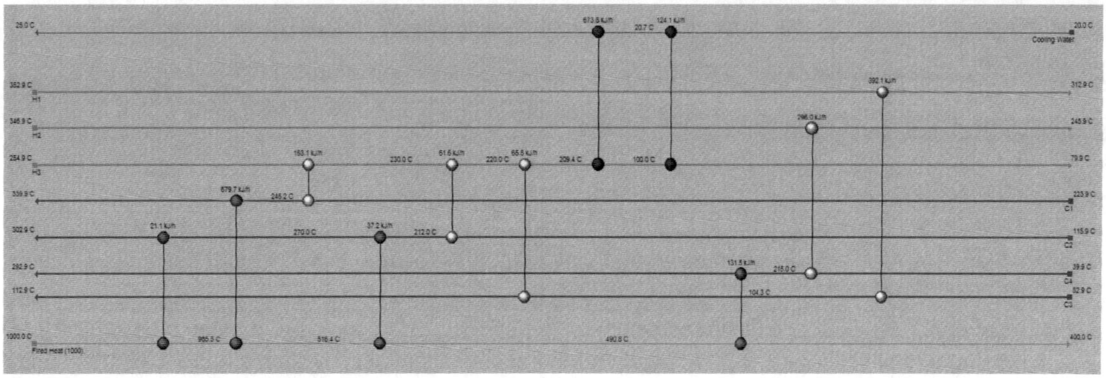

Figure 2 - Grid Diagram of the HEN synthesized for area target in Hysys for Example 2

The heat exchanger network synthesized by [13]shows seven units above the pinch position while five units are below the pinch position. This is slightly different from that synthesized in Hysys which shows six unit above and five units below the pinch point. The fewer number of heat exchanger (units) used by Hysys resulted in the HEN synthesized by Hysys having a cheaper annual cost alongside a smaller area compared to the HEN synthesized by [13]. It should also be noted that the eleven heat exchanger units used by the HEN synthesized using Hysys compared to the twelve used by [13] represent a percentage difference of 8.3 %. These differences are illustrated in Table 5.

Table5- Difference between the result obtained by [13]using area – target NPLand that obtained with Hysys

	AREA TARGET		
	NLP	HYSYS	% DIFFERENCE
NUMBER OF UNIT ABOVE PINCH	7	6	14.29
NUMBER OF UNIT BELOW PINCH	5	5	0
TOTAL NUMBER OF UNIT	12	11	8.33
TOTAL AREA	$188.9m^2$	$106.4m^2$	43.67

Hysys was able to obtain a smaller area with fewer numbers of heat exchanger units. That is, an area of $106.4m^2$ was obtained with eleven heat exchangers; while an area of $188.9m^2$ was obtained with twelve heat exchangers.

It can be seen that application of AspenHysys software in solving example two has resulted in a better minimum network area as compared to the result reported by [13] and obviously a cheaper capital cost, because fewer number of heat exchanger units were used.

4.0CONCLUSION.

Modified pinch technology which is embedded in AspenHysys has been shown in this work to be an effective tool in the optimization of the area of a heat exchanger network as depicted by the two examples solved using this technique.

REFERENCES

1. Ahmad. S., and Linnhoff, B., 1984. Overall cost targets for heat exchanger networks. IChemE Annual Research Meeting, 31 c.
2. Ahmad, S., Linnhoff, B., & Smith, R.1990. Cost optimum heat exchanger networks to targets and design for detailed capital cost models. Computers & Chemical Engineering. 14, 751-767.
3. Azeez ,S, Isiafiade, N. and Fraser, D.(2011). Supply and target based superstructure synthesis of heat and mass exchanger networks. Modeling simulation and applied optimization (ICMSAO), 4th International Conference, University of Technology, Malaysia. IEEE Conference Publication. Pp 1 – 6.
4. American Institute of Chemical Engineers. Atlantic City, NJ.

5. Anozie, A. N. and Odejobi 0. J. (2007). "Evaluation of Heat Exchanger Network Design and Energy Efficiency in the crude distillation units of Nigerian oil refineries", presented at the 37th Annual Conference of the Nigerian Society of Chemical Engineers, pp 1-10.
6. Ahmad (2007), optimal heat exchanger network synthesis with the detailed heat transfer equipment design. Comput. Chem. Eng., 31, 1432-1448.
7. Applied Thermal Analysis, 23, 141-151. Ravagnani, M. A. S. S. and Caballero, J.
8. Bertsimas. D.andsitsiklis. J., 1997. Introduction to linear optimization Athena Scientific, Massachusetts.
9. B.K. Piagbo (2011), Using simulation modelling to improve the performance of chemical engineering processes: retrofitting of the Port Harcourt refinery heat exchanger network (HEN) as case study, MSc. Thesis, Teesside University: Middlesbrough England, 2011.
10. B. Linhoff and E. Hindmarch, The pinch design method for heat exchanger networks, Chemical Engineering Science, 38 745-763.
11. B. Linhoff, Pinch analysis — A state-of-the-art overview, Chem. Eng. Res. Des, 71(A) (1993), 503-522.
12. C.A. Floudas, A.R. Ciric and I.E. Grossmann, Automatic synthesis of optimum heat exchanger network configurations, AICHE Journal, 32(1986) 276-290.
13. Colberg, R. and Morari, M. Area and capital cost targets for heat exchanger network synthesis with constraint matches and unequal heat transfer coefficients, computer chemical engineering, vol. 14, No 1, 1-22, 1990.
14. Delaby, O, 1989. Minimum area in heat ex-changer networks. Master's thesis, University of Manchester Institute of Science and Technology, UK.
15. El-Halwagi. M. M.andanousiouthakis. V.(1989) Synthesis of mass-exchange networks. AlChE Journal. 35. 1233-1244.
16. El-Halwagi, M. M., and Anousiouthakis, V. (1990). Automatic synthesis of mass-exchange networks with single-component targets. Chemical Engineering Science. 45(9). 2813-2831.
17. El-Halwagi. M. M..&Manousiouthakis, V. (1990b). Simultaneous syn-. thesis of mays-exchange and regeneration networks. AlChE Journal. 36(8), 1209-1219.
18. Hallale, N., & Fraser, D. (2000) Capital and total cost targets for mass exchanger networks. Part 1: simple capital cost models. Computers & Chemical Engineering, 23(11), 1661-1679.
19. Gundersen, T. and Naess, L. (1998), The synthesis of cost optimal heat exchanger networks—An industrial review of the state of the art. Computes Chem. Eng., 126, 503-530.
20. Gupta. A.andanousiouthakis, V. (1993). Minimum utility cost of mass-exchange networks with variable single component supplies and tar-gets. Industrial & Engineering Chemistry Research. 32. 1937-1950.

21. Nishimura, H. (1980). A theory for the optimum synthesis of heat ex-change systems. Journal of Optimization Thecny Applications, 30, 423-450.

IOP Conf. Series: Earth and Environmental Science **173** (2018) 012004 doi:10.1088/1755-1315/173/1/012004

Production and characterization of electrical transformer insulating fluid from blending Castor and Neem seed oils

Kenneth Onyeka Okafor* and Joseph Onyebuchi Okafor

Department of Chemical Engineering, Faculty of Engineering and Engineering Technology, Federal University of Technology, Minna, Niger State, Nigeria

Corresponding author E mail: mackent2025@gmail.com

Abstract. Castor and neem seed oils were first purified separately through the process of acid degumming, neutralization, bleaching, filtering, washing and drying. The refined castor and neem seed oils was mixed thoroughly at a ratio of 40:60, 50:50 and 60:40 to make up weight of 100g each. The blended ratios were synthesized separately through epoxidation and transesterification reactions to satisfactorily remove glycerol and unsaturation and form an epoxy in the triglycerides of the mixture. The epoxidation process of the triglyceride significantly increased the viscosity whereby making the blended fluids more physically, chemically and thermally balanced. The blended oil esters produced were washed properly with distilled water and dried to purify the esters. The blended epoxy castor-neem oil methyl ester in ratio of 40:60, 50:50 and 60:40 (BEME46, BEME55 and BEME64) were characterized and the following results were obtained: the viscosities of BEME46, BEME55 and BEME64 at a lower temperature of 30^oC were 9.30 cSt, 9.45 cSt and 9.48 cSt respectively, while the viscosities at a higher temperature of 70^oC were 3.30 cSt, 3.32 cSt and 3.48 cSt respectively, using a glass capillary kinematic viscometer and hot water bath setup. The dielectric strength of BEME46, BEME55 and BEME64 were 30.55 kV, 30.75 kV and 32.02 kV respectively using a megger test set. The acid value or neutralization number of BEME46, BEME55 and BEME64 were 0.1683 mgKOHg^{-1}oil, 0.1405 mgKOHg^{-1}oil and 0.1405 mgKOHg^{-1}oil respectively. The densities were 0.8260 gcm^{-1}, 0.8251 gcm^{-1} and 0.8219 gcm^{-1} respectively, while the flash points were 165^oC, 168^oC and 160^oC respectively. BEME46, BEME55 and BEME64 were compared with Transformer Mineral oil obtained from APO Transmission station. It was revealed that the blended oil esters met the transformer standard specifications.

1. Introduction

Transformer oil plays an important role to keep distribution and power transformers active and operate efficiently. It keeps the transformer cool and works as an insulator between the transformer windings and body [11]. In power holding networks electrical insulation material like the transformer oil play a very essential role, they resist and prevent the conduction of electricity between two or more conducting materials [13]. It is suggested that synthesized ester is as good as mineral oil in many ways and a viable alternative fluid for cooling and insulating power transformers [1]. Mineral oil has shown over the century to be unreliable and not sustainable because of spill ups, high flammability which eventually leads to high transformer fire, high toxicity and less economic benefits [5][13][21]. In order to perform its function, electrical power transformers need an insulating liquid to function as a cooler and insulator [11]. There are a few characteristics that an insulating medium is required to have in order to function effectively in transformers. These characteristics include long term stability due to possible problems brought about by oxidation, contamination and excessive temperature which can affect the reliability of the equipment and the suitability of the oil in rheological aspect [11][21].

2. Materials and Methods

2.1. Materials

The materials and reagents were purchased from FIN Laboratory and Chemicals, Abuja. The materials used in the course of this study include the following: Citric acid ($C_6H_8O_7$, purity: 99.7%), Sodium sulphate anhydrous (Na_2SO_4), Hydrogen Peroxide (about 30% H_2O_2 GPR, assay 30-31%, BDH chemical ltd), Sodium hydroxide (NaOH, purity: 99.9%), Sulphuric acid (H_2SO_4, purity: 99.8%) Silica gel, Glacial acetic acid (CH_3COOH, purity: 99.8%), Toluene ($C_6H_5CH_3$, purity: 99.7%), Methanol (CH_3OH, purity: 99.8%), Filter paper (24cm, medium crystalline; Whatman, England), Phosphoric acid (H_3PO_4, purity: >85%), Oxolane (THF), Distilled water (purified in SHESTCO), Crude castor seed oil (purchased from NARICT, Zaria), Crude Neem seed oil (purchased from NARICT, Zaria), Coconut shell activated carbon (SHESTCO).

2.2. Crude Seed Oil Refining

The process steps for separately refining crude castor seed oil and crude neem seed oil includes degumming, neutralization, bleaching, and deodorization [15]. The following procedures were carried separately for castor seed and neem seed oil.

250g crude sample of oil was heated in a 500 ml conical flask to 70˚C. (6% volume) of 64% aqueous citric acid solution was added gently and mixed thoroughly with a magnetic stirrer for 15 minutes at 800rpm. The oil sample was decanted gradually. 4 ml of 8% Sodium hydroxide (NaOH) solution was gently added and the mixture magnetically stirred at 400 rpm for 15 minutes. The sample was washed in a separating funnel with warm distilled water, and then dried in vacuum oven at 85°C for 50 minutes to reduce the water content in the oil. 2 g silica gel was added to the mixture at 70°C and it was agitated rigorously for 30 minutes at 300 rpm to prevent it from settling out. Coconut shell activated carbon was then added and magnetically stirred for 30 minutes at 95°C. The sample was then filtered and then washed with distilled water 3 times. The sample was dried over anhydrous sodium sulphate Na_2SO_4. Then the sample was dried in vacuum oven at 85°C for 50 minutes to reduce the water content in the oil [15].

2.3. Epoxidation process of blending castor and neem oils

100g of oil was used for the epoxidation process.

2.3.1 (PAA) preparation

Glacial acetic acid CH_3COOH (13 g) was placed in a 250 ml flat bottom flask followed by the addition of toluene C_7H_8 (13.5 g). The mixture was continuously stirred with a magnetic stirrer and after 30 minutes, followed by the addition of 30% aqueous hydrogen peroxide H_2O_2 (38 g) and continuously stirred (1200rpm) for 1 hour to produce peracetic acid (PAA).

2.3.2 Epoxidation at ratios of 40:60, 50:50 and 60:40

Refined castor seed oil and refined neem oil in ratios of 40:60, 50:50 and 60:40 were mixed separately to make 100g and then was poured into a two-necked flask equipped with a reagent injector and a condenser and stirred continuously and then was heated to 70°C. 1 % Sulphuric acid H_2SO_4 (10 g) was poured into the sample and continuously stirred for 15minutes at a constant temperature of 55°C. Followed by the addition of the PAA mixture into the two-necked reactor flask using the reagent injector and added in drops. The mixtures were continuously stirred (1200 rpm) at a regulated temperature between 40°C and 65°C for 7 hours. Aliquot was taken out for FTIR analysis to determine the effect of reaction time on the product. The reaction mixture was transferred from the flat bottom flask into a 1 litre separating funnel and then allowed for an interface to form. Then the bottom layer was discarded and the upper layer was collected. The upper layer was washed with distilled cold water 3 times and following by washing with distilled warm water 5 times and then distilled cold water again 2 times. The sample was dried over anhydrous sodium sulphate Na_2SO_4. The sample was put in the vacuum oven at 105°C to remove toluene C_7H_8 from the sample. Aliquot was taken for FTIR analysis to determine the formation of epoxide in the product [1][16].

2.4. Transesterification process of blended epoxy alkyl esters carried out for ratio of 40:60, 50:50 and 60:40

100g of blended epoxidized sample of 40:60, 50:50 and 60:40 were introduced separately into the flat bottom glass equipped with a magnetic stir bar. The epoxidized sample was heated to 55°C on a hot plate and magnetically stirred. Followed by addition of Oxolane (Tetrahydrofuran) (5g) $(CH_2)_4O$ into the flask of epoxidized sample and stirred continuously. A mass of 1 g of sodium hydroxide NaOH (1.0 wt %) as catalyst was weighed. And the catalyst was completely dissolved in 66 ml of pure methanol CH_3OH to form sodium meth-oxide. The resulting mixture (Sodium meth-oxide) was then added to the sample (oil + THF) in a flat bottom flask and heated to 60°C on a regulated hotplate and magnetically stirred at a rotational speed of 800 rpm. The set up was equipped with a reflux condenser. The mixture was refluxed and stirred for 2 hours at 60 - 65°C. The mixture was transferred into a 500 ml separating funnel and kept in an oven for 50 minutes at a temperature of 85°C. The bottom layer containing the glycerol was discarded while the top layer containing epoxidized alkyl ester was collected. The collected layer containing epoxidized alkyl ester was washed with 0.01 M phosphoric acid H_3PO_4 4 times and then washed with distilled warm water 3 times; the emulsion was kept in oven for one hour for separation. The bottom layer which contains washed out of residual glycerol and sodium hydroxide NaOH was discarded, while the top layer containing purified epoxidized alkyl ester was collected. The purified epoxidized alkyl ester + THF + methanol were dried over anhydrous sodium sulphate Na_2SO_4. The purified epoxidized alkyl ester + THF + methanol was then placed in a rotary evaporator under reduced pressure of about 10 mbar and temperature of 60°C to remove the excess methanol. The sample was then put in the vacuum oven at 105°C to remove to Oxolane (THF) from the sample [1][2][5][18].

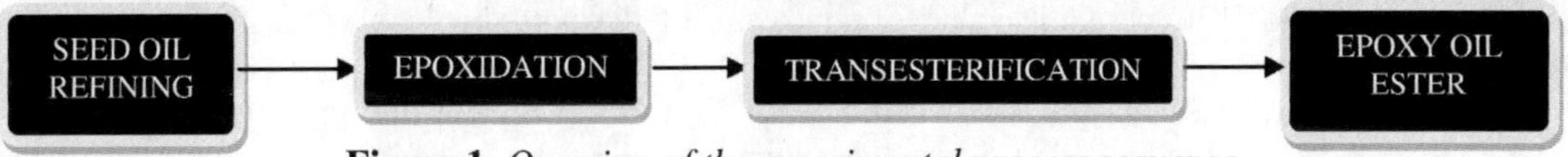

Figure 1. *Overview of the experimental process sequence*

2.5. Characterization of the blended epoxy castor-neem oil esters

2.5.1. Fourier Transform Infrared (FT-IR) Spectrometry: The FT-IR spectra were measured using a Thermo Scientific Nicolet iS5 FT-IR spectrometer. A drop of the sample was placed on the spectrometer measurement cell and a spectrum was obtained.

2.5.2. Dielectric Strength: The dielectric strength of the blended castor-neem samples were measured using the megger (OTS 60SX). The operation of the test set is simple and the results are displayed on a bright LED screen. The instrument is an automatic machine that can assess the quality of oil based on American Society for Testing and Materials (ASTM). The oil sample was placed between two electrodes with a 2.5 mm gap. A constant increasing voltage was applied until the oil discharges at a certain kV; the kV was recorded as the breakdown voltage [17].

2.5.3. Kinematic Viscosity: Viscosity of a fluid is a measure of its resistance to gradual deformation by shear stress or tensile stress. For liquids, it corresponds to the informal concept of "thickness". It was determined following ASTM D 445 standard specification. A suspended-level capillary viscometer was used to determine the kinematic viscosity. The viscometer is a U-shaped piece of glassware that was immersed into the hot water bath to regulate the varied temperature difference. This was multiplied by a constant specific for every viscometer (S3 viscometer used), to give a measure of the sample viscosity at 40°C, 50°C, 60°C and 70°C. The initial temperature was set at 30°C [1][7].

2.5.4. Flash Point: The flash-points of the samples were determined using an improvised method. Apparatus used were 50 ml conical flask, thermometer, heating mantle and a gas blower torch. A conical flask (50 ml) was filled with 5 ml of oil and equipped with a thermometer and then heated at as very low constant rate on the hot plate. A gas blower torch was brought to the mouth of the conical flask and gradually passed across the mouth of the conical flask after every degree increase in temperature as heating continued until a pop sound that ignited the fire was seen. The flash point was obtained when application of a test flame causes the vapour above the oil to ignite. The temperature was then recorded as the flash point of the sample. The procedure was repeated for the various samples [8][9].

2.5.5. Acid value and free fatty acid (FFA): The acid value of an acid or fat is the number of mg potassium hydroxide required to neutralize the free acid in (1 g) of the sample. The FFA is an aliphatic carboxylic acid which contain a long hydrocarbon; It can be hydrophilic or hydrophobic [14].
Procedure:- Mix 25ml diethyl ether with 25ml ethanol. Add 1ml of 1% phenolphthalein indicator solution. Neutralize with 0.1M potassium hydroxide solution. Dissolve 1 to 10g of the oil in the neutralized solvent mixture. Titrate with 0.1M potassium hydroxide solution. Shake constantly until a pink colour which persists for 115seconds is obtained [7][8][14][19].
Calculation:

$$Acid\ value = \frac{titration\ \ (ml) \times 5.61}{Wt\ of\ sample\ \ used}\ (\text{mg KOH/g}) \qquad (1)$$

Also acid value = 2 x Free fatty acid (FFA). (2)

FFA% is approximately ½ AV [8][14].

2.5.6. Peroxide value: The peroxide value is a measure of the peroxides contained in the oil [14].
Procedure:- The test should be carried out in subdue daylight. Weigh out 1g of oil into a clean dry boiling tube. Add 1g powdered potassium iodide and 10ml of the solvent mixture. Place the boiling water bath so that the liquid boils within 30seconds and allow to boil vigorously for more than 30seconds. Pour the content into a titration flask containing 20ml freshly prepared 5% potassium iodide solution. Wash the tube twice with 25ml portions of water and add the washings to the titration flask. Titrate with 0.002M sodium thiosulphate solution using starch as indicator (aml). Carry out a blank determination. Blank titre (bml) [8][9][19].

Calculation:

$$Peroxide\ value = \frac{2(a-b)}{Wt\ of\ oil}\ (\text{mEq/kg}) \qquad (3)$$

2.5.7 Saponification value: The saponification value of oil is the number of mg of potassium hydroxide to neutralize the fatty acids resulting from the complete hydrolysis of the sample [14].
Procedure:- Weigh 2g of oil (using a dropping pipette) into a flask. Add exactly 25ml of the alcoholic potassium hydroxide solution into the flask. Attach a reflux condenser and heat the flask on a water bath for 1 hour with occasional shaking. At the end of 1 hour remove the flask from the water bath and add 1ml of the 1% phenolphthalein indicator. Titrate while still hot with the standard 0.5M hydrochloric acid (aml). Carry out blank determination omitting oil (bml) [8][9][10][14].
Calculation:

$$Saponification\ value = \frac{(b-a)\times 28.05}{Weight\ of\ oil}\ (\text{mg KOH/g}) \qquad (4)$$

2.5.8 Iodine value: The iodine value of oil is the weight of iodine absorbed by 100 part by weight of the sample. The iodine value is a measure of the degree of unsaturation of the fatty acids in the oil. [14].

Procedure:- Weigh accurately about 0.2 – 0.5g of the oil (with the acid of a dropping pipette) into a glass stopper flat bottom flask of about 25ml. Add 10ml carbon tetrachloride to the oil and dissolve. Add exactly 20ml Wiji's solution and insert the stopper which has been moistened with potassium

IOP Conf. Series: Earth and Environmental Science **173** (2018) 012004 doi:10.1088/1755-1315/173/1/012004

iodide solution. Mix and allow to stand in a dark cupboard for 30minutes. Add 15ml 10% potassium iodide solution (freshly prepared), 100ml water and mix. Titrate the mixture with 0.1M standard sodium thiosulphate solution using starch as an indicator just before the end point (aml). Carry out a blank simultaneously omitting the oil (bml) [8][9][14][6].

Calculation:

$$Iodine\ value = \frac{(b-a)\times 1.269}{Weight\ of\ oil/fat} \quad (5)$$

Note: if b – a is greater than b/2, repeat the test using smaller amount of the sample. Record result in Wiji's.

2.5.9 Water content: This is a measure of the % moisture lost due to drying at a temperature of 105 degrees Celsius [14]. 2g of the sample was weighed (W_1) into preweighed crucible (W_0) and placed into a hot drying oven at 105 degrees Celsius for 3hours. The crucible were removed, cooled in a desiccator and weighed. The process of drying, cooling and weighing were repeated until a constant weight (W_2) was obtained [9][19][14]. The weight loss due to moisture was obtained by the equation 6:

$$Moisture\ (\%) = \frac{W_1 - W_2}{W_1 - W_0} \times 100 \quad (6)$$

Where: W_0 = Weight of the empty crucible (g). W_1 = Weight of the sample + empty crucible (g). W_2 = Weight of dried sample + empty crucible (g).

2.5.10 Density: The weight of a small beaker was determined using an electronic weighing balance. 2 ml of the oil was poured into it and the weight was noted [8][12].

Calculation:

$$Density = \frac{(Mass\ of\ oil)}{(Volume\ of\ oil\ weight)}\ (gcm^{-1}) \quad (7)$$

2.5.11 Boiling point: The boiling point of a liquid is the temperature at which the vapour pressure of the liquid equals the pressure surrounding the liquid and the liquid changes into a vapour. The oil was placed in a container with an inserted thermometer and heated to the time the first bubble was observed. The temperature was noted, the experiment was repeated twice and the average temperature calculated [8].

2.5.12. Specific gravity: An empty 10 ml pycnometer bottle was weighed. The 10ml pycnometer bottle was filled with water and weighed. The 10 ml pycnometer bottle was then filled with the same volume of oil as that of the water and weighed [19][19]. The specific gravity is calculated with equation 8 shown below.

Calculation:

$$Specific\ Gravity = \frac{W_3 - W_1}{W_2 - W_1} \quad (8)$$

where, W_1 = weight of empty bottle. W_2 = weight of bottle + water. W_3 = weight of bottle + sample oil.

2.5.13. Appearance or oil colour
The procedure was visual.

3. Results and Discussion

The FT-IR study of the blended epoxy castor-neem oil ester at ratio of 40:60, 50:50 and 60:40
The methylene vibration of CH_2 in which all groups of straight chain alkanes in the blended samples vibrate in Phase, this peaks at 722.29 cm^{-1} and 721.80 cm^{-1} respectively in Figure 3, 4 and 5. The samples of observed broad peak in BEME46, BEME55 and BEME64 were observed to peak between the region of the spectrum of 750 cm^{-1} and 1000 cm^{-1}. The carbon-carbon spectra peak wasn't present in the BEME46, BEME55 and BEME64 spectra [20].

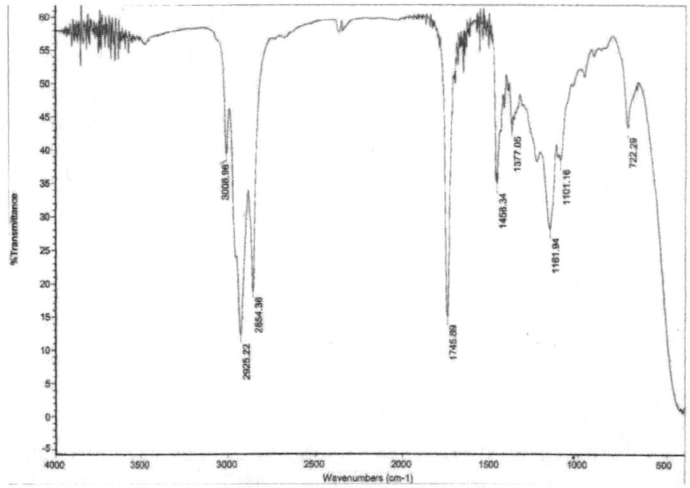

Figure 2: Fourier Transform Infrared FTIR Spectrum of Blended castor-neem oil methyl ester in ratio of 40:60 respectively (BEME46)

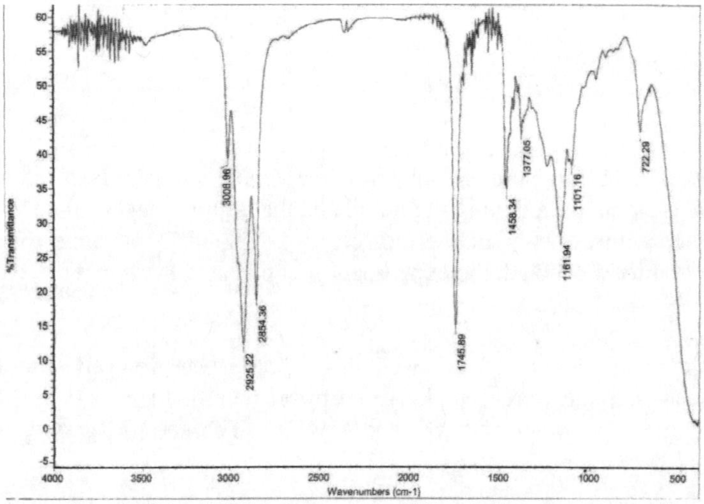

Figure 3: Fourier Transform Infrared FTIR Spectrum of Blended castor-neem oil methyl ester in ratio of 50:50 respectively (BEME55)

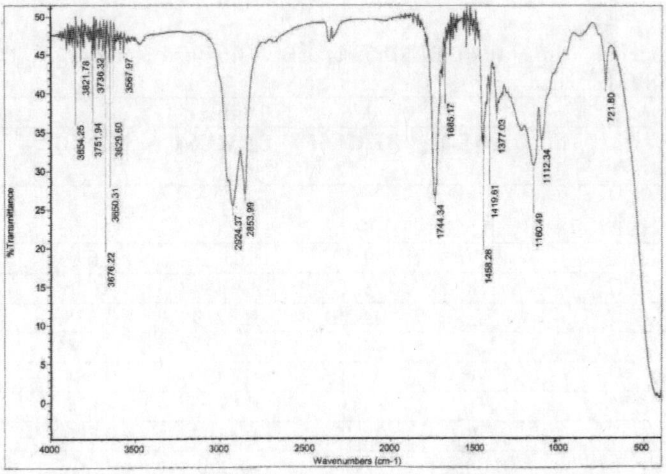

Figure 4: Fourier Transform Infrared FTIR Spectrum of Blended castor-neem oil methyl ester in ratio of 60:40 respectively (BEME64)

Table 1: Description of experimental samples

Sample	Description
BEME46	Blended epoxy castor-neem methyl ester {in ratio of 40:60}
BEME55	Blended epoxy castor-neem methyl ester {in ratio of 50:50}
BEME64	Blended epoxy castor-neem methyl ester {in ratio of 60:40}
MinOil	Transformer Mineral Oil obtained from AT3 Transmission Power Station

Table 2: Overall properties of BEME46, BEME55 and BEME64 and comparing with transformer oil specification of ASTM standard

Property	**Values**				
	BEME46	**BEME55**	**BEME64**	**MINOIL**	**Transformer oil ASTM Standard**
Dielectric Strength, *kV*, 27 °C	30.55	30.75	32.03	40.53	25 - 40
Viscosity, *cSt*, @ 30°C	9.30	9.45	9.48	16.03	9.3 – 27
Viscosity, *cSt*, @ 50°C	6.02	6.91	6.91	10.93	-
Viscosity, *cSt*, @ 70°C	3.30	3.32	3.48	5.46	-
Density, gcm^{-1} @ 27°C	0.8260	0.8251	0.8219	0.8888	0.55 – 0.89
Flash point, *°C*	165	168	160	138	Above 140
Acid value, $mg\ KOHg^{-1}oil$	0.1683	0.1405	0.1405	0.0846	0.01 – 0.3
Boiling point, *°C*	150	149	151	134	Above 120
Specific gravity	0.8491	0.8483	0.8502	0.8909	0.89 – 0.91
Saponification value, $mg\ KOHg^{-1}oil$	153.573	156.379	155.678	151.470	150 – 244
Peroxide value, mEq g^{-1}oil	8.240	8.000	9.115	4.440	5 – 10
Iodine value, g100g^{-1}oil	55.902	56.376	56.488	28.178	55 – 120
Free fatty acid, $mg\ KOHg^{-1}oil$	0.0842	0.0703	0.0703	0.0421	0.01 – 0.08
Moisture content, *%*	0.045	0.064	0.062	0.013	-
Appearance	Yellow tint	Yellow tint	Yellow tint	Yellow tint	White / yellow tint

*Transformer oil specification for ASTM standard was sourced from [8].

IOP Conf. Series: Earth and Environmental Science **173** (2018) 012004 doi:10.1088/1755-1315/173/1/012004

Table 3: Key properties of BEME46, BEME55 and BEME64 and comparing with transformer oil specification of ASTM standard

Key Property	Values				
	BEME46	**BEME55**	**BEME64**	**MINOIL**	**Transformer oil ASTM Standard**
Dielectric Strength, *kV*, 27 °C	30.55	30.75	32.03	40.53	25 - 40
Viscosity, *cSt*, @ 30°C	9.30	9.45	9.48	16.03	9.3 – 27
Viscosity, *cSt*, @ 70°C	3.30	3.32	3.48	5.46	-
Density, gcm^{-1} @ 27°C	0.8260	0.8251	0.8219	0.8888	0.55 – 0.89
Flash point, *°C*	165	168	160	138	Above 140
Acid value, $mg\ KOHg^{-1}oil$	0.1683	0.1405	0.1405	0.0846	0.01 – 0.3
Boiling point, *°C*	150	149	151	134	Above 120
Specific gravity	0.8491	0.8483	0.8502	0.8909	0.89 – 0.91
Moisture content, *%*	0.045	0.064	0.062	0.013	-
Appearance	Yellow tint	Yellow tint	Yellow tint	Yellow tint	-

*Transformer oil specification for ASTM standard was sourced from [8]

The refining process and the formation of an oxirane or epoxy in the molecular structure of BEME46, BEME55 and BEME64 that changed the double bonds had a significant effect by increasing the viscosity and dielectric strength of the blended oil esters. The transesterification synthesis of the blended epoxy oil esters that separated the free fatty acid FFA from the glycerol was observed to also increase the dielectric strength. The purification stage was observed to alter the acid value of the oils. The acid value decreased tremendously after the refining process. The observed and clear reduction in the acidity is as a result of neutralizing the oil with a base (sodium hydroxide NaOH) [5][15]. Transesterification of the blended epoxy oil esters decreased the acid value and by this reduced the FFA of the esters. The acidity of insulating fluid can be increased by thermal degradation of the fluid [13]. This is because oxidation products are often acidic in nature. Acids formed during ageing of mineral oil are classified into two; high molecular weight acids (HMWA) and low molecular weight acids (LMWA) [4]. The flash point BEME46, BEME55 and BEME64 has shown that the esters can safely be used even where the temperature is expected to be very high. Oils with flash point above 66°C are considered as safe oils. Viscosity of BEME46, BEME55 and BEME64 is observed to decrease with increase in temperature.

4. Conclusion

The electrical transformer insulating fluid from blending Castor and Neem seed oils have been produced and characterized. The refined castor and neem seed oils was mixed thoroughly at a ratio of 40:60, 50:50 and 60:40 to make up weight of 100g each.. The blended ratios were synthesized separately through epoxidation and transesterification reactions to satisfactorily remove glycerol and unsaturation and form an epoxy in the triglycerides of the mixture. The characterization of the blended epoxy castor-neem oil methyl ester in ratio of 40:60, 50:50 and 60:40 (BEME46, BEME55 and BEME64) revealed that the viscosities of BEME46, BEME55 and BEME64 at a lower temperature of 30°C were 8.30 cSt, 8.45 cSt and 8.48 cSt respectively, while the viscosities at a higher temperature of 70°C were 3.30 cSt, 3.32 cSt and 3.48 cSt respectively, using a glass capillary kinematic viscometer and hot water bath setup. The dielectric strength of BEME46, BEME55 and BEME64 were 30.55 kV, 30.75 kV and 32.02 kV respectively using a megger test set. However, the comparison of the BEME46, BEME55 and BEME64 with the Transformer Mineral oil obtained from APO Transmission station showed that the blended oil esters met the transformer standard specifications.

References

[1] Abdelmalika, A. A., Abbottb, A. P., Fothergill, J. C., Dodda, S. and Harris, R. C. (2010). Synthesis of a base-stock for electrical insulating fluid based on palm kernel oil; *Industrial Crops and Products*, vol. 33: 532-536.

[2] Aliyu, Jauro., and Momoh Haruna Adams. (2011). Production and Biodegradability of Biodiesel from Balanites Aegyptiaca Seed Oil; by Korean Chemical Society, in *Journal of the Korean Chemical Society*, vol. 55 (4), pp. 681-683.

[3] Al-Widyan, M. I., and Al-Shyoukh A. O. (2002). Experimental evaluation of the transesterification of waste palm oil into biodiesel; *Journal of Bioresource Technology*, vol. 85, no. 3, pp. 253-255.

[4] Azam, Mohibbe M., Amtul Waris, and Nahar N. M. (2005). Prospects and potential of fatty acid methyl esters of some non-traditional seed oils for use as biodiesel in India; Elsevier Publication, *Journal of Biomass & Bioenergy*, vol.29, pp. 295-300.

[5] Barakos, N., Pasias S. and Papayannakos N. (2008). Transesterification of triglycerides in high and low quality oil feeds over an HT2 hydrotalcite catalyst; *Journal of Bioresource Technology,* vol. 99, no. 11, pp. 5038-5041.

[6] Dhellot, J.R., Matouba, E., Maloumbi, M.G., Nzikou, J.M., Ngoma D.G Safou., Linder, M., Desobry, S and Parametier, M. (2006). Extraction, chemical composition and nutritional characterization of vegetable oils: Case of Amaranthus hybridus (Var 1 and 2) of congo Brazzaville. African Journal of Biotechnology; vol 5(11), pp. 1095-1101.

[7] Fares, N., Jabri Karoui, Sifi, S., and Abderrabba, M. (2016). Physical chemical and sensory characterization of olive oil of the region of Kairouan. *Journal of Material and Environmental Science*. Vol. 7(6), pp. 2148-2154.

[8] Garba, Z. N., Gimba, C. E., and Emmanuel, P., (2013). Production and Characterization of biobased transformer oil from jatropha curcas seed, *Journal of physical science*. Vol. 24(2), pp 49-61.

[9] George Anastopoulos, Ypatia Zannikou, Stamoulis Stournas and Stamatis Kalli-geros. (2009). Transesteriifcation of vegetable oils with ethanol and characterization of the key fuel properties of ethyl esters. *Journal of energies*, vol. 2, pp. 362-376.

[10] Giuseppe Toscano and Eleonora Maldini (2007). Analysis of the physical and chemical characteristics of vegetable oils as fuel. *Journal of Agric engineering* – Riv. di Ing. Agr. Vol. 3, pp 39-47.

[11] Hosier, L., Vaughan A. S. and Montjen F. A. (2006). Ageing of biodegradable oils for high voltage insulation systems; *IEEE CEIDP*, USA, pp. 481-482.

[12] Huseyin Sanli, Mustafa Canakci, and Ertan Alptekin. (2011). Characterization of waste frying oils obtained from different facilities. World renewable energy congress-sweden. *Bioenergy Technology*, pp. 479-484.

[13] Kejun, Q., Chengke Z., Allan, M. and Yue Y. (2011). Modeling of Load Demand Due to EV Battery Charging in Distribution Systems; *Power Systems, IEEE Transactions on,* vol. 26, pp. 802-808.

[14] Kelle Henrietta Ijeoma and Udeozo Ifeoma Prisca. (2005) Characterization of the Chemical properties of some selected refined vegetable oils commonly sold in Nigeria. *British Journal of Applied Science Technology*. Vol. 5(6), pp. 538-546.

[15] McDonnell, K. P. Ward, S. M. and Timoney, D. J. (1995). Hot water degumming of rapeseed oil as a fuel for diesel engines. *Journal of Agriculture and Engineering Resources*; vol. 60(1), pp. 7–12.

[16] Miao, S. D., Zhang S. P., Su Z. G., and Wang P. (2013). Synthesis of bio-based polyurethanes from epoxidized soybean oil and isopropanolamine; *Journal of Applied Polymer Science*, vol. 127 (3): pp 29-36.

[17] Perrier, C. and Beroual, A. (2009). Experimental Investigations on Insulating Liquids for Power Transformers: Mineral, Ester, and Silicone Oils, *IEEE Electrical Insulation Magazine*. Page 5-10.

[18] Rashmi, Kumari., Vinod Kumar Jain, and Sushil Kumar. (2012). Biodiesel production from seed oil of Cleome viscose L; by National Institute of Science Communication and Information Resources, CSIR, in Indian; *Journal of Experimental Biology*, vol. 50, pp. 502-510.

[19] Sidohounde, Assou., Nonviho, G., Djenontin, T. Sebastien., Agbangnan, P., Paris, C., and Sohounhloue, C.K. Dominique. (2014). Physio-chemical Characterization of Vegetable Oil and Defatted Meal from Two Varieties of Cyperus esculentus from Benin. *Chemical Journal*; vol. 04. Issue 01, pp.1-7.

[20] Stuart, B. (2004). *Infrared Spectroscopy: Fundamentals and Applications*. Wiley. pp 1-102

[21] Suwarno, M. I. (2006). Study on the Characteristics of Jatropha and Ricinnus Seed Oils as Liquid Insulating Materials; *IEEE CEIDP*, USA, pp. 162-164.

Rutherford Backscattering Spectrometry (RBS) method for the Determination of Elemental Constituent of Tropical Wood Matrices from Western Nigeria

Onumejor C. A.[1], Balogun F. A.[2], Akinpelu A. [1], Arijaje T. E.[1], Usikalu M. R.[1]
Department of Physics, Covenant University, Ota, Nigeria
Centre for Energy Research and Development, Obafemi Awolowo, University, Ile-Ife, Nigeria

charity.onumejor@covenantuniversity.edu.ng

Abstract. The elemental composition analysis and its concentration levels in percentage were done on several wood samples using Rutherford Backscattering Spectrometry (RBS) of Ion Beam Analytical (IBA) methods. A total of twenty seven (15) tropical hardwoods from south western Nigerian environment were analyzed and six (6) major elements (C, O, N, Na, S, Cl) were identified. Carbon has the highest percentage constituents, ranging from 58.30% to 73.46%, in all the samples with sample 4 (*Albizia gummifera*) having the highest concentration of carbon as 73.46% and sample 7 (*Anadelphia afzeliana*) having the least carbon quantity as 58.30%.Nitrogen N ranged from 1.21% - 7.76%, Oxygen O 17.96% - 34.83%, Sodium Na 0.36% - 1.08%, Chlorine Cl 0.11% - 0.46%, and Sulfur S 0.13% to 0.59%. C was above the overall wood estimated constituent 50%, O below estimated value of 40%. Cl and S were within estimated range as trace element percentage range. The variation in elemental composition of wood used in the experiment shows that wood species have thesame elemental compositional trend as some human tissues and could be harness for use as tissues equivalent material for application in medical radiation dosimetry.

Keywords: RBS method, Elemental composition of wood, Wood composition, Tissue equivalent material

1. Introduction

Ion Beam Analysis (IBA) consists of several methods of studying materials based on the interaction between accelerated charged particles (ions) and the bombarded materials (sample). IBA techniques include: PIXE (particle induced X-ray Emission) based on atomic fluorescence, PIGE (Particle Induced Gamma ray Emission) based on nuclear reaction, NRA (Nuclear Reaction Analysis) based on nuclear reaction, RBS (Rutherford Backscattering Spectrometry) based on nuclear scattering. RBS is an ion beam scattering techniques that is used primarily for thin film composition analysis. It is typically the choice method for other quantitative composition analysis as well, for example for optical coatings, elemental characteristics, depth profiling and other applications where control of film composition is critical. RBS can detect low Z elemental constituent in materials and can measure hydrogen H and all other elements except helium Hi, litium Li and beryllium Be. In this study, we are concentrating on use of RBS which is arguably the first nuclear reaction ever recorded, for the

IOP Conf. Series: Earth and Environmental Science **173** (2018) 012005 doi:10.1088/1755-1315/173/1/012005

determination of elemental composition of biomass material (that is; several wood samples). Natural biomass materials have been known to contain elements like carbon, oxygen, nitrogen, sulfur and silicon etc. wood as a biomass are composed of Low Z (atomic number) element of which RBS method has been found to be highly suitable. Overall, wood has an elemental composition of about 50% carbon, 6% hydrogen, 44% oxygen, and trace amounts of several metal ions (Pettersen, 1984). The detailed chemical composition break down of wood varies from species to species, but when accounting for trace element it is approximately 50% carbon, 6% hydrogen, 42% oxygen, 1% Nitrogen and 1% other elements (mainly calcium, potassium, sodium, magnesium, iron, and manganese) by weight [1]. Most studies on elemental composition of biomass use digestions and other destructive analytical methods. There are rare records of studies using non-destructive analytical such as RBS. Hence, this study is aimed at using RBS method for the elemental analysis of select tropical woods from western Nigeria. Knowing the elemental composition of wood can tell a lot about the possible usefulness of such wood for various purposes. The common use of wood is for furniture and as fuel. Recently, wood is becoming relevant in the medical radiation field of study. Scientist [2,3] and few others has done some study on wood for dosimetric purposes and medical phantom material suitability in radiation oncology. Bradley [4] studied the photon attenuation studies on tropical hardwoods, Tajuddin [5] reported radiographic and scattering investigation on the suitability of Rhophoraspp hardwood as tissue-equivalent medium for dosimetric study. Sudin [6] study included Rhizophoraspp tropical hardwood as water-equivalent materials. Investigation continued on Rhizophoraspp hardwood. The usefulness of this named hardwood began from its elemental constituent study. Thus, the elemental study of wide range of woods will form a basis for further discovery of some other useful tropical wood in radiobiology. The present study will report elemental compositions of 15 tropical woods from Western Nigeria as basis for further radiological wood property study.

2. Materials and Method

2.1. Wood sample preparation

15 different wood samples were selected within the south-west region of Nigeria. Selection was based on abundance of the wood type, easy availability and cultivation possibility. Experts from Forest Research institute of Nigeria (FRIN) Ibadan were involved in the selection process. Selected samples were harvested in its timbers forms from different areas within the stated region. Identification of such tree to be harvested was done by trained personnel from the Forest Research institute of Nigeria (FRIN) Ibadan. The harvested samples were all transported from harvest zones to the Center for Energy Research and Development (CERD), at Obafemi Awolowo University Ile-Ife and stacked at the designated area within CERD space.

The harvested timbers thick back were peeled off immediately harvesting was completed. The peeled timber samples were sawed (cut) to smaller wood pieces and left to dry. Drying was done by construction of wood drying platform under open and direct sun rays. Sun drying of samples was from morning to dusk, at dusk; wood samples will be packed of the rancho (constructed hut or house for wood sample storage). Samples drying time was minimum of 100days. A Portion of each dried wood sample was pulverized and sieved with 2mm mesh size sieve. The sieved wood dust was scooped (about 20g per portion) into sample dispensary bags. The sample bags were properly labeled and bagged samples were then taken for pelletizing (pellet size of about 250 – 350mg). Each prepared (pelletized) sample was kept sealed in another labeled sample bag for elemental composition analysis. RBS (Rutherford Backscattering Spectrometry) ion beam analytical method was employed for sample elemental analysis. SIMNRA programing Software was used for RBS spectrum analysis of each wood sample.

2.2. Wood common names, scientific names and sample codes
The wood common names, scientific names and sample codes are as shown in table 1.

Table 1: Common, Botanical Name and Codes for Wood Samples

Sample Codes	Local or Common Names	Botanical Names
1	Afon	Treculiaafricana
2	Arere (Obeche)	Triplochitonscleroxylon
3	Aynure (white) or ayunre	Albizia zygia
4	Aynure or Ayunre (yellow)	Albizia gummifera
5	Cidrela	Cedrellaodorata
6	Efo-onibo	Basella alba
7	Bere	Anadelphia afzeliana
8	Gmelina	Gmelinaarborea
9	Ire	Funtumiasppelastic
10	Iroko	Milicia excels
11	Isin	Blahiasapida
12	Ita	Celtisspp
13	Kara koro	Irvingiagrandifolia
14	Kasia	Sennasiamea
15	Masonia	MansoniaAltissima

2.3. Description of accelerator facility for RBS
The NEC 5SDH 1.7 MV Accelerator machine at the Center for Energy Research and Development (CERD), ObafemiAwolowoUversity Ile-Ife was used for the Ion Beam Analysis (IBA). The IBA facility centred on this accelerator is equipped with a RF charge exchange ion source. The ion source is equipped to provide proton and helium ions. The accelerator has provision for five beam lines but is presently maintaining one that is equipped with a general purpose End Station for Particle Induced X-ray Emission (PIXE), Rutherford Backscattering Spectrometry (RBS), Elastic Recoil Detection Analysis (ERDA) and Particle Induced Gamma ray Emission (PIGE). Which are stationed at 135°, 165°, 30° and 225°angled position respectively, while the window for observing the beam position and size is at 0°.Calibrations were done using commercially available standards for biological analysis. These standards were selected such that they provide a representation of the entire spectrum needed for the analysis. In calibrating the system, the 0.6 - 3.5MeV energy ranges of the 4He+ particles was used to bombard the standard samples.

3. Results and discussion
A total of 6 elements of interest were entered into the SIMNRA software setup for RBS spectra processes. The percentage weighted fractions for each of these elements are as shown in table 2.

IOP Conf. Series: Earth and Environmental Science **173** (2018) 012005 doi:10.1088/1755-1315/173/1/012005

Table 2: Element concentration of wood samples in %

Elements/sample ID	C	N	O	Na	Cl	S
1	64.73	2.0	32.31	0.49	0.14	0.33
2	65.10	2.27	31.48	0.54	0.18	0.42
3	64.48	3.33	30.96	0.65	0.18	0.38
4	73.46	7.50	17.96	0.37	0.11	0.59
5	70.00	7.76	21.11	0.68	0.26	0.43
6	69.01	6.48	22.90	0.84	0.32	0.45
7	58.30	5.05	34.83	0.89	0.42	0.51
8	70.51	7.54	20.91	0.36	0.11	0.56
9	66.13	1.97	30.51	0.73	0.41	0.34
10	67.91	5.87	24.76	0.79	0.34	0.33
11	63.96	4.02	30.21	1.08	0.39	0.34
12	67.17	2.36	26.41	1.02	0.46	0.57
13	64.18	4.16	30.19	0.72	0.35	0.40
14	67.02	4.66	26.76	0.76	0.38	0.48
15	64.03	1.21	33.79	0.63	0.21	0.13

3.1. Elemental analysis

A total of 6 elements of dosimetric interest were inputted into RBS spectra processing software for identification in samples. This was done for each of the 15 different wood pellets. The elements identified are C, N, O, Na, Cl and S. The results of the elemental compositions for the studied samples are presented in table 2. The concentrations of all the detected elements in the samples were found to vary slightly, but had the same concentration trend as shown in the graphical plots of element concentration against samples in table 2. Figure 1-6 shows the comparison of each element concentration in all the studied samples. Major points of variation were noted as are clearly seen on the graphs. The concentrations of Carbon C was found to be within the range of 58.30% -73.46%, Nitrogen N ranged from 1.21% - 7.76%, Oxygen O 17.96% - 34.83%, Sodium Na 0.36% - 1.08%, Chlorine Cl 0.11% - 0.46%, and Sulfur S 0.13% to 0.59%. C was above the overall wood estimated constituent 50%, O below estimated value of 40%.Cl and S were within estimated range as trace element percentage range. The results tallies with wood elemental constituents trend as reported by WPGS [1,7] and Pettersen, [8,9]. These detected elements were so called elements of dosimetric interest because they covered about 99% of the elemental constituents of the various human tissues and organs. Other detectable elements are in trace levels. According to Nelson [8,10], almost 99% of the mass of the human body is made up of six elements: oxygen, carbon, hydrogen, nitrogen, calcium, and phosphorus. Only about 0.85% is composed of another five elements: potassium, sulfur, sodium, chlorine, and magnesium. All 11 are necessary for life. The remaining elements are trace elements.

IOP Conf. Series: Earth and Environmental Science 173 (2018) 012005 doi:10.1088/1755-1315/173/1/012005

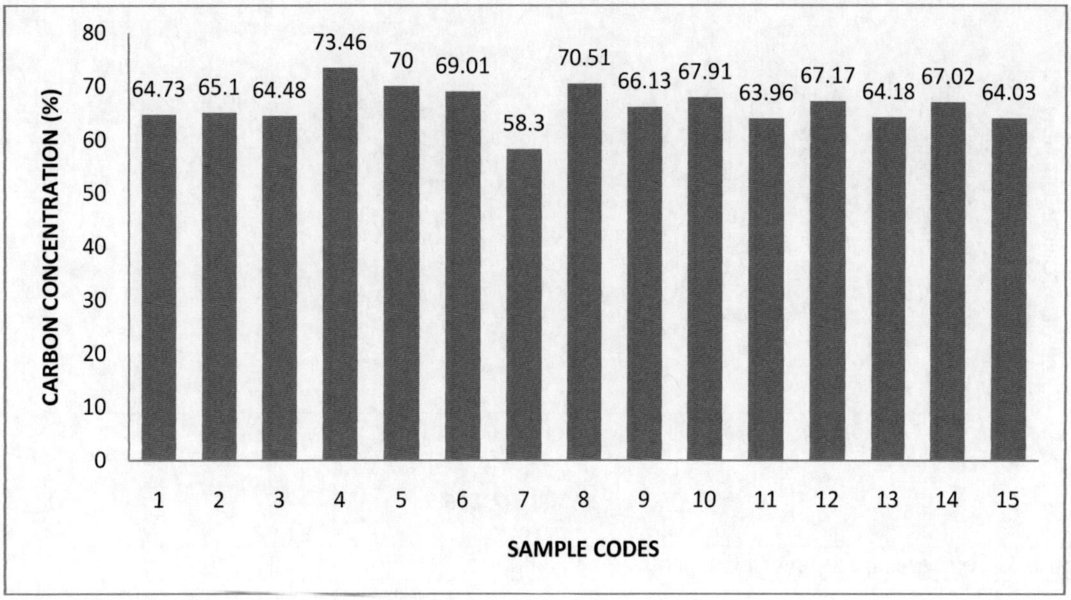

Figure 1: Comparison of Carbon Concentration in Samples

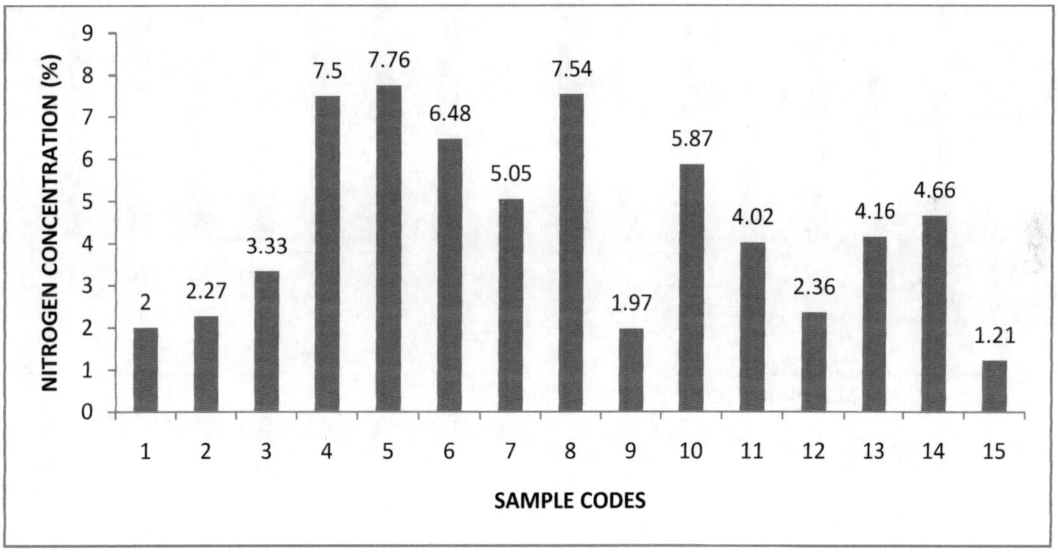

Figure 2: Comparison of Nitrogen Concentration in Samples

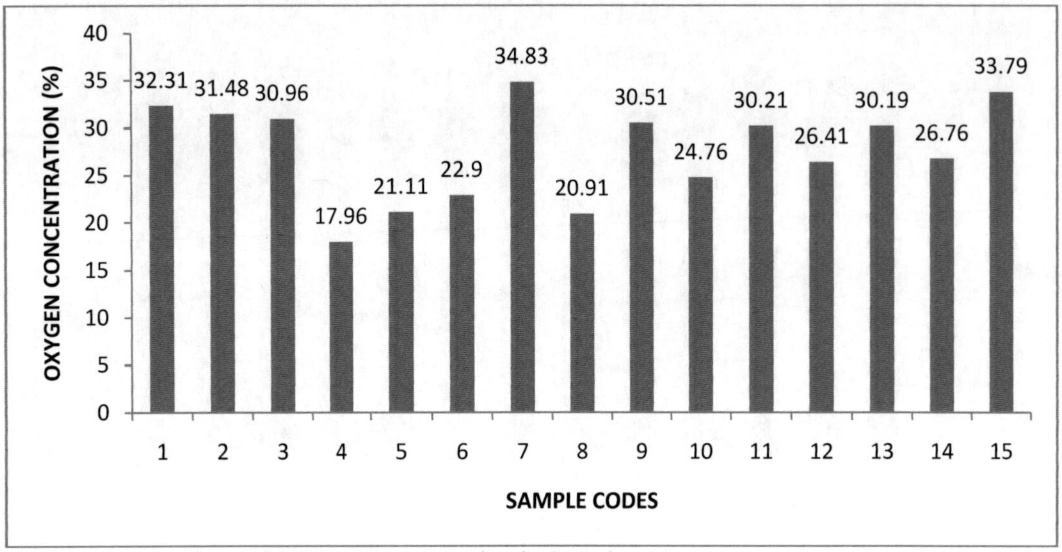

Figure 3: Comparison of Oxygen Concentration in Samples

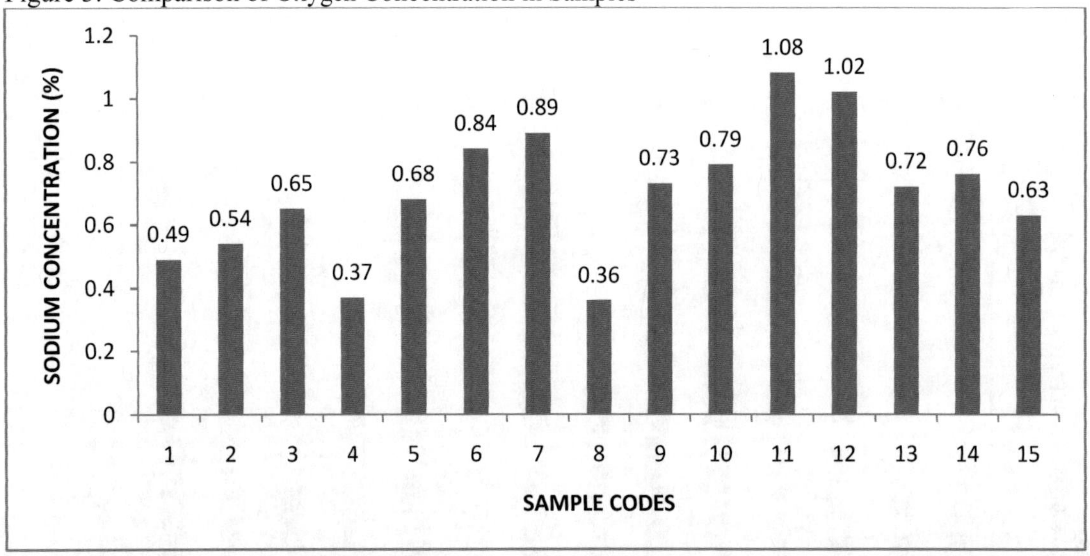

Figure 4: Comparison of Sodium Concentration in Samples

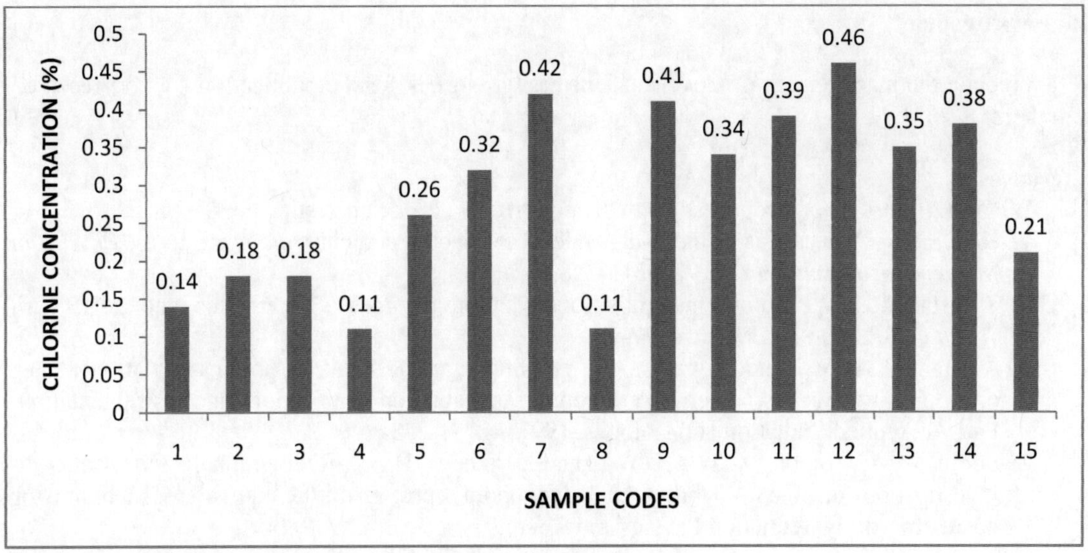

Figure 5: Comparison of Chlorine Concentration in Samples

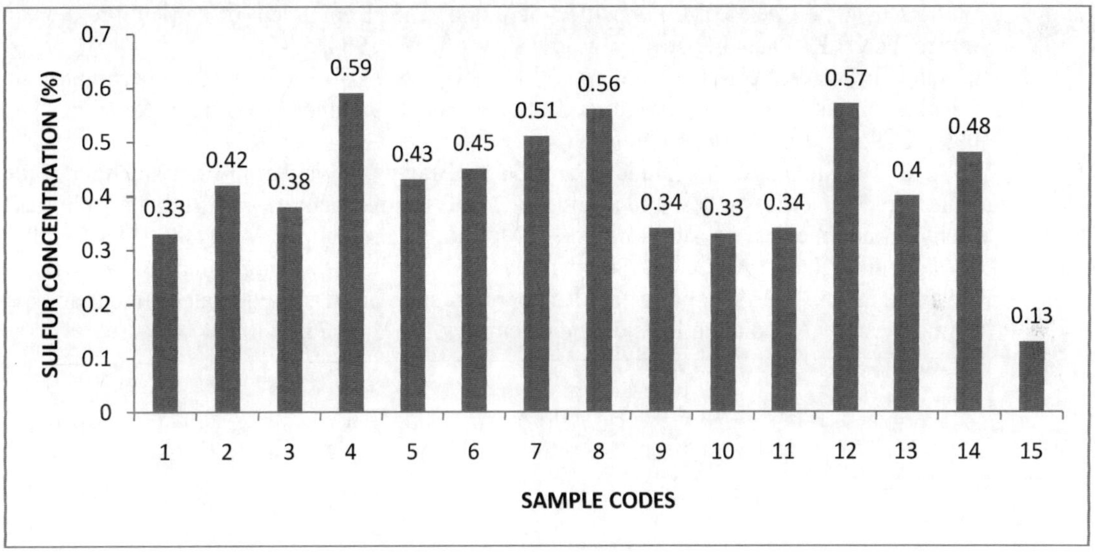

Figure 6: Comparison of Sulfur Concentration in Samples

4. Conclusion

RBS method has been used to determine the elemental constituent of several wood samples. The methods capability to detect element with low atomic number, has been harnessed as the idea non-invasive method for determination of low Z elemental matrix in materials samples such as wood. Looking at the composition for the wood samples, it was seen that the elemental composition of the wood followed the same trend as stated in literature. The knowledge of the elemental composition of wood can be harnessed as substitute material for animal and human biomas in application in health physics for radiological studies, medical application, radiation physics, radiology, nuclear engineering, radiation dosimetry, phantoms, radiation protection and space research.

Acknowledgement

We appreciate the management of Covenant University, Nigeria for the publication support received from them.

References

[1] WPGS (2015) - Wood Properties Growth and Structure 2015. treetesting.com.
[2] WHITE, D. R. An analysis of the Z-dependence of photon and electron interactions. *Physics in Medicine and Biology, 22,* 1977b219-228.
[3] C. Constantinou., "Phantom materials for radiation dosimetry," I. liquids and gels*British Journal of Radiology,* 1982, *52*217-224.
[4] D. A. Bradley, A. A. Tajuddin, C. W. A. C. W. Sudin, and S. Bauk, " Photon attenuation studies on tropical hardwoods," International Journal of Radiation Applications and Instrumentation. Part A. Applied Radiation and Isotopes, 1991, 42771-773.
[5] Tajuddin, A.A., Sudin, C.W.A.C.W., and Bradley, D.A., Radiographic and scattering investigation on the suitability of Rhizophora spp. as tissue-equivalent medium for dosimetric study, Radiation Physics and Chemistry, 1996, 47 739-740 .
[6] Sudin, C.W.A.C.W., Tajuddin, A.A., and Bradley, D.A., Evaluation of tissue-equivalent media for dosimetric studies, Proceeding of local seminar activities on radiation physics, biophysics and medical physics, Universiti Malaya, Kuala Lumpur, Malaysia, 1988. pp. 71-80.
[7] R.C. Pettersen, "The Chemical Composition of Wood" U.S. Department of Agriculture, Forest Service, Forest Products Laboratory, Madison, WI 53705.1984, 57-58.
[8] Nielsen and H. Forrest "Ultratrace minerals". Authors: USDA, ARS Source: Modern nutrition in health and disease / editors, Maurice E. Shils ... et al..Baltimore : Williams & Wilkins, c. 1999, 1283-303, Issue Date: 1999
[9] C. W. A. C. W. Sudin, A. A. Tajuddin, and D. A. Bradley, "Evaluation of tissue-equivalent media for dosimetric studies, Proceeding of local seminar activities on radiation physics, biophysics and medical physics," Universiti Malaya, Kuala Lumpur, Malaysia, 1988, 71-80.
[10] A. A. Tajuddin, C. W. A. C. W. Sudin,and D. A. Bradley, "Radiographic and scattering investigation on the suitability of Rhizophora spp. as tissue-equivalent medium for dosimetric study," Radiation Physics and Chemistry,1991, 47 739-740.

Construction of a portable cost effective temperature and humidity measuring device

A P Aizebeokhai[1], I O Ekumatalor[1], K D Oyeyemi[1] and N L Obafemi[1]

[1]Department of Physics, Covenant University, Nigeria

ahzegbobor.aizebeokhai@covenantuniversity.edu.ng; tommysin16@gmail.com

Abstract. Temperature and humidity are the most significant weather parameters in human development. The monitoring of temperature and humidity manually has been characterized with inaccurate measurements due to human errors. This project describes the construction of a portable cost effective device for measuring temperature and humidity; it is based on wireless technology. The measuring device obtains temperature and humidity data by using DHT22 sensor. The hardware used for the project includes Arduino Uno, Liquid Crystal Display (LCD) and DHT22 sensor; the software employed is the Arduino IDE. C programming language was used and the libraries of each component were installed in the Arduino IDE. A program code in the Arduino IDE (sketchbook) was created to enable the DHT22 sensor, LCD and Arduino work. The DHT22 was initialized at 5 V. The readings were gathered serially and transferred to a computer which communicates with the Arduino and displays the values on the LCD for human interface. The device was tested at different locations and different times for two consecutive days. The results show that the cost effective, portable device is able to detect and record temperature and humidity effectively.

Keywords: Arduino IDE, Temperature, Humidity, C programming, DHT22

1. Introduction

Weather is the given condition of a place or an environment at a particular time; these conditions include temperature, humidity, and fog. Weather forecasting is a process in which different techniques are employed in predicting the future state of the atmosphere in a given place. Weather plays a fundamental role in human lives. It is of vital importance to different people in various ways. It controls the distribution of rain on earth. All living organisms on earth require water to survive; human beings require fresh water for drinking. Also, rain water is required for agriculture; the farmers are able to know when to plant and what precautions to take. Weather conditions and events such as hurricanes, tornadoes, flood, and drought has a huge impact on human civilization. An example of this is hurricane Katrina, which took place on the Gulf of the US killing nearly two thousand people and displacing almost one million people. Air transportation also depends on Weather as temperature and humidity can determine the behaviour of a plane in the air. It also plays an important role in the shipping industry because they can be used to predict hurricanes and other storms, although ship construction has become more advanced to terrible weather conditions, hurricanes can still damage or even capsize some ships.

Weather is varied by the difference in air pressure in different locations. This difference depends on the direction of the sun at a specific spot. This spot is determined by latitude from tropics in pressure and temperature. Weather helps in keeping track of different climate patterns which include temperature, humidity, rainfall, wind speed, wind direction. Weather forecasts is predicted by collecting quality data about the state of the atmosphere in a given place using sensors to obtain values or data and trying to manipulate a deep understanding of how the atmosphere would evolve on that day. These temperature and humidity device would be carried out by using an Arduino microcontroller. Weather forecasting performed manually has turned out to be a major problem as inaccurate data is generated and it leaves a negative impact. The construction of an Arduino based weather station will help in predicting weather conditions accurately for farmers, weather based industries and so on. With this monitoring system, data about their field would be readily available with them. These would not only save time but also labour.

In weather monitoring, conditions such as temperature, humidity and rainfall or precipitation can be measured using sensors [1]. The advancement of technology has helped in making small and reliable electronic sensors capable of monitoring environmental conditions efficiently [2] [3]. [4] developed a monitoring system where sensors are used in measuring indoor weather and environment based on the conditions mentioned. Combination of these sensors with data acquisition system has proved to be a better approach for temperature and relative humidity monitoring [5]. [1] [6] introduced wireless sensing micro system for environmental monitoring, using capacitive-based sensors. [7] [8] introduced the usage of surface acoustic waves (SAW) devices as temperature sensor and pressure sensor respectively. These systems, however, are quite expensive and complex in nature as some of them require the use of on-chip transmitter circuit and involve fabrication processes. The aim of this research is to construct a portable, light weight, cost effective and low power consumption device for measuring temperature and humidity. The essence is to develop a device that is capable of providing in real-time temperature and humidity data at maximum speed without delay.

2. Methodology

The methodology adopted in this project involves the construction of a device for measuring temperature and humidity using a DHT22 sensor. The data observed or measured by the sensor is transmitted to a computer by an Arduino. The values are then displayed on a Liquid Crystal Display (LCD) for human interface [9]. The flow chart of the device is shown in Figure 1a. The principle (mode of operation) of this device explains the function of each component and their outputs. The circuit diagram is presented in Figure 1b. It was constructed using DHT22 which is a temperature and humidity sensor. The sensor was initialized by supplying +5 V. A program code in the Arduino IDE (sketch book) was created to enable the DHT22 sensor, liquid crystal display and Arduino works. The readings are gathered serially into a computer which communicates with the Arduino and displays the values on a Liquid Crystal Display (LCD) for human interface.

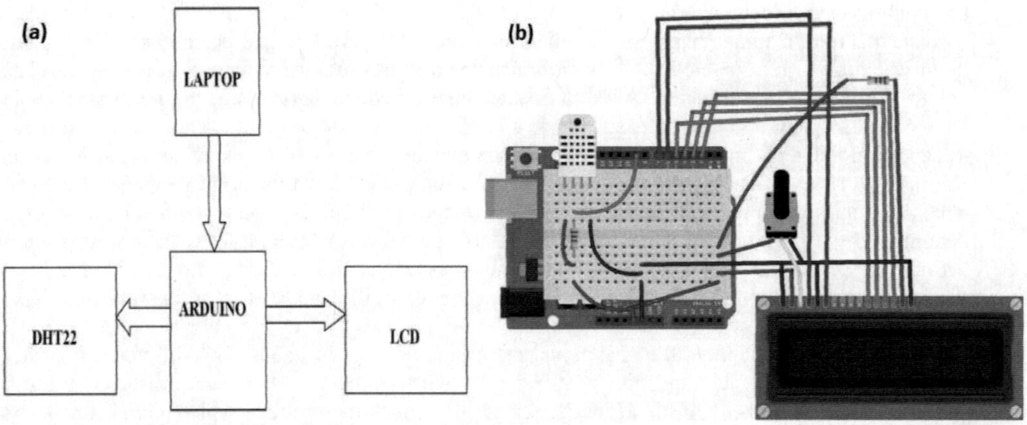

Figure 1: (a) Flow chart for the construction of the device and (b) Circuit diagram used to implement humidity and temperature device.

3. Results and Discussion

Multiple tests were carried out to check the performance of the hardware and software of the device. The tests were carried out at different stages during the development of the device in order to determine the efficiency of each segment of the device. These include the circuit for the temperature and humidity device and the results when tested in selected environments. The readings obtained are stored in files (Microsoft word). The files containing the readings were copied in Microsoft word Excel, where graphs are plotted to give a perfect view of the temperature and humidity for each specified location. The circuit of the temperature and humidity device consists of liquid crystal display, DHT22, Arduino Uno and potentiometer. The device was first implemented on a bread board as shown in Figure 2(a) to detect the working state. This allows us to know if the Dht22 sensor is reading values; it is first cross checked on the serial monitor of the Arduino IDE as shown in Figure 2(b).When it was confirmed that the DHT22 sensor is reading values, the Arduino Uno was then plugged into the computer before being transferred to a Vero board where it was soldered in order to reduce the size of the circuit as shown in Figure 2(c). Since portability is the main objective of this research, the final prototype of this device was constructed to fit within a small plastic enclosure as shown in Figure 3. The device was tested at different locations within the Covenant University campus and data acquired involving both temperature and humidity are presented in Table 1-5.

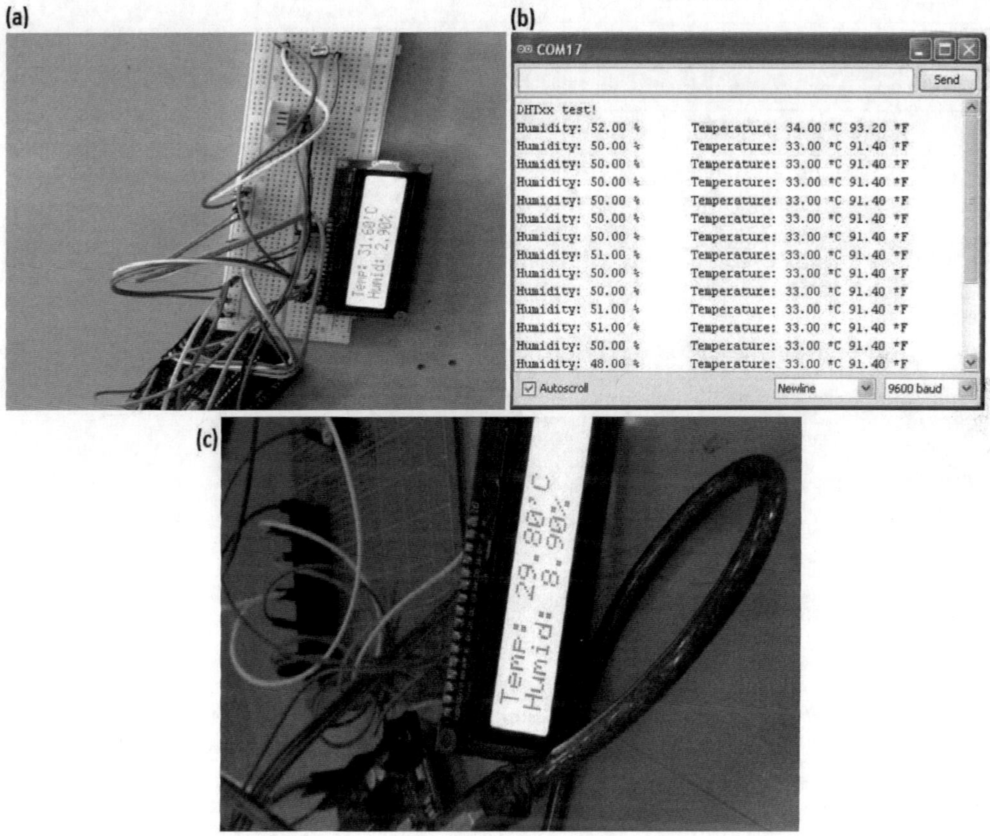

Figure 2: (a) Implementation of temperature and humidity values on the serial monitor of the Arduino IDE, (b) Temperature and humidity test circuit on breadboard and (c) Temperature and humidity test circuit on Vero board.

IOP Conf. Series: Earth and Environmental Science **173** (2018) 012006 doi:10.1088/1755-1315/173/1/012006

Figure 3: Temperature and humidity device in operation.

Table 1: Table of temperature and humidity in the Chapel

Date	**Time(m)**	**Temperature($^{\circ}c$)**	**Humidity (%)**
02/04/2018	9:00am	30.00	1.60
02/04/2018	9:10am	32.60	2.10
02/04/2018	9:20am	33.00	3.50
02/04/2018	9:30am	35.10	5.60
02/04/2018	9:40am	35.60	5.80
02/04/2018	9:50am	37.60	7.30

Table 2: Table of temperature and humidity in Physics Laboratory

Date	**Time(m)**	**Temperature(℃)**	**Humidity (%)**
02/04/2018	8:00am	30.40	4.10
02/04/2018	8:10am	35.00	6.60
02/04/2018	8:20am	37.50	8.50
02/04/2018	8:30am	37.60	9.60
02/04/2018	8:40am	38.00	9.90
02/04/2018	8:50am	42.30	8.10

Table 3: Table of temperature and humidity in Basket Ball Court

Date	**Time(m)**	**Temperature(℃)**	**Humidity (%)**
02/04/2018	10:05am	32.70	1.40
02/04/2018	10:10am	30.50	1.80
02/04/2018	10:15am	29.50	2.70
02/04/2018	10:20am	35.40	3.70
02/04/2018	10:25am	35.90	3.80
02/04/2018	10:30am	36.40	4.90

Table 4: Table of temperature and humidity in Lecture Theather1

Date	Time(m)	Temperature(°C)	Humidity (%)
02/04/2018	10:45am	30.70	4.60
02/04/2018	10:50am	31.60	1.90
02/04/2018	10:55am	31.70	2.20
02/04/2018	11:00am	32.50	2.40
02/04/2018	11:05am	33.70	4.80
02/04/2018	11:10am	32.59	5.80

Table 5: Table of temperature and humidity in Cafeteria 1

Date	Time(m)	Temperature(°C)	Humidity (%)
03/04/2018	3:00pm	30.00	1.20
03/04/2018	3:05pm	32.90	1.30
03/04/2018	3:10pm	32.80	2.00
03/04/2018	3:15pm	37.80	5.00
03/04/2018	3:20pm	38.80	4.00
03/04/2018	3:25pm	32.80	5.10

4. Conclusion

The construction of the temperature and humidity device was successfully implemented using cheap components. These components include: Arduino Uno, DHT22 and potentiometer. The sensor transmits temperature and humidity wirelessly to the Arduino which then displays them on the Liquid Crystal Display. The characteristics of the device after testing include low power consumption, small size, cheap cost, efficiency and high accuracy. The materials used for the project can be acquired locally. The device is suitable for all kinds of short distance conditions of wireless data acquisition and transmission.

Acknowledgements

The authors wish to appreciate the Covenant University Centre for Research, Innovation and Discovery for the conference support sponsorship.

References

[1] Ong K G Grimes C A Robbins C L and Singh R S 2001 Design and application of a wireless, passive, resonant-circuit environmental monitoring sensor. *Sensors and Actuators* **93** pp 33–43.

[2] Ayara W A Omotosho T V Usikalu M R Singh M S J and Suparta W 2017 Development of a solar charged laboratory bench power supply. *Journal of Physics: Conference Series* **852(1)** pp 012044.

[3] Falayi E O Usikalu M R Omotosho TV Ojoniyi O S and Akinwumi S A 2017 Impact of meteorological parameters over Covenant University, Ota. *Journal of Physics: Conference Series* **852(1)** pp 012012

[4] Odlyha M Foster G M Cohen N S Sitwell C and Bullock L 2000. Microclimate monitoring of indoor environments using piezoelectric quartz crystal humidity sensors. *Environmental Monitoring* **2** pp 127–131.

[5] Moghavvemi M Ng K E Soo C Y and Tan S Y 2005 A reliable and economically feasible remote sensing system for temperature and relative humidity measurements. *Sensors and Actuators* **117** pp 181–185.

[6] DeHennis A D and Wise K D 2005 A wireless micro system for the remote sensing of pressure, temperature and relative humidity. *Journal of Micro Electromechanical Systems* **14(1)** pp 12-22.

[7] Buff W Plath F Schmeckebier O Rusko M Vandahl T Luck H and Muller F (1994) Remote sensor system using passive SAW sensors. In Proceedings of IEEE Ultrasonic Symposium. Cannes, 585p.

[8] Roneel V S 2014 Development of a remote automatic weather station with a pc-based data logger. International Journal of Hybrid Information Technology **7(1)** pp 233-240.

[9] Nagendra D 2017 Monitoring environmental parameters. Humidity and temperature using Arduino based microcontroller and sensors. 2p.

Design and Construction of LED Matrix Display

Arijaje. T. E[1], Azi. S. O[2], Akinpelu. A[1], Onumejor. C. A[1], Omeje. M[1], Ogunrinola. I. E[1]

[1] Covenant University, Ota,Ogun State, Nigeria. [2] University of Benin, Benin City, Nigeria

theophilus.arijaje@covenantuniversity.edu..ng

Abstract. This paper presents a compilation of the design and construction of a programmable scrolling matrix display that scrolls the following characters "WELCOME TO PHYSICS DEPARTMENT". Dot Matrix Display based on the logic output signals of a digital signal processor (DSP) and microcontroller or other logic device. The Perspex board measuring 0.3x1.5 meters was acquired and holes of 4mm were drilled for the Perspex in order to allow the LED measuring 5mm in diameter fit tightly into the drilled holes. The circuit diagram was gotten after careful design and calculation had been done. The complete work was designed, simulated and routed with proteus and after the design, construction and testing, the device (programmable scrolling matrix display) was ascertained to be working satisfactorily.

Keywords: LED's; Microcontroller; Scrolling Matrix Display; Digital signal processor

1. Introduction

Analog display of information has long been in use in various forms some of which include sign posts for showing direction or caution signs, bill board for displaying prices and exchange rates, score boards for displaying scores in a game or even large boards for advertisement purposes. This involved arranging some form of lighting in a particular style to depict what is to be displayed. This style has one major challenge; no flexibility which possess numerous problems for the display. Due to this major setback in the analog format of display for signposts andWrite-up, it is important to solve this challenge digits and automatic electronic display system [1]. This is to enable the viewers to get enough information displayed in attractive colours [2]. The extent at which dissemination of information has reached in Nigeria and the world at large will be maintained if the information to be displayed is burnt into the microcontroller especially the ones that have to do with bigger firms. This will bring about mutual relationship between firms and its consumers. Matrix display of information is based on the logic input signals processor (DSP) and or other logic device.

LED matrix display operates on what is known as persistence of vision (POV). One LED only is displayed at every point in time and your eye combine the lit LEDs to get the image. This is known as the persistence or permanency of vision (POV) and this happens when the eyes sees images that changes at a very fast rate of 10 scans per second. A flickering effects is observe when the scan increases to 20 scans per second but when it is increase to 30 scans per second the effect is observe to be smooth.

In this work, the display works at a rate greater than 100 scans per second and the eye observe the display to be steady. If we reduce the scan rate, each of the LEDs can be seen as the efficiency of the microcontroller clock is reduced. In this work, we will designed and construct the electronic circuit for driving the LED matrix and also program the microcontroller and demonstrate its operation on the matrix display board.

2. Materials & Methods

2.1. Design process and implementation

In the design of the matrix display circuit, very little calculation is required due to the fact that most of the integrated circuits (ICs) used in this circuit operate with digital signals and require little or no transformation[5, 6]. Nevertheless, the calculations made in the cause of design are outlined as follows

2.2. Calculation for Voltage Regulator Parameters

Vout = LED regulated voltage
Vin(max) = LED maximum input voltage
Iload(max) = LED maximum load current for this work
Vout = 5V
Vin(max) = 5V
Iload(max) = 3A.
The inductor L, output capacitor (C_{out}) and input capacitor (C_{in}) are 680μH, 1000μf and 100μf respectively. [7]

2.3. The Current Limiting Resistors to the LEDs

Practically, a voltage (V_{DD}) = 2volts can considerably bias an LED to be driven in the forward direction. Diodes require 20mA from a source for full brightness. [7]

To limit the current flowing through the LED, it must have a resistor connected in series in other to prevent the LED from damaging. In this project a block of LED is used which consists of four LEDs (two parallel connected and two series connected). The resulting block is terminated to act as a single unit

For the series EDs,

$$V_{DD} = 2 \text{ x } 2v = 4\text{Volts} \tag{1}$$

For paralleled LEDs

$$I_{DD} = 20mA \text{ X } 2 = 40mA \tag{2}$$

In the circuit, driving voltage V_S = 12volts is applied to the line driving the LEDs. This is too high for the LEDs and hence, the current limiting resistors are needed whose value is calculated below

$$\frac{\mathbf{12v-4v}}{\mathbf{40mA}} \tag{3}$$

$$= \frac{\mathbf{12-4}}{\mathbf{0.04}} = 200\Omega \tag{4}$$

Considering safety, it is advisable to use resistor with higher value for each LED block to reduce the current driving them. 270Ω resistor is chosen as current limiting resistor for each block of the LEDs.

Hence the current that will drive the LED block from this current limiting resistor will be;

$$\frac{12v-4v}{270\Omega} = 30\text{mA} \quad (5)$$

30mA is a reasonable current value to drive a block of four (4) super bright LEDs requiring a maximum current of 40mA current.

2.4. *Construction Procedure*

2.4.1. The LED Matrix Board. The Perspex board measuring 0.3x1.5 meters was procured. The points for the placement of LEDs were carefully identified and evenly marked. Holes of 4mm were drilled in the marked points in the Perspex in order to allow the LED measuring 5mm in diameter fit tightly into the drilled holes. 980 holes were drilled for 980 LEDs on the Perspex.

The LEDs were tested and the faulty (including those with poor brightness) LEDs were promptly replaced. The LEDs were grouped in four and interconnected to form a single lighting. LEDs were wired such that all the cathodes along a row share a common terminal and all the anodes along a column shared a common terminal.

The arrangement of the dot matrix are the LED matrix which have a common anode for LEDs in a row and in each of the , the anode is common and the LED with a common cathode each row are also common.

*2.4.2. The Circuit.*The circuit diagram was gotten after careful design and calculation had been done. Most of the components were gotten from the local electronics store. The workability of all the discreet components and integrated circuits were verified before they were all used in the project.[8, 9]The complete project was designed, simulated and routed with Proteus and found to be working satisfactorily, before the actual components were then assembled on a printed circuit board. All the IC sockets were first soldered on the PCB before all other component were soldered after being tested.

2.4.3. Software Design. The software required for displaying characters "WELCOME TO PHYSICS DEPARTMENT" on the LED matrix. The hex code is downloaded on to the PIC18F4620 microcontroller and is verified by interfacing the microcontroller pins to the pins of the board. For any given character, a corresponding pattern of LED is generated and is used to display the character at run time. The equivalent data bytes required for the Display of "WELCOME TO PHYSICS DEPARTMENT" character were computed and stored at internal ram address and were transmitted sequentially by monitoring the TI flag of the register.

3. Results and Discussion

In displaying the world "WELCOME TO PHYSICS DEPARTMENT", we start by displaying the word 'WE'. The D-1 is selected, meaning D_1 is drag low and other columns are unselected columns by obstructing their paths by dragging D_2 through D_5 pins to logic high. By so doing, the first column is made active while the LED's in this rows R-1 through R-7 of this are turn on, which is done by the application of forward bias to row (1, 5, 7) until we get to the information to be displayed which is 'WELCOME TO PHYSICS DEPARTMENT', where D- means column and D- means row.. Consequently, a quick scan across the column greater than 100 times per second and the LEDs are turn on in each of the row of that column, the is graphics is seen to be still by stability of vision or observation.

The written program was examined and the result was used to test its functionality by executing the program with some chosen inputs before it was burnt into the microcontroller..
We then burnt the code which we have written in C compiler into the PIC18F4620, and later we assemble the control circuit and fix connectors and also the serial port to pc. Now we setup the software into computer and when we type the messages in the computer we get the display.
The tables below show the codes for the display 'WELCOME TO PHYSICS DEPARTMENT'.

Table 1: Shows the digital code for the alphabet 'WELCOME TO PHYSICS DEPARTMENT'

R/D	D_6	D_5	D_4	D_3	D_2	D_1
R-1	1	1	0	1	0	1
R-2	0	1	0	1	0	1
R-3	0	1	0	1	0	1
R-4	1	1	0	1	0	1
R-5	0	1	0	1	0	1
R-6	0	1	0	1	0	1
R-7	1	1	0	1	1	1
R-1	1	1	0	0	1	0
R-2	0	1	0	0	1	0
R-3	0	1	0	0	1	0
R-4	0	1	0	0	1	0
R-5	0	1	0	0	1	0
R-6	0	1	0	0	1	0
R-7	1	1	0	1	1	0
R-1	1	1	0	1	1	0
R-2	1	1	0	1	1	0
R-3	0	1	0	1	1	0
R-4	0	1	0	1	1	0
R-5	0	1	0	1	1	0
R-6	0	1	0	1	1	0
R-7	0	1	0	1	1	0
R-1	1	0	1	1	0	1
R-2	0	0	0	1	0	1
R-3	0	0	0	1	0	1
R-4	0	0	1	1	0	1
R-5	0	0	0	1	0	1
R-6	0	0	0	1	0	1
R-7	0	0	1	1	0	1
R-1	0	1	1	0	1	1
R-2	0	1	1	0	0	1
R-3	0	1	1	0	0	1
R-4	0	0	1	0	0	1
R-5	0	0	1	0	0	1
R-6	0	0	1	0	0	1
R-7	0	0	1	0	0	1
R-1	1	1	0	1	0	1
R-2	0	1	0	1	0	1

R-3	1	0	0	0	1	0
R-4	1	0	0	0	1	0
R-5	1	0	0	0	1	0
R-6	0	0	0	0	1	0
R-7	1	1	0	0	1	0
R-1	0	1	1	0	1	0
R-2	0	0	1	0	1	0
R-3	0	0	1	0	1	0
R-4	0	0	1	0	1	0
R-5	0	0	1	0	1	0
R-6	0	0	1	0	1	0
R-7	0	1	1	0	1	0
R-1	0	1	1	0	1	1
R-2	0	1	1	0	0	1
R-3	0	1	0	0	1	0
R-4	0	1	0	0	1	0
R-5	0	1	0	0	1	0
R-6	0	1	1	0	1	0
R-7	0	1	1	0	1	1
R-1	0	1	1	0	1	1
R-2	0	1	1	0	0	1
R-3	0	1	1	0	0	1
R-4	0	0	1	0	1	1
R-5	0	0	1	0	0	1
R-6	0	0	1	0	0	1
R-7	0	0	1	0	1	1
R-1	0	1	1	0	1	1
R-2	0	1	1	0	1	1
R-3	0	1	1	0	1	1
R-4	0	1	1	0	1	1
R-5	0	1	1	0	1	1
R-6	0	1	1	0	1	1
R-7	0	1	1	0	1	1
R-1	1	1	0	1	1	1
R-2	0	1	0	0	1	0
R-3	0	1	0	0	1	0
R-4	0	1	0	0	1	0
R-5	0	1	0	0	1	0
R-6	0	1	0	0	1	0
R-7	0	1	0	0	1	0
R-1	1	0	1	1	0	1
R-2	1	0	0	1	0	1
R-3	1	0	0	1	0	1
R-4	1	0	1	1	0	1
R-5	1	0	0	1	0	1
R-6	1	0	0	1	0	1
R-7	1	0	1	1	0	1

R-1	0	1	1	1	1	1
R-2	0	0	1	0	1	1
R-3	0	0	1	0	1	1
R-4	0	0	1	0	1	0
R-5	0	0	1	0	1	0
R-6	0	0	1	0	1	0
R-7	0	0	1	0	1	0

[10, 11, 12]

In table 1 above, one pin source for the current for only one LED across each row at a time while the column pin will sink the current from more than one LED. For instance, to display the word "WE", C1 column will sink the currents six (6) LEDs. An external transistor arrangement is required since the microcontroller I/O pin cannot sink much electric current. In digital electronics, 1 means ON while 0 means OFF. When powered on the LED, it displays "WELCOME TO PHYSICS DEPARTMENT. Only the areas that have 1 are light up while the areas that are 0 are OFF as shown in the table above.

4. Conclusion

The matrix display of light is made up of LEDs arranged in a rectangular array in such a way that by turning on or off, picture or image or characters can be displayed. With the help of the dot matrix controller, information can be turn into signals thereby turning on and off lights in the matrix in other to turn on the required display. Hence the need to expand research and development in the LED matrix display technology. All thanks the advent of microcontroller which has made it possible for information to be displayed in any format.

Acknowledgment

The authors thank Covenant University for conference support

References

[1] B. Amolgajre "Led display literature review. Light emitting diode, Incandescent light bulb, halogen lamp,"2013
[2] M.Archer, "advanced digital electronics, oxford University press," 2000
[3] B. Borat, " A Course in Digital Processing," 6th edition, Wiley 1997
[4] M. A. Mazidi. "the Microcontroller and Embedded System," Second Edition, Pearson, 2012
[5] fL. Chris "Field application," Engineer cooper electronic technologies.2013
[6] T. Edward, *"Principles of information display,"* 3rd edition, graphic press. *Electrical Design News,* Volume 32, Issues 11-15, 1990, Rogers Publishing Company, 1987.
[7] J. Greenfield, " Digital Electronic Design," 4th edition, Oxford University press. 1999
[8] C Kesparis, "Electrical and Computer Engineering," 7th edition, central Florida USA. 2003
[9] C. W. Lu and K. J. Hsu "A high-speed low-power rail-to-rail Column driver for AMLCD application IEEE journal of solid-state circuit." Vol.39, No 8. 2004
[10] M.J. Bell. " An LED column driver using a switch capacitor DAC," IEEE Journal of solid-state circuit capacitor. Vol. 40, No 12, pp2756-2763, 2005, ISSN 0018-92
[11] J. Lim & A. Oppenheim, ¬"Advanced Topics in Digital Processing," 2nd edition, Prentice Hall limited, New Delhi. 1988
[12] A. Staveren & A. Roermond, A "Structured Electronic Design," 5th edition, springer publishers. 2000

IOP Conf. Series: Earth and Environmental Science **173** (2018) 012008 doi:10.1088/1755-1315/173/1/012008

Kolanut pod husk as a biobase catalyst for fatty acid methyl ester production using Thevetia peruviana (Yellow oleander) seed oil

E. U. Osakwe [1*], I. J. Ani[1], U. G. Akpan[1], M. A Olutoye[1]

[1]Chemical Engineering Department, Federal University of Technology, PMB 65 Minna, Niger State, Nigeria.

*Corresponding author. Email: uzeziosakwe@yahoo.com

Abstract. The potential of kolanut pod husk as a biobase catalyst for the conversion of Thevetia peruviana (yellow oleander) seed oil to fatty acid methyl ester (FAME) in transesterification reactions was investigated. Oil was extracted from Thevetia peruviana (yellow oleander) seeds and a yield of 60 % was obtained. Kolanut pod husk was ashed in a muffle furnace at 600 °C and characterized. In producing the FAME, three concentrations (0.5, 1.0 and 1.5 wt %) of kolanut pod husk ash (KPHA) were investigated in transesterification reactions. Methanol to oil ratio was kept constant at 6:1 while temperature and time were varied at 40-60 °C and 60-120 min respectively. The design of experiments was used in the transesterification step to determine the effects of process variables on the yield of Thevetia peruviana methyl esters (TPME). The result showed the TPME was consistent with ASTM specifications for biodiesel. The best yield of 84.50 % was obtained for 1.5 wt % KHPA at a temperature of 60 °C and a time of 90 min. XRD and elemental analysis indicated that the catalytic effect of KHPA resulted from the presence of potassium and the microstructural formation from calcination at 600 °C. The study highlights the possibility of reducing the cost of production for FAME by using non edible vegetable oil and kolanut pod husk, an agricultural waste as a biobase catalyst in transesterification reactions.

Key words: Kolanut pod, Fatty Acid Methyl Esters, Transesterification, Yellow Oleander

1. Introduction

Energy is an important tool needed for global development; the use of fossil fuel to meet energy demands has however brought with it prosperity as well as some negatives [1]. Energy sources from fossil fuel amount to about three quarters of the world's energy consumption [2] and this high reliance on it as a non renewable resource has led to its great depletion. The continuous and increasing need for energy resources and the desire to bridge the gap of insufficiencies has resulted in an obvious alternative: Biodiesel [3].

Biodiesel is attractive and a natural alternative to fossil fuel because it is a renewable fuel, it is environmental friendly; it is also biodegradable and non toxic [4]. Best of all it possess very similar characteristics with present day diesel fuel from fossils [5]. The very wide range of feedstock for biodiesel production is also an added advantage. Feedstock can be sourced from edible and non-edible vegetable oil, animal fats, waste cooking oil and more recently algae [5]. The production of biodiesel is achieved by the transesterification of triglyceride using alcohol with the aid of a catalyst [6]. Demand for biodiesel has greatly increased over the years; this has generated concern for its sustainability mainly because major industrial production of biodiesel is from food sources: starch and sugar [7]. However Research into new generation bio-fuels from cellulose by enzymatic conversion and fermentation as well as non-edible plants and plant parts are currently on going. It is believed that in the near future, commercialization of biodiesel production from non-edible sources will take center stage.

Thevetia peruviana as a crop is one of such non-edible sources of feedstock for biodiesel production. It produces seeds which contain up to 65 % of non-edible vegetable oil [8]. It belongs to the order apocynales of apocyanaceae family [9]. It is a small tree believed to have its origin in tropical America: Mexico, Brazil and West Indies [10]. It also found in tropical parts of the world including Nigeria. It is a plant that grows well under varying climatic conditions; in arid zones as well as in areas with high rainfall. The evergreen plant starts to flower after 18months of planting and grows to a height of 3-4m. It flowers and fruits all year round producing about 400- 800 fruits per year [11]. The fruits are green in colour and become reddish black when fully mature and ripe; each mature fruit contains one to four seeds in its kernel.

In addition to using non edible oil in biodiesel production, another way to help reduce the cost of production is utilizing agricultural wastes as heterogeneous catalysts in place of homogenous ones [12]. The ashes of theses agricultural wastes are known to contain oxides of potassium and oxides of other elements [13] which have been reported to serve as effective bio-based catalysts for transesterification reactions [12].

In this paper, we investigate the use of kolanut pod husk ash as a bio-based catalyst for the transesterification of Thevetia peruviana seed oil. We also optimized the process variables (Temperature, reaction time and catalyst amount) for the transesterification reaction using response surface methodology (RSM).

2. Materials and methods

2.1. *Materials*

Thevetia peruviana seeds were handpicked at a residential premise in the Federal Capital Territory (F.C.T) Water Board Staff Quarters, Lower Usuma Dam Area, Abuja. Kolanut pods were sourced from Suleja market in Niger State, Nigeria. All chemicals and solvents used were of analytical grade, and were used without further treatment.

2.2. *Methods*

2.2.1. Thevetia peruviana seed oil extraction. The fleshy outer layers of the Thevetia peruviana seeds were peeled off and the inner part inner part sundried for 7 days then it was shelled and decorticated. The seeds were then sundried for another 14 days until a constant weight was reached. The seeds were ground to powder using a manual grinding machine and stored in an air tight plastic bag. The Soxhlet apparatus and petroleum ether as extraction solvent were used for this work. 250 ml of petroleum ether was delivered into the soxhlet extractor. 25 g of powered Thevetia peruviana seed on Whatman No1 filter paper was placed in the thimble of the soxhlet extractor. The heating mantle was set at a specified temperature and the extraction was carried out for a period of 3 h. The extraction solvent in the solvent-oil mixture was recovered by distillation.

2.2.2. Characterization of Thevetia peruviana seed oil. The Thevetia peruviana seed oil obtained was analyzed to determine its physical and chemical properties. Using ASTM standard methods (1983), density, specific gravity, viscosity, acid value, iodine value and saponification value were determined.

2.2.3. Preparation of kolanut pod husk catalyst. The Kolanut pods were manually peeled and the kolanut pod husks collected were washed with distilled water. They were then cut into smaller pieces to aid faster drying. The cut pieces were spread on a drying mat for continuous sun drying; which was carried out for 30 days until a constant weight was achieved. At the end of the drying period the kolanut pod husks were placed in a furnace where it was combusted at a temperature of about 600°c for 3 h (until white ash was formed) and allowed to cool for another 2 h. The kolanut pod husk ashes (KPHA) formed were collected and stored in air tight plastic bags.

2.2.4. Characterization of KPHA. To determine the specific properties of the Kolanut Pod Husk Ash, various analyses were carried out. X-ray Diffraction (XRD) Analysis helped to show crystalline phases and x-ray diffraction patterns. To determine the surface area and pore size distribution, Brunauer, Emmett and Teller (BET) Analysis were performed. While elemental composition of the KPHA was determined using atomic absorption spectrophotometer.

2.2.5. Fatty acid methyl esters production. The biodiesel samples were prepared using transesterification reactions. KHPA Catalyst load of 0.5%, 1%, 1.5 % were utilized in the transesterification processes with methanol-to-oil ratio of 6:1 and reaction temperature of 40, 50 and 60°C. The reaction time used was varied from 60 to 120 min. A constant mixing speed of 400 rpm was utilized for all runs. The Thevetia peruviana methyl esters formed were then separated from the glycerol using a separating funnel after washing with distilled water. The washed Thevetia peruviana methyl esters were then allowed to air dry for 2 days. Air dried Biodiesel produced were characterized. Biodiesel yield was determined using equation 1.

$$\textbf{Biodiesel yield} = \frac{\textbf{Volume of biodiesel produced}}{\textbf{Volume of oil used}} \quad (1)$$

2.2.6. Experimental design for transesterification by Response surface methodology (RSM). The Thevetia peruviana methyl ester production was optimized using Response surface methodology, The Box-Behnken design with three factors and three coded levels having five centre points and 12 factorial points was used for a Quadratic, second order response surface.

Table 1 represents the optimization process variables and their code levels. It gives the three process variables used in the transesterification process, showing the symbols assigned t each variable and the three range and levels.

Table 1
Experimental range and levels of independent variables for Box-Behnken design

Variables	**Symbols**	**Range and Levels**		
		-1	0	1
Temperature (°C)	A	40	50	60
Time (min)	B	60	90	120
Catalyst (wt %)	C	0.5	1	1.5

The design expert version 7.0.0 (trial version) software was utilized in the regression analysis of actual experimental data.

2.2.7. Biodiesel Analysis. The biodiesel produced were analyzed for their density, viscosity, specific gravity and acid values according to ASTM standards. The biodiesel samples were also analyzed by Gas Chromatograph- Mass Spectrometer.

3. Results and discussion

3.1. Thevetia peruviana seed oil extraction and properties

The Thevetia peruviana seed oil produced was light gold in colour; it had a sweet smell and was liquid at room temperature. Analysis of the Thevetia peruviana oil produced determined some of its physical and chemical properties. Characteristics such as density, specific gravity, viscosity, acid value, iodine value and saponification value were determined, as shown in Table 2 below

IOP Conf. Series: Earth and Environmental Science **173** (2018) 012008 doi:10.1088/1755-1315/173/1/012008

Table 2
Physiochemical properties of Thevetia peruviana seed oil

Property	**Value**
Density @ 40 °C (g/cm^3)	0.895
Acid value (mg KOH/g)	1.089
Free fatty acid (mg KOH/g)	0.545
Saponification Value	341.599
Specific Gravity @ 40 °C	0.901
Viscosity @ 40 °C (Cp)	32.620
Oil Content (wt %)	59.66
Moisture content of seed cake (wt %)	2.797
Iodine Value (g I_2/100g)	70.600

The oil content of 60% obtained from the Thevetia peruviana seeds indicates that the seeds are rich in oil. This value is however lower than the value reported by Masime et al. [8] (66%) but can be considered to be ideal. It is ideal because for a seed to be seen as commercially viable with regards to large scale oil extraction it should possess at least 40% in oil content [14]. The low value for free fatty acid content of the extracted oil obtained in this research is an indication of its good resistance to hydrolysis. The acid value and free fatty acid values of 1.0893 and 0.5446 respectively were lower than the values reported by Yarkasuwa et al. [14] and Ogunneye et al. [15] but higher than that reported by Deka et al. [16]. The viscosity is also an important parameter in determining if an oil type would produce good biodiesel. The value obtained from this study shows a workable viscosity that would not prolong the time needed for complete transesterification of the oil.

3.2. Characterization of KPHA catalyst

The elemental compositions of KPHA are presented in Table 3. The major elements present were potassium, calcium, magnesium, silicon, phosphorous, sulphur, chlorine, aluminium and sodium with percent mass fraction 33.914, 25.382, 12.398, 7.335, 7.329, 4.921, 3.529, 1.959 and 1.740 respectively. The other elements were present in trace quantities.

X-ray diffraction analysis was used to identify the crystalline phase composition and crystalinity of KPHA as expressed in Fig 1. It showed the presence of Calcium Carbonate, 6-Aminopurine trihydrate, Calcium Hydroxide and Nitrogen. The KPHA surface area, pore size and pore volume were also determined. The specific surface area was found to be 398.547 m^2/g, pore volume 0.203 cm^3/g and the pore diameter 2.138 nm.

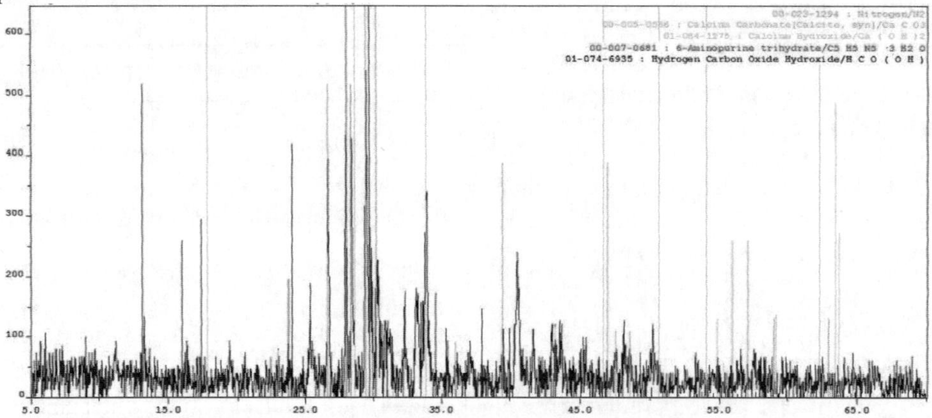

Figure 1. XRD chromograph of KPHA

From the pore diameter obtained the material is mesoporous. The high surface area and mesoporous nature attributes to the good performance of the catalyst.

Table 3
Elemental composition of KPHA

Element	**% mass fraction**
Potassium, K	33.914
Calcium, Ca	25.382
Magnesium, Mg	12.398
Silicon, Si	7.335
Phosphorous, P	7.329
Sulphur, S	4.921
Chlorine, Cl	3.529
Aluminum, Al	1.958
Sodium, Na	1.740
Iron, Fe	0.540
Manganese, Mn	0.407
Zinc, Zn	0.400
Others	0.148

3.3. Optimization of fatty acid methyl esters using RSM.

Table 4 below is a representation of the Ben-Behnken design showing the 17 standard runs randomly generated to minimize errors in the observed response. As shown in the table, the actual values obtained from the transesterification reactions and the predicted values generated are very similar which implies that the mathematical model generated (a quadratic regression model) best describes the process.

Table 4
Box-Behnken design for fatty acid methyl ester production

Standard run	**A**	**B**	**C**	**Actual Value**	**Predicted Value**	**Residual**
1	-1	-1	0	52	52.19	-0.19
2	1	1	0	72	72.56	-0.56
3	-1	1	0	57	56.44	0.56
4	1	1	0	79	78.81	0.19
5	-1	0	-1	55	55.50	-0.50
6	1	0	-1	73	73.13	-0.13
7	-1	0	1	59	58.88	0.13
8	1	0	1	84.5	84.00	0.50
9	0	-1	-1	65	64.31	0.69
10	0	1	-1	68	68.06	-0.063
11	0	-1	1	70	69.94	0.063
12	0	1	1	76	76.69	-0.69
13	0	0	0	69	68.94	0.060
14	0	0	0	68.7	68.94	-0.24
15	0	0	0	69.2	68.94	0.26
16	0	0	0	68.5	68.94	-0.44
17	0	0	0	69.3	68.94	0.36

A: Temperature; B: Time; C: Catalyst load

The Analysis of variance (ANOVA) test was used in the evaluation of the adequacy of the model. The results show that the quadratic regression model expressed using Equation 2 gives the best description of the process.

$$Yield = +68.94 + 10.69 * A + 2.62 * B + 3.56 * C + 0.50 * A * B + 1.88 * A * C + 0.75 * B * C - 2.91 * A^2 - 1.03 * B^2 + 1.84 * C^2 \quad (2)$$

where A is the reaction temperature, B is the reaction time and C is the catalyst load.

The Model F-value of 336.05 implies the model is significant. Values of "Prob > F" less than 0.0500 indicate model terms are significant. In this case A, B, C, AB, AC, BC, A^2, B^2 and C^2 are significant model terms. Values greater than 0.1000 indicate the model terms are not significant. The standard deviation as obtained was 0.61 while the mean value was 67.95. The R-Squared value evaluated in experimental design is used to determine the amount of variation around the mean described by the model by adding factors to the model. The R^2 value evaluated for this work was 0.9977 meaning 99.77% of the variability can be accounted for. The Pred R-Squared value of 0.9688 is in reasonable agreement with the Adj R-Squared of 0.9947 Also the signal to noise ratio of 67.548 was obtained indicating adequate signal as a ratio greater than 4 is desirable.

Using the design expert software to solve the regression equation represented as equation 2, the optimum values for the process variables needed for the production of Thevetia peruviana methyl esters from Thevetia peruviana seed oil were obtained. They are reaction temperature 60 °C, reaction time 90 min and Catalyst load 1.5 wt %. The predicted yield of Thevetia peruviana methyl ester conversion under optimal condition was 84 %, this was however validated to be 84.12 % by redoing the experiment three times.

Table 5
Summary of the test of significance for regression coefficients and ANOVA

Source	**Sum of Squares**	**Df**	**Mean Square**	**F Value**	**p-value Prov>F**	
Model	1140.44	9	126.72	336.05	< 0.0001	significant
A-Temperature	913.78	1	913.78	2423.36	< 0.0001	
B-Time	55.13	1	55.13	146.19	< 0.0001	
C-Catalyst	101.53	1	101.53	269.26	< 0.0001	
AB	1.00	1	1.00	2.65	0.1474	
AC	14.06	1	14.06	37.29	0.0005	
BC	2.25	1	2.25	5.97	0.0447	
A^2	35.59	1	35.59	94.40	< 0.0001	
B^2	4.49	1	4.49	11.90	0.0107	
C^2	14.29	1	14.29	37.91	0.0005	
Residual	2.64	7	0.38			
Lack of fit	2.19	3	0.73	6.45	0.0517	Not significant
Pure error	0.45	4	0.11			
Cor Total	1143.08	16	-			

Figure 2a illustrates the three dimensional plot of the relationship between temperature and time on the Thevetia peruviana seed oil conversion to biodiesel. The plot shows an increase in temperature significantly affects the conversion of oil to biodiesel. The three dimensional plot as a function of temperature and catalyst load is represented in Figure 2b The significant influence of the variables of temperature and catalyst load on oil to biodiesel conversion is well defined in the plot. Figure 2c represents the plot of reaction time and catalyst load, showing interactions between the reaction time and catalyst load.

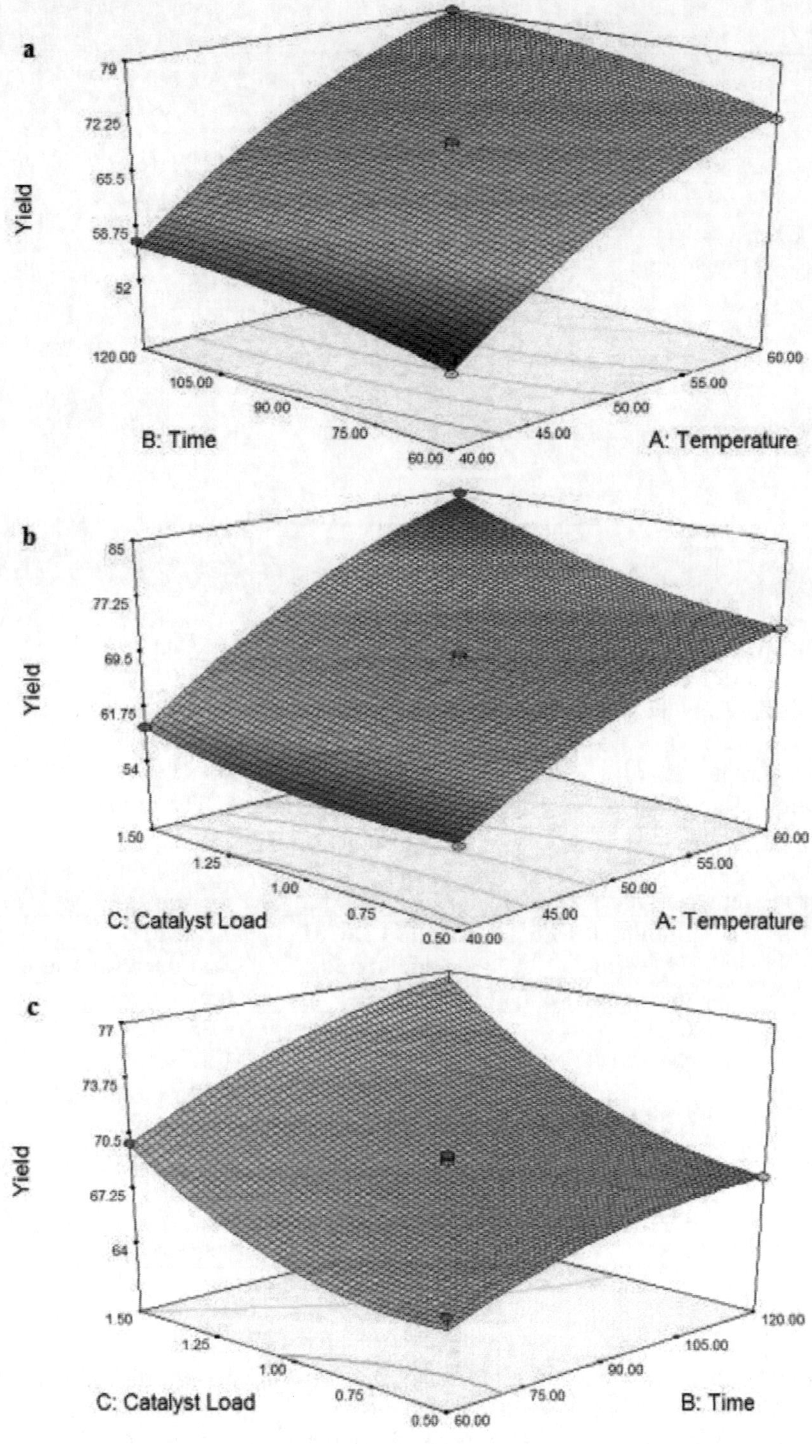

Figure 2. Three dimensional plot of Thevetia peruviana methyl ester yield

Figure 3 represents the plot of predicted values versus actual values obtained. The plot shows the similarity between the actual values obtained and the values predicted by the model, proving the accuracy of the model

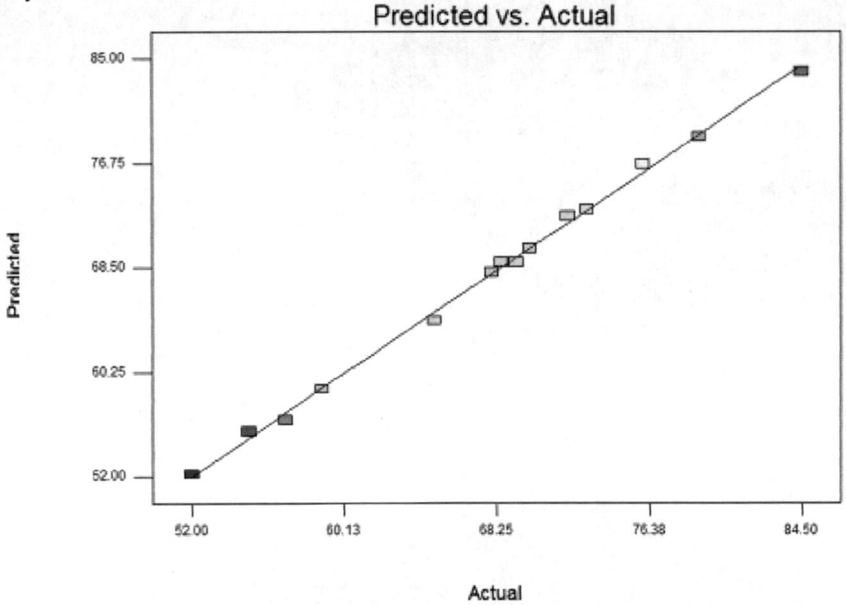

Figure 3. Plot of predicted values against actual values of biodiesel yield

3.4. *Characterization of fatty acid methyl ester produced.*

The summary of the fuel properties of the biodiesel produced from the KPHA catalysed transesterification reaction from Thevetia peruviana seed oil is shown in Table 6. The properties fell within the range of values specified by ASTM D6751 and ASTM D975 standards.

Table 6
Properties of biodiesel from Thevetia peruviana seed oil

Properties	**Biodiesel from Thevetia peruviana seed oil**	**ASTM D6751 Standard for biodiesel**	**ASTM D975 Standard for Diesel fuel**
Density (g/m^3)	0.872	0.88	0.85
Kinematic Viscosity 40 °C (mm^2/s)	5.4	1.9 - 6	1.3 - 4.1
Saponification Value (meq/kg)	170.8	-	-
Acid value ($mgKOHg^{-1}$)	0.2	<0.8	-
Iodine Value (gI_2 $100g^{-1}$)	84	-	-
Flash point (oC)	168	100-170	60-80
Cloud point (oC)	-5.6	-	-
Pour point (oC)	0.8	-15 to 10	(-35) to -15

4. Conclusion

The potential of the developed KPHA as an effective bio-base catalyst for biodiesel production has been demonstrated in this work. The availability of kolanut pod husk as an agricultural waste means the catalyst can be sourced with little or no cost. The successful optimization of process variables (reaction temperature, reaction time and catalyst load) in the transesterification reaction using response surface methodology to obtain a biodiesel yield of 84.5 % emphasizes the fact that Thevetia peruviana seed oil and kolanut pod husks can serve as cost effective feedstock for biodiesel production.

References

[1] Cottrell J, Fortier F, Schlegelmilch K, (2015). Fossil Fuel to Renewable Energy: Comparator Study of Subsidy Reforms and Energy Transitions in African and India Ocean Island States. United Nations Office for Sustainable Development, Incheon, Republic of Korea, January 2015.

[2] Etim E.E., (2012). The prospects of biofuels in Complementing the Nigeria's energy needs. International Journal of Environment and Bioenergy. 4(2), 74-85.

[3] Gashew A., Lakachew A., (2014). Production of Biodiesel from non edible oil and its properties. International Journal of Science, Environment and Technology. 3(4), 1544-1562

[4] Basumatary S, (2013). Transesterification with heterogeneous catalyst in the production of biodiesel: a review. Journal of Chemical and Pharmaceutical Research. 5(1), 1-7.

[5] Okoro J.O, Tanure A.A, (2015). A review on the use of non-conventional heterogeneous catalysts for biodiesel synthesis. International Journal of Petroleum Engineering. 2(4), 128-138.

[6] Betiku E., Akintunde A.M., Ojumu T.V, (2016). Banana peels as a biobase catalyst for fatty acid methyl esters production using Napoleon's plume (Bauhinia monandra) seed oil: A process parameters optimization study. Energy. 103(2016), 797-806.

[7] Ruth L, (2008). Bio or burst? The economic and ecological cost of biofuels. Embo Reports 9(2), 130-133.

[8] Masime J.O., Ogur E., Mbatia B., Aluoch A.O., Otieno G. (2017). Optimization and thermodynamics of the extraction of yellow oleander seed oil using soxhlet extractor. Journal of scientific Achievements. 2(1), 43-50.

[9] Dhoot S.B., Jaju D.R., Deshmukh S.A., (2011). Extraction of Thevetia peruviana seed oil and optimization of biodiesel production using Alkali-catalysed methanolysis. Journal of Alternative Energy Sources and Technologies. 2(2), 8-16.

[10] Jabar J.M., Lajide L., Bakare I.O., Oloye M.T. (2016). Extraction and characterization of vegetable oil from Thevetia peruviana and Jatropha Curcas seeds. FUTA Journal of Research in Sciences. 12(1), 73-80.

[11] Ibiyemi S.A., Fadipe V.O., Akinremi O.O., Bako S.S., (2002). Variation in oil composition of Thevetia Peruvians Juss "Yellow olender" Fruit, Journal of Applied Science and Environmental Management. 6 (2), 61-65. ISSN 1119-8362

[12] Amos O., Ogunniyi D.S., Odetoye T.S. (2016). Production of Biodiesel from Parinari polyandra B. Seed Oil using Bio based Catalysts. Nigerian Journal of Technological Development. 13(1), 26-30.

[13] Azeez S. O., Agbagunde G. S. (2016). Production and Characterization of liquid detergents from some agricultural waste products. Nigerian Journal of Technology. 35(1), 60-65.

[14] Yarkasuwa C. I., Wilson D., Michael E. (2013) Production of Biodiesel from Yellow Oleander (Thevetia peruviana) Oil and its Biodegradability. Journal of the Korean Chemical Society, 57(3), 377– 381.

[15] Ogunneye, A.L., Banjoko, O.O., Jabr, J.M., Shomoye, O.F. (2014). Extraction and characterization of oil from Thevetia peruviana seeds (yellow oleander) using nhexane as a solvent, Proceedings of the International Conference on Science, Technology, Education, Arts, Management and Social Sciences, isteams Research Nexus, Afe Babalola University, Ado Ekiti, Nigeria, 195–202.

[16] Deka D.C., Basumatary, S., (2011). High quality biodiesel from yellow oleander (Thevetia peruviana) seed oil. Biomass Bioenergy, 35, 1797–1803.

Comparative Analysis of Saturated and Under-saturated oil Viscosity Correlations using Statistical Tools, Niger Delta Case Study

Ogunkunle Temitope Fred, Abraham, V. Damilola., Afolabi Richard Oyekunle and Oriola Opedamola

Department of Petroleum Engineering, Covenant University, Ota, Nigeria

temitope.ogunkunle@covenantuniversity.edu.ng

Abstract. Crude oil viscosity is one of the most important fluid properties that affects fluid flow behavior; either in pipeline hydraulics or in the porous media (reservoir). Viscosity is a vital physical property that plays a major role in the petroleum industry, the production processing and transportation of oil due to influence on the flow through porous rock, oil wells, multiphase flow through tubing and piping system. Therefore, the need for accurate determination of viscosity for oil and gas applications cannot be overemphasized. Numerous empirical correlations exist in literature for predicting crude oil viscosity but their accuracy is limited based on range of conditions of application, composition of the crude used in developing the correlation, specific range of data and experimental conditions. In the present work, experimental data of oil viscosity from different samples of Nigerian oil reservoirs were statistically compared with correlation predicted viscosity using the most common viscosity empirical correlations. Validity and accuracy of these empirical models has been confirmed for both saturated and under-saturated Niger Delta oil samples. It was observed that for under-saturated oil viscosities, Elshawarky & Alikhan's correlation gave a better prediction based on the Absolute average percentage error and standard deviation while for the case of saturated oil viscosities Chew and Connally proved to be the closest to the experimental results.
Keywords: Viscosity, Statistical analysis, Crude oil

1. Introduction

Fluid sampling is a vital aspect of the petroleum engineering. Fluid sampling enables the understanding of the behaviour of various reservoir fluids as they exhibit distinct behaviour based on their composition/properties. Viscosity is an important aspect of fluid sampling as it affects the fluid flow behaviour. Viscosity is generally known as an intensive property of a fluid that causes an internal resistance of the fluid to flow[1]. Crude oil viscosity, a Newtonian fluid is a vital physical property that plays a major role in the petroleum industry, the production processing and transportation of oil due to its influence on the flow through porous rock, oil wells, multiphase flow through tubing and piping system[12]. Crude oil viscosity plays a major controlling and determining role in the successful implementation of secondary recovery process, EOR processes and reservoir simulation modeling. Optimum reservoir management and sound design facilities is hinged on reliable evaluation of viscosity data[1].

Viscosity is often determined using PVT analysis, however, this laboratory determination of viscosity is expensive and time consuming. An accurate correlation that can reliably determine the viscosity of crude oil in the Niger Delta region will reduce the costs and time involved in the determination of the viscosity of crude oil. There are currently a large number of existing viscosity correlations that has gained acceptance in the industry, such include [1, 3, 4, 9, 16 & 17], but most of which were differs in terms of regions of the world and different compositions of crude oils, specific range of data and different experimental conditions. Therefore, this posed a serious accuracy and precision concern in the applicability of these correlations for the Niger Delta, where the properties of the fluid may differ from the conditions at which those models were developed. Determination of the most accurate viscosity correlation from the already existing correlations that can accurately represent the viscosity conditions in the Niger Delta will serve as a means for further prediction of the viscosity of crude oils in this region. The purpose of this study is to provide a comparative analysis of existing saturated and under-saturated oil viscosity correlations to determine an accurate oil viscosity correlation that can best fit the oil viscosity conditions in the Niger Delta. Where quick estimates of oil viscosity are required in the Niger Delta, the selected viscosity correlations can be applied.

2 METHODOLOGY

The determination of the performance of the correlation involved the use of the available correlations to predict (estimate) the experimental database (measured) for the reservoir oil viscosity property. Comparative analysis on the two values (measured and predicted) was carried out to evaluate the efficiency of the correlation[12]. The performance of the existing correlations in this research- saturated oil, under-saturated oil viscosities were evaluated by using statistical tools to evaluate the accuracy of the correlation with the Niger Delta experimental viscosity

2.1 Correlation Selection and Crude Oil Viscosity Determination

This research utilizes various viscosity correlations (saturated and under-saturated) to predict the viscosities involved in this research- measured by PVT analysis in the Niger Delta region considered. The PVT input data of the Niger Delta is to be used and from the application of this input data, comparative analysis of the viscosity of crude oils from the various viscosity correlations will be determined. Correlations within the same range of data specified with the Niger Delta were tested. Extensive range of correlations were applied in this research which involved the early viscosity correlations and the recently developed correlations for screening purposes, correlations with an insignificant coefficient of determination value were removed. Correlations within the range of coefficient of determination between 0.9-1 and correlations developed for Niger Delta were then selected for further evaluation using statistical tools. 10 correlations were then selected for under-saturated oil viscosity and 9 were selected for saturated oil viscosity.

2.2 Comparative Analysis of the Selected Correlations and the Experimental Data

Comparative analysis to measure the agreement between the computed and measured viscosity data was done using the statistical tools. These statistical tools assist in making judgments on the variations in the computed results from the selected correlations and the experimental viscosity in determining performance of the selected correlations. The statistical tools applied in this research for the comparative analysis are outlined below:

2.3 Average Absolute Percentage Relative Error, AARE

This is a measure of the relative absolute deviation of the estimated values of the oil viscosity determined from the various correlations against the experimental values of the oil viscosity. It indicates the relative absolute deviation in percentage from the experimental values; the lower the value of the average absolute percentage relative error, the more accurate the correlation. E_r is expressed as: $AARE = \frac{1}{n}\sum_{i=1}^{n}|E_i|$

Where E_i is relative deviation of the estimated (predicted) value from an experimental value and is defined as:

$$E_i = \left[\frac{(\mu)_{exp} - (\mu)_{est}}{(\mu)_{exp}}\right] \times 100 \quad (1)$$

Where n=number of samples

2.4 *Standard deviation using Statistical Package for Social Science (SPSS)*

The standard deviation (S_r) is a measure of the percent relative spread or dispersion of the oil viscosity data distribution:

$$S_r = \sqrt{\frac{1}{n-1}\sum_{i=1}^{n}[\mu_i - \mu_{mean}]^2} \quad (2)$$

Where n=number of sample
μ_i= oil viscosity for each sample
μ_{mean} =mean oil viscosity
The standard deviation is generated from the paired sample test on SPSS. The lower the value of standard deviation means a smaller value of scatter. The difference between the standard deviation of the experimental data from the standard deviation using the various correlations indicates the measure of the relative spread from the experimental data. The accuracy of the correlation is determined by the value of the difference in standard deviation. Smaller value indicates better accuracy.

2.5 *Coefficient of Determination, $\mathbf{R^2}$(Graphically)*

The coefficient of determination is a simple statistical parameter that tells how the model/correlation fits the data and thereby represents a measure of the utility of the model. The R^2 in this study will be determined by making cross plots of the experimental oil viscosity against the predicted oil viscosities from the correlations. The closer the value of the R^2 is to 1, the better the model/correlation for oil viscosity fits the data acquired from the Niger Delta. $R^2$is expressed as:

$$R^2 = 1 - \frac{SSE}{SS_{yy}} \quad (3)$$

SSE measures deviations of experimental values from their predicted values:

$$SSE = \sum_{i-1}^{n}(\mu_{experimental} - \mu_{calc})^2 \quad (4)$$

SS_{yy} measures deviations of the experimental values from their mean value

$$SS_{yy} = \sum_{i-1}^{n}(\mu_{experiemental} - \mu_{mean})^2 \quad (5)$$

3 Materials

This research involves viscosity input data for bottom-hole samples taken from the Niger Delta region. The bottom-hole samples had been flash separated to obtain solution gas oil ratio, gas relative density and oil API gravity. Viscosity data had been obtained by rolling ball viscometer for various pressures and temperatures. The viscosity of the reservoir fluid data consists of different data sets based on the classifications of the viscosity of the crude oil. Four wells were utilized for the analysis.

The total pressure points for the four wells are 26 points for under-saturated oil viscosity and 23 points for the saturated oil viscosity. The range of data for the four wells is outlined at the various pressures.

Table 1: Fluid property

PVT property	Range
Oil Viscosity above Bubble Point (cp)	0.2-13
Oil Viscosity below Bubble Point (cp)	0.2-35
Bubble Point Pressure (psia)	1100-2400
Solution Gas Oil Ratio ($\boldsymbol{R_s}$)	27-1200
Relative Oil Density	0.6-0.9
Relative Gas Density	0.5-0.95
Pressure (psia)	165-6100

4. RESULTS AND DISCUSSION

The generated results from the comparative analysis using statistical tools are outlined below. These results include the calculated oil viscosities using correlations, graphical analysis, average absolute relative error, and standard deviation from the paired samples test using Statistical Package for Social Science (SPSS) to support the evaluation of the accuracy of the correlations in selecting the correlation with the least error from the overall analysis of these results generated from statistical measures.

4.1 Graphical Analysis

Graphical analysis was carried out by constructing plots of experimental oil viscosities against the calculated oil viscosity to show the coefficient of determination.

4.2 Comparative Analysis Using Average Absolute Relative Error:

The average absolute relative error analysis for the Under-saturated and Saturated oil viscosity was determined and ranked according to its increasing error. From table 2: Saturated Oil Viscosity ranking based on average absolute relative error, Chew & Connally[6] shows the lowest average absolute relative error of 7.4% with Kahn[10] showing the largest error of 18.1% due to the inverse viscosity trend.

From table 3: Under-saturated Oil Viscosity ranking based on average absolute relative error, Elsharkawy te al[7] shows the lowest average absolute relative error of 3.985% with Vasquez & Beggs[15], showing the largest error of 15.568%.

Table 2: Saturated Oil Viscosity ranking based on average absolute relative error.

RANKING BASED ON AVERAGE ABSOLUTE RELATIVE ERROR	
Correlations	**AARE(%)**
(Chew & Connally, 1959)	7.413
(Elsharkawy, A. M.; Alikhan, A. A., 1999)	10.715
(Naseri, Nikazar, & Mousavi Dehgani, 2005)	12.843
(Beggs & Robinson, 1975)	17.683
(Kahn, 1987)	18.075

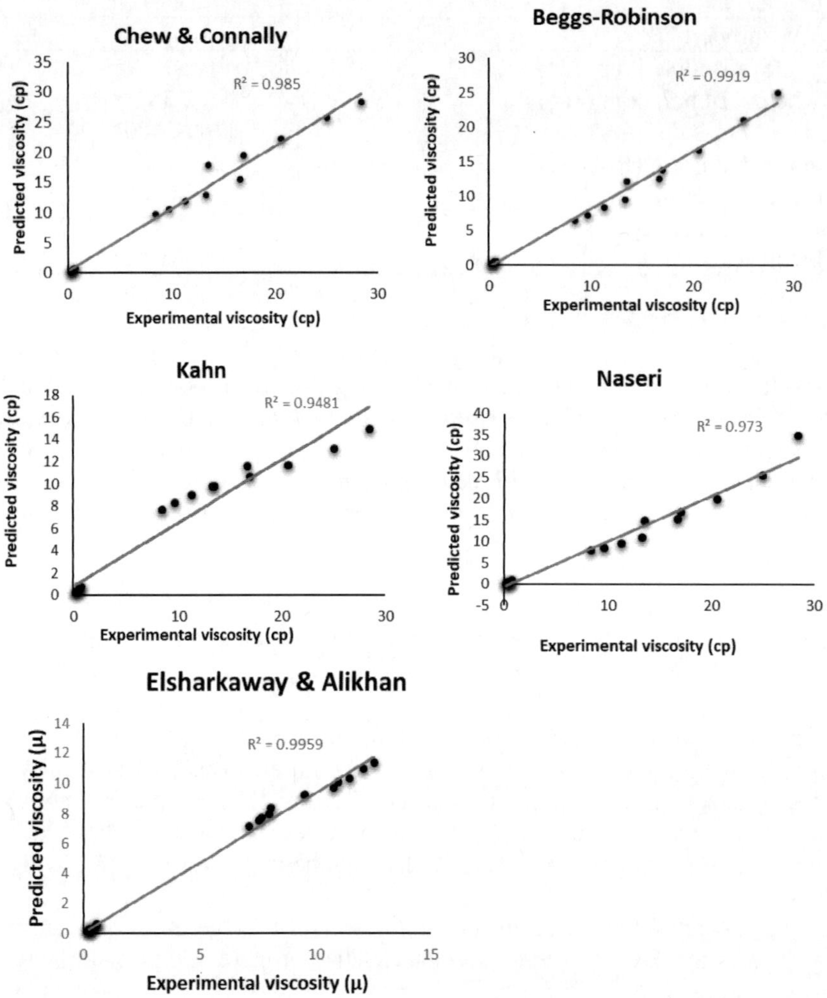

Figure 1: Predicted versus experimentally measured saturated oil viscosity

Table 3: Under-saturated Oil Viscosity ranking based on average absolute relative error.

RANKING BASED ON AVERAGE ABSOLUTE RELATIVE ERROR	
Correlations	AARE (%)
(Elsharkawy, A. M.; Alikhan, A. A., 1999)	3.985
(Kahn, 1987)	5.845
(Bergman & Sutton, 2006)	5.876
(Kartoatmodjo & Schmidt, 1994)	9.232
(Vasquez & Beggs, 1980)	15.568

4.3 Paired Sample T- Test Results Showing Standard Deviation:

The database for the T-test is a combination of all the four wells into one database in order to have a generalized result for the comparative analysis. Conclusion will then be drawn from the generalized result.

From Table 4: Mean and Standard Deviation difference between experimental and the correlations, the following can be inferred; Elsharkawy et al.[7] shows the least deviation from the experimental standard deviation and Vasquez & Beggs[15] with the largest deviation from the experimental standard deviation.

From Table 5: Mean and Standard Deviation difference between experimental and the correlations, the following can be inferred; Chew & Connally[6] shows the least deviation from the experimental standard deviation and Isehunwa[9] with the largest deviation from the experimental standard deviation.

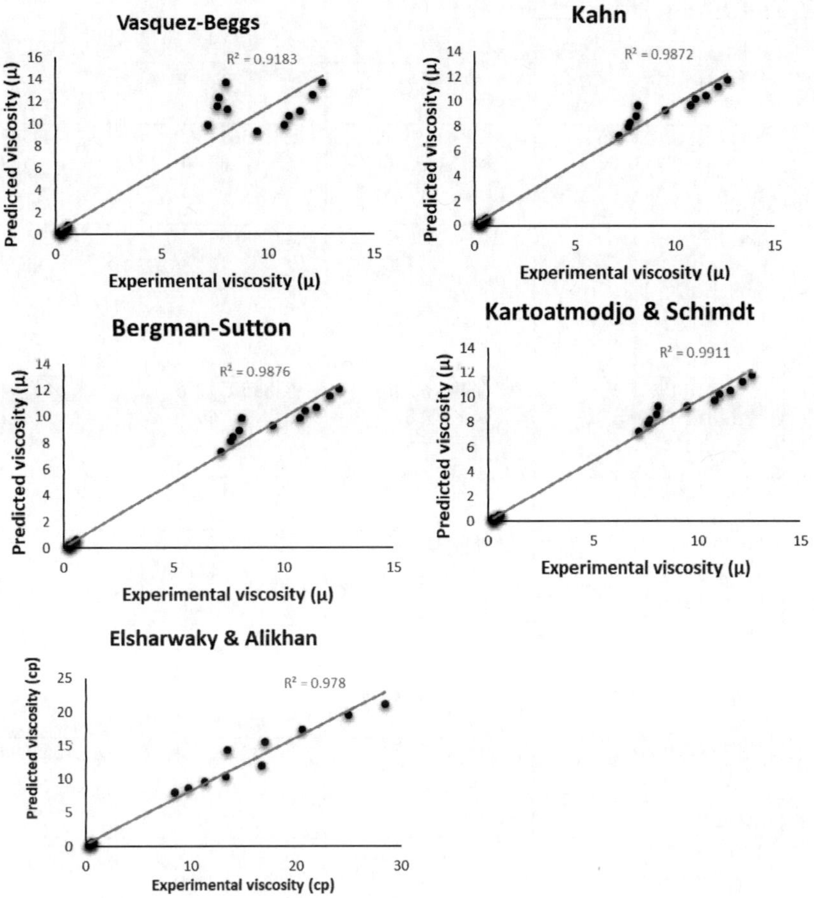

Figure 2: Predicted versus experimentally measured under-saturated oil viscosity

IOP Conf. Series: Earth and Environmental Science **173** (2018) 012009 doi:10.1088/1755-1315/173/1/012009

Table 4: Mean and Standard Deviation Viscosity difference between experimental and the correlations (Under-saturated)

		Mean	Std. Deviation	Std. Error Mean
Pair 1	EXPERIMENTAL - VASQUEZ-BEGGS	-.784756428660258	1.710361166063740	.335429421578858
Pair 2	EXPERIMENTAL – ELSHARKAWAY and ALIKHAN	.219552006327480	.445489691776270	.087367716595063
Pair 3	EXPERIMENTAL - ALMEHAIDEB	.196444835888013	.538955389554956	.105697848011412
Pair 4	EXPERIMENTAL - BEAL	.117411922695281	.495610162295201	.097197149564543
Pair 5	EXPERIMENTAL - A. HEMMATI	.098989621520830	.568479975871874	.111488095771440
Pair 6	EXPERIMENTAL - BERGMANSUTTON	-.025524181371737	.538428884345962	.105594591844694
Pair 7	EXPERIMENTAL - ISEHUNWA	.009546904592505	.568830858731586	.111556909561773
Pair 8	EXPERIMENTAL - PETROSKY	.594228977072582	1.012854083292956	.198637028274343
Pair 9	EXPERIMENTAL - KAHN	.056202147479362	.563537214473395	.110518740509060
Pair 10	EXPERIMENTAL – KARTOATMODJO and SCHIMDT	.089798186019134	.471647622100586	.092497708793482

Table 5: Mean and Standard Deviation viscosity difference between experimental and the correlations (Saturated)

		Paired Differences		
		Mean	Std. Deviation	Std. Error Mean
Pair 1	EXPERIMENTAL - CHEW and CONNALLY	-.547797568282084	1.249623996892070	.286683406309259
Pair 2	EXPERIMENTAL - ISEHUNWA	-3.301615830486082	6.964456908801230	1.597755992742538
Pair 3	EXPERIMENTAL - BEGGS.ROBINSON	1.675032861855476	1.691537249728671	.388065259515895
Pair 4	EXPERIMENTAL - ABU-KHAMSIN and ALMARHOUN	2.911398170272748	3.916882204135443	.898594405346276
Pair 5	EXPERIMENTAL - KAHN	3.021394464540832	4.261791577850468	.977722042435944
Pair 6	EXPERIMENTAL – KARTOATMODJO and SCHMIDT	1.846997304642271	2.424607483914645	.556243105270335
Pair 7	EXPERIMENTAL – ELSHARKAWAY and ALIKHAN	1.469701008100454	2.255290248479358	.517399067446014
Pair 8	EXPERIMENTAL - NASERI	-.011853718903746	1.790547072091324	.410779670573617
Pair 9	EXPERIMENTAL - A. HEMMATI	3.527028085172082	4.544927300432159	1.042677832016995

5. Conclusion and Recommendation

In this research, the aforementioned objectives were achieved using statistical tools for a Niger Delta case study. From the results of the analysis, the following conclusions can be drawn:

- Using the correlations from the literature for various regions of the world, deviations from the experimental viscosities were obtained for the Niger Delta case study indicating that the correlations are geographically dependent.

From the comparative analysis for the under-saturated oil viscosities correlations using statistical tools such as average absolute relative error and standard deviation as the screening criteria. Elshawarkay & Alikhan proved to be the best correlation from the assessment showing the lowest average absolute error of 3.985% and lowest standard deviation of 0.4455 in the combined wells from the SPSS.

❖ From the comparative analysis for the saturated oil viscosities correlations using statistical tools such as average absolute relative error and standard deviation as the screening criteria. Chew & Connally correlation proved to be the best correlation from the assessment showing the lowest average absolute error of 7.413% and lowest standard deviation of 1.2496 in the combined wells from the SPSS.

However, the following recommendations are given in order to extend this study. The selected viscosity correlations should be applied in prediction of saturated and under-saturated oil viscosities when field viscosity data is unavailable. The performance of the selected viscosity correlations should be evaluated with other available viscosity data from the Niger Delta.

Acknowledgment

The authors thank Covenant University for conference support.

References

1. *Abu-Khamsin , S. A., & Al-Marhoun , M. A. (1991). DEVELOPMENT OF A NEW CORRELATION FOR BUBBLE-POINT. The Arabian Journal for Science and Engineering, 16, 99-106.*
2. *Almehaideb, R. A. (1997). Improved PVT correlations for UAE crude oils. SPE middle east oil show CONF (pp. 109-120). Mamamah, Bahrain: SPE.*
3. *Beal, C. (1946). The Viscosity of Air, Natural Gas, Crude Oil and its Associated Gases at Oil Field Temperatures and Pressures. TRANS AIME, 94-112.*
4. *Beggs, H. D., & Robinson, R. J. (1975). Estimating the Viscosity of Crude oil systems. Journal of petroleum technology, 1140-1141.*
5. *Bergman, D. F., & Sutton, R. P. (2006). Under-saturated Oil Viscosity Correlation for Adverse conditions SPE 103144. 2006 SPE annual techinal conference San Antonio. Texas: SPE.*
6. *Chew, J., & Connally, C. A. (1959). A Viscosity Correlation for Gas- Saturated Crude Oils. Trans (pp. 23-25). AIME.*
7. *Elsharkawy, A. M.; Alikhan, A. A. (1999). "Models for Predicting the Viscosity of Middle East Crude Oil". Fuel, 78, 891-903.*
8. *Hemmati-Sarapardeh, A., Khishvand, M., Naseri, A., & Mohammadi , A. H. (2013). Toward reservoir oil viscosity correlation. Chemical Engineering Science ELSEVIER, 53-68.*
9. *Isehunwa, O. S., Olamigoke, O., & Makinde, A. A. (2006). A Correlation to Predict the Viscosity of Light Crude Oils. 31st Annual SPE International Technical Conference and Exhibition in Abuja (pp. 1-5). Abuja: Society of Petroleum Engineers.*
10. *Kahn, S. A. (1987). Development of viscosity correlations for crude oils . Texas: university of texas .*

11. *Kartoatmodjo, F., & Schmidt, Z. (1994). Large data bank improves crude physical property correlation. Oil and Gas Journal, 4, 51-55.*
12. *Muhammad, Z. K., & Mohd, D. N. (2013). Developed Correlation to Estimate The Malaysian Crude Oil Density and Viscosity. Lahore.*
13. *Naseri, A., Nikazar, M., & Mousavi Dehgani, S. A. (2005). A correlation approach for prediction of crude oil viscosities. Journal of petroleum science and engineering, 163-174.*
14. *Petrosky, G. E. (1990). PVT correlations for Gulf of Mexico Crude oils. Lousiana: University of southern Louisiana.*
15. *Vasquez, M., & Beggs, H. D. (1980). "Correlation for Fluid Physical Property Predictions". JPT, 32, 968-970.*
16. *Glaso, O., (1980). Generalized Pressure-Volume-Temperature Correlations. Journal of Petroleum Technology. Pp. 785 - 795.*
17. *Ikiensikimama, S. S., & Ogboja, O. (2009). New Bubble-Point Pressure Empirical PVT Correlation. In Proceedings 33rd annual SPE international technical conference and exhibition in Abuja, Abuja, Nigeria.*

Exergy and Energy Analysis of Solid Oxide Fuel Cell Fuelled Using Methanol Propane, and Butane

Emordi* C.O., Abdulkareem A.S., AzeezO.S., Afolabi S.

P.M.B 65, Chemical Engineering Department, Federal University Of Technology, Minna, Nigeria

emordichukwuma@gmail.com

Abstract. Exergy and Energy analysis used the principle of conservation of mass and energy, in connection with thermodynamics law in designing and analysing of thermal system This Paper focus on Energy and Exergy analysis of Solid Oxide Fuel Cell (SOFC) system by computer simulation. A comprehensive effect of Energy and Exergy analysis on SOFC using Methanol, Propane, and Butane fuel system with the aid of Thermolib simulation toolbox was investigated. From the configurations simulated for the three different fuel sources, the data produced were used for thermodynamic analysis. The result obtained showed that Butane configuration has the highest energy efficiency of 50.3% .while Methanol and Propane system has an energy efficiency of 46.5% and 47.7% respectively. The total energy produced by Methanol, Propane and Butane Fuel System are 273.66 KW, 234.67 and 263.92 kW respectively. While that required are 69.45 KW, 0.062 KW and 4.972 KW respectively. This shows that the highest energy requiring and producing system is the Methanol Fuelled system. The principle of energy conservation was met in all configurations. The Exergy analysis indicated that around equipment such as Reactor, lambda burner and SOFC stack there is a change between the inlet and outlet chemical Exergy. However the chemical Exergy for Pumps, Heat exchangers, compressor and Mixer remains the same because no chemical reaction occurs in them. In addition, equipment such as pump and compressor gives higher Exergetic efficiency than others. In terms of overall loss work, Methanol, Propane, and Butane system are 422.2 KW, 247.7KW, 195.7 KW respectively. The Overall System Exergetic efficiency of the three fuel systems are 44.2%, 49.3%, and 46.7% respectively. This shows that Propane system has the highest Exergetic efficiency and least irreversibility. While Methanol fuel system has the lowest Exergetic efficiency and most irreversibility. Exergy and energy efficiency favours the choice of propane fuel system. This is because Propane is the most Exergy efficiency system.

Keywords: Solid oxide fuel cell, Exergy analysis, energy analysis, Methanol, Propane, Butane.

1. Introduction

Continued industrial growth as a result of the developments in science and technology has improved comfortability and human standard of living. It has also led to the increase in demand for energy [1]. World stocks of fossil fuels continue to fall, as it is the main source of energy, leading to shortfalls in supply and creating serious environmental degradation as they introduce foreign materials in the atmosphere. This concern has led several countries to review their energy policy and sought radical measures to remove waste [2].it has steered the search for other sources of energy that are more effective than the usual heat engine with no or slight emissions of pollutant [3, 4]. According to [5, 6], among several possible fuels, fuel cells (FCS) were recognized as one of the best, and the potential of clean energy technologies to meet all the constraints of energy supply, the economic development and environmental sustainability.

According to [1], fuel cells (FCs) are energy device that converts chemical energy to electrical energy and heat by reduction and oxidation reaction on the anode and the cathode of the cell using hydrogen or hydrogen-rich fuel and oxygen from the air. Merging both the benefits of internal combustion engines in their continuous operation and simplicity of use, and highly efficient and low-noise operation of batteries, fuel cells thus seems to be the perfect energy-alternative.

Hydrogen is the perfect fuel for fuel cell. Fuel cell powered with hydrogen derived from a renewable resource would emit nothing but gaseous water [3, 7, 8]. Despite the fact that hydrogen is the most abundant element on Earth, there are a number of obstacles limiting its direct use [9]. Hydrogen does not occur naturally on Earth, and as such, it is derived from substances notably: natural gas, gasoline, methanol, ammonia and propane [10]. This makes hydrogen expensive. Hydrogen is 14 times as light as air and as a result of this small volumetric density and molecular size, moving even a small amount of hydrogen is very expensive [11]. Hence, the transport and storage of hydrogen is considered as unfeasible. Another major drawback for the direct use of hydrogen is that no hydrogen infrastructure currently exists.

A good analysis should be done to the investigations of the energy and exergy efficiency using various hydrogen fuel sources. The use of energy analysis only to determine the effectiveness of a thermodynamic system can be misleading. Exergy analysis identifies the primary source of loss and offers more accurate information about the effectiveness of a system.

[12-15] describe the technique to carry out Exergy analysis and enhance the understanding of this technique. All the necessary equations to determine exergy analysis, including exergy calculation for heat exchangers, combustion chambers, compressor and gas turbines are explained in their research. In addition, the procedure for the calculation of the second law efficiency of each component and the overall system was illustrated.

[16], in their study centered on an integrated internal-reforming SOFC gas turbine (IRSOFC-GT) power system fed with natural gas. From the simulation results, IRSOFC-GT energy system could reach a net electrical efficiency beyond 59% and system efficiency beyond 79%.

[17] said that the Exergy analysis of a hybrid SOFC show that the SOFC stack was responsible for the largest Exergy loss This high rate of exergy loss originated in the ineffectiveness of chemical and electrochemical reactions that occur in SOFC stack. In addition, it shows the catalytic burner where anode off-gas flow was burned, has high exergy destruction. They concluded in the end that in designing hybrid energy systems, special emphasis ought to be given to the part with the greatest exergy loss.

[18] in his study gives an understanding from his investigation of operating various fuels on an integrated SOFC reformer system. The simulated results show that significant issues in thermal management may arise from the use of different fuels in the same integrated fuel cell reformer system.

Another study by [1] on the Thermo-economic performance of PEMFC shows that methane-fueled configuration requires less amount of energy compared to methanol configuration. The result of the analysis of the exergy of both configurations revealed greater exergy efficiency around important equipment such as pump and compressor. Based on exergy efficiency, methanol-configuration is better than methane system. Economic analysis revealed that methane system has low capital costs and slightly higher annual costs against methanol system.

[20-22] analyse various integrated system based on Exergy destruction calculation and second law efficiencies. They also show different ways achieved detailed physical and chemical Exergy calculations for both SOFC and PEMFC system.

The results of Thermo-economic study of a hybrid SOFC [7] using Thermolib with Ethanol and methane fuel shows that when the system is powered with ethanol, the solid oxide fuel cell stack constitute around 29% of the overall exergy loss. For methane configuration, the equipment having the greatest loss of exergy is CO2 compressor, which constitute around 51 percent of the overall exergy loss. The findings also show that the turbine has the greatest exergy efficiency in the two configurations. The operation of equipment in the two systems reveals that methane has more component with high exergy efficiency, while ethanol configuration has more component with high irreversibility. The simulated results show a total exergy efficiency of the methane and ethanol systems to be 22.3% and 24.6% respectively with a total loss of work of 1066.36 kW and 782.33 kW respectively. This Indicate that ethanol fuel configuration yield the greatest exergy efficiency and the highest percentage of irreversibility in relation to methane fueled system.

The use of Thermodynamic analysis in SOFC system with Thermolib using methanol, propane and butane as fuels have not been explored. This study presents a comparative analysis of three different fuels namely: Methanol, propane and Butane on the thermo-economic performance of SOFC. This study provides information essential for the design of profitable conversion systems.

2. Material and Method

2.1. Selection of SOFC Configurations

The configuration proposed by [1] was augmented by the SOFC-DEMO configuration of Thermolib software from which improved configurations for Methanol, Propane, and Butane fuel were derived.

2.2. Description of Process

After the process configurations for the three fuels have been selected, they are modeled and simulated with the aid of trail version 5.30 of Thermolib software. The parameters are set to attain a power yield of 200KW for each of the configurations. After simulation, the results obtained were used for Energy and Exergy analysis for the three systems.

The process description for Methanol fuel as shown in Figure 2.1 is the same with that of Propane and Butane fuel except that in Propane and Butane system there is an additional WGS reactor that converts CO to H2 and CO2.

For the Propane and Butane system modeling presented in Figure 2.2 -2.3, the Inlet fuel is mixed with steam and pump through heat exchanger2 to heat it up to the required temperature for the reaction. At the reformer, the Fuel is converted to H2 and CO. The reformer outlet is passed to a water-gas shift (WGS) reactor where CO is transformed to additional H2 and CO2 [23-25]. The stream from the WGS reactor is sent to the SOFC anode while compressed air is set to the cathode of the SOFC. It is assume that 85% of the hydrogen reacts with oxygen in the fuel cell to generate electricity. The SOFC outlet containing unused hydrogen is sent to the afterburner where it is burnt. So as to reduce the energy requirement of the system, the afterburner outlet is split into two streams: one is sent to heat exchanger 2 and the other to the reforming reactor to provide heat for steam reforming process.

2.3. Modelling and Simulation of Selected Configurations

After the process configurations for the three fuels have been selected, they are modelled using Thermolib, which is a MATLAB/Simulink's toolbox. Thermolib is developed for modelling and simulation of thermodynamic systems. A detailed description on using Thermolib to model the selected configuration is given by [1]. The parameters are set to reach a power productivity of 200KW for each of the configurations. After simulation, the results obtained were used for Exergy and economic analysis for the four systems. The simulated configurations for the three fuels are show in figure 2.1-2.3.

IOP Conf. Series: Earth and Environmental Science **173** (2018) 012010 doi:10.1088/1755-1315/173/1/012010

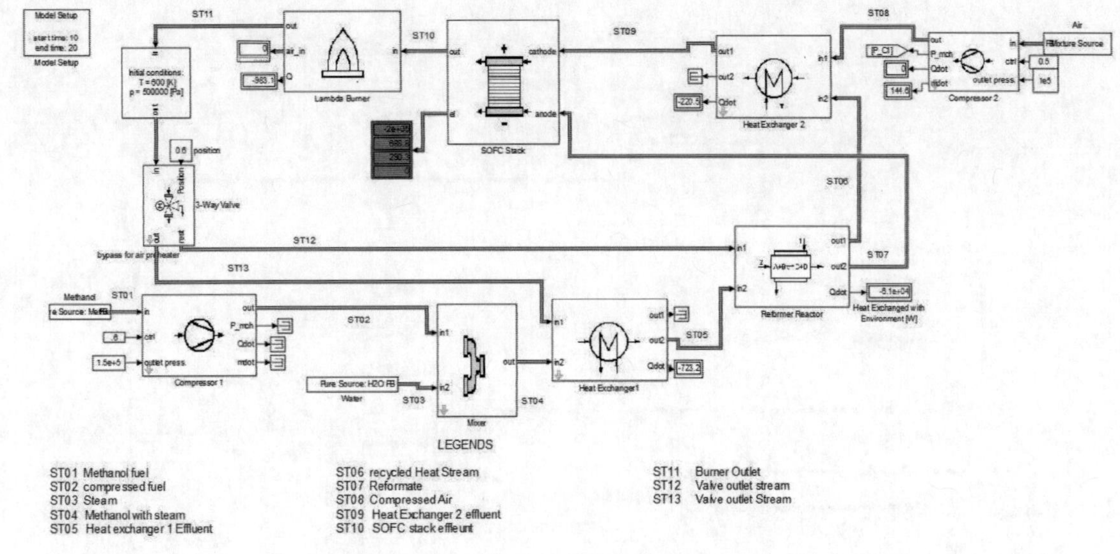

Figure 2.1 Methanol SOFC System

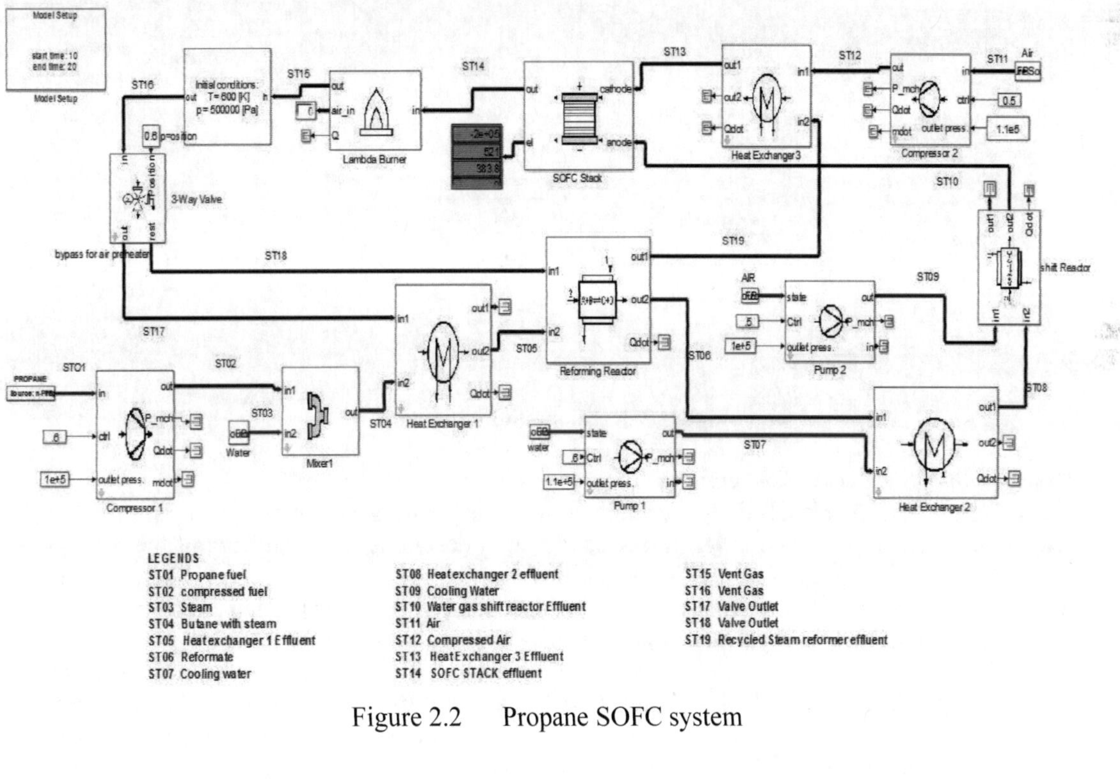

Figure 2.2 Propane SOFC system

IOP Conf. Series: Earth and Environmental Science **173** (2018) 012010 doi:10.1088/1755-1315/173/1/012010

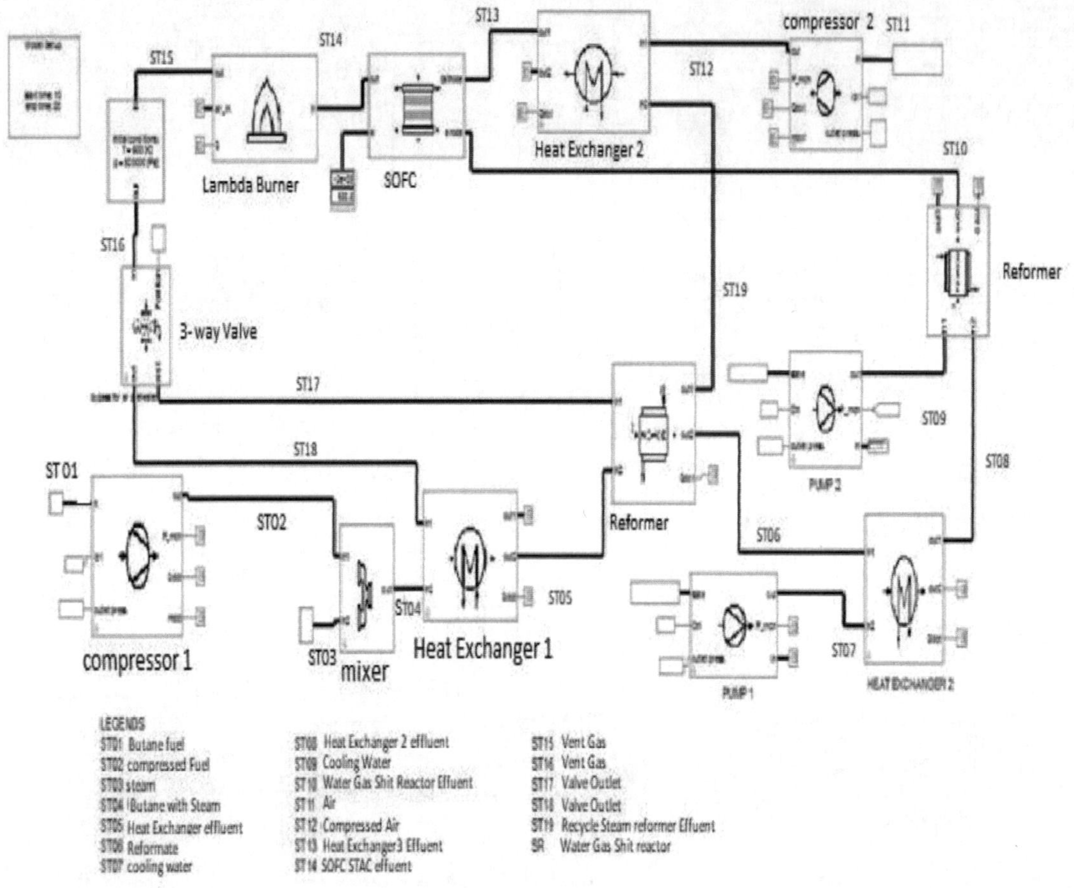

Figure 2.3 Butane SOFC system

2.4. Selection of SOFC Configurations

The configuration Exergy analysis is a technique of using the principle of conservation of mass and energy, in connection with thermodynamicssecond law in designing and analyzing of thermal system [24]. Exergy is the highest quantity of work that can be achieved when a system is taken to equilibrium (mechanical, thermal and chemical) with the environment [5]. The parameters determine in the exergy analysis are physical exergy and chemical exergy, loss of work and exergetic efficiency.

According to [24], the pressure P = 1.01 bar, temperature To = 298.15 K, and environment composition of 0.92% argon, 3.03 % water, 0.03 % carbon dioxide, 20.35 % oxygen, and 75.67 % Nitrogen are assumed as reference. Unlike energy, Exergy is not generally conserve but destroyed by irreversibility within a system. Without magnetism, electricity, nuclear and surface tension effects, [9] defines specific molar Exergy (KJ/mol) as the summation of kinetic Exergy

(1)

The kinetic and potential Exergy is assume negligible in this study. Hence the Exergy equation is reduced to only physical and chemical Exergy as given by [15] in equation (2)

(2)

IOP Conf. Series: Earth and Environmental Science **173** (2018) 012010 doi:10.1088/1755-1315/173/1/012010

Thermal Exergy, which is also known as physical Exergy, is the maximum amount of work, which can be obtained from a stream of material brought to the environmental state from its initial state but exchanging heat only with the environments thermal reservoir [24, 25]. The chemical Exergy according to [25] is the total amount of work, which can be acquired from a stream of material brought from the environment (restricted dead) state to the total dead (unrestricted) state while exchanging only materials and heat with the environment. The equation for the chemical and thermal Exergy is specified by [21, 24, 26] in equation 3 as

(3)

The lost work (LW) around each piece of equipment with stream inlet Exergy flow (n_(i.) 〖ex〗_i), outlet exergy flow (n_(o.) 〖ex〗_o), work flow Ẇ (kJ/s), and utility heat duties Q (kJ/s), was evaluated with equation (4) given by [26].

(4)

Also Exergetic efficiency is calculated as follow [1, 26]

(5)

Table 2.2 shows the simulated parameters for the selected configurations.

Table 2.2 Simulation Parameters for the Selected Configurations

Parameter	Methanol Configurations	Propane Configurations	Butane Configurations
Fuel utilization	0.85	0.85	0.85
Compressor 1 power required (Watt)	0.2	0.2	0.2
Compressor 2 power required (Watt)	24100	33.9	1060
Pump 1 power (Kw)	-	0.1923	0.193
SOFC Temperature (^{o}C)	850	850	850
SOFC Pressure (bar)	1	1	1
SOFC Area (cm^2)	323	323	323
Number of SOFC cell	800	800	800
Reformer Pressure	20	1	1.2
Reformer Temperature	650	603	612
Water gas Shift Reactor Pressure	-	1	1
Water gas Shift Reactor Pressure	-	302	411

3. Result and Discussion

3.1. Energy Analysis

Table 3.1 - 3.3 shows the flow of energy in each equipment of Methanol, Propane, and Butane configuration respectively. The results presented in the Tables shows that the equipments requiring energy are the compressors and pumps with Compressors having the highest energy requirement. The results show that the energy available in the outlet stream is higher than in the inlet stream wit exception to pumps and compressors because they are energy requiring equipment. The Tables also reveal that the SOFC stack is the highest energy producing equipment. Methanol Fuel System has the highest energy requirement which can be attributed to the High energy required by the Methanol compressor. Note that despite the fact that Propane and Butane has an extra unit (WGS Reactor), the Energy required by the two system which is 0.062 KW and 4.972 KW respectively is less than 69.45KW required by Methanol system. This is also validated by [1]. The energy produced by Propane Fuel system is 234.67 KW which is the least while that produce by Methanol is 273.66 KW being the highest energy produced among the four system. The first law of Thermodynamic was satisfied in all system.

Table 3.1. Energy Flow for Methanol System

Equipment	Energy of inlet stream (Kw)	Energy of Outlet Stream (Kw)	Required Energy (Kw)	Energy Produced (Kw)
Compressor 1	105.1	69.45	69.45	-
Mixer	665.2	665.2	-	-
Heater 1	1051.6	1052.5	-	0.92
Reforming Reactor	687.07	758.5	-	71.43
Compressor 2	0.1	0.1	-	-
Heater 2	342.27	342.5	-	0.24
SOFC Stack	428.7	628.7	-	200
Lambda Burner	627.6	628.7	-	1.07
3 Way Valve	628.7	628.7	-	-

Table 3.2 Energy Flow for Propane System

Equipment	Energy of inlet stream (Kw)	Energy of Outlet Stream (Kw)	Required Energy (Kw)	Energy Produced (Kw)
Compressor 1	19.600	19.550	0.060	-
Mixer	362.500	362.500	-	-
Heater 1	405.500	406.400	-	0.870
Reforming Reactor	403.210	429.600	-	26.390
Compressor 2	0.100	0.100	-	-
Heater 2	586.400	586.400		0.260
Heater 3	190.230	190.400	-	0.170
Shift	246.800	246.800	-	-

Reactor				
Pump 1	3.100	3.100	-	-
Pump 2	164.800	164.780	0.002	-
SOFC Stack	156.300	356.300	-	200.000
Lambda Burner	356.300	363.280	-	6.980
3 Way Valve	363.280	363.280	-	-

Table 3.3. Energy Flow For Butane System

Equipment	Energy of inlet stream (Kw)	Energy of Outlet Stream (Kw)	Required Energy (Kw)	Energy Produced (Kw)
Compressor 1	21.170	16.200	4.970	-
Mixer	1191.800	1191.800	-	-
Heater 1	1914.500	1915.330		.830
Reforming Reactor	1553.300	1609.340		56.040
Compressor 2	.100	.100	-	-
Heater 2	1237.900	1238.240		.340
Heater 3	536.500	536.710		.210
Shift Reactor	1078.300	1078.300	-	-
Pump 1	3.100	3.100	-	-
Pump 2	164.802	164.800	0.002	-
SOFC Stack	989.800	1189.800		200.000
Lambda Burner	1189.800	1196.300	-	6.500
3 Way Valve	1196.300	1196.300	-	-

Table 3.4 shows the overall energy efficiency of the four Fuel SOFC system. From the result presented in Table 3.4, Butane fuel System is the most energy efficient system having an efficiency of 50.3%. However, the quality and direction of energy flow is not considered. Though Butane Fuel system has the most energy efficiency, it does not mean that it has the least energy degradation. Since no information is available on the degradation of energy occurring in each process.

Table 3.4 Energy Efficiency

Parameter	Energy Efficiency
Methanol	46.5
Propane	47.7
Butane	50.3

3.2. Energy Analysis

Table 3.1 - 3.3 shows Table 3.5 - 3.7 shows the behavior of physical, chemical Exergy, and the total Exergy flow forMethanol, Propane, and Butane fuel system respectively. The result presented indicate that around equipment such as Reactor, lambda burner, and SOFC stack there is a change between the inlet and outlet chemical Exergy. This is as a result of chemical reaction occurring in them, which alters the chemical composition of the outlet stream. From the Propane and Butane fuel results shown in table 3.6 - 3.7 the SOFC inlet has the highest chemical Exergy flow of about 14.7% and 14.4% respectively of the total chemical Exergy flow. This is because of the high standard chemical Exergy of hydrogen in the streams. Lambda burner outlet has the highest physical Exergy owning to the combustion reaction, which gives off heat energy. From the results, pumps and compressor have very low physical Exergy because the streams conditions are or almost same with the reference temperature and pressure and the flow rate were very low. Also note that the Physical Exergy for Propane increase by 10.2% and Butane Fuel system increased by 24.7%. The increase in Physical Exergy in propane and Butane system is due to the high increase in Physical Exergy at the SOFC stack component. Although there is a rise in the Physical Exergy of the two systems, the overall Exergy of the system decrease by 19.6% and that of Butane by 2.9% respectively.

Table 3.5 Equipment Exergy Flow for Methanol System

Equipment	Physical Exergy In (KW)	Chemical Exergy In (KW)	Total Exergy In (KW)	Physical Exergy Out (KW)	Chemical Exergy Out (KW)	Total Exergy Out (KW)
Compressor 1	0	424.8	424.8	10.7	424.8	435.5
Mixer	26.8	427.4	454.2	26.5	425.9	452.4
Heater 1	182.2	437.4	619.6	98.8	437.4	536.2
Reactor	234.4	433.6	668	43.9	434.7	478.6
Compressor 2	0	16.1	16.1	0	16.1	16.1
Heater 2	12.7	25.9	38.6	7.3	20.9	28.2
SOFC	41.5	447.9	489.4	312.9	81.8	249.1
Lambda Burner	249.1	81.8	330.9	155.3	19.2	174.5
3 Way Valve	155.3	19.8	174.5	154.7	19.2	173.9

Table 3.6 Equipment Exergy Flow for Propane System

Equipment	Physical Exergy In (KW)	Chemical Exergy In (KW)	Total Exergy In (KW)	Physical Exergy Out (KW)	Chemical Exergy Out (KW)	Total Exergy Out (KW)
Compressor 1	0	402	402	0	402	402
Mixer	0	403.6	403.6	0	403.6	403.6
Heater 1	120.3	414.6	534.9	46.4	414.6	461

Reactor	156	413.9	569.9	51.6	431.8	483.4
Pump 1	0	30.1	30.1	0	30.1	30.1
Heater 2	16.9	433.9	450.8	12.5	433.9	446.4
Pump 2	0	0.75	0.75	0	0.75	0.75
Shift Reactor	12.5	433.2	445.7	14.6	434	445.6
Compressor 2	0	0	0	0	0	0
Heater 3	34.7	39	73.7	9.1	39	48.1
SOFC	17.2	464.2	481.4	162	88	250
Lambda Burner	162	88	250	169	22.6	191.6
3-Way Valve	169.2	22.6	191.8	167	22.6	189.6

Table 3.7 Equipment Exergy Flow or Butane System

Equipment	Physical Exergy In (KW)	Chemical Exergy In (KW)	Total Exergy In (KW)	Physical Exergy Out (KW)	Chemical Exergy Out (KW)	Total Exergy Out (KW)
Compressor 1	0	380.8	380.8	1.9	380.8	382.7
Mixer	10.4	387.3	397.7	10.4	387.3	397.7
Heater 1	148.4	410.7	559.1	133.8	410.7	544.5
Reactor	178.4	405.5	583.9	83.3	443.3	526.6
Pump 1	0	26.9	26.9	0.0	26.9	26.9
Heater 2	36.3	433.6	469.9	34.8	433.6	468.4
Pump 2	0	0.75	0.75	0.0	0.8	0.8
Shift Reactor	36	432.9	468.9	31.0	431.7	462.7
Compressor 2	0	0.013	0.013	0.0	0.0	0.0
Heater 3	47	37.3	84.3	32.1	37.3	69.4
SOFC	61.6	458.6	520.2	181.0	88.1	269.1
Lambda Burner	181	185.1	366.1	228.6	26.0	254.6

IOP Conf. Series: Earth and Environmental Science **173** (2018) 012010 doi:10.1088/1755-1315/173/1/012010

3-Way Valve	228.3	26	254.3	228.2	26.0	254.2

The result of the exegetic performance (Exergy loss and Exergy efficiency) is presented in Table 3.8-3.10. Table 3.8 - 3.10 show the Exergetic efficiency and loss work for Methanol, Propane, and Butane. The calculations for Exergy loss takes into account the work required or Produced and the heat supply or loss. The results show that there are components with 100% Exergetic efficiency such as pumps, mixers and compressors. This high efficiency and low exergetic irreversibility in compressors and pumps clearly show that the equipment such as compressors and pumps that use electricity as a source of energy have high efficiency because it has good energy. This is validated by [1, 24]. It is observe that the Lambda Burner in all systems has the highest degradation of energy (Exergy loss). This could be ascribed to high temperature change of the equipment, therefore having the lowest Exergy efficiency of 53%, 79.4% and 71.3% for Methanol, Propane and Butane Fuel System respectively. This agrees with the findings of previous work by [1].

Table 3.8 Exergetic Performance Analysis Result for Methanol System

Equipment	Loss Work(Kw)	Exergetic Efficiency (%)
Compressor 1	13.4	96.8
Mixer	1.8	99.6
Heater 1	82.5	86.7
Reactor	118.0	82.3
Heater 2	0.0	100.0
Compressor 2	10.2	73.6
SOFC Stack	40.3	91.8
Lambda Burner	155.4	53.0
3 Way Valve	0.6	99.7

Table 3.9 Equipment Exergetic Performance Analysis Results for Propane System

Equipment	Loss Work(Kw)	Exergetic Efficiency (%)
Compressor 1	0.0	100.0
Mixer	0.0	100.0
Heater 1	73.0	86.4
Reactor	60.1	89.5
Pump 1	0.0	100.0
Heater 2	4.1	99.1
Pump 2	0	100.0
Shift Reactor	0.1	100.0
Compressor 2	0.0	100.0
Heater 3	25.4	65.5
SOFC	31.4	93.5

IOP Conf. Series: Earth and Environmental Science **173** (2018) 012010 doi:10.1088/1755-1315/173/1/012010

Lambda Burner	51.4	79.4
3-Way Valve	2.2	98.9

Table 3.10 Equipment Exergetic Performance Analysis Results for Butane System

Equipment	Loss Work(Kw)	Exergetic Efficiency (%)
Compressor 2	2.0	99.5
Mixer	0.0	100.0
Heater 2	13.8	97.5
Reforming Reactor	1.4	99.8
Heater 1	0.0	100.0
Heater 3	1.2	99.8
Shift Reactor	0.0	100.0
Compressor 1	6.2	98.7
Pump 1	0.0	100.0
Pump 2	14.7	82.6
SOFC Stack	51.1	90.2
Lambda Burner	105.2	71.3
3 Way Valve	0.1	100.0

Table 3.11 shows the overall loss work, energy and Exergy efficiency of Methanol, Propane, and Butane system. Among the three fuels, Methanol has the highest loss work of 422.2 KW while Butane has the least loss work of 162.7 KW. From the table methanol has the least Exergetic efficiency of 44.2% while Butane has the highest Energy efficiency. The result also reveals that propane which has the second highest Energy efficiency after Butane Fuel has the highest Exergy efficiency. Hence, from the result propane fuel is most favourable. Hence energy analysis alone is not sufficient to give the best possible process part, instead a combination of both energy and Exergy analysis will give a better and actual performance of a system.

Table 3.11 Overall Loss work, Energy Efficiency, and Exergy Efficiency

System	Overall Loss Work (Kw)	Overall Energy Efficiency	Overall Exergetic Efficiency
Methanol	422.2	46.5	44.2
Propane	247.7	47.7	49.3
Butane	195.7	50.3	46.7

4. Conclusion

The exergy and energy analysis on SOFC system using different fuels was successfully carried out. SOFC system configurations for Methanol, Propane and Butane fuel system was design using Thermolib. The configuration was successfully simulated to give power output of 200KW using Thermolib. From the results obtained, Exergy analysis was done on the different fuel SOFC system. From the result obtained, Butane Fuel system gave the highest Energy efficiency; Propane Fuel system has the highest Exergy efficiency. Although Butane system is the most Energy effective system when compared with the other three, Nevertheless a compromise between Exergy and energy efficiency favours the choice of propane fuel system. This is because Propane is the most Exergy efficiency system.

References

[1] Suleiman, B., Abdulkareem, A.S., Musa, U., Mohammed, I.A., Olutoye, M.A., and Abdullahi, Y.I. (2016): Thermo-economic analysis of proton exchange membrane fuel cell fuelled with methanol and methane. Energy Conversion and Management, 117, 1-3.

[2] Yannay. C., Luis. E., Mayra M., Elena. M., Dewulf. J. (2010): Energy and Exergy Analysis of an Ethanol Fueled Solid Oxide Fuel Cell Power Plant, 3-6.

[3] Abdulkareem, A.S. (2009). Design and Development of Proton Exchange Membrane From Synthetic Rubber And Carbon Nanocells for PEMFC. PhD Thesis, University Of Witwatersrand, South-Africa, 1-35.

[4] Hussain, M.M., Dincer, I. and Li, X., (2004): Energy and Exergy Analysis of an Integrated SOFC Power System, ASME FORUM, 10, 3-7.

[5] Colpan, O.C. (2009): Thermal Modelling of Solid Oxide Fuel Cell Based Biomass Gasification Systems, PhD Thesis, Carleton University Ottawa, Ontario, Canada, 1-55.

[6] Lan, R. and Tao, S. (2014): Ammonia as a suitable fuel for fuel cells. Frontiers in Energy Research, 2, 35, 2-4.

[7] Afolabi, S.A., 2kareem, A.S., Bilyaminu, S., and Yenkwo, K.C. (2015). Exergy and Economic Analyses of a Hybrid Solid Oxide Fuel Cell by Computer Simulation. Proceedings of the World Congress on Engineering, 2, 1-6.

[8] Siefert, N.S. (2014): Experimental and Thermo-Economic Analysis of Catalytic Gasification and Fuel Cell Power Systems. Thesis Submitted in Partial Fulfillment of the Requirements for the Degree of Doctor of Philosophy, Carnegie Mellon University Carnegie Institute Of Technology.

[9] Jamalabadi, M.Y.A., Hooshmand, P. and Broumand, B. (2004): Economic and Environmental Modeling of A Mgt-SOFC Hybrid Combined Heat And Power System For Ship Applications. Middle-East Journal of Scientific Research, 22 (4), 3.

[10] Cheddie. D. (2012): Ammonia as a Hydrogen Source for Fuel Cells: A Review. Hydrogen Energy – Challenges and Perspectives, 5-23.

[11] Barelli . L., Bidini. G., Gallorini. F., Ottaviano. A. (2011): An Energetic–Exergetic Analysis of a Residential CHP System Based on PEM fuel cell: Applied Electricity, Science direct,88 3-6.

[12] Meyer, L., Castillo, R., Buchgeister, J., and Tsatsaronis, G. 2009 . Application Of Exergoeconomic And Exergoenvironmental Analysis To An SOFC System With An Allothermal Biomass Gasifier. Int. J. Of Thermodynamics, 12 (4), 2-6.

[13] Jubeh. N. M. (2005): Exergy Analysis and Second Law Efficiency of a Regenerative Brayton Cycle with Isothermal Heat Addition. Entropy, 7, 172-187.

[14] Baheta. A., Gilani. S. (2011): Exergy Based Performance Analysis of a Gas Turbine at Part Load Conditions. Journal of Applied Sciences, 11, 4-9.

[15] Sreeramulu. M., Gupta. A., Srinivas. T. (2011): Exergy Analysis of Gas Turbine – Fuel cell

based combined Cycle Power Plant. International Journal of Engineering Science and Technology, 3, 227-236.
[16] Chan. S.H., Ho. H.K., Tian. Y. (2002). Multi-level modeling of SOFC–gas turbine hybrid system. International Journal of Hydrogen Energy, 28, 9-10.
[17] Zabihian, A. and Fung, F. (2009): A Review On Modeling Of Hybrid Solid Oxide Fuel Cell Systems. International Journal Of Engineering (IJE), 3(2), 2-8.
[18] Yi. Y., Rao. A. D., Brouwer. J., Samuelsen. G.(2005). Fuel flexibility study of an integrated 25kW SOFC reformer system. Journal of Power Sources, 144, 67-76.
[19] Arsalis, A. (2007). Thermo-economic Modelling and Parametric Study of Hybrid Solid Oxide Fuel Cell – Gas Turbine – Steam Turbine Power Plants Ranging from 1.5 MW to 10 MW. MSc thesis submitted to the faculty of the Virginia Polytechnic Institute and State University in partial fulfillment of the requirements for the degree of Master of Science in Mechanical Engineering, 55-120.
[20] Hussain. M.M., Baschuka. J.J., Li. X., Dincer. I. (2005): Thermodynamic analysis of a PEM fuel cell power system. International Journal of Thermal Sciences, 44, 3-11.
[21] Hotz, N., Lee, M., Grigoropoulos, C.P., Senn S.M., and Poulikakos, D. (2006): Exergetic Analysis Of Fuel Cell Micropower plants Fed By Methanol. International Journal Of Heat And Mass Transfer, 49, 1-2.
[22] Ang. S., Fraga. E., Brandon. N.P., Samsatli. N.J., Brett. D. (2011). Fuel cell systems optimization Methods and strategies. International Journal of Hydrogen Energy, 36, 678-703.
[23] Von Spakovsky. M. R., Nelson. D. J., Ellis. M. W., Olsommer, B., and Ogburn, M. J. (2000).A Multi- / Inter-Disciplinary Approach To Fuel Cell System Development: The U.S. DOE GATE Center For Automotive Fuel Cell Systems At Virginia Tech, Journal on Society of Automotive Engineers, 1, 1.
[24] Douvartzides, S., Coutelieris, F., Tsiakaras, P. (2004): Exergy Analysis of A Solid Oxide Fuel Cell Power Plant Fed By Either Ethanol Or Methane. Journal of Power Sources, 49, 3-4.
[25] Yunus, A.C. and Michael, A.B. (2003): Thermodynamics: An Engineering Approach, Fifth Edition, chapter 8.
[26] ICIS Pricing. ("n.d"). Petrochemical, energy and Fertilizer Market Intelligence. Retrieved from www.icis.com/about/price-reports.

Propensity for Peptides as nutraceuticals: a survey of the Nigerian Market

Omini Joy J, Omotosho Omolola E

Department of Biochemistry, Covenant University, Nigeria
joy.omini@covenantuniversity.edu.ng

Abstract. Over time, scientists have studied the potential of peptides as nutraceuticals by measuring their bio functionalities. The aim of this study was to ascertain the potential of isolated peptides as nutraceuticals in Nigeria using market availability. A market survey was carried out to determine the willingness of citizens who possess buying power, to purchase this product. The survey was conducted using survey monkey. 54.17% of the sample population declared unlikelihood to purchase these bioactive peptides, 20.83% of this group had no idea, of which most were men. Of the 45 and above age group, 55.5% declared high likelihood for purchase of nutraceutical peptides. On the assumption that only the aging group subscribes to nutraceutical peptides, there is little prospect of seeing these peptides being utilized to their maximum potential in Nigeria for health maintenance. Hence enlightenment of the public is necessary to increase acceptability and prospect of bioactive peptides as nutraceuticals.

Keywords: Netraceuticals, Bioactive, Peptides, Market survey, Nigeria

1. Introduction

Nutraceuticals or functional foods have received increased attention and wide acceptance over the years due to their health promoting effects [1]. This reflected in a finance report by the global nutraceutical industries report [2] that the global nutraceutical market should reach $285.0 billion by 2021 from $198.7 billion in 2016 at a compound annual growth rate (CAGR) of 7.5%, from 2016 to 2021. Bioactive peptides however, have not received as much attention as other nutraceuticals because they are often considered to be proteins, whereas they differ in size, structure and function. A lot of studies have confirmed their diverse functionalities which include antidiabetic, cholesterol lowering, antihypertensive, skin repair, anticancer effects among others. These studies have also shown the promising prospects of these peptides as nutraceuticals based on their bioactivity. However, no study has examined the prospects of peptides as nutraceuticals based on market acceptability. Hence the need for this market survey to measure the potential efficacy of peptides if used as nutraceuticals.

Nature provides the most abundant source for bioactive peptides, from living things and their products. Peptides are generally present in various quantities in food of plant or animal origin, and they are released as a result of proteolytic hydrolysis by enzymes participating in the fermentation, aging or ripening process. For example *Oryzasativa,* upon digestion *in vivo* yields anti-inflammatory, antioxidant and immunomodulatory peptides [3].

Isolation and concentration of these peptides in a medium is important to maximise their bioactivities. This is because they may be obtained in minor, insignificant quantities from food sources. They also cannot be stored in the body and so to exert significant effects, they need to be taken in large quantities. Enzyme hydrolysis and fermentation have been the most practiced methods for production of bioactive peptides from protein sources [4]. Hydrolysis by enzymes involves subjecting a target protein to cleavage by a protease at optimum temperature and pH. One or more crude or purified proteolytic enzyme can be used to break down the protein containing material to a solution containing short peptide sequences called a protein hydrolysate [5]. Every proteolytic enzyme yields peptides as the end product of the reaction.

Microbial fermentation on the other hand involves culturing of the selected microbial cells on a protein-containing substrate. The growing cells secrete various proteolytic enzymes into the medium that hydrolyse proteins to yield peptides at different phases of their life cycle. The health benefits of these peptides are numerous. Antihypertensive peptides can be derived from protein-containing foods and used for treatment of high blood pressure or hypertension. These peptides inhibit Angiotensin I converting enzyme (ACE) making them non-functional. ACE cleaves angiotensin I to angiotensin II-bradykinin inhibitor which inhibits bradykinin [6]. Bradykinin is a strong vasodilator that dilates and expands the blood vessels to modulate blood pressure. Therefore inactivation of bradykinin increases blood pressure. However by inhibiting ACE, bradykinin is active. Fermented protein containing foods are a source of peptides that lower blood pressure.

Reactive oxygen species have been linked to many diseases which are usually as a result of tissue damage and cell death. These species can however be cleared from the system by antioxidants, which capture the species within their structure. Tryptophan, histidine, phenylalanine, methionine, tyrosine, valine have been identified as components of antioxidant peptides [7]. The ability of a peptide to scavenge radicals in circulation is conferred by the peculiarity of its side chain structure which stabilizes electron deficient radicals by easily donating protons to them. Antimicrobial peptides have cytotoxic effects on a vast array of microbial cells. The presence of cationic or lipophilic amino acid residues at the exposed region of peptides increases the ability of AMPs to interact with microbes. AMPs attack bacteria by drilling holes through the cell membrane or by interacting with protein or sugar moieties within the cell cytosol [8]. Cationic amino acids disrupt viruses, yeasts or bacteria by mechanisms such as micro pinocytosis.

Peptides also have antidiabetic functions which they exert by inhibiting the activity ofα-Amylase. Furthermore, they were observed to act by reducing starch absorption from the gastrointestinal tract through inhibition of cleavage of large starch moieties into simpler permeable ones, thus restricting absorption of starch. Examples of antidiabetic peptides include Cumin seed peptides [9]. Some peptides have cholesterol lowering effects. Cholesterol accumulation in the coronary artery could diminish blood flow and consequently, supply of oxygen to the heart and prompt diseases of the cardiovascular system [10]. Cholesterol lowering peptides act by reducing cholesterol solvency in lipid vectors, and restraining cholesterol uptake in some monolayer cells. Bioactive peptides confer selective cytotoxicity of malignant tissues through various means. A particular peptide HVLSRAPR obtained from *S. platensis* hydrolysates demonstrated solid inhibitory action against HT-29 cancerous cell division yet indicated little restraint against ordinary liver cells [11]. The peptides prompted cell arrest in the S stage by heightening p21 and p27 expression while diminishing cyclinA production. Moreover, the peptides cleaved caspase 3, down regulated Bcl-2, PARP and caspase 9 activities yet up regulated p53 and Bax articulation.

2. Materials and methods

2.1. Survey Design

A mail survey was developed to assess the awareness and market power of peptide nutraceuticals for health benefits. The survey consisted of one section which included one question and was designed for all individuals who received the questionnaire, whether or not they had used the product previously. This questionnaire covered the willingness of respondents to purchase isolated peptides. The questionnaire consisted of a multiple choice question. One best answer was requested for each question. Percentages were calculated using the number of respondents as the denominator and therefore, summed up to 100%.

2.2. Sampling

The survey was conducted on line in spring 2018 on a sample of two hundred and fifty individuals, representative of the Nigerian population for gender, age (20 and above), and region of residence. The reference population taken into account for the direct survey is represented by the working class residents of the states in Nigeria and different works of life.
The survey was sent out using survey monkey.

2.3. Data analysis

Data were summarized based on gender and also pooled together. It was presented as pictorial results for the sampling frame, Microsoft excel and Survey monkey were used to analyse the results into charts.

3. Results and Discussion

250 respondents made up the survey. About 30% of these respondents were over 45 years of age. Overall, about 8% of all respondents declared high unlikelihood to purchase isolated peptides as nutraceuticals. Equal percentage (25%) of the sample size showed willingness for purchase and unwillingness for purchase. 20.83% of the total respondents declared high likelihood to subscribe to isolated peptides as nutraceuticals while another 20.83% had no idea entirely about nutraceuticals and peptides as shown in figure 1. The outcome of the survey was also grouped into male and female categories. This was done to analyse the interdependence of opinion on the gender of the respondents. It was observed that while 33.33% of the female population declared high likelihood to purchase these nutraceuticals, no male in the total sample population absolutely opted (very likely) to purchase peptides as nutraceuticals. With regards to average likelihood to purchase the peptides, about 27% of females and 22% of males responded. About 44% of males and 6% of females had no idea whatsoever about the nutraceuticals and as such they do not know if they would likely purchase isolated peptides or not. 22% of male and 27% of females declared unlikelihood to purchase the peptides. Finally, 11% of males and 6% of females said they were very unlikely to purchase the peptides as nutraceuticals. When the results were grouped according to age groups, it was observed that 55.5% of the 45 and above age group, declared willingness to buy the nutraceuticals, while 22% opted to not purchase the product. Of the 20-45 age groups, a total of 46.66% declared willingness to purchase while 54.44 declared unlikelihood to buy even with purchasing power.

From the survey conducted on an excerpt of the Nigerian market population, the success of bioactive peptides as nutraceuticals can be predicted. The prospect of peptides as nutraceuticals depends on its bioactivity and largely also on the willingness of the population to purchase it for use. Since 2002, [12] reported that nutraceuticals (including isolated nutrients) represent the fastest growing segment of the food industry. However Nigeria seems to be lagging behind as observed in this survey where a fifth of the total sample population had no idea whether or not they will purchase these peptides as nutraceuticals. This may have been due to ignorance of the subject matter. Of these, 80% were shown to be men. People will normally buy protein-containing foods which will be digested into peptides in the gut, but to maximize the functionality of these peptides, they have to be isolated and concentrated in a medium. Hence the population has to be first aware of nutraceuticals as a whole for the prospects to be actualized. The potential of peptides as nutraceuticals in the Nigerian market depends highly on the desire to buy this product. Again, 20.83% of the population declared high likelihood while 25% declared average likelihood to purchase peptides. This does not correlate with the recent financial predictions by Global Nutraceuticals Industry Report that the global nutraceuticals market should reach $285.0 billion by 2021 at a compound annual growth rate (CAGR) of 7.5%, from 2016 to 2021. This low percentage however may be as a result of nutraceuticals or peptides not being accepted in Nigeria yet. The financial prediction above ought to cut across every nutraceutical product and every nation, and its actualization is also dependent on Nigerians involvement. A total of 45.83% positive responses were harvested while the negative responses came from 54.17% of the population. Judging from the marginal difference between these two, it can be implied that the future of peptides as nutraceuticals in the Nigerian market is not so promising. The comparison of male and female responses showed significant differences in the responses to the choices-no idea and very likely declared by the different genders, which suggests that propensity for this product may be gender dependent. Males have less interest and are more unlikely to purchase these peptides than women. The live Nigerian population clock as at 20-03-18, 13:20 reported a male population of 50.6%, which is higher (although not significantly), than that of women. Based on the census, the men should have a higher determining power on how successful peptides will be as nutraceuticals in the Nigerian market. Thus the prospect of peptides as nutraceuticals in the Nigerian market is not promising as men have shown less likelihood to purchase them. The outcome was also grouped as 20-45 and 45 and above. Many bioactivities of peptides such as antidiabetic effect, antihypertensive effect, cholesterol lowering effect, anti-cancer effect, skin repair effect, etc. are important for a particular age group. This is due to the fact that age-related diseases are usually suffered by the 45 and above age group. Based on this, it is expected that this age range should show more willingness for purchase of this nutraceutical product. The result was observed to corroborate with the expected outcome, as over half of the 45 and above age group opted to purchase this nutraceutical.

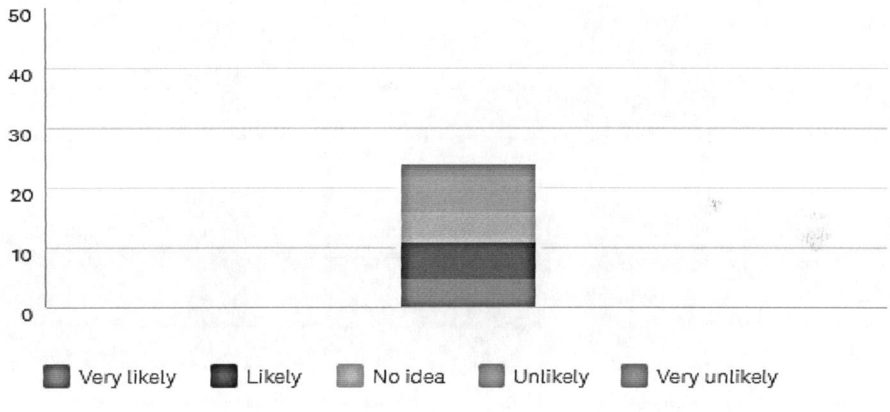

ANSWER CHOICES	RESPONSES
Very likely	20.83%
Likely	25.00%
No idea	20.83%
Unlikely	25.00%
Very unlikely	8.33%

Figure 1: Chart representation of the general survey outcome.

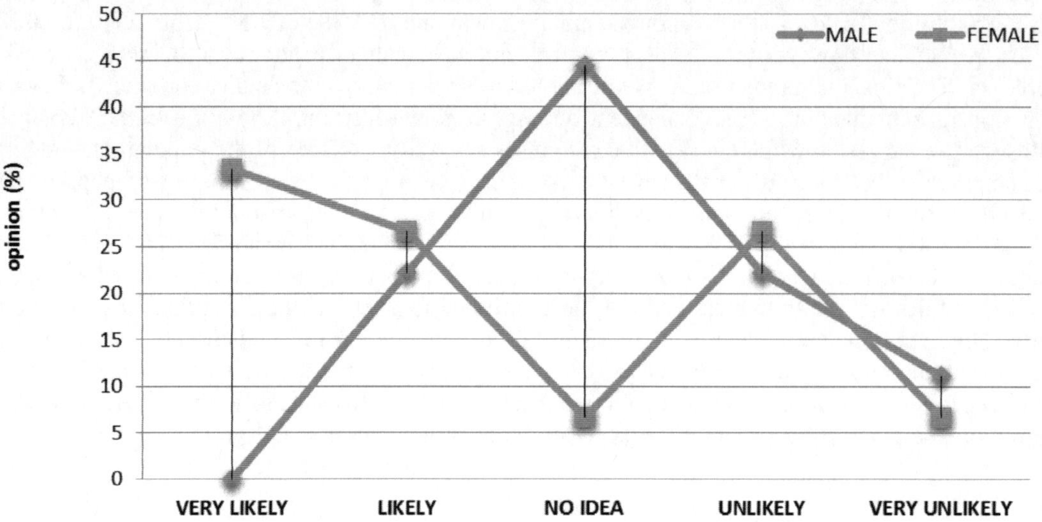

Figure 2: Graph representation of difference in male and female responses.

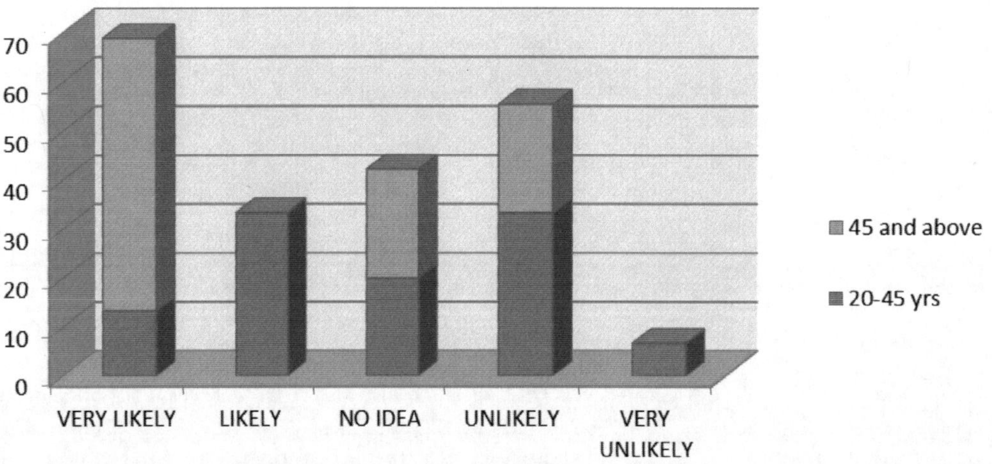

Figure 2: Graphical representation of the outcome of different age groups.

4. Conclusion

Based on the outcome of the major target market in Nigeria alone, the prospect of peptides being used as nutraceuticals in Nigeria looks promising. However, the functions of peptides are not limited to those linked with age-related diseases and in order for bioactive peptides to fulfil and maximize its potential in the Nigerian market, subscription has to increase. Establishing its functionality is not sufficient to determine its prospect. Citizen's awareness of the importance, efficacy and benefits of nutraceuticals as compared to synthetic drugs has to be increased.

Acknowledgements
We are very grateful to the Covenant University Centre for Research, Innovation and Discovery for the conference support sponsorship.

References

[1] Khalid, G., Singh, K. and Rifat J. (2015). Nutraceuticals and Functional Foods: The Foods for the Future World. *Critical Reviews in Food Science and Nutrition,* 56**,** 2617-2627.

[2] Global Nutraceuticals Industry Report: 2017-2021 - Analysis of the Multi-billion Functional Food and Beverage Industries - Research and Markets (2016).https://www.prnewswire.com/news-releases/global-nutraceuticals-industry-report-2017-2021---analysis-of-the-multi-billion-functional-food-and-beverage-industries---research-and-markets-300422477.html. AccessedTuesday, 20 March 2018 at 7 p.m.

[3] Ferri, M., Graen-Heedfeld, J., Bretz, K., Guillon, F., Michelini, E., Calabretta, M.andTassoni, A. (2017). Peptide fractions obtained from rice by-products by means of an environment-friendly process show in vitro health-related bioactivities. *PloS one*, 12(1), 170-174.

[4] Pihlanto, A. (2006). Antioxidative peptides derived from milk proteins. *InternationalDairy Journal*, 16(11), 1306-1314.

[5] Kristinsson, H. G., &Rasco, B. A. (2000). Fish protein hydrolysates: production, biochemical, and functional properties. *Critical Reviews in Food Science and Nutrition,* 40(1), 43-81.

[6] Daskaya-Dikmen, C., Yucetepe, A., Karbancioglu-Guler, F., Daskaya, H., &Ozcelik, B. (2017). Angiotensin-I-converting enzyme (ACE)-inhibitory peptides from plants. *Nutrients,* 9(4), 316.

[7] Matsui, R., Honda, R., Kanome, M., Hagiwara, A., Matsuda, Y., Togitani, T. andTerashima, M. (2018). Designing antioxidant peptides based on the antioxidant properties of the amino acid side-chains. *Food Chemistry,* 245, 750-755.

[8] Aktories, K. (2011). Bacterial protein toxins that modify host regulatory GTPases. *Nature Reviews Microbiology,* 9(7), 487.

[9] Siow, H. L. and Gan, C. Y. (2016). Extraction, identification, and structure–activity relationship of antioxidative and α-amylase inhibitory peptides from cumin seeds (*Cuminumcyminum*). *Journal of Functional Foods,* 22, 1-12.

[10] Singh, B. P., Vij, S. andHati, S. (2014). Functional significance of bioactive peptides derived from soybean. *Peptides*, 54, 171-179.

[11] González-Montoya, M., Hernández-Ledesma, B., Silván, J. M., Mora-Escobedo, R.and Martínez-Villaluenga, C. (2018). Peptides derived from in vitro gastrointestinal digestion of germinated soybean proteins inhibit human colon cancer cells proliferation and inflammation. *Food Chemistry,* 242, 75-82.

[12] Andlauer, W. and Furst, P. (2002). Nutraceuticals: a piece of history, present status and outlook. *Food Research International,*35, 171-176.

Antiplasmodial Activity shown by Secondary Metabolites Extracted from the Seeds of*Pentaclethramacrophylla*Benth

Grace I. Olasehinde[*b], Raphael C. Mordi[*a], Arnold N. Eluagwule[a], Hassana O. Jonathan[a], Joan I. Ayo-Ajayi[a], Adebola P. Okedere[a], Daniel O. Uchenna[b], OreoluwaR. Oluseye[b], Diji-Geske I. Ruth[b], OlabodeOnile-ere[b]and Dominic E. Azuh[c]

[a]Department of Chemistry, [b]Department of Biological Sciences and [c]Department of Economics and Development Studies, Covenant University, Canaan Land, Km 10 Idiroko Road, Ota, Ogun State, Nigeria

grace.olasehinde@covenantuniversity.edu.ng; ruth.diji-geske@stu.cu.edu.ng

Abstract. Oil extracts from the African oil bean seed (*Pentaclethramacrophylla*Benth.) was analyzed for its phytochemical and mineral content and proximate, physicochemical and anti-microbial analyses were also performed. Phytochemical analysis showed the presence of tannins, saponins, quinones, terpenoids, phenols and coumarins in the oil sample. Mineral determination of the cotyledon showed the presence of iron (Fe) (with the highest concentration), Cu, Zn, Mn, Cr, Pb and Cd; while proximate analysis gave the following result: moisture (14.2%), ash content (1.5%), crude fibre (4.9%), crude proteins (12.8%), oil contents (4.9%), and carbohydrate (61.8%). GC-MS analysis of the partitioned petroleum ether and chloroform fractions of the oil revealed the presence of 9-Octadecenoic acid, 9,12-Octadecadienoic acid and their methyl esters,cis-9-Hexadecenal among the many components of the oil extract. Physicochemical analysis of the oil indicateda saponification value (148.67 mg KOH/g), peroxide value(8.0 meq/g), iodine value (10.41 mg iodine/g) and free fatty acid (8.98 mg KOH/g). The need for the development of new drugs for malaria led to our study of the antiplasmodial activity of the oil from the seeds of *Pentaclethramacrophylla*. Toxicological studies were carried out to determine the $LD_{50}$with chloroquinediphosphate as positive control and normal saline as negative control. Using the Peter's 4 day suppressive test a parasite inhibition rate of 47.72% (25 mg/kg), 63.63% (50 mg/kg) and 61.36% (100 mg/kg) on day 4 after treatment was recorded. A 95.45% chemo-suppression was observed for animals treated with 10 mg/kg chloroquine. This resultis an indication that the extract had appreciable signs of chemosuppression.

Keywords: *Pentaclethramacrophylla*, phytochemical analysis, antimicrobial analysis, GC-MS, Plasmodium, secondary metabolites, chemo-suppression, malaria

1. INTRODUCTION

In rural Africa, people are exposed to a variety of plants, fruits and seeds, some of which are taken as food because of their nutritional value and some for treatment of various ailments because of their medicinal properties (Opara*etal*., 2013). *Pentaclethramacrophylla*Benth. also known as oilbean tree but locally called the African oil bean tree is one such plant whose seeds are eaten by a group of people in Southeastern Nigeria and the extracts from its leaves and stem bark have been used for treatment of ailments such as diarrhea, leprosy wounds and convulsion and has been used as an anthelmintic (Fern, 2014; Akindahunsi, 2004). The seeds are gotten from perennial legume tree called *Pentaclethramacrophylla*, Benth and also commonly known as the oil bean tree. It has also been suggested that consumption of the seeds increases the hemoglobin value; increases oxygenation of tissue; enhances specific hormones and quickens the production of red blood cells (Okpala, 2015).

Antimalarial drug resistance has emerged as one of the greatest challenges facing malaria control today (Olasehinde*et al*., 2015). Based on the emergence of Plasmodium falciparum resistance to artemisinin derivatives (ART) in Cambodia, which is currently threatening the world's malaria control and elimination efforts (Ariey*et al*., 2014), development of new drugs will be a welcome initiative. There are several approaches to discovering new agents, of which discovery from natural sources is a highly desirable approach.

A high proportion of the global population still depends upon the herbal/traditional drugs for their health care. Because of their ready availability, believe in their safety and effectiveness, rural people rely on these plants for their health care issues andneeds. *Pentaclethramacrophylla*Benthis widely used as food and for treatment of some ailments and is widely abundant in Nigeria; hence this study is focused on investigating its properties and harnessing them for the treatment of malaria. The investigation was done by extracting oil from the seed cotyledon with ethanol and petroleum ether as solvents.

The oil extracted from the cotyledon and seed cover was tested for its antimicrobial and antimalarial activities. These were determined along with its chemical composition using Gas chromatography–mass spectrometry (GC-MS). The cytotoxicity and in vivo toxicity; evaluation of in vitro and in vivo antiplasmodial activities as well as the determination the LD50 and chemo-suppressive properties of ethanolic extracts against *Plasmodium bergheiberghei*were considered.

2. MATERIALS AND METHODS

2.1 Plant Material

The African oil bean (*Pentaclethramacrophylla*) seeds were obtained from Oyingbo market, EbuteMetta, Lagos, Lagos State, Nigeria.The seeds were authenticated at the Federal Research Institute of Nigeria (FRIN), Ibadan. It was assigned a voucher No: Pm/Bio/H806 with Ascension code: OgCUNo807 Thecover of the seeds of *Pentaclethramacrophylla* were manually removedwith knife and cut into small sizes of about 5 mm. These were then stored in a refrigerator until required.

2.2 Plant Extraction

The extraction was performed by taking a weighed amount of the seed and placing it in a thimble of the Soxhlet extractor. The extraction was performed exhaustively with the appropriate solvent (ethanol or petroleum ether).After the extraction, the solvent was removed using a rotary evaporator. The oil collected was keptin the refrigerator for further analysis. The oil percentage extract was calculated as:

$$\%\ oil\ content = \frac{weight\ of\ oil\ extracted}{weight\ of\ seed\ sample} \times 100$$

The ethanolic oil extract was partitioned between chloroform, petroleum ether and distilled water. About seventeen (17) grams of ethanolic sample extract was first partitioned between water (50 mL) and petroleum ether (50 mL) in a 250 mL separating funnel. The water fraction was then poured back into the separating funnel and 50 mL of chloroform was added to the mixture and the chloroform fraction was obtained. Both the petroleum ether and chloroform fractions were allowed to evaporate and analyzed by GC/MS.

2.3 Gas Chromatography/ Mass Spectroscopy

The GC-MS analysis was carried out at Shimadzu Training Center for analytical instruments (STC) Lagos. The GC-MS instrument used for the analysis was GCMS-QP2010SE SHIMADU JAPAN. The machine comprises of injector and gas chromatograph interfaced to mass spectrophotometer. The gas

IOP Conf. Series: Earth and Environmental Science **173** (2018) 012012 doi:10.1088/1755-1315/173/1/012012

chromatography condition was: column oven temperature (60°C); injection temperature (200°C); injection mode split ratio (1:1); helium as carrier gas flow rate (1.56 mL/min); the system was temperature programmed from 60°C(at 10°C/min) to 160°C(held for 2min) then at (10°C/min) to 250°C; the injection volume was 0.5 µL. Mass spectrophotometer condition: Ion source temperature 200°C; Interface Temperature 250°C solvent cut time 4.5 min and acquisition was in the scan mode.

2.4 Physicochemical and Proximate Analysis

Physicochemical and proximate analyses were performed using the standard methods as described in the literature. Antimicrobial analysis was performed according to the method previously described by Olasehinde*et al*., (2016).

2.5 In vivo Antimalarial Activity

2.5.1 Experimental Animals

Experimental animals used in this study were the Swiss pre-infected albino mice (*Musmusculus*) of either sex, weighing between 18 – 25 g. All animals were housed in standard cages and kept in the animal house, at Covenant University, Ota, Ogun state, Nigeria. The animals were fed with a balanced standard pellet diet and maintained under standard laboratory conditions (24 – 28oC), standard light cycle and water ad libitum according to the NRC guide, for the care of laboratory animals(Institute for Laboratory Animal Research, 2011).They were allowed to acclimatize for five days before the commencement of the experiment. Ethical approval for use of animals was obtained from the BIOSREC, Covenant University.

2.5.2 Parasites

The chloroquine sensitive Plasmodium bergheiNK65 rodent parasite used to assess the antimalarial potential of *Pentaclethramacrophylla*was obtained from the National Institute for Medical research (NIMR), Lagos, Nigeria.

Each mouse was inoculated with 2 mL of plasmodium infected blood, containing about 0.5*103 of parasitized erythrocytes intra-peritoneally. This volume was prepared by determining the percentage parasitemia and the erythrocyte count of the donor mice and diluting 15 – 20 drops of blood, from the tail tip of the infected albino mice with 25 mL of Phosphate buffered solution (PBS). The animals were then left to attain maximum parasitemia for seven days.

2.5.3 Acute Oral Toxicity

Lorke's method was used to evaluate the acute toxicity of the ethanol extract of *Pentaclethramacrophylla*, which then allowed us todetermine the LD_{50}(Lorke, 1983).

Twelve mice were selected and divided into three groups of four mice each. The groups of mice were administered intra-peritoneally, with doses of 25 mg/kg, 50 mg/kg and 100 mg/kg respectively. The mice were observed for signs of toxicity, which included paw licking, salivation, stretching of the whole body, weakness, sleep, respiratory disease, coma and death in the first four hours and subsequently daily for 7 days. The animals were fasted for 24 h before the commencement of the experiment.

The number of deaths in each group within 24 hours was recorded and oral median lethal dose LD_{50} was calculated using this formula:

$$LD_{50} = \sqrt{minimum\ toxic\ dose \times maximum\ tolerated\ dose}$$

2.5.4 In Vivo Schizontocidal Activity 4-Day Peters' Test

The Peter's 4-day suppressive test against chloroquine sensitive *Plasmodium berghei*NK65 infection in mice was employed to test for suppressive activity(Peters, 1968). Adult Swiss Albino mice weighing 20 – 25 g were inoculated by intra-peritoneal injection with standard inoculum of

*Plasmodium bergheiberghei*with 1×107 infected erythrocytes. The mice were randomly divided into 3 groups of 4 mice each for the *P. macrophylla*extract for 25 mg/kg, 50 mg/kg and 100 mg/kg. Two control groups were used; the positive control was treated daily with 10 mg of chloroquine/kg, while the negative control group was given 2 mL/kg normal saline. The body weight of each mouse was measured on the first day, and the fifth day, while the body temperature was taken before and after drug administration (D0),and monitored daily to the fifth day.

On each day of the experiment, blood was collected from the tail of each mouse and a smear was made on the microscope slide to make a film(Saidu et al., 2000). The blood films were fixed with methanol, stained with 10% Giemsa at pH 7.2 for 10 minutes and parasitemiaexamined.Parasitemia was determined by Light microscopy using a 100× objective lens and the following equation was used to calculate the average percentage chemo-suppression:

$$\%\ chemosuppression = \frac{A - B}{A} \times 100$$

Where: A = Average parasitemia in the negative control group; B = average parasitemia in the test group.

Parasitemia estimation was done by allowing the slide to dry in a vertical position and examined under a light microscope at ×100 magnification and determining the number of parasites per field on a slide.

3. RESULT AND DISCUSSION

Percentage oil extracted from the seeds of P. macrophylla was very low; ethanol (4.89%)and petroleum ether (2.75%). This low oil content is a confirmation of the result reported by (Sofowora*et al.*, 2013) but contrasts significantly with values reported by other workers(Odoemelam, 2005; Oguekе*et al.*, 2010; Oyeleke*et al.*, 2014) whose values were obtained in the process of fermenting the seed cotyledon.

The results of the proximate analysis, phytochemical screening and physicochemical analysis of the oil extract are presented in tables 1, 2 and 3 respectively. Table 2 shows the presence of family of compounds whose activities are well known and have been reported on many occasions(Chung *et al.*, 1998; Ntie-Kang *et al.*, 2014; Oyeleke*et al.*, 2014).

Table 1: Results of proximate analysis of seed

Proximate Analysis	Moisture	Ash Content	Crude Fiber	Proteins	Oil Content	Carbohydrate
Determined Values (%)	14.16	1.53	4.85	12.79	4.89	61.78

Table 2: Results of phytochemical screening of seed oil

Component	Tannins	Saponins	Quinones	Terpenoids	Phenols	Coumarins
Observation	+	+	+	+	+	+

Table 3: Results of the physicochemical properties

Test	Saponification value (mg-KOH/g)	Free fatty acid (FFA) (mg-KOH/g)	Peroxide value (meq/kg)	Iodine value (mg-iodine/g)	pH	
	148.665	8.98	8.0	10.4058	4.57 (ethanol extract)	4.20 (petroleum ether extract)

Table 4:Resultsof Mineral composition in seed

Minerals	Iron	Copper	Zinc	Manganese	Chromium	Cadmium	Lead
Concentration (mg/L)	3.0271	1.4102	1.0703	0.4663	0.0568	0.0075	0.0278

The minerals analyzed for are as shown table 4 above. The minerals found present apart from sodium (Na), potassium (K) and calcium (Ca), which were not analyzed for are iron (Fe), zinc (Zn), manganese (Mn), copper (Cu), chromium (Cr), cadmium (Cd) and lead (Pb). The values obtained for these minerals compare very well with those obtained byOyelekeet *al*., (2014). These minerals have all been reported as essential for the body to function properly(Dana *et al*., 2014; Ikhuoriaet *al*., 2008). GC-MS analysis was carried out on the crude extract, petroleum ether and chloroform fractions to determine the components of the oil. The chromatograms of the petroleum ether and chloroform fractions are shown in Figures 1 and 2 below.

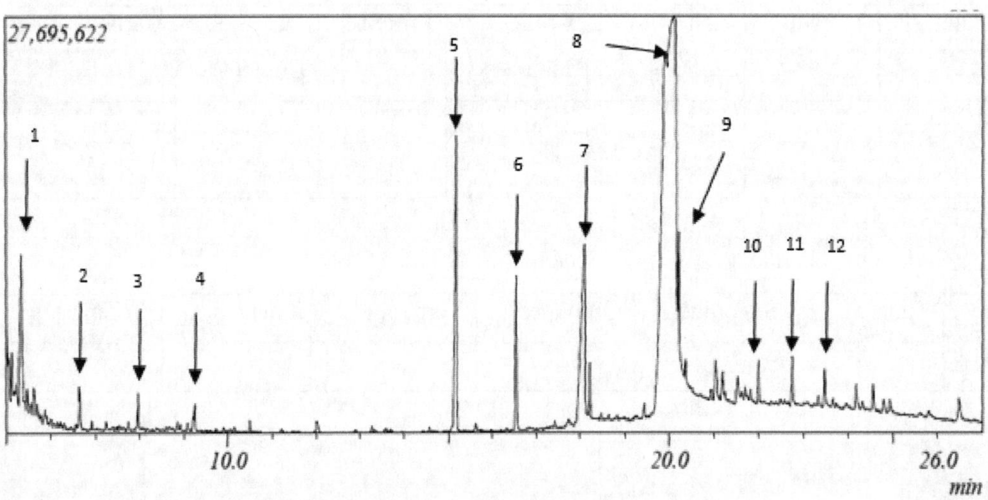

Figure 1: Chromatogram of the petroleum ether fraction

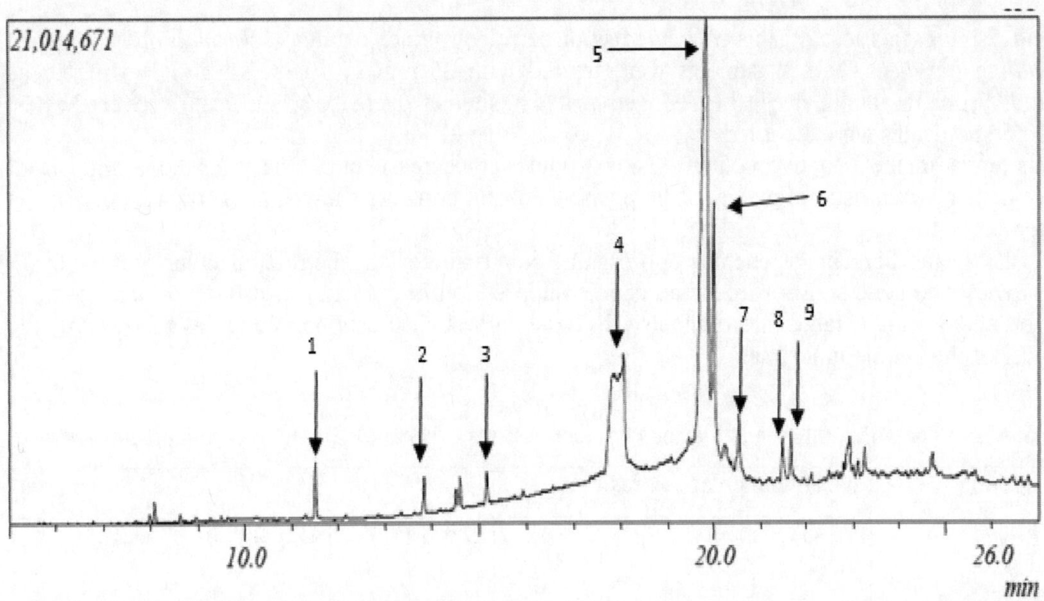

Figure 2: Chromatogram of the chloroform fraction

The identity of numbered peaks in the chromatograms was confirmed by comparison of their mass spectra with mass spectra of the authentic compound from literature. For the petroleum ether fraction, the following compounds were identified as numbered:**1** – Undecane (5.338); **2** – dodecane (6.648); **3** – tridecane (7.965); **4** – tetradecane (9.246); **5** – isobutyl, butyl glutaric acid (15.180); **6** – dibutyladipate (16.547); **7** – n-hexadecanoic acid (18.104); **8** – 9,12-octadecadienoic acid (19.985); **9** – Octadecanoic acid (20.205); **10** – 2-methylhexacosane (21.988); **11** – pentatricontane (22.744).

For the chloroform fraction, the following compounds were identified as numbered: **1** – Vanillin (11.510); **2** – 5-oxo-Pyrrolidinecarboxylic acid-ethyl ester(13.829); **3** – 1-(1-cyclohexen-1-yl)-Pyrrolidine (15.165); **4** – n-Hexadecanoic acid (17.834); **5** – 9-Octadecenoic acid (19.857);**6** – octadecanoic acid (20.014);**7** – 9,12-Octadecadienoic acid (20.525); **8** – 2-hydroxy cyclopentadecanone (21.477); **9** – Eicosanoic acid (21.665). The number in parentheses after the name of each compound is the retention time for that compound in minutes.

The anti-plasmodial activity of petroleum ether and ethanolic extracts of seeds of *Pentaclethramacrophylla*is presented in tables 4, 5 and 6. It confirmed that petroleum ether seed extracts of *Pentaclethramacrophylla*is non-toxic as high as 100 mg/kg, Table 4.

Table 5: Acute toxicity test forEthanol and petroleum ether extract of seeds of *Pentaclethramacrophylla*

Group	Average Weight (g)	Dose	Survival at 24 h	Survival at 48 h	Survival at 72 h
Group 1	26 g	25 mg/kg	Active and healthy		
Group 2	25 g	50 mg/kg			
Group 3	25 g	100 mg/kg			

IOP Conf. Series: Earth and Environmental Science **173** (2018) 012012 doi:10.1088/1755-1315/173/1/012012

Table 5 represents test organisms treated with petroleum ether extract of *P. macrophylla*against *Plasmodium berghei*. The test subjects were treated with 25 mg/kg, 50 mg/kg and 100 mg/kg as indicated. From the table, it can be deduced that parasitemia decreased significantly over the test period, and no deaths were recorded.

Parasitemia in the negative control (Table 5)did not decrease but rather showed a significant increase as is characteristic of malaria. The positive control however showed a significant decrease in parasitemia.

Table 6shows the percentage chemo-suppression over time for the petroleum ether extract of *P. macrophylla*. The table shows percentage chemo-suppression of parasites at different concentrations for different days. In the table chloroquine showed the highest chemosuppression rates followed by the 100mg/kg of the petroleum extract.

Table 6: Mean Daily Parasitemia at Various Concentrations ofEthanol Extract of *P. macrophylla*

Concentration (mg/Kg)	Mean Percentage Parasitemia				
	Pre-Treatment (%)	Day 0 (%)	Day 1 (%)	Day 2 (%)	Day 3 (%)
25	11.00 ±1.00	9.00±1.00	7.50±0.5	5.00±0.0	3.50±0.05
50	11.00±1.00	8.50±0.5	6.00±0.0	4.50±0.5	2.50±.050
100	13.00±1.00	10.50±0.5	8.00±0.0	5.50±0.5	2.00±0.00
Chloroquine (Positive Control)	15.00±0.00	10.00±0.00	6.50±0.5	4.00±1.0	1.00±1.00
Normal Saline	12.00±1.00	15.00±1.00	17.50±1.5	19.00±1.0	22.00±0.00

Values are expressed as mean ±SEM

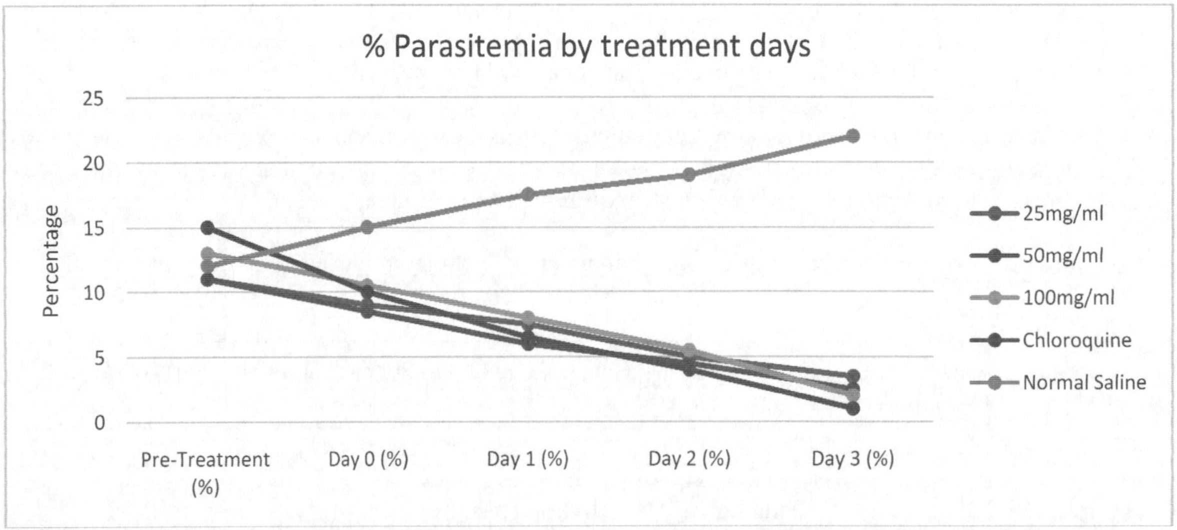

Figure 3: Percentage parasitemiaat Different Concentrations of Ethanol Extract of *P. macrophylla*

Table 7: Mean Daily Parasitemia at Various Concentrations of Petroleum Ether Extract of *P. macrophylla*

Concentration (mg/Kg)	Mean Percentage Parasitemia				
	Pre-Treatment (%)	Day 0 (%)	Day 1 (%)	Day 2 (%)	Day 3 (%)
25	22.5 ±3.89	17.0±0.71	13.5±1.06	14.0±0.35	11.5±0.35
50	19.00±0.71	15.50±0.35	15.0±0.71	10.00±0.00	8.0±0.00
100	19.50±0.35	17.00±0.00	11.5±2.47	17.00±5.66	8.50±1.06
Chloroquine (Positive Control)	15.00±0.00	10.00±0.00	6.5±0.35	4.00±0.71	1.00±0.71
Normal Saline	12.00±0.71	15.00±0.71	17.5±1.06	19.0±0.71	22.00±0.00

Values are expressed as mean ±SEM

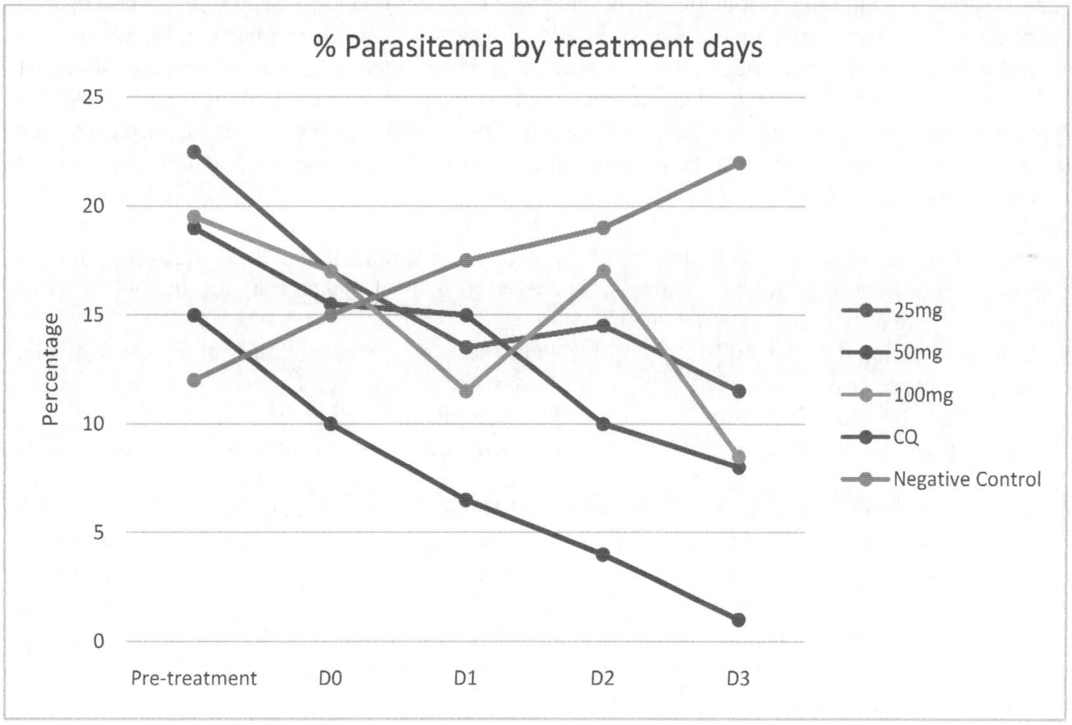

Figure 4: Percentage parasitemia at Different Concentrations of Petroleum Ether Extract of *P. macrophylla*

IOP Conf. Series: Earth and Environmental Science **173** (2018) 012012 doi:10.1088/1755-1315/173/1/012012

Table 8: Percentage Chemo-suppression of the Ethanoland Petroleum Ether extracts of *P.macrophylla*seed

Drug	Dose	Percentage chemo-suppression (%) Ethanol	Petroleum Ether
Physiological saline	0.5mL	-	-
P.macrophylla	25mL/kg	84.09	47.72
	50mL/kg	88.63	63.64
	100mL/kg	90.90	61.36
Chloroquine Phosphate	10mg/kg	95.45	95.45

Discussion

The *in-vivo* antiplasmodial activities of crude ethanolic and petroleum ether extracts of seeds of *P.macrophylla*Benth of the family *Fabacaea*with voucher No: Pm/Bio/H806 and Ascension code: OgCUNo807 were determined in this study. In vivo models are usually employed in anti-malaria studies because they take into account, the possible pro drug effect and probable involvement of the immune system in eradication of the pathogen(Waako *et al*., 2005). During early infections, Peter's 4 day suppressive test was used to evaluate schizontocidalactivity. In all methods, determination of percent of inhibition of the parasitemia was the most reliable parameter. A mean parasitemia level that is ≤90% of that of the vehicle treated animals (negative control), usually indicates that the test compound is active(Peter and Anatoli, 1998). In the 4 day suppressive test, seed extracts ofP.macrophyllasignificantly reduced parasitemia in animal models in a dose dependent manner, with these drugs showing anti-malarial activities compared to that of the standard drug tested. There were no deaths or any signs of toxicity observed after oral administration of single doses up to 100mg/kg of the extracts of P. macrophylla. This data indicates that the medium lethal dose (LD_{50}) should be 100mg/kg for both extracts, for test animals. There were no significant changes in the behavior, cutaneous breathing of both male and female mice.

The $LD_{50}$was calculated to be 707.11mg/kg for Petroleum ether seed extract of P. macrophylla. The $LD_{50}$is the concentration at which 50% of the test subjects will die. This shows that the extract becomes toxic at 707.11mg/kg. Also for P.macrophylla, the LD_{50} was 707.11mg/kg

The seed extract of P.macrophylla exerted dose dependent chemo-suppressive effect against Plasmodium bergeibergei. The petroleum ether extract caused a 47.72%, 63.64% and 61.36% chemo-suppressions at concentrations 25mg/kg, 50mg/kg and 100mg/kg respectively by the 4th day of the test, while the standard drug chloroquine caused chemo-suppressions of 95.45%. The ethanolic extracts of P.macrophyllaseed at dosing concentrations of 25mL/kg, 50mL/kg and 100mL/kg body weight of mice yielded 84.09%, 88.63% and 90.90% inhibition respectively as against 95.45% for the standard drug, chloroquine. The observed higher efficacy of Chloroquine may in part be due to non-selectivity of the extract or slow absorption and poor bio-availability of the extract. This is common with medicinal plant extracts(Adzu and Haruna, 2007).

P. Bergeibergei has been utilized in studying the activity of potential anti-plasmodials in-vivo in rodents (Pedroni *et al*., 2006)and it produces diseases similar to those of human plasmodial infection (Kumar et al., 2008). Agents with suppressive activity against P. Bergeibergeiwere known for their anti-plasmodial activity(Elufioye and Agbedahunsi, 2004).

4. CONCLUSION

This study has shown the efficacy of *P. macrophylla*as an antimalarial and as such validates its use in traditional settings. Future research should focus on identifying the active fractions of the plant.

References

[1] Adzu, B., Haruna, A.K., 2007. Studies on the use of Zizyphus spina-christi against pain in rats and mice. African Journal of Biotechnology 6, 1317–1324. doi:http://dx.doi.org/10.5897/AJB2007.000-2183

[2] Ajaiyeoba, E., Ashidi, J., Abiodun, O., Okpako, L., Ogbole, O., Akinboye, D., Falade, C., Bolaji, O., Gbotosho, G., Falade, M., Itiola, O., Houghton, P., Wright, C., and Oduola, A., 2005. Antimalarial Ethnobotany: In Vitro. Antiplasmodial Activity of Seven Plants Identified in the Nigerian Middle Belt. Journal of Pharmaceutical Biology 42, 588 – 591.

[3] Akindahunsi, A.A., 2004. Physicochemical studies on African oil bean (*Pentaclethramacrophylla* Benth.) seed. Journal of Food Agriculture & Environment 2, 14–17.

[4] Anosa, G. N., Udegbunam. R. I., Okoro, J. O., Okoroafor, O. N., 2014. In vivo antimalarial activities of Enantiapolycarpa stem bark against *Plasmodium berghei berghei* in mice. Journal of Ethnopharmacology 153, 531-534. doi: 10.1016/j.jep.2014.02.022

[5] Ariey, F., Witkowski, B., Amaratunga, C., Beghain, J., Langlois, A.-C., Khim, N., Kim, S., Duru, V., Bouchier, C., Ma, L., Lim, P., Leang, R., Duong, S., Sreng, S., Suon, S., Chuor, C.M., Bout, D.M., Ménard, S., Rogers, W.O., Genton, B., Fandeur, T., Miotto, O., Ringwald, P., Le Bras, J., Berry, A., Barale, J.-C., Fairhurst, R.M., Benoit-Vical, F., Mercereau-Puijalon, O., Ménard, D., 2014. A molecular marker of artemisinin-resistant *Plasmodiumfalciparum* malaria. Nature 505, 50–55. doi:10.1038/nature12876

[6] Charles Chukwumah, O., Inyang, I., Kelechi Angelina, I., 2013. KINETICS OF FERMENTATION OF OIL BEAN SEED. Annals. Food Science and Technology 14.

[7] Chung, K.T., Wong, T.Y., Wei, C.I., Huang, Y.W., Lin, Y., 1998. Tannins and human health: a review. Critical reviews in food science and nutrition 38, 421–64. doi:10.1080/10408699891274273

[8] Dana, S., Prusty, D., Dhayal, D., Gupta, M.K., Dar, A., Sen, S., Mukhopadhyay, P., Adak, T., Dhar, S.K., 2014. Potent antimalarial activity of acriflavine *in vitro* and *in vivo*. ACS chemical biology 9, 2366–73. doi:10.1021/cb500476q

[9] Devi, P. B., Vijayabharathi, R., Sathyabama, S., Malleshi, N. G., and Priyadarisini, V. B. 2014. Health benefits of finger millet (Eleusine coracana L.) polyphenols and dietary fiber: a review. J. Food Sci. Technol 51, 1021–1040. doi: 10.1007/s13197-011-0584-9

[10] Elufioye, T.O., Agbedahunsi, J.M., 2004. Antimalarial activities of Tithonia diversifolia (Asteraceae) and Crossopteryx febrifuga (Rubiaceae) on mice *in vivo*. Journal of Ethnopharmacology 93, 167–171. doi:10.1016/j.jep.2004.01.009

[11] Ene, A. C. Atawodi, S. E., Ameh, D. A., Kwanashie, H. O. and Agomo, P. U. (2009) In vivo antiplasmodial effect of chloroform extracts of Artemisia maciverae Linn and Artemisia maritima Linn. African Journal of Biotechnology 8, 6612 – 6616.

[12] Akwaji P.I.*, Eyam E. O.and Bassey R. A. (2017). Ethnobotanical Survey of Commonly Used Medicinal Plants in Northern Cross River State, Nigeria. World Scientific News;70(2):140-157

[13] Grimberg, BT., and Mehlotra, RK., 2011. Expanding the Antimalarial Drug Arsenal—Now, But How? Pharmaceuticals 4: 681 – 712. doi:10.3390/ph4050681

[14] Ikhuoria, E.U., Aiwonegbe, A.E., Okoli, P., Idu, M., 2008. Characteristics and Composition of African Oil Bean Seed (*Pentaclethramacrophylla* Benth). Journal of Applied Sciences 8, 1337–1339. doi:10.3923/jas.2008.1337.1339

[15] Institute for Laboratory Animal Research, 2011. Guide for the Care and Use of Laboratory

Animals: 8th Ed., Guide for the Care and Use of Laboratory Animals. doi:10.2307/1525495

[16] Kumar, S., Das, S.K., Dey, S., Maity, P., Guha, M., Choubey, V., Panda, G., Bandyopadhyay, U., 2008. Antiplasmodial Activity of [(Aryl)arylsulfanylmethyl]Pyridine. Antimicrobial Agents and Chemotherapy 52, 705–715. doi:10.1128/AAC.00898-07

[17] Lorke, D., 1983. A new approach to practical acute toxicity testing. Archives of Toxicology 54, 275–287. doi:10.1007/BF01234480

[18] Ntie-Kang, F., Onguéné, P.A., Lifongo, L.L., Ndom, J.C., Sippl, W., Mbaze, L.M., 2014. The potential of anti-malarial compounds derived from African medicinal plants, part II: a pharmacological evaluation of non-alkaloids and non-terpenoids. Malaria journal 13, 81. doi:10.1186/1475-2875-13-81

[19] Odoemelam, S.A., 2005. Proximate Composition and Selected Physicochemical Properties of the Seeds of African Oil Bean (*Pentaclethramarcrophylla*). Pakistan Journal of Nutrition 4, 382–383. doi:10.3923/pjn.2005.382.383

[20] Ogueke, C.C., Nwosu, J.N., Owuamanam, C.I., Iwouno, J.N., 2010. Ugba, the Fermented African Oilbean Seeds; its Production, Chemical Composition, Preservation, Safety and Health Benefits. Pakistan Journal of Biological Sciences 13, 489–496. doi:10.3923/pjbs.2010.489.496

[21] Olasehinde, G.I., Akinlabu, D.K., Owoeye, F.T., Owolabi, E.F., Audu, O.Y., Mordi, R.C., 2016. Phytochemical and Antimicrobial Properties of Oil Extracts from the Seeds of Ricinodendron heudelotii. Research Journal of Medicinal Plants 10, 362–365. doi:10.3923/rjmp.2016.362.365

[22] Olasehinde, G.I., Ojurongbe, D.O., Akinjogunl, O.J., Egwari, L.O., Adeyeba, A.O., 2015. Prevalence of Malaria and Predisposing Factors to Antimalarial Drug Resistance in Southwestern Nigeria. Research Journal of Parasitology 10, 92–101. doi:10.3923/jp.2015.92.101

[23] Oyeleke, G.O., Odedeji, J.O., Ishola, A.D., Afolabi, O., 2014. Phytochemical Screening and Nutritional Evaluation of African Oil Bean (*Pentaclethramacrophylla*) Seeds. Journal of Environmental Science, Toxicology and Food Technology 8.

[24] Pedroni, H.C., Bettoni, C.C., Spalding, S.M., Dalla Costa, T., 2006. *Plasmodiumberghei*: Development of an irreversible experimental malaria model in Wistar rats. Experimental Parasitology 113, 193–196. doi:10.1016/j.exppara.2005.12.017

[25] Peter, I.T., Anatoli, V.K., 1998. The current global malaria situation. Malaria parasite biology, pathogenesis, and protection. ASM Press. WDC 1, 1–22.

[26] Peters, W., 1968. The chemotherapy of rodent malaria, II. Annals of Tropical Medicine & Parasitology 62, 246–251. doi:10.1080/00034983.1968.11686556

[27] Saidu, K., Onah, J., Orisadipe, A., Olusola, A., Wambebe, C., Gamaniel, K., 2000. Antiplasmodial, analgesic, and anti-inflammatory activities of the aqueous extract of the stem bark of Erythrina senegalensis. Journal of Ethnopharmacology 71, 275–280. doi:10.1016/S0378-8741(00)00188-4

[28] Sofowora, A., Ogunbodede, E., Onayade, A., 2013. The role and place of medicinal plants in the strategies for disease prevention. African journal of traditional, complementary, and alternative medicines : AJTCAM 10, 210–29.

[29] Tarkang, P. A., Franzoi, K. D., Lee, S., Lee, E., Vivarelli, D., Freitas-Junior, L., Liuzzi, M., Nolé, T., Ayong, L. S., Agbor, G. A., Okalebo, F. A., and Guantai, A. N., 2014. "*In Vitro*Antiplasmodial Activities and Synergistic Combinations of Differential Solvent Extracts of the Polyherbal Product, Nefang" BioMed Research International,1. doi:10.1155/2014/835013

[30] Tarkang, P. A., Okalebo, F. A., Ayong, L. S., Agbor, G. A., and Guantai, A. N., (2014) Anti-malarial activity of a polyherbal product (Nefang) during early and established Plasmodium infection in rodent models, Malaria Journal; 13:456 . doi: 10.1186/1475-2875-13-456.

[31] Tepongning, R. N., Yerbanga, S. R., Dori, G. U., Lucantoni, L., Lupidi, G., et al. (2013) *In vivo*

Efficacy and Toxicity Studies on Erythrina senegalensis and Khayaivorensis Used as Herbal Remedies for Malaria Prevention in Cameroon. Eur J Med Plants3: 454 – 464. doi: 10.9734/EJMP/2013/3928

[32] Titanji, VPK., Denis Zofou, D., and Moses N. Ngemenya, MN., (2008) The antimalarial potential of medicinal plants used for the treatment of malaria in Cameroonian folk medicine, Afr. J. Traditional Complementary Alternative Medicine; 5: 302 – 321.

[33] Valderramos, S. G., Scanfeld, D., Uhlemann, A. C., Fidock, D. A., and Sanjeev K. S., 2010. Investigations into the Role of the *Plasmodiumfalciparum* SERCA (PfATP6) L263E Mutation in Artemisinin Action and Resistance. Antimicrobial agents and chemotherapy 54, 3842 – 3852. doi:10.1128/AAC.00121-10

[34] Waako, P.J., Gumede, B., Smith, P., Folb, P.I., 2005. The *invitro* and *invivo* antimalarial activity of Cardiospermum halicacabum L. and Momordica foetida Schumch. Et Thonn. Journal of Ethnopharmacology 99, 137–143. doi:10.1016/j.jep.2005.02.017

[35] Watts, D.L., 1990. Nutritiolal Interrelationships: minerals, vitamins, endicrines. Journal of orthomolecular Medicine 5, 11-19.

[36] Wu, M., Atchley, D., Greer, L., Janssen, S., Rosenberg, D., Sass, J., 2009. Dosed without prescription: preventing pharmaceutical comtamination of our nation's drinking water. Natural Resources Defense Council White Paper.60.

IOP Conf. Series: Earth and Environmental Science **173** (2018) 012013 doi:10.1088/1755-1315/173/1/012013

Cloud attenuation modelling for satellite network links performance improvement

O. M. Adewusi[2], T.V. Omotosho[1], M. L. Akinyemi[1], S. A. Akinwumi[1], O. O. Ometan[2]

[1]Department of Physics, College of Science and Technology, Covenant University, PMB 1023, Ota, Ogun State, Nigeria.
[2]Department of Physics, Lagos state University, Ojo, Lagos state, Nigeria.

madewusi@gmail.com, omotosho@covenantuniversity.edu.ng, marvel.akinyemi@covenantuniversity.edu.ng,oluwasayo.akinwumi@covenantuniversity.edu.ng, ometanfunmi@gmail.com

Abstract. All local climate zones need to develop relevant climatic models such as cloud attenuation model which can serve as correlative resources to global satellites applications and a means of determining local link margins for satellite networks services in localities. Effects of suspended water droplets (SWD) and suspended ice crystals (SIC) which constitute clouds are major concern in the design and successful operation of satellite communication system at frequencies above 3GHz because the hydrometeors reduces the services availability critically as frequency increases. This work is a review of cloud attenuation modelling, using as a case study typical research work been carried out, to develop cloud attenuation model for tropical Ota (6.7oN, 3.23oE), southwest Nigeria. Cloud attenuation modelling for a station involve obtaining clouds parametric system equation from their numerical representations, taking into consideration the climatic initial and boundary conditions. The numerical representations are collected climatological and radiometric data, derived from well-designed experiments in which cloud parameters measurements are carried out using radiances change measured by satellites and visual observations from the surface station(s) on land and ships in the ocean. The on-going research work begin with the study of performance of eight foundation cloud models at the station, through their detailed evaluation from cloud cover data and radiosonde data analysis, followed by analysis of collected over three years spectrum analyser signal attenuation data. Then modelling analysis of the station spectrum analyser data collected at computed elevation angle 56.18o to Astra 2(E-G) Satellite located at 28.2oE, which involve comparisons of the derived attenuation distribution curves, both of the spectrum analyser data and those of the existing cloud models for the station, with the derived attenuation distribution curves of output data generated by each run session of the station simulation equation program. The general representation of the station cloud attenuation model is stated and the specific model equation is been worked on.

1. Introduction

The increasing demand for satellite services and consequent increases in bandwidth requirements for current range of telecommunication applications have driven satellite service providers to continually seek wider bandwidth. The consequent increases in transmission frequency are phenomenally accompanied by directly proportional hydrometeors attenuation [1, 2]. The uniqueness of cloud attenuation among other hydrometeors is that on any given day about half of the Earth is covered by clouds, it is also more persistent and more stable in the atmosphere. Previous research works have established that transmission frequencies above 3 GHz have associated increase in signal propagation impairment in the troposphere due to atmospheric hydrometeors' dielectric relaxation among others, and that the impact of cloud attenuation at Ku band (12/14) GHz, Ka band (20/30) GHz and above have consequently been established as quite significant [3-6]. Operating at high frequency bands enables antenna elements size reduction, and moreover it becomes feasible to have multiple antennas not only at base stations, but also on mobile handsets – for producing sufficiently high signal to noise ratio [7, 8].

However satellite climatic models' algorithms need be correlated with accurately developed earth surface local climatic zones' models, as the algorithms requires various experimentally determined input parameters. For example on the Moderate Resolution Imaging Spectrometer (MODIS) which is a 36-bands scanning radiometer flying on the National Aeronautics and Space Administration (NASA)'s Terra and Aqua platforms, differences between model derived and measured clear sky radiances are mitigated with a radiance bias adjustment to avoid height assignment errors [9].

Measureable parameters of an electromagnetic signal are the signal amplitude, frequency, phase, bandwidth and polarization; while the cloud properties affecting radio propagation are the amount of cloud liquid water (SWD) and ice crystals (SIC), cloud top and base heights, and horizontal extent. The ten cloud types have been classified into three based on their normal range of vertical heights of existence in the atmosphere relative to the earth surface, namely – low, middle and high clouds [10]. Attenuation is reduction of the amplitude of a propagating signal with respect to distance travelled along its transmission path and cloud attenuation is due to relaxation losses in the molecules of the SWD. Dielectric relaxation is the inability of the molecules of the dielectric as water to become or remain aligned with the direction of the electromagnetic field applied to it by the signal. Relative to relaxation oscillator circuit, energy loss in the dielectric molecules occur through random interference of generated electromagnetic waves with the interacting signal, since such circuit generate electromagnetic waves by its periodic conduction and non-conduction cycles [2, 11-14].

The provision of broadband internet access through satellites to millions of commercial and individual users globally through small terminals as POS terminals and hand phones has inestimable socio-economic values. Thus improvement of the satellite services availability in every locality is fundamental. This requires studies at each local climate zones to develop relevant models as effective cloud attenuation model, which can be used to reliably estimateneeded parameters' values for global satellites applications. Numerous cloud models have been independently developed over the last eight decades based on Rayleigh scattering and Mie absorption theories, using empirical data and assumptions that cloud cover horizontally and vertically are uniform, and that clouds are non-precipitating. Foundation cloud models include Gun and East Model, Staelin Model, Slobin Model, Liebe et al. Model, Altshuler and Marr Model, Salonen Model, Salonen and Uppala Model and the ITU-R Model [15-20].

2. Cloud Cover Data Acquisition

This case study station's observatory is at a practical visibility height, having clearance altitude over high rise objects as buildings or trees that may obstruct the view of the clouds. The observation is conducted at fixed regular times daily – 9 A.M. and 3 P.M., considering local times of free or least obscurity or any fixed light interference. A digital camera with correct date and time setting is used to record a short video for each observation view. Each assessment view of the total amount of cloud entails determination of cloud amount in cloud layers between the cloud base and maximum vertical distance visible for each of the four sky quadrants during each of the fixed daily observation times. On the use of instrument to determine the cloud amount surrounding a station, a reliable directly measuring device is not common up to date, however multiple sensors method exist. Available devices for the multiple sensing methods are ceilometer, pyrometer and sky camera. Multiple ceilometers may be used in clustering technique to determine cloud amount, this procedure require more than three ceilometers with sky conditions multiple-ceilometer-algorithm. Similarly, the pyrometers in multiple sensing deployment configurations operate as passive remote sensing infrared thermometer.

In the satellite observations segment of cloud cover determination for a station, radiance changes measured by satellites are interpreted using climatic models' algorithms. For example the MODIS on the National Aeronautics and Space Administration (NASA) earth observing satellites system consisting of Terra and Aqua platforms, provide unique measurements for deriving global and regional cloud properties. Single spectra band approach by earlier satellites programs in estimating global cloud cover variations often under estimate thin or semi-transparent clouds as cirrus. MODIS multispectral observation capacity at high spectra resolution facilitated seasonal and annual changes studies of clouds, particularly the semi-transparent clouds [9].

3. Cloud Cover Data Analysis

The cloud cover of a station is a graphical relationship that shows the monthly variation of the percentage of cloud amount on the station's sky dome, thus daily total cloud amount need be accurately estimated by visual method or by an efficient cloud estimating instrument, and the monthly average total is computed through each year.Typical monthly record of the visual cloud data is shown inTable 1 and Figure 1 shows the cloud cover result – visual and satellite.At the Ota station, data from SRB, ISCCP, CERES, TERA-AQUA-MODIS, CALIPSO and CLOUDSAT satellites were downloaded and processed. Analyses of each set of satellite data were carried out and their output charts displayed in the figure 1 are used for corroborating the surface visual cloud cover result.

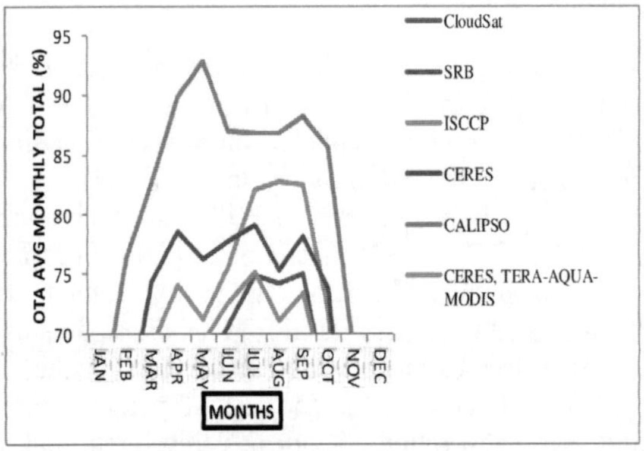

FIGURE 1: SATELLITES CLOUD COVER

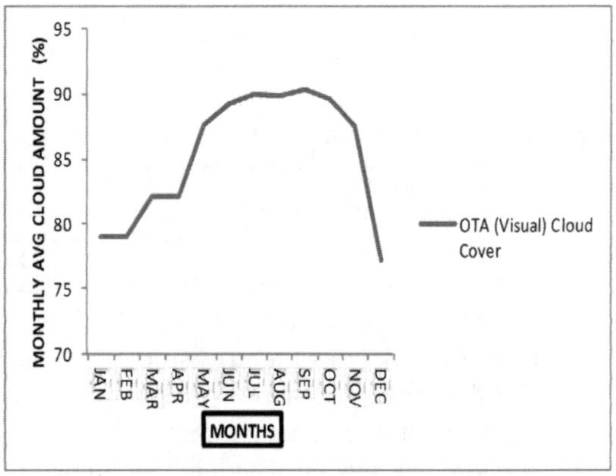

FIGURE 2: OTA - CU CLOUD COVER

4. Radiometric Data Acquisition

Radiometric devices commonly deployed on surface stations for measurement of propagation impairment includes Millimetre Wavelength Cloud Radar (MWCR), Microwave Radiometer (MWR), 9103 Spectrum Analyser, Laser Ceilometer and Micropulse Lidar, and Weather Surveillance Radar – 1988 Doppler (WSR – 88D). In a typical observation station, at least one of these listed devices will be required for radiometric data collection, and radiosonde data or satellite radiometric data will be needed for corroboration. In the Ota – CU station case study, fifty - three years radiosonde observations data was acquired, filtered and extracted to obtain each cloud layer's values for the required primary parameters – pressure (hPa), temperature (K) and calculated geopotential height (m); and every minute measurement and logged data of signal attenuation for over three years- using spectrum analyser, has been carried out.

Table 1: Typical Monthly Record of Visual data Measurement

	EAST (%)			SOUTH (%)			WEST (%)			NORTH (%)			CLOUD TYPES PRESENT			
DATE	A.M.	P.M.	AVG.	A.M.	P.M.	AVG.	A.M.	P.M.	AVG.	A.M.	P.M.	AVG.	TOTAL	E	S	W	N
1-Aug	24	23.5	23.75	23.5	23.5	23.5	23.5	22.5	23	23.5	23	23.25	93.5	Ns ; Sc	Ns ; Sc	Ns ; Sc	Ns ; Sc
2-Aug	24	23.5	23.75	23.5	23.5	23.5	23.5	24	23.75	24	24	24	95	Ns ; Ns, Sc	Ns ; Ns, Sc	Ns ; Ns, Sc	Ns, Cu ; Ns, Sc
3-Aug	23.5	24	23.75	23.5	24	23.75	23.5	23.5	23.5	23.5	23.5	23.5	94.5	Ns ; Ns,	Ns ; Ns,	Ns ; Ns,	Ns ; Ns,
4-Aug	23	20.5	21.75	22.5	22.5	22.5	22.5	23.5	23	23	22.5	22.75	90	Sc, Ns ;Sc	Sc, ;Sc	Sc, ;Sc, Ns	Sc, ;Sc
5-Aug	21	23	22	20	22.5	21.25	20.5	24	22.25	21.5	23.5	22.5	88	St, ;Sc	St, ;Sc	St, Ac ; Sc	St, Ac ; Sc
8-Aug	21	23	22	22	23.5	22.75	22	24	23	22	23	22.5	90.25	Sc ; Sc	Sc ; Sc	Ns, Sc ; Sc	Ns, Sc ; Sc
9-Aug	24	22	23	23.5	22.5	23	23	22.5	22.75	24	23	23.5	92.25	Ns, Sc ; Sc	Ns, Sc ; Sc	Ns, Sc ; Sc	Ns, Sc ; Sc
10-Aug	20	21	20.5	20.5	20.5	20.5	21.5	22.5	22	22.5	22.5	22.5	85.5	St ; Sc	St, Sc ; Sc, St	St, Sc ; Sc, Ns	St, Sc ; Cu, St
11-Aug	23.5	23	23.25	22.5	21.5	22	23	22.5	22.75	23	21	22	90	Ns, Cu ; Sc	Sc, ;Sc	Sc, ;Sc	Sc, ; Cb
12-Aug	23	21.5	22.25	23	21.5	22.25	22	20.5	21.25	21	20.5	20.75	86.5	Ns, Ac ; Cu, St	Sc ; Cu, St	Sc ; Sc	Ns,Ac,As ; Cu
15-Aug	24	24	24	23.5	23.5	23.5	24	23	23.5	23.5	23	23.25	94.25	Ns, Ac Cu; Ns	Ns, Ac Cu; Ns	Ns, Ac Cu; N	Ns, Ac Cu; Ns
16-Aug	24	23	23.5	23	22.5	22.75	23.5	22	22.75	24	22	23	92	Ns ; St	Ns ; St	Ns ; St	Ns ; St
17-Aug	23.5	23.5	23.5	22	23	22.5	22	23.5	22.75	23	23.5	23.25	92	Ns ; Sc	St, ;Sc	St, ;Sc	Sc, ;Sc
18-Aug	22	21.5	21.75	21	22	21.5	21	22.5	21.75	20.5	22	21.25	86.25	As,Ac ; Sc	As,Ac ; Sc	As,Ac ; Sc	As,Ac ; Sc
19-Aug	23	22.5	22.75	23	23.5	23.25	24	23.5	23.75	24	23.5	23.75	93.5	Sc, ; St, Sc	Sc, ; St, Sc	Sc, ; Ns	Sc, ; Ns
22-Aug	24	20	22	23.5	20.5	22	23.5	21	22.25	23.5	20.5	22	88.25	Ns ; St	Ns ; St	Ns ; St	Ns ; St
23-Aug	23.5	20	21.75	22.5	21	21.75	22.5	21.5	22	23	21	22	87.5	Ns, St	Sc	Sc	Sc
24-Aug	24	21	22.5	23.5	23.5	23.5	23	23	23	22.5	23	22.75	91.75	Ns ; Sc	Sc, ;Sc	Sc, ;Sc	Sc, ; Cu
25-Aug	20.5	23.5	22	20.5	23	21.75	20	23.5	21.75	20.5	23	21.75	87.25	As,Ac ; Ns, Cu	As,Ac ; Ns, Cu	St ; Cb	As,Ac ; Ns, Cu
26-Aug	22.5	23	22.75	23.5	23.5	23.5	23.5	23.5	23.5	23.5	23.5	23.5	93.25	Sc, ;Sc	Sc, ;Sc	Sc, ;Sc	Sc, ;Sc
29-Aug	22	20.5	21.25	22	21	21.5	22.5	21.5	22	23	21.5	22.25	87	St, ; Ac, Cu	St, ; Sc	St, ; Sc	St, ; St, Cu
30-Aug	20	24	22	20.5	20.5	20.5	21	23.5	22.25	21.5	23.5	22.5	87.25	St, ; Ns	St, ; Cb	St, ; Ns	St, ; Ns
31-Aug	20	20.5	20.25	20.5	21	20.75	21	21	21	20	20.5	20.25	82.25	St, ; St	St, ; St	St, ; St	St, ; St
AVG													89.913				

5. Radiometric Data Analysis

Attenuation distribution and statistics for each of the eight listed cloud models were computed from the processed radiosonde data. The station logged spectrum analyser data (SPAD) were extracted and analysed under two conditions - rainy and non-rainy days, using the climatological device's monthly data for each year. Then all the processed data over three years were integrated and the integrated data cumulative distribution curve obtained. The integrated cloud attenuation cumulative distribution curve has similar curve geometry with those of the 2014, 2015 and 2016 respectively. Each distribution curve is a result of raw data processing using Matlab and electronic spread sheet programming to implement several layers of required station conditions such as geographic, climatic and spectra conditions. The resulting cloud attenuation cumulative distribution curves at 12.5 GHz for each of the existing cloud models and that for the spectrum analyser data were compared in figures 3 and 4.

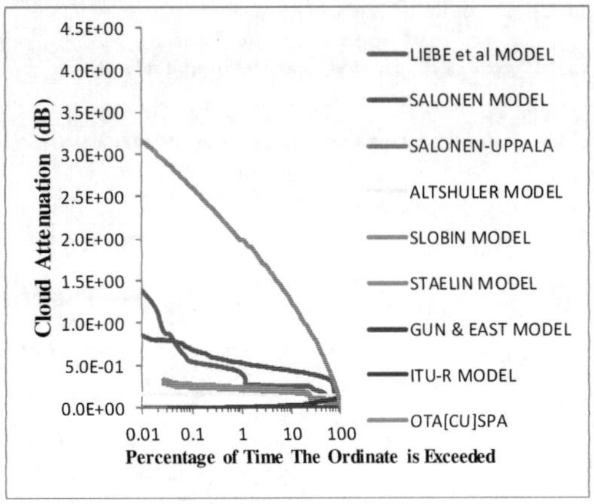

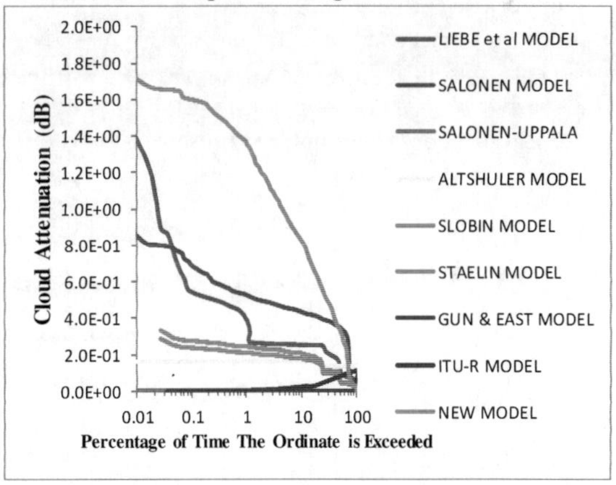

FIGURE 3: SPAD AND CLOUD MODELS DISTRIBUTION CURVES AT 12.5 GHz

FIGURE 4: CLOUD ATTENNUATION MODELS DISTRIBUTION CURVES AT 12.5 GHz

6. Cloud Attenuation Modelling Result

The SPAD integrated attenuation cumulative distribution curve approximates a straight line of slope -0.31011 and intercepts 3.089 dB, which appear as a resultant constructive interference of some pairs of periodic functions of comparable amplitudes. A possible general mathematical representation of these observations is equation (1).

$$Ac\ (WL, \theta, t, f) = f(x) = a\cos(x) + b\sin(x) = A\cos(x - \alpha) \qquad (1)$$

where A = (a2 + b2)1/2 and α = tan-1(b/a) in the range 0 < α < π/2; considering the variables of the cloud attenuation (Ac) - cloud liquid water content (W_L) of each cloud layer series (CLS) and their temperature (t); angle of elevation (θ) and frequency (f) of propagating radio signals; x is considered to be cloud layer specific attenuation coefficient (K_L) defined in the International Telecommunication Union Recommendation (ITU-R) P.840 – 4, corresponding to the frequency function g(f) in the Salonen – Uppala procedure. The definition is based on a mathematical model, using Rayleigh scattering and a double – Debye model for dielectric permitivity Ɛ (f) of water (ITU-R, 2009; Salonen – Uppala, 1991). Considering the relatively gentle slope of the SPAD line, the range of x should be 0 to π rather than 0 to π/2, and using the straight line analogue, a = - 0.31011, b = 3.089 dB, as directly measured from the graph; computed A = 3.08915 and α = - 1.560796 are typical values been used in the simulation equation (equation 2) program.

$$Ac = AW_L COS\ ((0.5K_L) + \alpha) \qquad 0 \leq KL \leq \pi \qquad (2)$$

Hence equation (2) is a general representation of the cloud attenuation model for the station measured data distribution (SPAD), having shown considerable geometric pattern agreement with the foundation models at various frequencies in Figures 4 – 8. The specific model equation is been worked out.

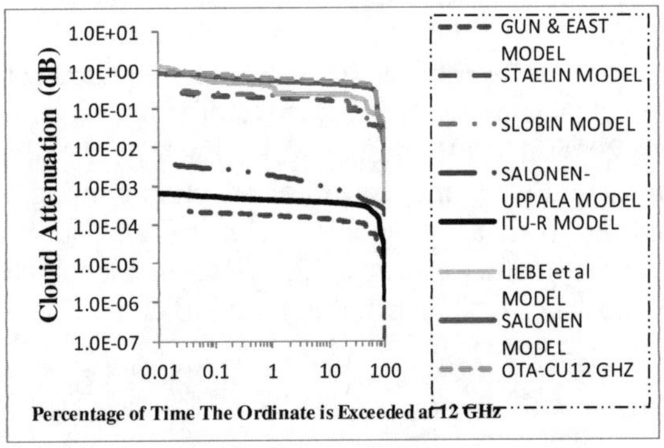

FIGURE 5: OTA-CU CLOUD MODEL AND FOUNDATION MODELS AT 12 GHZ

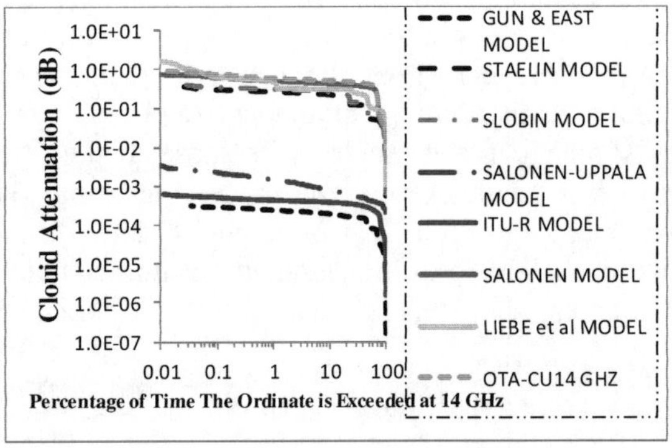

FIGURE 6: OTA-CU CLOUD MODEL AND FOUNDATION MODELS AT 14 GHZ

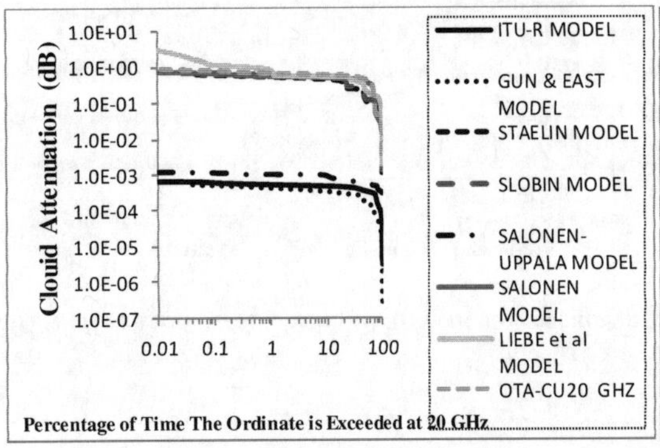

FIGURE 7: OTA-CU CLOUD MODEL AND FOUNDATION MODELS AT 20 GHZ

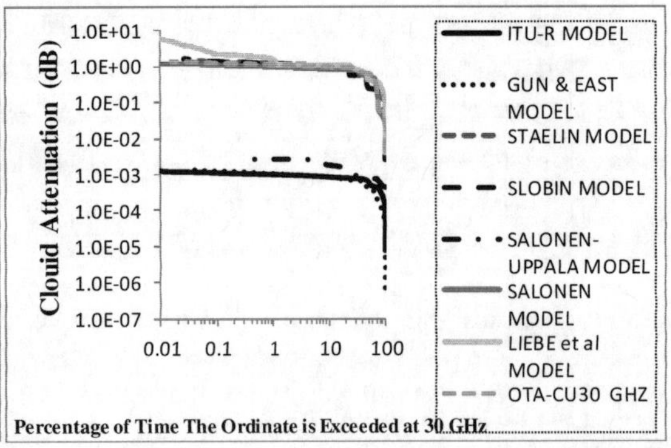

FIGURE 8: OTA-CU CLOUD MODEL AND FOUNDATION MODELS AT 30 GHZ

Conclusion

Cloud attenuation modelling begins with a well-designed experimental procedure for effective measurements and data collection scheme of defined climatological and radiometric parameters. Then graphical analysis of the primary data groups directly give some of the required results as the cloud cover of the station; and indirectly produce some other required results as the derived integrated cloud attenuation cumulative distribution curve from the annual distributions. The integrated cumulative distribution curve is used to develop required model equation using the climatic initial and boundary conditions. This model's chart is compared with those of existing cloud attenuation models as displayed by the figures 4 to 8 for correlation. However, a single global satellite based climatic model is desirable, where modular algorithm would be used to integrate the various climatic zones models.

Acknowledgment

The authors thank Covenant University for conference support.

References

[1] Allnut, J. E. (1989): Satellite - to - Ground radiowave propagation. Institution of Electrical Engineers;Electromagnetic Waves Series. Peter Peregrinus Ltd.
[2] Ippolito, L. J. (2008): Satellite Communication System Engineering; John Willey & Sons.
[3] Ali, M. S., Aduwati, S., Mandeep, J. S. and Ismail, A. (2013): Extracted atmospheric impairmentson earth-sky signal quality in tropical regions at Ku-band. Journal of Atmospheric and Solar-Terrestrial Physics, 104: 96-105.
[4] Antonio, M., Marina, B., Ermanno, F., Ondrej, F., Frank, S. M., Watson, P., Wrench, C. (2002): Propagation effects due to atmospheric gases and clouds; COST 255, ESA, Netherlands
[5] Das, S., Chakraborty, S. and Maitra, A. (2013): Radiometric measurements of cloud attenuation at atropical location in India; Journal of Atmospheric and Solar-Terrestrial Physics, 105: 7-100.
[6] NASA, (2014): The CloudSat Downlink; National Aeronautic and Space Administration, 7: 1-6
[7] Giovanni, E. C. ed. (2007): Information Technology Transmission Processing and Storage – Digital satellite Communication. Springer, ISBN 0-387-34649-X. 313 – 314, 361 – 365.
[8] Bruce, R. E. (2004): Satellite Communication Applications Handbook, 2nd Edition. Artech House, Inc., Norwood. 1- 58053 - 490 – 2.
[9] Menzel, W. P., Frey, R. A., Zhang, H., Wylie, D. P., Moeller, C. C., Holtz, R. E., Maddux, B., Baum, B. A., Strabala, K. I. and Gumley, L. E. (2008): MODIS Global Cloud – Top Pressure andAmount Estimation: Algorithm Description and Results. Journal of Applied Meteorology and Climatology, American Meteorological Society, 47, 1175 – 1198.
[10] Warren, S. G. and Hahn, C. J. (2002): Climatology; Encyclopaedia of Atmospheric Sciences, Academic Press.
[11] Debye, P. (1929): Polar Molecules; Chemical Catalogue Company. NY, USA.
[12] Cole, S. C. and Cole, R. H. (1941): Dielectric Relaxation in Glycerol, Propylene, Glycol and n – propanol; Chem. Phy., 9: 341-351.
[13] Omotosho, T.V., Mandeep, J. S. and Mardina A. (2013): Cloud Attenuation Studies of the Six Major Climatic Zones of Africa for Ka and V Satellite System Design; Annals of Geophysics; 56 (6): 1-2.
[14] Eva, R. I., Gustavo A. S., Jose, M. R. and Pedro, G. P. (2015): Atmospheric Attenuation in Wireless Communication Systems at Millimeter and THz Frequencies; IEE Antenna and Propagation Magazine, 57 (1): 48-61.
[15] Gerace, G. C. and Smith, E. K. (1990): A Comparison of Cloud Models; IEEE Antennas and Propagation Magazine.
[16] Liebe, H. J., Manabe, T., and Hufford, G. A., (1989): Millimetre Wave Attenuation and Delay Rates Due to Fog/Cloud Conditions; IEEE Trans. Ant. Prop. 37 (12):1617-1623.
[17] Edward, E. A. and Richard, A. M. (1989): Cloud Attenuation at Millimeter Wavelengths; IEEETransactions on Antennas and Propagation, 37 (11).
[18] Salonen, E., Karhu, S., Jokela, P., Zhang, W., Uppala, S., Aulamo, H. and Sarkula, S. (1990): Studyof Propagation Phenomena, for Low – Availabilities; ESA/ESTEE/Final Report.
[19] Salonen, E. and Uppala, S. (1991): New prediction method of cloud attenuation; Electron, Lett 27,1106-1108.
[20] ITU-R (2013): Attenuation due to Cloud and Fog; International Telecommunication Union, Radio Propagation Recommendation, P. 840-6.

Interpolation for Aerosol Optical Depth at Four Wavelengths over Ilorin from Observed Aeronet Data

*Sa'id R.S and Garba,S.
Department of Physics,
PMB 3011, Bayero University, Kano

*Correspondence Author: rssaidu.phy@buk.edu.ng, 08033976393

ABSTRACT

The Aerosol Optical Depth (AOD), which is the integral part of the atmospheric extinction coefficient from the surface to the top of the atmosphere, is an important parameter for observation of visibility degradation due to atmospheric pollution, solar radiation extinction and other climate effects. The wavelength dependence of Aerosol Optical Depth (AOD) varies between different aerosol types because of their different physical and chemical characteristic. This paper presents new data points using interpolation for Aerosol Optical Depth (AOD) at four spectral wavelengths (440-870nm. The data used consist of measurements collected at Ilorin AERONET site (8°32´N, 4°34´E) for AOD and Angstrom exponent at the four wavelengths ($\alpha_{440-870}$). Angstrom empirical formula was used to compute new data points by interpolation. A Regression analysis was then used to estimate the relationship between the interpolated AOD and the observed AOD at the different wavelengths of 440nm, 500nm, 670nm and 870nm. The coefficients of determination were found to be 0.978, 0.980, 0.999 and 0.997 respectively, indicating that the interpolated data agrees well with the measured data.

Key words: Interpolation, Aerosol, AERONET, Aerosol Optical Depth (AOD)

...

1. Introduction

Atmospheric aerosols are solid fine particles or liquid droplets suspended in air or another gas. Atmospheric aerosols originate from the condensation of gases and from the action of the wind on the earth's surface [1]: [2]. Aerosols in the atmosphere have several important and environmental effects; they scatter and absorb visible radiation, thereby limiting visibility and affect the earth climate both directly (by scattering and absorbing radiation) and indirectly (by serving as nuclei for cloud formation) as reported by [3]. Certain atmospheric aerosols absorb solar radiation, specifically carbonaceous particles and mineral dust [4], this absorption affects the atmosphere by reducing the radiation budgets reaching the surface and increasing the temperature at the top. Scattering of solar radiation by aerosols increases the Earth Albedo because a fraction of the scattered light is reflected back to space. There are two types of aerosols; the Natural aerosols (Fog, Forest exudates, Geyser steam e.t.c) and anthropogenic aerosols (Haze, Dust, Particulate air pollutants, Smoke e.t.c).

Aerosols have the ability to influence earth radiation budget as well as their effects on health, air quality and clouds [5]. Particles which are produced as a result of human activities (Anthropogenic) are in general smaller in size, and naturally produced particles are generally bigger in size. Characterization of aerosol properties is important because of the rapid growth of both population and economic activities, which cause the anthropogenic aerosol emission rates to increase as a result of the increase in fossil-fuel combustion and biomass burning [6]. These pollutants directly affect the climate and at the same time increase the haze, fog, and cloudy conditions which decrease visibility, particularly under high atmospheric turbidity [7]; [4].

Atmospheric aerosols generally have a bimodal distribution, the smaller particles are referred to as the fine mode or accumulation mode aerosols, these particles have radii between 0.1 to 0.25 microns but the larger particles comprise of coarse mode with a radii ranging between 1.0 to 2.5 microns [8]. The behavior of dust aerosols are affected by their mineralogical composition [9].

2. Methodology

Interpolation method was used in this study to construct new data points within the range of a discrete set of known data points; the known data set was obtained from Ilorin AERONET site from 2005-2009 within the spectral range 440-870nm. The two Angstrom empirical relations used to compute the interpolated AODs are as follows:

$$\tau_2 = e^{-(\ell n\lambda_2 - \ell n\lambda_1)} + \tau_1 \quad (1)$$

$$\beta = \tau_\lambda \lambda^\alpha \quad (2)$$

where α is the Angstrom Exponent
τ_2 is AOD at wavelength 2
τ_1 is AOD at wavelength 1
λ is wavelength
β is the Atmospheric turbidity

The data of Angstrom exponent ($\alpha_{440-870nm}$) [10] and the Aerosol optical depth (τ_λ) were obtained at different spectral bands (440, 500, 670, and 870nm) from Ilorin AERONET (AEROSOL ROBOTIC NETWORK) site for period of five years (2005- 2009), AERONET is a network of ground-based aerosol monitoring station that provides long term, continuous and readily accessible public domain database of aerosol optical properties for aerosol research and characterization as reported by [11].

Several gaps, at times spanning many days often occur in the AOD data series, thus, making the use of monthly or annual averages reliable than the daily averages for evaluating long term trends with the data, in this analysis the usage of monthly and annual data of the Angstrom exponent and AOD was used. The AOD data in this study was computed in three data quality levels (1.0, 1.5, and 2.0), level 2.0 was used which is cloud screen and quality assured. Interpolation method was used to construct new data points of AOD within the range of discrete set of known AOD data points which were measured by the solar extinction AERONET Algorithm, the known data points at different wavelength (440, 500, 670, and 870nm) from 2005-2009 and the interpolated data points were computed in this study using the following relationship;

$$\alpha = -\left(\frac{\ell n\tau_2 - \ell n\tau_1}{\ell n\lambda_2 - \ell n\lambda_1}\right) \quad (3)$$

where α is the Angstrom Exponent

τ_2 Is AOD at wavelength 2
τ_1 Is AOD at wavelength 1
$\boldsymbol{\lambda}$ is wavelength
Now by cross multiplying equation (3),

$$\alpha(\ell n\lambda_2 \_ \ell n\lambda_1) = -(\ell n\tau_2 \_ \ell n\tau_1) \qquad (4)$$

Rearranging equation (4)

$$\alpha(\ell n\lambda_2 \_ \ell n\lambda_1) = (\ell n\tau_1 \_ \ell n\tau_2) \qquad (5)$$

$$\ell n\tau_2 = -\alpha(\ell n\lambda_2 \_ \ell n\lambda_1) + \ell n\tau_1 \qquad (6)$$

Taking natural log of side of equation (6)

$$\tau_2 = e^{-(\ell n\lambda_2 \_ \ell n\lambda_1)} + \tau_1 \qquad (7)$$

Equation (7) was used in this study to calculate the new data points (interpolated AODs) at different spectral bands (440, 500, 670 and 870nm) from 2005-2009. The interpolated AOD(τ_2) data was calculated, for each year from 2005-2009 the procedure was repeated at different wavelengths (λ_1 *and* λ_2) from 440-870nm and the new constructed data set (interpolated AOD (τ_2) was tabulated.

3. Results and Discussion

3.1 Average AODs for Observed and Interpolated AERONET Data Sets

Figure 1 shows the trend between different spectral channels (440, 500, 670, & 870nm) for averages of AOD for 2005-2009; the average AOD for four spectral wavelengths was plotted for five years against month from January-December.

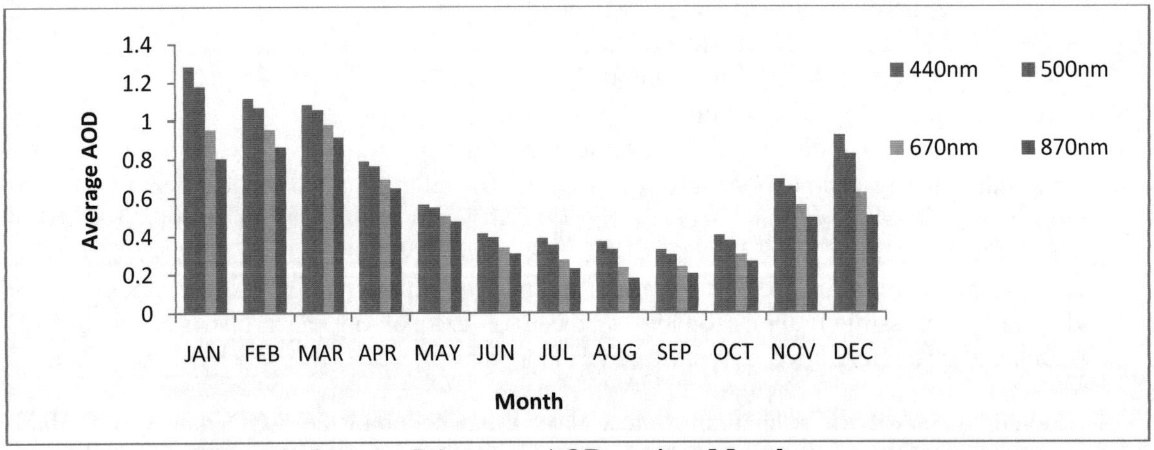

Figure1: Observed Average AOD against Month

The average AOD is highest in January with a value of 1.2826 at 440nm within five years period of study, AOD has a value of 1.1796 at 500nm still in January then followed by 1.1188 in February at 440nm and this shows that aerosol particles in the Atmosphere are more influential at shorter wavelength than at longer wavelength. It has been shown that the lower the wavelength the higher the AOD [12]: [13]. The lowest Average AOD in this trend is in August with a value of 0.1774 at 870nm this is the time when rainfall is at its peak, there is washing away of aerosol which result in lower AOD. The Average interpolated AOD was also plotted as shown in Figure 2

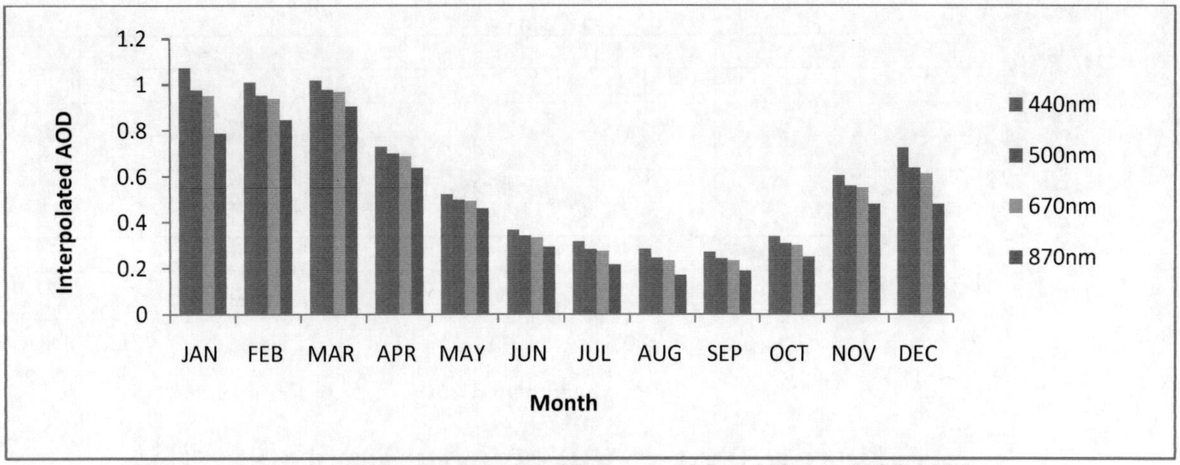

Figure 2: Average Interpolated AOD versus Month from 2005-2009

A plot of the measured and interpolated AOD was done at the four wavelengths to observe the trend as shown in figure 3. From the figure, it can be seen that for all the wavelengths, the interpolated values are slightly below the measured values. The only exception, where there is a close agreement of the two is at 675nm. In the months of June, July, August, September and October, the agreement of the two values is at the two wavelengths of 500 nm and 675 nm

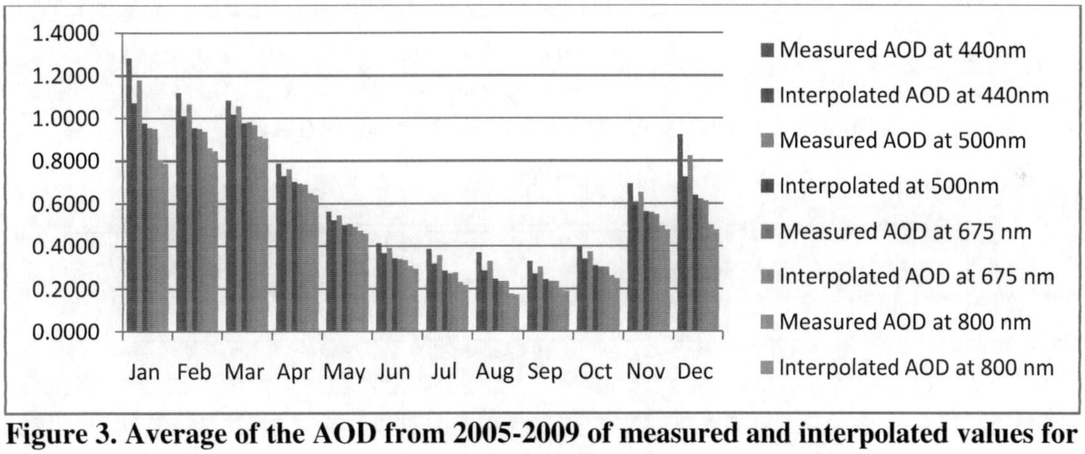

Figure 3. Average of the AOD from 2005-2009 of measured and interpolated values for the four wavelengths

3.2 Regression Analysis between Interpolated and Observed AODs

The regression analysis was done in this research to estimate the statistical relationship between the interpolated and observed AODs, the following graphs 4a, 4b, 4c and 4d shows the statistical relationship between the two AODs in which the average AODs for interpolated data set against ground-based measured data set at the four wavelengths: 440, 500, 670 and 870nm from 2005-2009 were plotted.

IOP Conf. Series: Earth and Environmental Science **173** (2018) 012014 doi:10.1088/1755-1315/173/1/012014

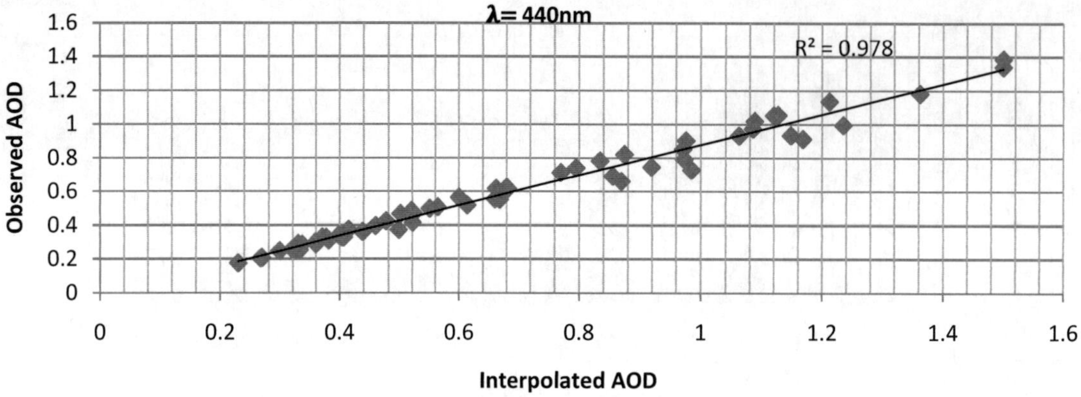

Figure 4a: Observed AOD versus Interpolated AOD

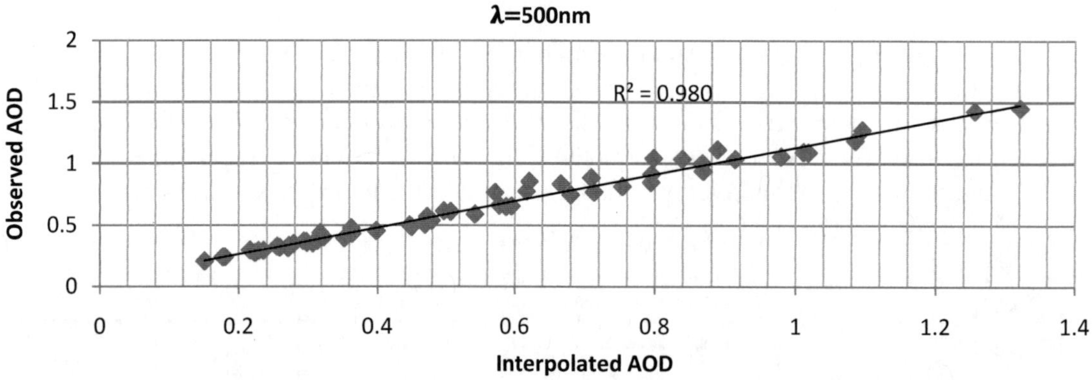

Figure 4b: Observed AOD versus Interpolated AOD

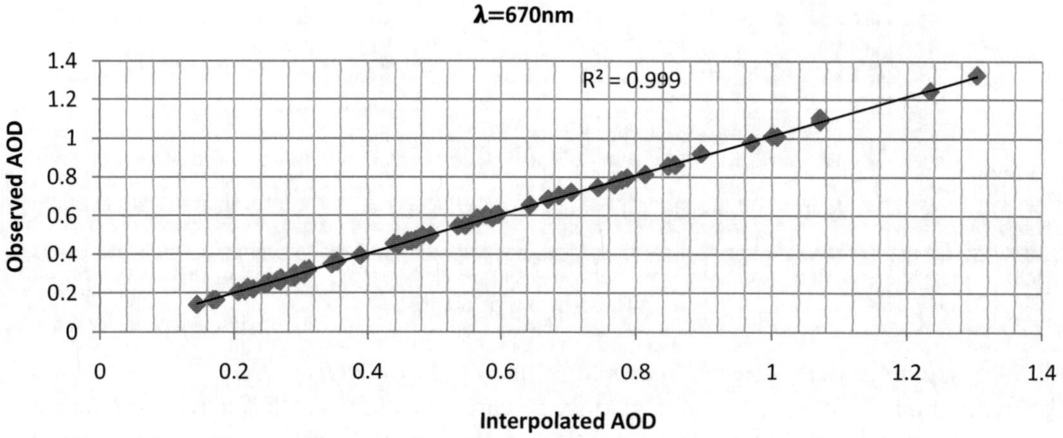

Figure 4c: Observed AOD versus Interpolated AOD

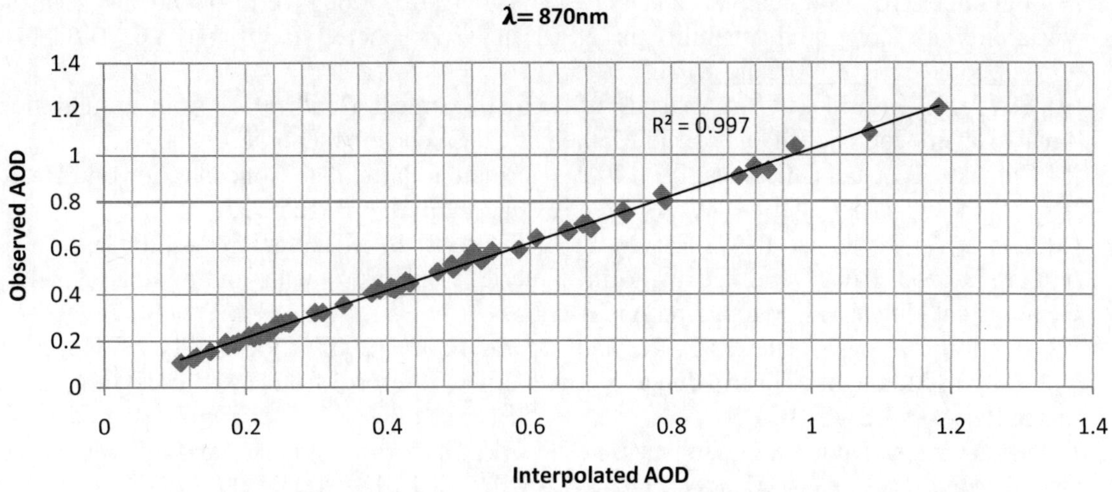

Figure 4d: Observed AOD versus Interpolated AOD

The regression in figure 4a, 4b, 4c and 4d shows that at wavelength 440nm the coefficient of determination is very close to 1 with a value of 0.978, at 500nm a value of 0.980 was obtained and at wavelength of 670nm and 870nm, the coefficient of determination produced values of 0.999 and 0.997 respectively. This indicates good relationship between the interpolated data set and the observed data set because all the data points in the four graphs plotted in this study are connected via the regression line in the manner that the distance from the line to the points plotted were the smallest. The regression analysis revealed that, the higher the values the more related the observed AOD and interpolated AOD are. Looking at the four values obtained in the study which revealed a good relationship at different wavelengths, this shows that, the interpolation method used in the study can be applied to construct new data points (interpolated data) wherever there are missing data in the ground-based AERONET measurements.

4. Conclusion

In conclusion, the data obtained from observation of AOD in Ilorin (8°32´N, 4°34´E) AERONET site from 2005-2009 at wavelengths of 440-870nm shows that, the method employed in the study of constructing new data points (interpolated) within the range of discrete known data sets (observed data) revealed a significant relationship between the two AODs. Although, the interpolated AOD is slightly lower than the measured value at the selected wavelengths for the average AOD of the years 2005-2009, however, there is a good agreement at 500 nm and 675 nm for the months of June to October.

References

[1] Oyem, A.A. and Igbafe, A.F. (2010) Analysis of Atmospheric Aerosol Loading over Nigeria. *Environmental Research Journal*, **4**, 145-156.

[2] O'Neill, N., T. Eck, A. Smirnov, B. N. Holben, and S. Thulasiraman (2003), Spectral Discrimination of coarse and fine mode optical depth, J. Geophys. Res., 108 (D17), 4559, doi:10.1029/2002JD002975.

[3] Balarabe, M., Abdullah, K. and Nawawi, M. (2015) Long-Term Trend and Seasonal Variability of Horizontal Visibility in Nigerian Troposphere. *Atmosphere*, **6**, 1462-1486. http://dx.doi.org/10.3390/atmos6101462

[4] Sa'id, R.S. (2011). "An Analysis of Aerosol Optical Depth Data from around Ilorin Area". Bayero Journal of Physics and Mathematical Science. VOL.4. NO.1.

[5] Chineke, T.C. and Chiemeka, I.U. (2009) Harmattan Particulate Concentration and Health Impacts in Sub-Sahara Africa. *The African Review of Physics*, **3**, 125-135.

[6] Cox S. J, Stackhouse P. W, Harvey V. L, Pierce R. B, Miskolczi F. M (2003) Dust and biomass aerosol radiative effects at Ilorin, Nigeria, from instruments and models. American Geophysical Union, Fall Meeting,pp A12B-0095.

[7] Anuforum, A. C., Okeke. P. N. and Opara, F. O. (2005): Variability of Horizontal Visibility in Maiduguri, Nigeria due to Harmattan Dust Aerosol from 1971 to 2000. Nig. J. of Space Res. 1, 1. 121-130.

[8] Schuster GL, Dubovik O, Holben BN. 2006. A° ngstro¨m exponent and bimodal aerosol size distributions. *J. Geophys. Res.* **111**: D07207. Doi: 10.1029/2005JD006328.

[9] Sa'id, R. S. and Kashimbila M. M. (2012) Scattering Phase Functions of Constituents of Mineral Dust Aerosols. Journal of the Nigerian Association of Mathematical Physics. Vol. 20, March, 2012. Pp. 331-342.

[10] Angstrom A. (1964). The parameters of atmospheric turbidity. *Tellus* **16**: 64–76.

[11] Holben, B.N., Eck, T.F., Slutsker, I., Tanré, D., Buis, J.P. Setzer, A., Vermote, E., Reagan, J.A., Kaufman, Y.J. Nakajima, T., Lavenu, F., Jankowiak, I. and Smirnov, A. (1998). AERONET—A federated instrument network and data archive for aerosol characterization. *Remote Sens. Environ.* 66: 1–16.

[12] Kaskaoutis, D.G., Kambezidis, H.D., Hatzianastassiou, N., Kosmopolos, P.G. and Badariahnath, K.V.S. (2007) Aerosol Climatology: On the Discrimination of Aerosol Types over Four AERONET Sites. *Atmospheric Chemistry and Physics Discusion*, **7**, 6357-6411. http://dx.doi.org/10.5194/acpd-7-6357-2007.

[13] Adeyewa, Z.D., Balogun, E.E., (2003). Wavelength Dependence of aerosol optical depth and the fit of the Angstrom law. Theor. Appl. Climatol 74, 105–122.

Analysis and comparison of tropospheric scintillation prediction models at Covenant University

S. A. Akinwumi[1], T. V. Omotosho[1], M. R. Usikalu[1], T. A. Adagunodo[1], M. O. Adewusi[2], O. O. Ometan[2]
[1]Department of Physics, College of Science and Technology Covenant University PMB 1023, Ota, Ogun state, Nigeria
[2]Department of Physics, Lagos state University, Ojo, Lagos state, Nigeria

oluwasayo.akinwumi@covenantuniversity.edu.ng

Abstract. Knowledge of tropospheric scintillation is an important phenomenon in the design of satellite communication system. One year (January 2015-December 2015) scintillation data extracted from Astra 2E/2F/2G Satellite link measurement installed at Covenant University, Ota (Lat: 6.7 oN, Long: 3.23 oE) southwest Nigeria, at an elevation angle of 59.9o and a frequency of 12.245 GHz was used in this study. The analysis and the result were compared with some reputable scintillation prediction models so as to obtain best performance model for Ota region. From the result, it was discovered that the Karasawa model gives the lowest percentage error rate for both fade and enhancement of about 0.57% at 0.1 percentage of time and 6.93% at 0.01 percentage of time respectively and therefore was best found fit for the prediction of propagation impairment for the region. However, the model should be tested further using higher frequency band such as Ka and V bands to confirm the accuracy of the model. The information provided in this study will help in fade margin for antenna sizing and performance needed for satellite communication link in the region.

Keywords: Tropospheric scintillation, Attenuation prediction, Satellite communication, Ku band, Performance evaluation

1. INTRODUCTION

Tropospheric scintillation occurrences transpires to be one of the key signal impairments that affect earth-satellite path [1-6] in modern earth-space communication systems, most importantly at higher frequency bands (both uplink and downlink). The influence of scintillation on radio wave signal transmission cannot be over highlighted due to its constant variation in phase and amplitude which affect signal power [7-8]. At short microwave or millimeter-wave bands, scintillation intensity increase with decrease in elevation angle, antenna size and with increase in frequencies [9]. Yet, the tropospheric scintillation is an intricate occurrence on earth-space transmission path, which includes the presence of gasses [10], and margin level up to altitudes of about 20 km within the tropical troposphere [11].
These disparities in turn transform the amplitude and phase of the received electric field. In actual fact scintillation intensity is a variable that can be influenced by atmospheric conditions [8], and the subsequent variability in tropospheric scintillation strength has a significant impact on the statistics of the scintillation process [12]. Also, tropospheric scintillation is acknowledged to show a robust correlation with some climatic parameters such as temperature, pressure and humidity [7].

2. METHODS AND DATA ANALYSIS

The tropospheric data site is located at the Covenant University, Ota from Astra 2E/2F/2G beacon satellite 12.245 GHz (Lat: 6.7 ^{o}N, Long: 3.23 ^{o}E, Elev. Angle: 59.9^{o}) at a sample rate of 1 second. The data for this study were measured from January 2015 to December 2015. However, the non-rainy days were separated from rainy days for the analysis by using Davis automatic weather station and spectrum analyser at rain rate 0 mm/h observed for non-rainy events, meanwhile the rain rate above 0 mm/h were removed from the corresponding days and time data within the period of observation because the rain rate above 0 mm/h indicate the presence of rain. A monthly average data were used as reference data signal level and was subtracted from daily measured received data signal level in other to obtained the non-rainy attenuation on every one minutes for each clear-sky day [1]. Subsequent filtering procedure result in data which consists of positive (enhancement) and negative (fade) scintillation amplitude that is above or below the mean level.

Also, the ground tropospheric scintillation measured data were compared with some of the existing scintillation prediction models. In this study, the considered models that predict the variance of signal log-amplitude are: ITU-R model [13], Karasawa model [14], Otung model [15], Van de Kamp model [16]. Finally, performance evaluation of each of the tropospheric scintillation was tested based on the fractional percentage error as presented in Eq. (1).

$$\text{Error } (\varepsilon) = \frac{predicted\ -measured}{measured}\ x\ 100\% \qquad (1)$$

Table 1: Percentage error of the models

% of time	ITU-R	Fade			Enhancement		
		Kasarawa	**Otung**	**Van de Kamp**	**Kasarawa**	**Otung**	**Van de Kamp**
0.01	55.77558	28.02516	190.9347	-39.1306	6.92731	71.48973	-31.7537
0.1	22.36777	0.570562	45.80179	-54.7107	-10.2135	1.607013	-50.2639
1	-2.97606	-20.2577	-25.8888	-66.6068	-24.1086	-51.4064	-64.6062
10	-33.2918	-51.3273	-56.6163	-78.5266	-44.9347	-98.2782	-77.5358

3. RESULT AND DISCUSSION

Table 1 shows the tropospheric scintillation prediction models with statistical comparison between fade (negative) and enhancement (positive). It can be observed that Karasawa model gave the lowest percentage error for fade of about 0.57% at 0.1% of time and 28.03% at 0.01% of time respectively. The next model to that of Karawasa is ITU-R, which is of about 22.37% at 0.1%. Same was observed for enhancement at 0.01% of time followed by ITU-R. However, at 1 and 10% of time, ITU-R is observed to have the least percentage error. Figure 1 demonstrates the assessment of the ground data in Ota and several models prediction for both fade and enhancement, in order to understand the limits of each predicted model and the degree of validity in Ota region.

For fade signal scintillation amplitude, Karasawa model (0.1%) and ITU-R model (1%) showed a very close agreement with the ground data values in Ota almost for all the percentage of time predicted. This was followed by Otung model, which slightly deviated from other models at 0.01%. The Van de Kamp model variance from other models may be attributed to the scintillation measurement during the presence of heavy clouds and rainfall. However, enhancement signal scintillation amplitude also shows a close relationship between measure data and Karasawa predicted model at all level of percentage of time (most especially at 0.1%). This closeness may be because the model was developed during non-rainy period with strong contribution from water vapour obtain from surface temperature and relative humidity [6]. No result for ITU-R model in enhancement because the model was only designed to yield result for signal fade.

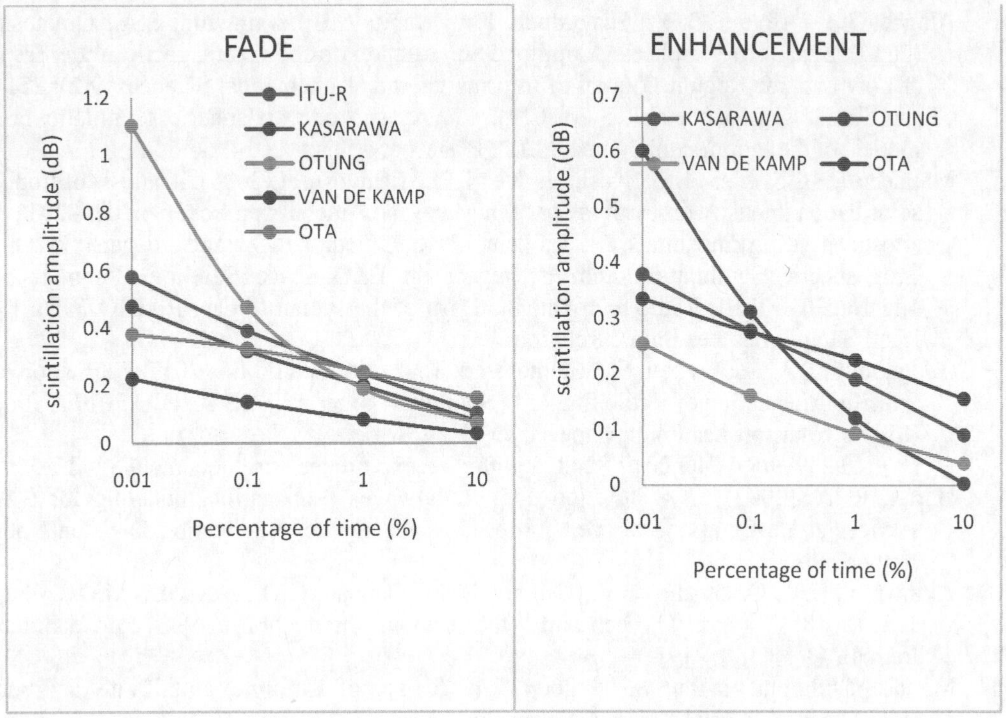

Fig. 1: Comparison of Ota data with Four existing models for fade and enhancement

4. CONCLUSION

Evaluation of four existing clear-sky scintillation models namely: ITU-R, Karasawa, Otung and Van de Kamp models have been presented in this study. This existing models were compared with the ground data in Ota obtained from Astra 2E/2F/2G satellite beacon at 12.245 GHz located at Covenant University. The ground measurements from Ota have confirmed that Karasawa model gave the best prediction for tropospheric scintillation intensity for Ota and its environment.

ACKNOWLEDGMENT

The authors are grateful to Covenant University for sponsoring the publication of this research

REFERENCES

[1] Omotosho,T. V.,Akinwumi,S. A., Usikalu, M. R., Ometan,O.O.,Adewusi,M. Oand Abdullah, M. (2017). Analysis of non-rainy attenuation on earth-space path in Ota,Southwest, Nigeria. IOP Conf. Series: Journal of Physics: Conf. Series 852 (2017) 012039 doi:10.1088/1742-6596/852/1/012039.

[2] Akinwumi,S. A., Omotosho,T. V., Usikalu, M. R.,Adewusi, M. Oand Ometan, O. O. (2016). Atmospheric gases attenuation in West Africa. 2016 IEEE Radio and Antenna Days of the Indian Ocean, RADIO 2016; Hotel Le Recif Saint-Gilles Les Bains; Reunion.

[3] Afolabi, O.O., Adagunodo, T.A., Ayorinde, T.T., Rabiu, A.B., Akinwumi, S.A., Usikalu, M.R. (2017). Impact of Magnetic Activity on Occurrence of Ionospheric Amplitude Scintillation over Lagos, Nigeria. IEEE RADIO 2017 at Protea Hotel by Marriott Sea Point, Cape Town, South Africa. DOI: 10.23919/RADIO.2017.8242229.http://ieeexplore.ieee.org/document/8242229

[4] Alagbe G.A., Afolabi O.O., Adagunodo T.A., Rabiu A.B., Akinwumi S.A., Omotosho T.V. (2017). Study of Ionospheric Amplitude Scintillation during Geomagnetic Activities of 2012 at Low Latitude Region. Journal of Informatics and Mathematical Sciences, 9(2): 251 – 256.

[5] Cheng,Y. C. and Mandeep,J. S. (2014). Comparison of tropospheric scintillationprediction models of the Indonesian climate. Earth, Planets and Space 2014, 66:64.

[6] Mandeep,J. S., Syed, S. H., Kiyoshi, I., Kenji,T., Mitsuyoshi, I.(2006). Analysis of tropospheric scintillation intensity on earth to space in Malaysia. Amer J App Sci 3(9):2029–2032.

[7] Omotosho, T. V., Akinwumi,S. A., Usikalu, M. R., Ometan,O. O., and Adewusi, M. O. (2016). Tropospheric Scintillation and its Impact on Earth-Space Satellite Communication in Nigeria. 2016 IEEE Radio and Antenna Days of the Indian Ocean, RADIO 2016; Hotel Le Recif Saint-Gilles Les Bains; Reunion.

[8] Adagunodo, T.A., Akinwumi, S.A., Omotosho, T.V., Akinyemi, M.L. 2017. Estimation of Specific Attenuation of Radio Signal in Southwest Nigeria. IEEE RADIO 2017 at Protea Hotel by Marriott Sea Point, Cape Town, South Africa. DOI:10.23919/RADIO.2017.8242225.http://ieeexplore.ieee.org/document/8242225

[9] Harris, R. A. (2002). COST Action 255, Radiowaves propagation modeling for SATCOM services at Ku-Band and above, European Space Agency publication Division Noordwijk, Netherlands.

[10] Akinwumi, S.A., Omotosho, T.V., Usikalu, M.R., Ometan, O.O., Adewusi, M.O., Adagunodo, T.A. (2018). Study of Oxygen and Water Vapour Attenuation in West Africa. Interciencia Journal, 43(1): 180 – 191.

[11] Mandeep, J.S. and Hassan, S. I. (2004). Comparison of 1-minute rainfall rate distribution for tropical and equatorial climates. Space Commun 19:193–198.

[12] Gibson, J. D. (2002). The Communication Handbook, second edition, CRC Press, LCC, 59-1 to 59-16, 2002.

[13] ITU-R; Recommendation P.618, 2012. Propagation data and prediction methods required for the design of Earth-space telecommunication system.

[14] Karasawa Y., Yamada M., Allnutt J. E. (1988). A new prediction method for tropospheric scintillation on Earth-space paths. IEEE Trans Ant Prop 36(11): 1608-1614.

[15] Otung I. E., (1996). Prediction of tropospheric amplitude scintillation on a satellite link. IEEE Trans Ant Prop 44(12): 1600-1608.

[16] Van De Kamp M. M. J. L., Tervonen J. K., Salonen E. T., Poirares B. J. P. V. (1999). Improved models for long-term prediction of tropospheric scintillation on slath paths. IEEE Trans Ant Prop 47(2): 249-260.

Review on Li-Fi: an advancement in wireless network communication with the application of solar power

W. A Ayara, M. R Usikalu, M. L Akinyemi, T. A Adagunodo and K. D Oyeyemi

Department of Physics, Covenant University, PMB 1023, Ota, Nigeria

williams.ayara@covenantuniversity.edu.ng

Abstract. Light Fidelity, commonly referred to as Li-Fi is a technology that was introduced by a German physicist called Herald Hass. This technology in its own uniqueness considers the use of light (Light Emitting Diode, LED to be specific) as its medium of delivering high speed communication that complies with standard IEEE 802.15.7 bidirectional communication standard. This optical means of communications is also known as visible light communication. It is safer when compared with Wi-Fi network. This paper will focus on Li-Fi applications, comparison with existing technologies like Wi-Fi and the inclusion of an alternative source of power i.e., solar power, as a means of effectively optimizing on the availability and accessibility of the benefits in places where erratic or total power outage is a common trend.
Keywords: Li-Fi, solar power, Visible Light Communication (VLC), Wi-Fi

1. Introduction

Light Fidelity, which is uniquely referred to as Li-Fi is a technology that was introduced first by a German Physicist called Herald Hass from the University of Edinburg. It is a Visible Light Communication (VLC) that deals with the use of LED as light source for the transmission of data at high speed. Due to the ever increasing demand, which optimally will be on the rise for wireless data communication, compelled the wireless communication industry to respond to this challenge by considering the radio spectrum above 10 GHz (mm-wave communication) since the available radio spectrum that is below 10 GHz (cm-wave communication), has become increasingly insufficient as a result of massive data being generated daily [1]. The technology provided by Li-Fi comes with much larger spectrum for transmission when it is compared with the conventional methods that employ wireless communications that rely on radio waves. The working principle that guides this technology is that data can be transferred through the use of LED light by varying light intensities faster than what the human eyes is able to perceive. The Visible Spectrum, which is a part of the electromagnetic spectrum that is not greatly used is what Li-Fi technology employs.

Because Li-Fi as a technology on its own is already a great discovery, to have data transfer rate that is comparable to fibre optics makes this new technology a major issue and an area that is worth further research. Imagine having a light source, that provides not only networking facilities but also provides illumination with data transfer at astonishing speeds [2]. The possibilities provided by Li-Fi are comparably endless, and there are greater potentials yet ahead as more researches are conducted. Soon, Li-Fi will become an everyday technology. Hence the importance of introducing or adding solar power to its operation becomes a welcome and a desirable need. This is promising, considering the low energy consumption level of LED related devices under which LI-Fi is categorized.

Presently, researches have shown that Light Fidelity Li-Fi is gaining more ground in the area of high speed data transmission. For example data transmission speed is over 500 megabytes per second [3]. In data transmission, the consideration of Speed and security is a major issue. Data that are transmitted with the use of Wi-Fi are susceptible to security challenges since it can penetrate through walls easily and because of this reason hackers are able to access documents which ordinarily they are not given access to. Li-Fi on the other hand due to its characteristic nature does not penetrate walls and so provides more security when compared to Wi-Fi [4]. The major component that makes up Li-Fi communication is the white LED, which acts as a communication source and a silicon photodiode which shows good response to visible light. A data rate of greater than 100Mbps is possible by the high speed LEDs.

1.1 HISTORY AND WORKING PRINCIPLE

The idea of Li-Fi was first put forward and used by Professor Harald Haas from University of Edinburgh, United Kingdom. This was done at the TED global talk in 2011. Li-Fi technology was given a place of "one of 50 best inventions of 2011" on TED world site on the internet [5]. The very first project that gives explanation of the idea of Li-Fi was the D-Light project, which stands for "data through illumination" [6]. In October 2011, industries players formed the Li-Fi consortium, whose aim is to propagate the awareness of high-speed optical wireless systems and with it conquer the limited amount of space available for radio-based wireless spectrum by exploiting a completely different part of the electromagnetic spectrum. [6].

The principle of operation of Li-Fi technology is implemented by using white LED light bulbs for illumination powered by constant current. In order to achieve data transmission, the LED bulb is subjected to very fast variations of the current at extremely high speeds. This switches on and off such that the human eyes cannot detect thereby producing digital signal of '1'for the LED in 'on' state, or a signal of '0' for the LED in 'off' state [**2**]. In order to be able to send and receive data, some LEDs and a controller that codes data into those LEDs and for receiving data, an Image Sensor, and a Photodiode which is used as a detector, are needed to makes up the basic components requirement as shown in Figure 1. The LED bulb is designed to have a micro-chip whose work is to process the data received. Figure 2 shows summary of the working principle of a Li-Fi system. To transmit data therefore, a single LED or multi LED maybe required. On the side of the receiver, a photo detector is installed, which convert the light signal into electric signals which is delivered to the device connected to it. A voltage regulator and level shifter circuits are needed on both sides to convert or maintain a stabilized voltage level between the transmitter and receiver [7].

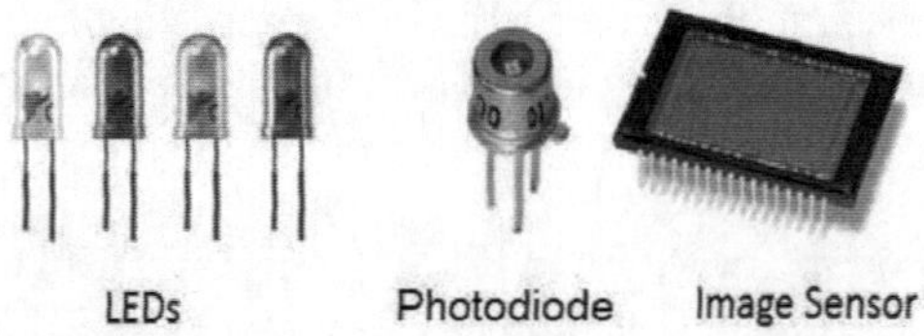

Figure 1: Basic component requirement [6]

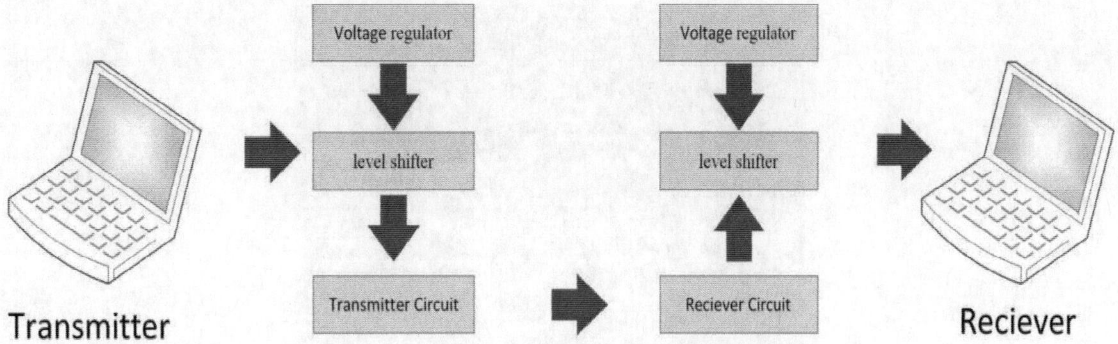

Figure 2: Li-Fi Principle of operation [6]

2. LI-FI APPLICATIONS

The usefulness of LI-Fi technology is now advancing into several fields of human endeavours. A few of the areas are as discussed in the next subsections.

2.1. Hospital

Theatre where surgical work are carried out do not permit the use of Wi-Fi because of its inherent radiation and the interferences it can cause to monitoring equipment like Magnetic Resonance Imaging (MRI) scanners during its use. This challenge is solved with the use of Li-Fi owing to the fact that light is one of the most glaring fixtures in the surgical room. Moreover, Li-Fi has 10,000 times the spectrum of Wi-Fi [2,8,9].

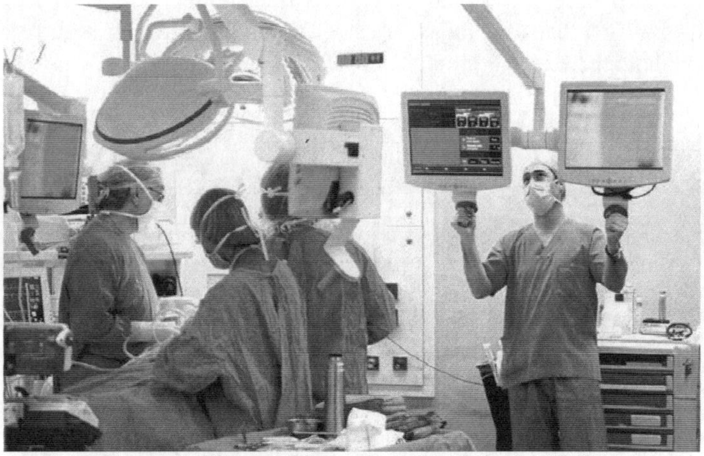

Figure 3: The working principle of the Li-Fi [6].

2.2. Smart Street Lighting

These days, majority of street lights are LED based, thus deviating from the usual high pressure lamps. These LED lamps easily find application by their conversion to Li-Fi hotspots that serves both as light source and for data transfer in communication [4].

Figure 4: Smart lighting with Li-Fi [6].

2.3. Mobile Device Communication
The mode of communication has been made much easier by the introduction of mobile devices. These smart mobile devices easily communicate electronically with Li-Fi technology to provide very impressive data transfer with high security [2,4].

2.4. Education
The application of Li-Fi technology in learning is very promising. It affords both the instructor and the student a base to communicate on an individual platform. It therefore makes it very easy to progressively monitor individual student [13]. And because of its unique features the download of tutorials and other materials for learning becomes easy.

2.5. High Risk Environments
Places like petrochemical plants, power plants etc., will find the technology of Li-Fi as a safe haven from the application of Wi-Fi that uses radio wave for communication in high risk areas because of their high inflammable tendencies.

Figure 5: The use of Li-Fi in power plant [6, 10].

2.6. Aviation
Li-Fi can find application in the aviation sector especially in the cabin and at the control tower where the need to avoid interference while using radio frequency related devices on-board the cabin is of very great importance [4, 7, 9, and 10].

Figure 6: Li-Fi in air plane [6].

2.7. Deep water Communications
In order to overcome the challenge of signal strength reduction in deep water communication, the technology of Li-Fi provides a way out for an effective under water communications [4, 9, and 10].

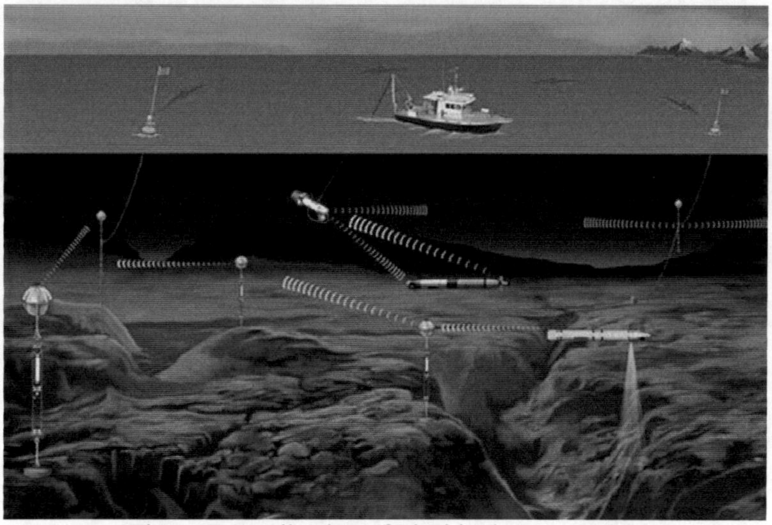

Figure 7: Application of Li-Fi in deep water [6].

3. COMPARING LI-FI AND WI-FI

Table 1 shows compared features between Li-Fi and Wi-Fi in tabular form. A close look at the table is suggestive of the importance that Li-Fi can provide high benefit in transfer of data in terms of cost,

IOP Conf. Series: Earth and Environmental Science **173** (2018) 012016 doi:10.1088/1755-1315/173/1/012016

speed and safety of data. This section is meant to show only the basic difference that exist between the use of Li-Fi, and Wi-Fi network [11].

Table1. Comparing Light Fidelity (Li-Fi) and Wi-Fi

Features	Li-Fi	Wi-Fi
Identification	Popularly called Li-Fi	Generally referred to as Wi-Fi
Mode of operation	Data transmission in Li-Fi is achieved using light from LED bulbs.	Data transmission in Wi-Fi is achieved using radio waves.
Possible Interference	Do not have any observed possible interference as with Wi-Fi	There are cases of interference from close by similar devices
Operating Frequency of the technology	Made up very wide spectrum. Above 100Ghz.	Low when compared with Li-Fi
Distance coverage	It is known to cover about 10 meters	Have potential to cover about 32 meters (WLAN 802.11b/11g), vary based on transmission power and antenna type
Hard ware System requirement	1. LED Lamp driver, 2. LED bulb 3. Photo detector, ete., will make up complete Li-Fi system.	Requires 1. Routers. 2. Installed subscriber devices like (laptops, PDAs, desktops) as stations.

4. THE PLACE OF SOLAR POWER IN LI-FI TECHNOLOGY

Availability and accessibility of power is a vital factor for optimum performance of any known technology. Solar power is readily available and it is a renewable source of energy [16,17]. Every known technology is aimed at being able to perform given task more efficiently. Hence availability of power and the ease of accessibility cannot be compromised when such new technology like Li-Fi will need to be deployed to developing world where erratic power supply and total power outage is an issue that is being contended with [15,17].

4.1 THE LI-FI/SOLAR POWER COMPONENTS

The desire is that the system will be purely direct current (DC) based configuration. Hence the need for a power inverter may not be necessary except for cases that may uniquely require the inclusion of an inverter system. The component that will make up the system will include; 12 or 24 V deep cycle battery(ies) – to be determined by the system requirement, Solar charge controller, DC combiner box, Inverter (optional), Solar cell modules – system size to determine number of panels (solar cell modules) and Li-Fi system configuration. Figure 8 shows block diagram of a proposed arrangement for the implementation of solar power with Li-Fi system.

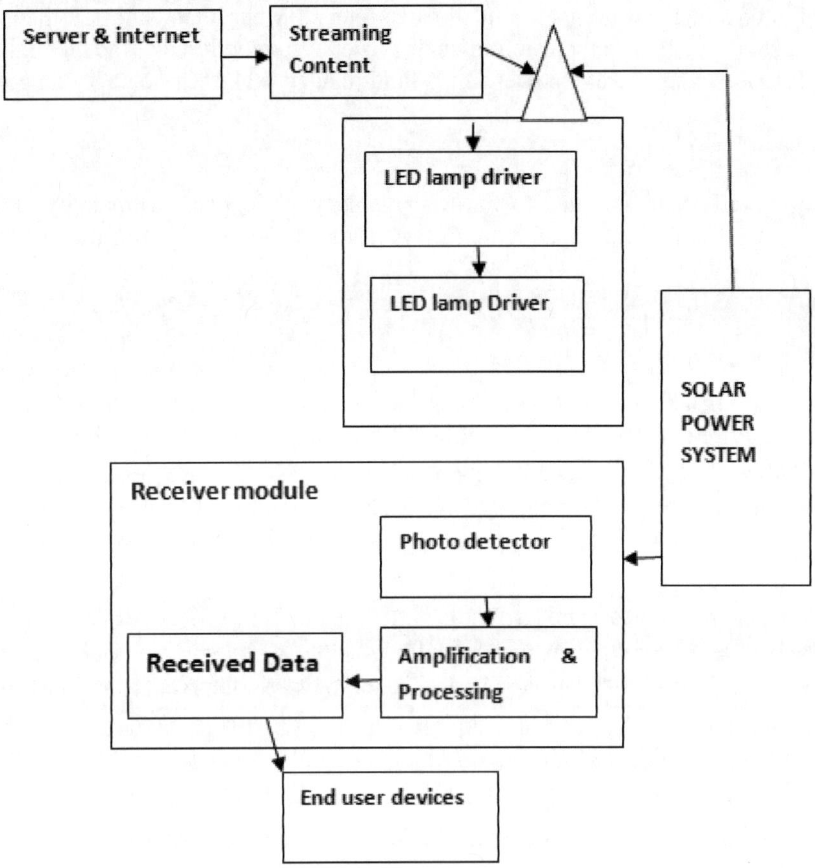

Figure 8: Block diagram of a proposed arrangement for the implementation of solar power with Li-Fi system.

4.2 CHALLENGES OF LI-FI

Li-Fi though new compared with Wi-Fi, its great prospect is not without some challenges. Such challenges are listed below. Research works are on towards overcoming these challenges. Overcoming them will further give Li-Fi more edge above Wi-Fi.

- Effective data transmission will require that both transmitter and receiver are appropriately placed in same line of sight.
- Access to the internet is lost, whenever there is a disruption with the light source. This makes internet access to be dependent of light source.

- Since visible light cannot penetrate brick walls, simply walking in front of LED source can disrupt connectivity between transmitter and receiver.
- Dealing with the effect of changing weather conditions is a concern if the system is meant to operate outdoors.

5. CONCLUSION

The need to communicate and doing it in a most friendly, safe and convenient way is on the increase daily. Huge amount of data is generated continuously, and this results in the gradual reduction of available space on the electromagnetic spectrum upon which the Wi-Fi network depends. Solving this challenge opens up the Li-Fi technology that has greater space provided by the visible light spectrum and capable of producing data transfer rate greater than 10Mbps that is faster than any average broadband connection [1,12]. The inclusion of solar power system will make the technology available and accessible in places where erratic power supply and regular total power outage is a concern.

ACKNOWLEDGEMENT

The authors do express their gratitude to the Management of Covenant University and also to the University's Research Centre (CUCRID), Ota, Nigeria for sponsoring the publication of this article.

REFERENCES

[1] H. Harald, Y. Liang and W. Yunlu, "What is LiFi," *Journal Of Lightwave Technology,* 2015.

[2] C. Shubham, A. Shalabh and N. Asoke, "Scope and Challenges in Light Fidelity(LiFi) Technology in Wireless Data Communication," *International Journal of Innovative Research in Advanced Engineering (IJIRAE),* vol. 2, no. 6, pp. 1-9, 2015.

[3] www.studymafia.org, "Li-Fi Technology," www.studymafia.org, 2015.

[4] S. Ashmita, "A Comparative Study and Analysis on Li-Fi and Wi-Fi," vol. 150, no. 6, pp. 43-48, 2016.

[5] (Feb. 2013), Seminar projects and [Online]., "http://seminarprojects.com/Thread-li-fi-light-fidelity-the-future-technology-in-wireless-communication".

[6] E. Abdulhafid and Z. Nizar, "Overview Li-Fi Technology," Syrian Arab Republic, 2015.

[7] P. N. Dharmrajsinh, M. V. Khushbu and B. H. Jay, "LI-FI Technology – A Visible Light Communication," *International Journal Of Engineering Development And Research,* 2014.

[8] wikipedia. [Online]. Available: http://en.wikipedia.org/wiki/Li-Fi. [Accessed 18 July 2016].

[9] M. Mrs. Padmini, P. Jyoti, P. Sonu and k. Minu, "A Review On LiFi : The Green WiFi," *International Research Journal of Engineering and Technology,* vol. 3, no. 3, pp. 99-103, 2016.

[10] P. Subarna, S. Md and A. Dr. Jeyasekar, "Li-Fi Technology –Next Gen Data Transmission through Visible Light Communication," *International Journal of Innovative Research in Science, Engineering and Technology,* vol. 4, no. 11, pp. 144-151, 2015.

[11] P. Yusuf, "The Next Generation of Wireless Communication Using Li-Fi (Light Fidelity) Technology," *Journal of Computer Networks,* vol. 4, no. 1, pp. 20-29, 2017.

[12] M. Mutthamma, "A survey on Transmission of data through illumination- Li-Fi," *International Journal of Research in Computer and Communication Technology,* vol. 2, no. 12, 2013.

[13] T. Kirti and T. Kirti, "Smart Class Using Li-Fi Technology," *The International Journal Of Engineering And Science*, vol. 3, no. 7, pp. 16-18, 2014.

[14] A. A. Adeyinka, J. N. Samuel and I. U. Dike, "Design and Implementation of a Virtual Machine Video Conferencing Application," *International Journal Of Engineering And Computer Science,*

vol. 3, no. 3, pp. 5054-5063, 2014.

[15] S. N. John, R. E Okonigene, B. A. Akinade, I. Kasali. and A. A. Adewale "Developed Model to Control Congestion on Converge Network," *International Journal Of Innovative Technology & Creative Engineering,* vol. 1, no. 8, pp. 1-5, 2011.

[16] W. A. Ayara, T. V. Omotosho, M. R. Usikalu, M. S. J. Singh and W. Suparta, "Development of a solar charged laboratory bench power supply," in *Journal of Physics: Conference Series*, Malaysia, 2017.

[17] M. E. Emetere, M. L. Akinyemi and E. B. Edeghe, "A Simple Technique for Sustaining Solar Energy Production in Active Convective Coastal Regions," *International Journal of Photoenergy,* p. 11, 2016.

Influence of Meteorological Variables on UHF Radio Signal: Recent Findings for EBS, Benin City, South-South, Nigeria

K.E. Ukhurebor[1] and O.J. Umukoro[2]

[1] Department of Physics, Edo University, Iyamho, Edo State, Nigeria
[2] Department of Physics, Federal University of Petroleum Resources, Effurun, Nigeria

ukhurebor.kingsley@edouniversity.edu.ng; ukeghonghon@gmail.com

Abstract. In this study the influence of some meteorological variables on Ultra High Frequency (UHF) radio signals from Edo Broadcasting Service (EBS) in Benin City, South-South, Nigeria located within Latitude 6°20’17″N and Longitude 5°37’32″E was investigated. The measurements of the radio signal strength from EBS Television, transmitting at 743.25 MHz UHF and some meteorological variables (air temperature, atmospheric pressure and relative humidity) using the Digital Community-Access/Cable Television (CATV) analyzer and a self designed cost effective portable weather monitoring system were carried out simultaneously at a residential building within Benin City, in order to ascertain the influence of these meteorological variables on the signal strength. The measurements were done every eight hours (between 7am-8am, 3pm-4pm and 11pm-12am respectively) for a period of one year (2017). From the obtained results, it was inferentially observed that the radio signals was having inverse relationship with the air temperature, atmospheric pressure and the relative humidity; on the assumption that for any of the aforementioned meteorological variables, the others were observed constant. Statistically, the signal strength and these meteorological variables were having correlation values of -0.94, -0.92 and -0.96 for the air temperature, atmospheric pressure and the relative humidity respectively. These results obtained from this study should be taken into account by the management of radio communication systems for enhancement and planning purposes.

Keywords: Troposphere; Radio Signals; Air temperature; Atmospheric pressure; Relative humidity

1. Introduction

Air temperature, atmospheric pressure, relative humidity and most of the other meteorological variables have great influence on the electromagnetic waves that are propagated in the troposphere (lower atmosphere), although some other factors like the different components that made up the troposphere, also have much influence as well. But it is noteworthy that the variations in these meteorological variables are the ones actually affecting the components that made up the troposphere, which in turn have these influence on the electromagnetic waves [1-3]. The variations in these meteorological variables cause the refractive index of the air in the troposphere to differ from one place to another [3-5]. The bending in the path of the electromagnetic waves which is as a result of the non-homogenous spatial distribution of the refractive index in the air have contentious effects like multipath fading and interference, attenuation due to diffraction on the terrain obstacles which is also called the radio holes [3, 6, 7].

The propagation pattern of radio signals from the radio transmitter to the radio receivers which is controlled by the regions of the troposphere where they go through is of great significance in the management and planning processes in radio communication systems. Radio links in any radio communication system are exposed to weather variability, which could cause severe degradation in their system performance [6, 8]. It is therefore, imperative to explore the various factors affecting the radio link quality in these radio communication systems in order to take the necessary measures and adaptation options to alleviating and controlling its effects.
In the past few years, some researchers have tried to evaluate the impacts of some meteorological variables such as air temperature, atmospheric pressure, relative humidity, rain, wind, fog, dew point temperature, light intensity, solar irradiance, wet snow and evaporating duct on microwave frequency bands, to this end, some useful results and models have been postulated by few of these researchers, as a useful step to having more reliable signal communication systems [6, 8-14]. It is noteworthy, to the best of our knowledge from existing literatures that most of the research findings and models on the recent influence of air temperature, atmospheric pressure, relative humidity and the other meteorological variables on UHF signal propagations from television (TV) transmission stations for tropical climate in developing nations, like Nigeria are below the belt. In this study the measurement results of some meteorological variables (air temperature, atmospheric pressure and relative humidity) and the radio signal strengths from EBS TV transmission station were done every eight hours for a period of one year (January – December, 2017) simultaneously at a residential building in Benin City the capital of Edo State, South- South, Nigeria using a self designed cost effective portable weather monitoring system and the Digital Community-Access/Cable Television (CATV) analyzer respectively. The main aim of this research is to analysis the recent influence of these three meteorological variables (air temperature, atmospheric pressure and relative humidity) on the UHF signal propagation from EBS TV transmission station, so as to fashion how it can be used for effective radio link margins/link budgets in tropical rain forest regions for planning and improvement purposes in radio communication systems.

2. Materials and Methods

Statistical data of meteorological variables (air temperature, atmospheric pressure and relative humidity) and the radio signal strengths from EBS TV station transmitting on 743.25 MHz (UHF) measurements were done for every eight hours (between 7am-8am, 3pm-4pm and 11pm-12am respectively) for a period of one year (January – December, 2017) simultaneously at a residential building in Benin City, the capital of Edo State, South- South, Nigeria which is situated around 40 km north of the Benin River and 320 km by road east of Lagos and located within Latitude 6°20′29″N and Longitude 5°37′55″E with an elevation of about 88 m (288 ft) above sea level. It has an approximate population of 1,125,058 making it one of the largest and most ancient cities in South-South, Nigeria. The city operates on the West Africa time zone [15]. The coordinates are given in the latest version of the World Geodetic System (WGS 84) coordinate reference system, which is used in mapping and navigation, including the Global Positioning System (GPS) satellite navigation system and the canonical form of latitude and longitude representation uses (°), (′) and (″) for degrees, minutes and seconds respectively.

The measurements of the signal strength was done using the Digital Community-Access/Cable Television (CATV) analyzer (receiver) with more than 20 channels, spectrum 46 – 870 MHz, connected to a domestic receiver antenna of about 5 m height; while, the measurements of the meteorological variables were done using a self designed cost effective portable weather monitoring system for the aforementioned time simultaneously. The weather monitoring device was designed in

IOP Conf. Series: Earth and Environmental Science **173** (2018) 012017 doi:10.1088/1755-1315/173/1/012017

such a way that it can be used remotely and the measured meteorological variables are displayed on a user friendly Liquid Crystal Display (LCD) in numerical digital values for air temperature (^oC), atmospheric pressure (mbar), relative humidity (%) and light intensity (lux) which can be send to the computer via the programmed micro SD card or/and through the serial port; the Arduino SD card module. The user has the option of choosing how often the meteorological variables will be logged, measured, recorded, displayed and stored. The acquired meteorological variables are analyzed and the LCD displays the values respectively. In addition, the meteorological variables for each measurement are saved on the micro SD card in Excel format on a separate file with each file created with a file name that corresponds to the date and time when the meteorological variables were acquired. The users also have the option to stop the meteorological variables acquisition process at anytime by interrupting the routine. The details of the weather monitoring system design and implementation including its validity are contained in [16].

To ascertain the recent influence of these respective meteorological variables on the signal strength; each of the other meteorological variables which would have effect(s) on the inference on the result of that respective meteorological variable were observed constant, and we only made use of the average monthly measured signal strength together with the average monthly measured meteorological variables of air temperature, atmospheric pressure and relative humidity for the analysis of this research.

3. Results and Discussion

The results of each of the average monthly measured air temperature, atmospheric pressure and relative humidity with the average monthly measured signal strength of EBS TV station, transmitting on 743.25 MHz UHF were analyzed statistically in order to ascertain their influence and relationship. Table 1 shows the monthly average measurements of the signal strength together with the monthly average measured air temperature, atmospheric pressure and relative humidity, all through the year 2017.

Table 1: The Monthly Average Measurements of the Signal Strength and Meteorological Variables for 2017

Month	Average Signal Strength (dBm)	Air Temperature (^0C)	Atmospheric Pressure (mbar)	Relative Humidity (%)
January	13.50	28.20	1001.70	42.70
February	10.50	31.80	1002.50	48.20
March	11.40	30.10	1004.60	68.70
April	12.20	29.30	1007.10	85.20
May	13.40	28.40	1005.70	92.10
June	15.90	26.30	1005.90	92.40
July	16.10	25.00	1004.70	95.30
August	15.10	26.60	1003.90	78.60
September	16.00	25.50	1005.40	78.40
October	16.20	25.70	1005.60	68.90
November	13.10	27.80	1004.70	68.20
December	15.30	26.50	1002.90	58.30
Total Average	14.06	25.28	1005.00	73.08

In Figure 1 the direction of the residential building where the CATV analyzer and the weather monitoring device was stationed away from the EBS-TV transmission station is shown; while in Figures 2, 3 and 4 the relationship between the signal strength and the air temperature, atmospheric pressure and relative humidity are graphically shown respectively.

IOP Conf. Series: Earth and Environmental Science **173** (2018) 012017 doi:10.1088/1755-1315/173/1/012017

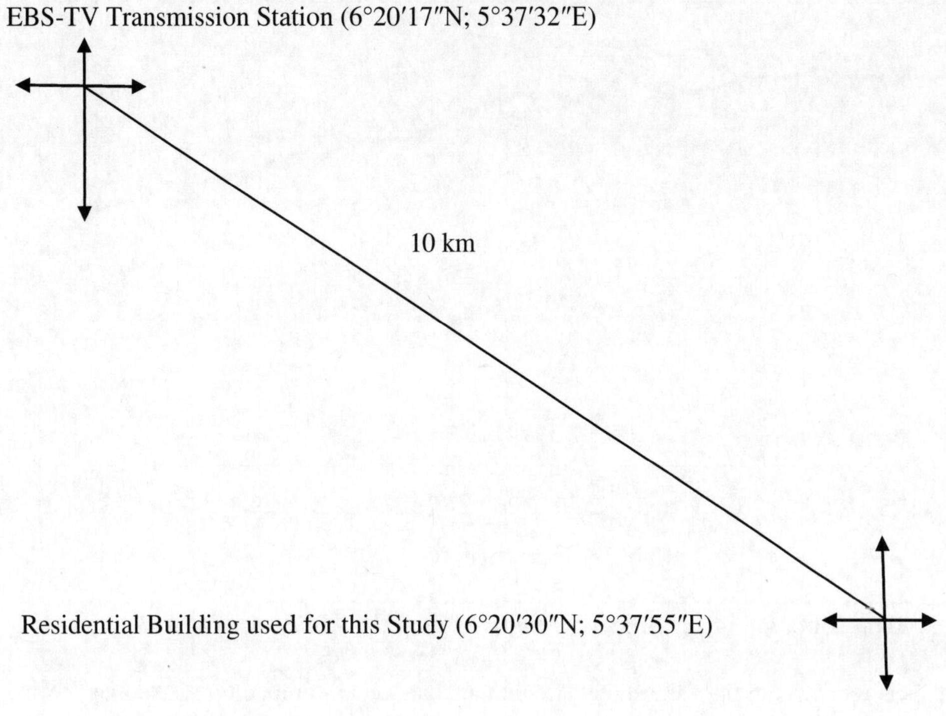

Fig. 1: Diagrammatic Description of the Direction of EBS-TV Transmission Station and the Residential Building used for this Study

In Fig. 1 the direction of the residential building where the CATV analyzer and the weather monitoring device was stationed away from the EBS-TV transmission station is shown. From EBS-TV transmission station to the residential building was about 10 km. We also took into account of the fact the distance from the transmission station can also affect the signal strength.

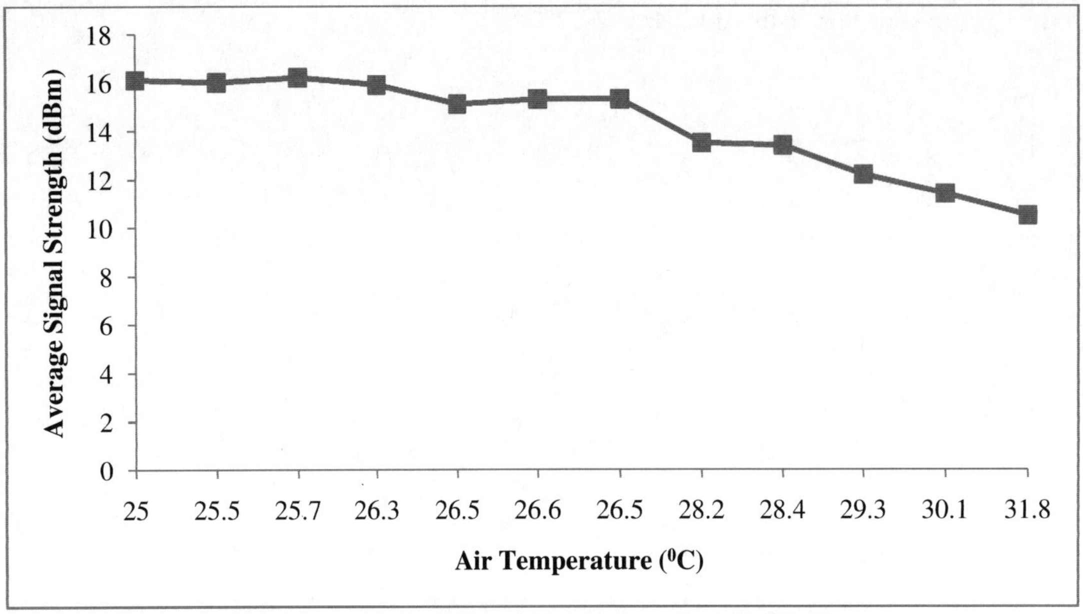

Fig. 2: Graphical Representation of the Signal Strength against the Air Temperature

In Fig. 2 the graphical representation of the signal strength against the air temperature is shown. It was observed that the signal strength decreases as the air temperature increases slightly. Statistically, the correlation between the signal strength and the air temperature was having a value of -0.94. This implies that, the signal strength is inversely proportional to air temperature, on the assumption that the other meteorological variables are observed constant. The increase in air temperature causes degradation of signal strength, this is true because there is a collision between increasing raining particles of light from the sun as air temperature which is the measure of temperature at different levels of earth's atmosphere increases with the radio signals (since both waves are electromagnetic waves), this attenuates the signal strength.

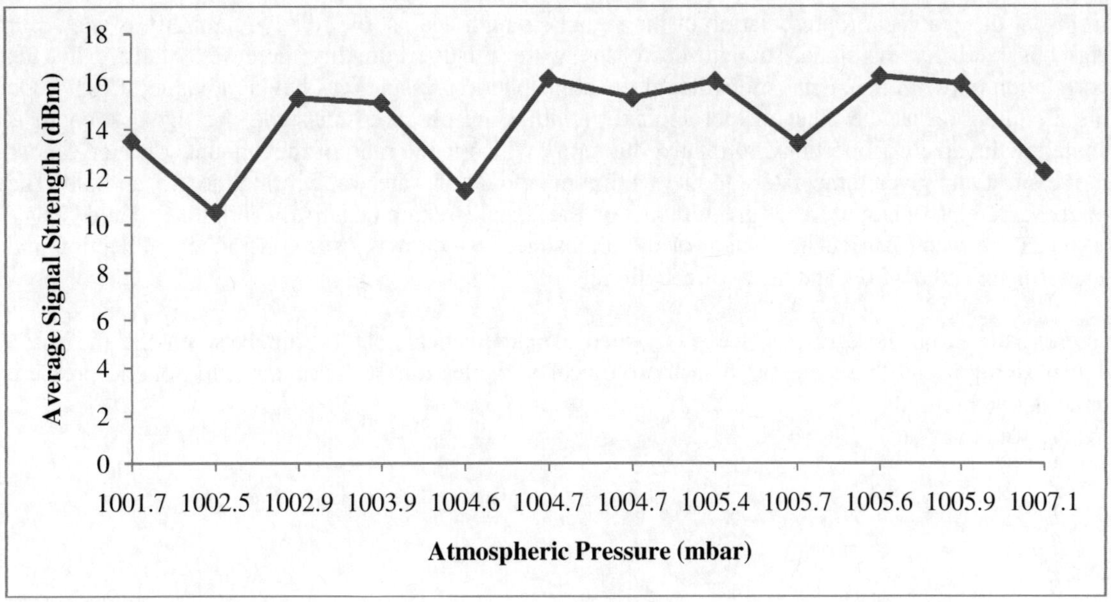

Fig. 3: Graphical Representation of the Signal Strength and the Atmospheric Pressure

In Fig. 3 the graphical representation of the signal strength against the atmospheric pressure is shown. It was observed that the signal strength decreases as the atmospheric pressure increases. Statistically, the correlation between the signal strength and the atmospheric pressure was having a value of -0.92. This implies that, the signal strength is somehow inversely proportional to atmospheric pressure, on the assumption that the other meteorological variables are observed constant. The atmospheric pressure impacts negatively on signal strength. Atmospheric pressure which is the force exerted on unit surface area on the earth by the weight of earth's atmospheric air above its surface has a negative influence on the signal and that the signal has to overcome it as its travels through the atmospheric channel.

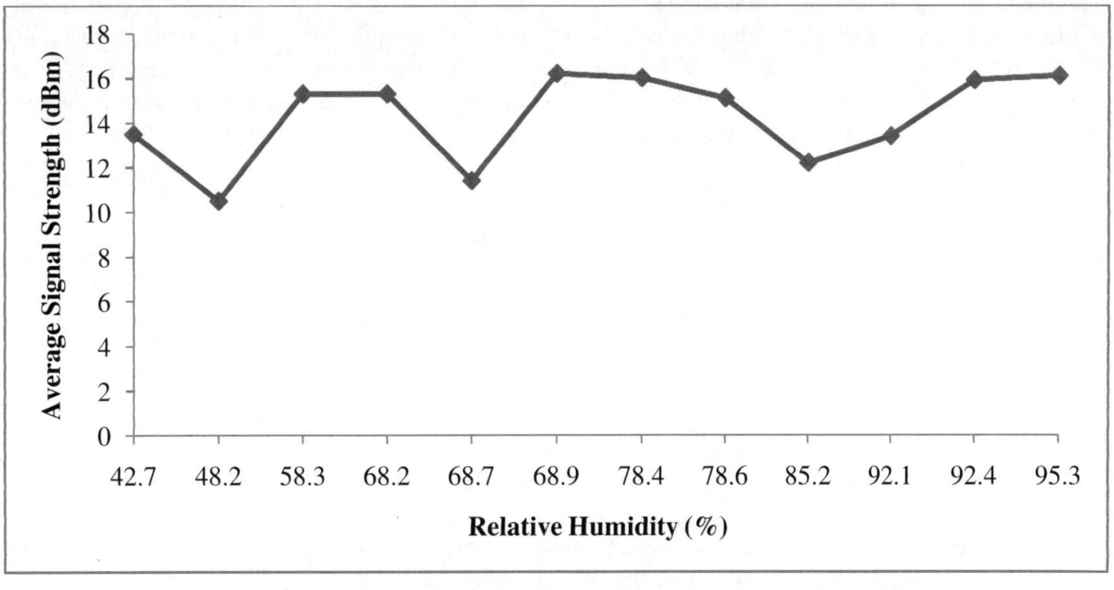

Fig. 4: Graphical Representation of the Signal Strength and the Relative Humidity

In Fig. 4 the graphical representation of the signal strength against the relative humidity is shown. It was observed that the signal strength decreases as the relative humidity increases. Statistically, the correlation between the signal strength and the atmospheric pressure was having a value of -0.96. On the assumption that the other meteorological variables are observed constant, the signal strength is somehow inversely proportional to relative humidity which is the ratio of the amount of water vapour in the air at any given temperature to the maximum amount of water vapour that the air can hold. The increase in relative humidity has great impact on the signal strength negatively and this is could be as a result of the water particulate content of the atmosphere which may cause diffraction, reflection and scattering of radio waves and hence attenuation.

Inferentially from these results, we established a mathematical relationship between the measured signal strength and these measured meteorological variables (air temperature, atmospheric pressure and relative humidity).
Where we have that;

$$S \propto \frac{1}{T} \quad (1)$$

$$S \propto \frac{1}{P} \quad (2)$$

$$S \propto \frac{1}{H} \quad (3)$$

Combining and resolving Eqn. (1), (2) and (3) mathematically, we have;

$$S = \frac{K}{TPH} \quad (4)$$

$$K = \boldsymbol{S} \times \boldsymbol{T} \times \boldsymbol{P} \times H \quad (5)$$

This implies that; $S_n \times T_n \times P_n \times H_n = S_{n+1} \times T_{n+1} \times P_{n+1} \times H_{n+1}$ (6)

Where, S, T, P, H and K are the signal strength, air temperature, atmospheric pressure, relative humidity and the proportionality constant respectively. The constant K is an important factor that needs to be quantified and n is in sequence of 1, 2, 3 … nth.

The influences of these meteorological variables on the signal strength were observed higher during the months with much rainfall (rainy season), compared to the months with lesser rainfall (dry season) for the period under consideration (2017), and among these three meteorological variables used for this study; air temperature was having the highest influence on the signal strength, this affirm the fact that air temperature is the most widely measured meteorological variable because of it crucial nature and its influence on other meteorological variables [15].

4. Conclusion

In this study the relationship and influence of air temperature, atmospheric pressure and relative humidity on the radio signals from EBS TV station transmitting on 743.25 MHz (UHF) in Benin City, South-South, Nigeria was analyzed. The results obtained from this study will go a long way in providing assistance for the management of radio communication systems in improving their quality of service within the studied area and beyond.

The following conclusions were inferentially reached from the results obtained:

- The air temperature, atmospheric pressure and relative humidity were inversely proportional to the signal strength during the studied period (2017).
- The influences of these meteorological variables were observed higher during the months with higher rainfall (wet season) during the studied period (2017).

- The air temperature was having the highest influence on the signal strength compared to the other two meteorological variables used for this study.
- Further studies should be carried out, using more meteorological variables and more radio signal service providers over a long period of time, in order to have more comprehensive results.

5. Acknowledgement
The authors are grateful to the management and staff of Edo Broadcasting Cooperation, Benin City, Nigeria for their assistance. We also appreciate the assistance and contributions of Prof. Augusto José Pereira Filho of the Institute of Astronomy, Geophysics and Atmospheric Sciences, University of Sao Paulo, Brazil.

References
[1] Korak shaha, P.N (2003): The Physics of the Earth and its Atmosphere, *John and Sons Inc., New York, USA.*
[2] Agbo, G.A; Okoro, O.N; Amechi, A.O (2013): Atmospheric Refractivity over Abuja, Nigeria, *International Research Journal of Pure and Applied Physics, Vol. 1, No. 1, pp 37-45.*
[3] Ukhurebor, K.E; Azi, S.O (2018): Review of Methodology to Obtain Parameters for Radio Wave Propagation at Low Altitudes from Meteorological Data: New Results for Auchi Area in Edo State, Nigeria, *Journal of King Saud University – Science (DOI:10.1016/j.jksus.2018.03.001).*
[4] Ayantunji, B.G; Okeke, P.N; Urama, J.O (2011): Diurnal and Seasonal Variation of Surface Refractivity over Nigeria, *Progress in Electromagnetics Research, Vol. 30, pp 201-222.*
[5] Okoro, O.N; Agbo, G.A (2012): The Effect of Variation of Meteorological Parameters on the Tropospheric Radio Refractivity for Minna, *Global Journals Inc. (USA), Vol. 12 Issue 2, pp 1.*
[6] Alam I; Mufti N; Shah S.A.A; Yaqoob M (2016): The Effect of Refractivity on Propagation at UHF and VHF Frequencies, *International Journal of Antennas and Propagation, 2016, pp 1-8.*
[7] Martin, G; Vaclav, K (2011): Atmospheric Refraction and Propagation in Lower Troposphere, *Electromagnetic Waves, Prof. Vitaliy Zhurbenko (Ed.), ISBN: 978-953-307-304-0, In Tech (www.intechopen.com).*
[8] Amajama, J (2016): Impact of Weather Components on (UHF) Radio Signal, *International Journal of Engineering Research and General Science, Vol. 4, Issue 3, pp 481-494.*
[9] Yeeken, O.O; Michael, O.K (2011): Signal Strength Dependence on Atmospheric Particulates, *International Journal of Electronics and Communication Engineering, 4(3), pp 283–286.*
[10] Karmakar P.K; Maiti, M; Bhattacharyya, K; Angelis, C.F; Machado L.A.T (2011) Rain Attenuation studies in the Microwave Band over Southern Latitude, *The Pacific Journal of Science and Technology, Vol. 12, No. 2, pp 196-205.*
[11] Meng, Y.S; Lee, Y.H; Ng, B. C (2009): The Effects of Tropical Weather on Radio Wave Propagation over Foliage Channel, *IEEE Transactions* on Vehicular *Technology, Vol. 58, No. 8, pp 4023-4030.*
[12] Olsen R.L; Segal, B (1992): New Techniques for Predicting the Multipath Fading Distribution on VHF/UHF/SHF Terrestrial Line of Sight Links in Canada, *Canadian Journal of Electrical and Computing Engineering, Vol.17, No. 1, pp 11-23.*
[13] Waheed-uz-Zaman M; Yousuf Zai M.A.K (2010): Study the Effects of Radio Waves Propagation under Sea at Pakistan Coastal Zones, *Journal of American Science, (2010), pp 413-418.*
[14] Xiaofeng Z, Sixun H (2010): Influence of Sea Surface Roughness on the Electromagnetic Wave Propagation in the Duct Environment, *Journal of Radio Engineering, Vol.19, No. 4, pp 601-605.*
[15] Ukhurebor, K.E; Batubo, T.B; Abiodun, I.C; Enoyoze, E (2017a): The Influence of Air Temperature on the Dew Point Temperature in Benin City, Nigeria, *Journal for Applied Science and Environmental Management, Vol. 21(4); pp 657-660.*
[16] Ukhurebor, K.E; Abiodun, I.C; Azi, S. O; Otete, I; Obogai L.E (2017b): A Cost Effective Weather Monitoring Device, *Archives of Current Research International, 7(4): pp 1-9.*

Theoretical study on the corrosion inhibitor potential of moxifloxacin for API 5L X-52 steel in acidic environment

M E Ikpi[1] and F E Abeng[1]

[1]Corrosion and Electrochemistry Research Laboratory, Department of Pure and Applied Chemistry, University of Calabar, P.M.B. 1115, Calabar-Nigeria.

E-mail: fidelisabeng@yahoo.com

Abstract. Density functional theory (DFT) is used to obtain the energy of an electron in the ground state which in turn enables the prediction of the inhibition efficiency of a molecule in the corrosion of metals. The electronic and molecular structure relationship of moxifloxacin acting as corrosion inhibitor on the corrosion of API 5L X-52 steel in acidic media was studied using DFT with B3LYP basis set. Quantum chemical parameters calculated from the Density functional theory model were as follows; E_{HOMO}, E_{LUMO}, electronegativity χ, global hardness η, electrophilicity index ω, the number of electrons transferred from the inhibitor molecule to the metal ΔN and energy of the back-donation ($\Delta E_{back\text{-}donation}$). The outcome of the analysis suggests that the studied molecule functions as an effective additive that slows down the corrosion process, and the data corroborates experimental results earlier reported.

1. Introduction

Corrosionof metals commonly occur in most chemical manufacturing, construction and oil and gas industries. Corrosion is the destruction of a material caused by the environment in which the material resides [1]. The industrial cost involved in preventing the problem of corrosion is high and so there is considerable drive from industry to find cheap and effective additives to control the problems of corrosion [2]. The chemical components in medicinal drugs act as inhibitive substances in corrosion prevention. Several reports on the use of pharmaceutical drugs as corrosion inhibitor has been documented using experimental measurements such as gravimetric and electrochemical methods [3-15].

As a contribution to the current interest, the present study aims at broadening the application of Density Functional Theory (DFT) for metallic corrosion inhibition. Moxifloxacin is utilized for the study as it presents the basic characteristic of an effective corrosion inhibitor, namely the presence of nitrogen and oxygen, and in addition is economically friendly and cheap. The reactivity of the molecule is affected by the ground state spatial arrangement of the molecule, and the shape and structure of both its highest occupied molecular orbital (HOMO) and its lowest unoccupied molecular orbital (LUMO).

The inhibition efficiency of the studied compound was previously reported [1, 2] using potentiodynamic polarization measurements, gravimetric measurements and electrochemical impedance spectroscopy. Theoretical chemical parameters to be calculated include energy of highest occupied molecular orbital E_{HOMO}, energy of the lowest unoccupied molecular orbital E_{LUMO}, electronegativity χ, ionization energy I, electron affinity A, global hardness η, global softness σ, the number of electrons transferred from the inhibitor molecule to the atom of the metal ΔN, energy of back-donation $\Delta E_{back\text{-}donation}$ and electrophilicity index ω.

2. Methodology

One of the computational chemistry tools that offer the basics for interacting many chemical concepts employed in diverse fields of chemistry is DFT. To investigate the consistency between theoretical and experimental data, quantum chemical calculations were carried out on a completely optimized geometrical structure of the compound. The optimization was performed by B3LYP functional at level of DFT. The density functional theory is regarded as an important method to examine the interactions at the inhibitor/surface interface. The chemical structure of moxifloxacin and the result of the geometry optimization of the studied compound is illustrated in Figure 1 and Figure 2 respectively.

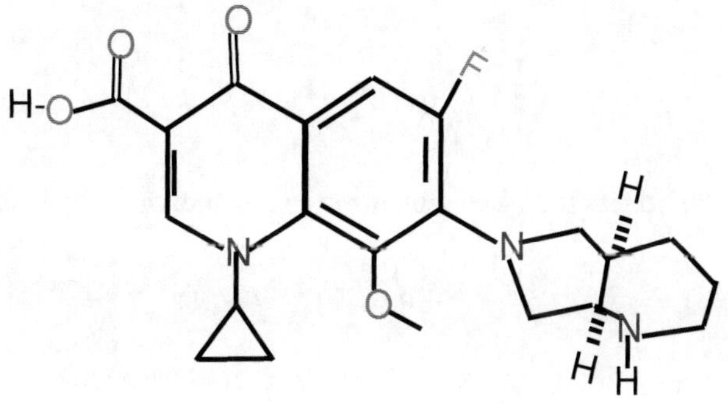

Figure 1. Chemical structure of moxifloxacin.

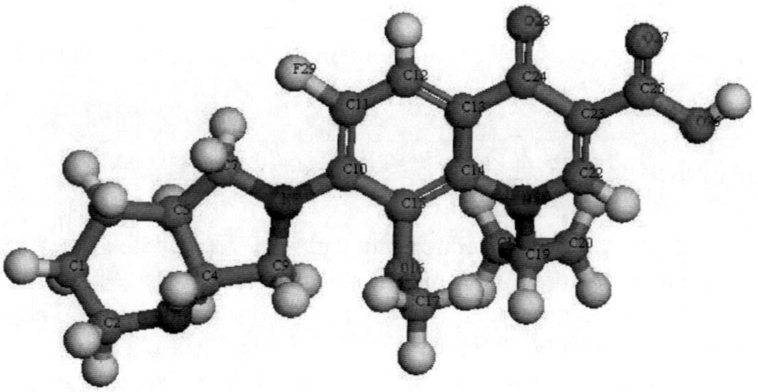

Figure 2. Optimized structure of moxifloxacin.

2.1. Theoretical and computational details

Parr, Donnelly, Levy and Palke [14, 16] related the first derivative of the electronic energy with respect to the number of electrons to the chemical potential μ. The mean value of ionization energy I and electron affinity A as well as the negative of electronegativity χ are given in (1) as finite difference version of chemical potential [9].

$$-\mu = \chi = {}^{\partial E}/_{\partial Nele} \approx {}^{(I+A)}/_{2} \qquad (1)$$

The principle of Koopmans theorem explains how E_{HOMO} and E_{LUMO} of the molecule are associated to I and A respectively. The principle considers the negative of HOMO energy as I and the negative of LUMO energy as A [9, 17]. The electronegativity χ is also expressed in terms of orbital energy as shown in (2).

$$\chi = {-(E_{HOMO} + E_{LUMO})}/{2} \tag{2}$$

The global hardness η (3) can be expressed as the second derivative of energy with respect to the number of electrons and may also be estimated from the values of the ionization energy and electron affinity of the studied compound.

$$\eta = \frac{1}{2}\left({\partial^2 E}/{\partial^2 Nele}\right) \approx {(I - A)}/{2} \tag{3}$$

The reciprocal of global hardness indicates global softness σ shown in (4).

$$\sigma = {1}/{\eta} \approx {2}/{(I - A)} \tag{4}$$

Equation (5) defines electrophilicity index ω as a degree of electrophilic strength of a molecule [15, 17].

$$\omega = {\chi 2}/{2\eta} = {(I + A)2}/{4(I - A)} \tag{5}$$

The number of electrons transferred ΔN can be computed based on Pearson theory [18], as expressed in (6).

$$\Delta N = {\chi_{Fe} - \chi_{inh}}/{2\sum \eta_{Fe} + \eta_{inh}} \tag{6}$$

The electronegativity of iron and the inhibitor molecule are represented as χ_{Fe} and χ_{inh} respectively, while η_{Fe} is the global hardness for iron atom and η_{inh} the global hardness for the inhibitor molecule. Theoretically, χ_{Fe} and η_{Fe} are assigned the values 7.0 eV and 0 respectively. The power of a molecule to pull to itself electrons in a covalent bond is identified in the electronegativity χ and global hardness η parameters [15, 17]. Equation (6) shows that the number of electrons transferred is related to the differences in χ and the sum of the hardness parameter [9].

Gomez et al. [17] developed a framework of charge transfer for donation and back donation of charges. Electron back-donation as illustrated in (7) is a process that explains the interaction between the metal surface and inhibitor molecules.

$$\Delta E_{Back-donation} = -{\eta}/{4} \tag{7}$$

When the energy of the back-donation is less than zero ($\Delta E_{Back\text{-}donation} < 0$), and the hardness is greater zero $\eta > 0$, it implies that there is transfer of charge to the molecule which is succeeded by back-donation to unoccupied d orbital of the metal atom and such behaviour would reflect a good inhibition process [9, 15, 17]. Equation (8) and (9) give expressions for electron donating ω^- and electron accepting ω^+ powers respectively.

$$\omega^- = {(3I + A)2}/{16\,(I - A)} \tag{8}$$

$$\omega^+ = {(I + 3A)2}/{16\,(I - A)} \tag{9}$$

A large ω^+ value reflects an enhanced capacity of the molecule to accept charge. On the other hand, a small value of ω^- is indicative of a good electron donor. Equation (10) can be used to match together ω^- and ω^+; the electron donating and electron accepting power.

$$\Delta\omega^{\pm} = \omega^+ - (-\omega^-) = \omega^+ + \omega^- \tag{10}$$

3. ResultsandDiscussion

Chemical reactivity as described by the Frontier Molecular Orbital Theory (FMOT), has it that the interaction between the inhibitor molecule and the metal occurs through the transfer of the electron from the HOMO of the inhibitor to the vacant d-orbital of a metal (LUMO) [13].

3.1. Ionization energy

Chemical reactivity of molecules and atoms can be described from a knowledge of the ionization energy I. High stability and chemical hardness is suggestive of a high ionization energy and vice versa [9, 15]. The ionization energy value of the studied molecule has been presented in Table 1. It shows the molecule having a low ionization energy value which indicates high inhibition efficiency of the molecule. Generally, a small ionization energy of the range of -0.1 to 10 eV has been reported by [18-23], while I above 10 eV is considered a large type of ionization energy as reported by [24].

3.2. Electronegativity

As earlier mentioned, the absolute electronegativity depicts a molecule's capacity to pull on the electrons in a covalent bond. Table 1 shows the calculated value of electronegativity of the molecule which can quickly reach equalization, expecting realistic inhibition efficiency according to Sanderson's electronegativity equalization principle [17].

Table 1. Calculated quantum chemical parameters of the studied compound.

E_{HOMO} (eV)	E_{LUMO} (eV)	I (eV)	A (eV)	χ (eV)	η (eV)	Σ (eV^{-1})	ω^- (eV)	ω^+ (eV)	$\Delta\omega^{\pm}$ (eV)	ΔN (e)	$\Delta E_{Back\text{-}donation}$
-4.613	-2.173	4.613	2.173	3.393	1.220	0.819	6.567	3.174	9.741	1.478	- 0.303

3.3. Hardness and Softness

Another quantum chemical parameter that measures molecular reactivity and stability is global hardness η. It is a signifier of the resistance towards the polarization of the electron cloud of the molecules, atoms or ions when these systems undergo a small perturbation. Hard molecules have a large energy gap value (ΔE_{LUMO}-E_{HUMO}) greater than 10 eV [24] whereas a small energy gap of within -0.01-10 eV is attributed to a soft molecule [7, 18-23]. Our study reveals that the inhibitor molecule can be considered a soft molecule implying that it is reactive as it readily gives electrons to the metal. This means that the studied molecule would offer a good inhibition performance.

3.4. Number of electrons transferred

The calculated value for the sum of electrons transferred is determined using (6) and is listed in Table 1. Several authors have opined that if the number of electrons transferred is less than 3.6 ($\Delta N < 3.6$), such inhibitor has a good electron donating ability to the empty d-block orbital [9, 13, 15]. It was revealed from the study that the molecule has value of 1.478 as the number of electrons transferred, meaning it would be expected to function well as a good corrosion inhibitor. Ikpi et al. [1] confirmed the efficacy of this molecule using electrochemical corrosion test methods.

3.5. Back-donation

Table 1 also illustrates the value of $\Delta E_{Back\text{-}donation}$ calculated from (7). During charge transfer process, back donation charges also occur, which influence the bonding interaction between the inhibitor molecule and the surface of the metal. According to Gomez et al. [17], when $\Delta E_{back\text{-}donation}$value of inhibitor is less than zero and the value of hardness of the same inhibitor is greater than zero, it signifies that the molecule is favoured energetically. The results therefore correspond with the theory of Gomez et al. [17].

3.6. Electrophilicity index

Electrophilicity index ω measures the stabilization in energy after a system accepts further amount of electron charge from the environment [15]. The electrophilicity index ω is further defined in (5), (8), (9) and (10). The net electrophilicity index is given as $\Delta\omega^{\pm}$, whereω^{-} represents the electron donating power and ω^{+} the electron accepting power. Their values are presented in Table 1. The value of ω^{-} makes the inhibitor an excellent electron donor whereas ω^{+} value makes the studied compound capable of accepting charge.

3.7. Mulliken atomic charge

The diagrams of HOMO and LUMO of the investigated compound are shown in Figure 3. The HOMO reflects dense electron cloud while LUMO shows lack of electron dense cloud. To explain the mechanism of inhibition of the molecule we have to employ mulliken atomic charges (Table 2).

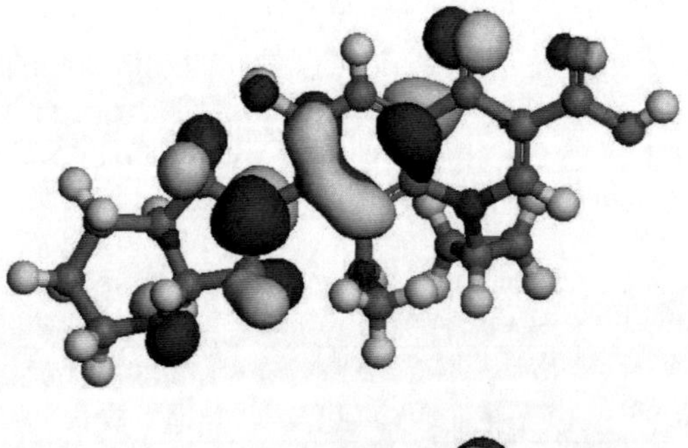

(a)

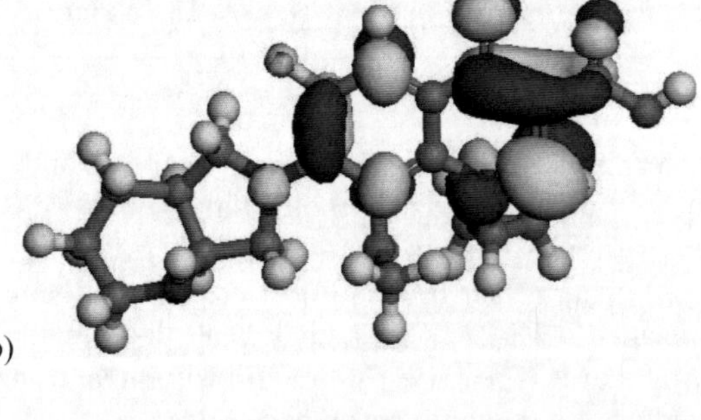

(b)

Figure 3. Frontier molecular orbital diagrams of moxifloxacin, (a) HUMO and (b) LUMO.

The ease at which an atom donates electrons to the unoccupied d-orbital of the metal is dependent on the atomic charge of the constituent atoms of the molecule that is more negative. Table 2 reveals that oxygen and nitrogen atoms exhibit more negative charges. These atoms can provide electrons to the surface of the metal thereby forming a coordinate complex as shown in Figure 4. Nitrogen and

Table 2. Mulliken atomic charges on moxifloxacin.

Atom	Charge	Atom	Charge	Atom	Charge
C1	-0.255	C10	0.138	C19	-0.273
C2	-0.140	C11	0.328	C20	0.096
N3	**-0.458**	C12	-0.103	**N21**	**-0.354**
C4	0.036	C13	-0.065	C22	-0.009
C5	-0.091	C14	0.113	C23	0.027
C6	-0.254	C15	0.239	C24	0.292
C7	-0.054	**O16**	**-0.570**	C25	0.532
N8	**-0.357**	C17	-0.156	**O26**	**-0.678**
C9	-0.059	C18	-0.239	**O27**	**-0.387**

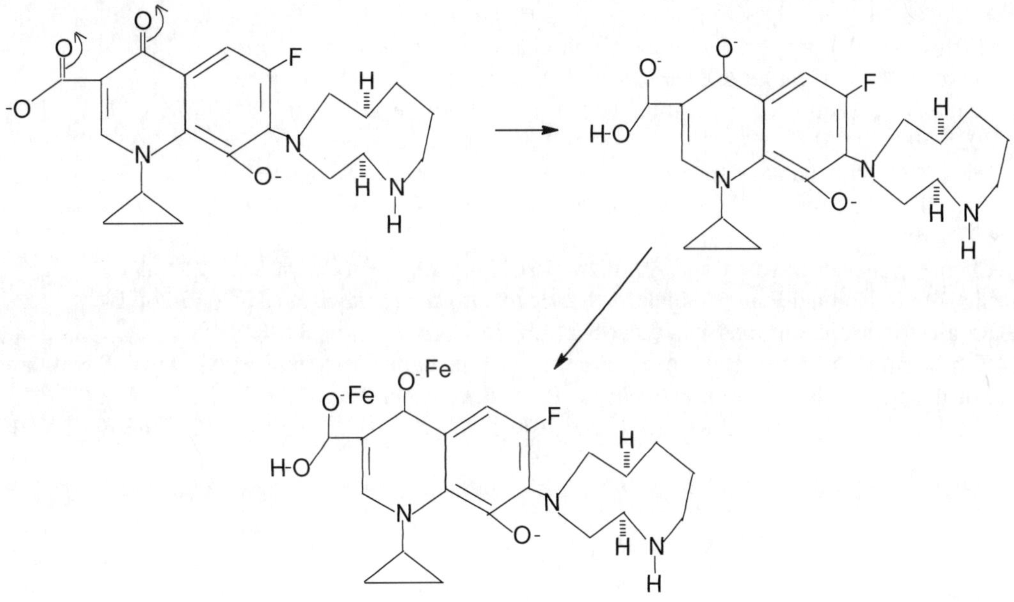

Figure 4. Corrosion inhibition mechanism of moxifloxacin on API 5L X-52 steel in 2 M HCl.

oxygen atoms are heteroatoms that have lone pairs of electron which can easily donate electrons to the empty d-orbital of the metal, enhancing adsorption onto the metal surface, facilitating inhibition of corrosion of the metal. The results reveal that adsorption of the studied molecule can easily occur at N3, N8, O16, N21, O26 and O27 atoms.

4. Conclusion

Density functional theory (DFT) has been used to study the inhibitory properties of moxifloxacin at B3LYP/6-31G basis set. The calculated quantum chemical parameters that describe the electronic activity of the investigated compound is in agreement with the inhibition efficiency results of this molecule earlier reported by [1, 2].The potentially good inhibition efficiency of the molecule reflects in thevalue of $E_{HOMO}$and the low value of the energy gap obtained from the computational analysis.

References

[1] Ikpi M E, Abeng F E and Obono O E2017*World News Nat. Sci.***9** 52-61
[2] Abeng F E, Ikpi M E, Uwakwe K and Ikpi G 2017*Int. Res. J. Pure Appl. Chem.***15** 1-12
[3] Ikpi M E, Abeng F E and Okonkwo B O 2017*World News Nat. Sci.***9** 79-90
[4] Ikpi M E and Abeng F E 2017*Int. J. Sci. Res.***6** 623-628
[5] Eddy N O and Ebenso E E 2010*Pigm. Resin Technol.***39**77-83
[6] Eddy N O and Odoemelam S A 2008*Adv. Nat. Appl. Sci.***2** 225-232
[7] Eddy N O, Stoyanov S R and Ebenso E E 2010*Int. J. Electrochem. Sci.***5** 1127-1150
[8] Akpan I A and Offiong N O 2014*Chem. Process Eng. Res.***26** 20-23
[9] Rajendran M, Kaerthika K, Kowsalya M and Devapiriam D 2016*Der Pharm. Chem.***8** 71-79
[10] Ikpi M E and Abeng F E 2017*Chem. Sci. Trans.***6** 569-576
[11] Eddy N O 2011*J. Adv. Res.***2** 35-37
[12] Ebenso E E, Isabirye D A and Eddy N O 2010*Int. J. Mol. Sci.***11** 2473-2498
[13] Wazzan N A and Mahgoub F M 2014*Open J. Phys. Chem.***4** 6-14
[14] Dariva C G and Galio A F 2014 Corrosion Inhibitors-Principles, Mechanisms and Applications *Developments in Corrosion Protection* ed M Aliofkhazraei (London: IntechOpen Ltd) chapter 16 pp 365-379
[15] Udhayakala P, Rajendiran T V and Gunasekaran S 2012*J. Adv. Scientif. Res.***3** 67-74
[16] Levine I N 2006 *Quantum Chemistry* (New Dehli: Prentice-Hall)
[17] Gomez B, Likhanova N V, Dominguez-Aguilar, Martinez-Palou R 2006*J. Phys. Chem.***110** 8928-8934
[18] Eddy N O, Ita B I and Ebenso E E 2011*Int. J. Electrochem. Sci.***6** 2101-2121
[19] Obot B I, Ebenso E E, Akpan A I, Gasen Z M and Afolabi A S 2012*Int. J. Electrochem. Sci.***7** 1978-1996
[20] Chen S, Scheiner S, Kar T and Adhikari U 2012*Int. J. Electrochem. Sci.***7** 7128-7139
[21] Shahabi S, Norouzi P and Ganjali M R 2015*Int. J. Electrochem. Sci.***10** 2646-2662
[22] Raja G, Saravanan K and Sivakumar S 2015*RASAYAN J. Chem.***8** 8-12
[23] Elmsellem H, Nacer H, Halaimia F, Aouniti A, Lakehal I, Chetouani A, Al-Deyab S S, Warad I, Touzani R and Hammouti B 2014 *Int. J. Electrochem. Sci.***9** 5328-5351
[24] Ameh P O, Eddy N O and Koha P U 2015*Chem. Sci. J.***6** 100. doi:10.4172/2150-3494.1000100

Room temperature Phytosynthesis of Ag/Co bimetallic nanoparticles using aqueous leaf extract of *Canna indica*

Anuoluwa Abimbola Akinsiku[1,*], Enock Olugbenga Dare[2,], Olayinka Oyewale Ajani[1], Joan Ayo-Ajayi[1], Olabisi Theresa Ademosun[1], Samuel Oluwakayode Ajayi[1]**

Department of Chemistry, Covenant University, Canaan Land, Nigeria
Faculty of Chemistry and Pharmacy,Institute for Organic Chemistry,Universitatsstr. 31. Regensburg 93053, Regensburg University, Germany

Corresponding authors: anu.akinsiku@covenantuniversity.edu.ng
Enock-Olugbenga.Dare@chemie.uni-regensburg.de

Abstract. Continuous quest for safe environment has led to 'green' approach of synthesizing nanoparticles. The method is easy, eco-friendly and cost-effective. Locally sourced medicinal plant *Canna indica* was used as capping/ stabilizing agent instead of toxic chemicals. Secondary metabolites in the plant extract acted as reducing agents. Optical measurements were carried out using Uv-vis spectrophotometer and photoluminescence (PL). Formation of core-shell was detected in the TEM micrograph, which was supported by two peaks observed in surface plasmon resonance shown in the Uv-vis spectra. FT-IR spectrophotometric analysis indicated the presence of some specific functional groups in the phytochemicals which were adsorbed on the surface of nanoparticles. This analysis depicted the presence of hydroxyl group (O-H stretching) with a broad strong peak at 3360 cm^{-1}, C-H stretching at 2938 cm^{-1}, C=C stretching at 1659 cm^{-1}, C=N stretching at 1557 cm^{-1} and C-O deformation at 1065 cm^{-1}. Crystalline phase of the nanoparticles was determined using X-ray diffraction (XRD). The plant-mediated green synthesized silver/cobalt nanoparticles are potential optical materials as a result of their broad absorption band and emission.

1. Introduction

Recently, nanoparticles and nanomaterials are products of a rapidly growing technology. Paradigm shift has led to the development of 'green' nanoparticles as they are biocompatible; form stable nanoparticle products and they possess less adverse effects on health and environment. Different fields such as nanobiotechnology, catalysis, chemical, bio-sensing, photo-catalysis among others explore metallic nanoparticles due to their unique optical, electronic, magnetic and chemical properties [1-5]. Biosynthesized metallic nanoparticles have been explored in biomedical for antimicrobial coatings, imaging and drug delivery applications. Catalytic application includes removal of environmental contaminants,they are also used as sensor in electrochemical application. Silver nanoparticles are applied as colorimetric sensor in chemical sensing as a result of their sharp localized surface plasmon resonance [6-7].The high surface energy and high ratio in the surface to volume characteristics possess by nanoparticles are also of great advantage [8].

Bimetallic nanoparticles (BNPs) involves combination of two different metals to develop novel properties of the two metals present. They are more important as they possess enhanced specific properties than their corresponding monometallic nanoparticles (MNPs). Features of BNPs include additional degree of freedom; higher surface area, directly proportional to their adsorption power which qualifies them as effective catalysts [9]. Moreover, synergistic effects of bimetallic nanoparticles have also made them to be highly active in catalysis; but, transition-metal nanoparticles are reported to be unstable (as a result of their high surface energy), self-aggregate, expensive and difficult to separate from reaction mixture [10-11].

From research, synergism has been reported to occur when BNPs are used for catalysis, as interaction between the two metals is key. Shapes or arrangements of BNPs depend on the synthetic method and thermodynamic equilibrium of the two metals (precursors). Studies have shown that BNPs with few tens and hundreds of atoms possess active surfaces, and they can serve as green catalysts, as this could solve the challenges encountered in transition-metal nanoparticles. Ismail *et al.* [12] used an adsorption method to synthesize Cu/Ag and Cu/Ni bimetallic nanoparticles loaded on ginger powder. Antibacterial and antibiofilm activities of Ag, Au and Ag/Au BNPs were carried out by Gopinath *et al.*[13] using leaf extract of *Gloriosa superba*.Plant-mediated green syntheses of silver nanoparticles and bimetallic Ag/Ni nanoparticles were carried out in our previous works [14-16]. However, few studies have been carried out on green synthesis of Ag/Co bimetallic nanoparticles. Pt/Ag, Pt/Rh, Ag/Ni, Au/Pt, Pd/Au, Ni/Au, Ag/Co BNPs synthesized using chemical and physical methods have been reported [17-21].

Unfortunately, use of toxic organic solvents and strong reducing agents during nanoparticles synthesis has posed great threat to the environment. Hence, development of eco-friendly and cheap nanomaterials is a necessity [22]. In view of the synergistic properties of BNPs, we report for the first time, phytosynthesis of Ag/Co BNPs at room temperature, using leaf extract of *Canna indica* as a reducing agent instead of toxic chemical. Secondary metabolites in the plant extract acted as capping/stabilizing agents for the newly formed nanoparticles. Optical measurements were carried out using Uv-vis spectrophotometer and photoluminescence (PL). Further characterizations were done using x-ray diffraction (XRD), transmission electron microscope (TEM) and Fourier transformed infra-red (FT-IR).

2. Experimental details

2.1. Materials and reagents

Cobalt chloride hexahydrate ($CoCl_2.6H_2O$), silver nitrate ($AgNO_3$)were purchased from Sigma-Aldrich Company, UK, and used as received. Distilled-deionized water, Whatman number 1 filter paper and fresh leaf extract of *Canna indica.*

2.2. Preparation of aqueous leaf extract

Fresh *C. indica* leaves were collected from Ogun State, Nigeria (Fig. 1). They were washed thoroughly with distilled de-ionized water (d-d water), after which 20 g of the homogenized leaves were mixed with 200 mL d-d water at room temperature. Filtrate from the mixture was collected using Whatman number 1 filter paper and preserved at 4°C for nanoparticle synthesis and phytochemical screening [23].

2.3. Synthesis Ag/Co BNPs at room temperature

Ag/Co bimetallic nanoparticle formation was prepared via phyto-reduction method with modification to previous study [12,16]. Plant extract (50 mL) was added to precursor mixture (250 mL) in a beaker i.e. equal molar concentration containing 125 mL of $AgNO_3$ and 125 mL $CoCl_2.6H_2O$. This reaction was carried out at varied concentrations (0.5-3.0 mM) of the precursor. Resulting mixture was stirred vigorously and left at 25°C, noting the initial colour. Metal ion reduction of Ag(I) ions to $Ag_s(0)$ and Co(II) to $Co_s(0)$ was monitored closely for any colour change. Appearance of surface plasmon resonance (SPR) was checked using double beam Uv-vis spectrophotometer by taking aliquot samples at different time intervals, placed in quartz cuvette, operated at a resolution of 1 nm so as to measure the absorbance.

2.4. Isolation of metallic Ag/Co nanoparticles

After appearance of SPR band, the reaction mixture was transferred into a Thermo Fisher Scientific Centrifuge (Thermo Electron LED), which was operated at 5000 rates per minute (rpm) for 30 minutes, so as to collect the nanoparticles. Ag/Co BNPs were purified by re-dispersing in d-d water in order to remove the organics present. It was centrifuged again for 10 minutes. The suspension was dried in an oven at 70°C, collected and kept for structural and morphological characterizations.

2.5.Characterization

2.5.1. Optical characterization

Optical properties of Ag/Co BNPs were carried out using Uv-vis spectrophotometer (Thermo Scientific GENESYS 10S model) at wavelength range between 250 and 800 nm to measure absorbance maximum. Distilled-deionized water was the blank used. Also, photoluminescence (PL) emission of Ag/Co BNPs was measured by Perkin-Elmer 55 spectrophotometer at room temperature. The sample was placed in 1 x 1 cm quartz cell.

2.5.2. Structural characterization

Morphology, particle size determination and detailed structure of the biosynthesized Ag/Co BNPs were determined using transmission electron microscope (Technai G2), coupled with energy-dispersive x-ray spectrometer (TEM-EDX) to determine elemental composition of the nanocluster. The equipment was operated at 200 KeV (accelerating voltage) and 20 μA current. Films on the TEM grids were left for 2 minutes to dry before taking any measurements, Ag/Co BNPs were prepared by drop-coating them onto carbon-coated copper TEM grids.

3. Results and discussion

Image of the utilized biodiversity and map of Nigeria indicating site of collection are shown in Figs. 1a and1b respectively.

Fig. 1 (a): Image of *Canna indica* (FHI No. 109928)

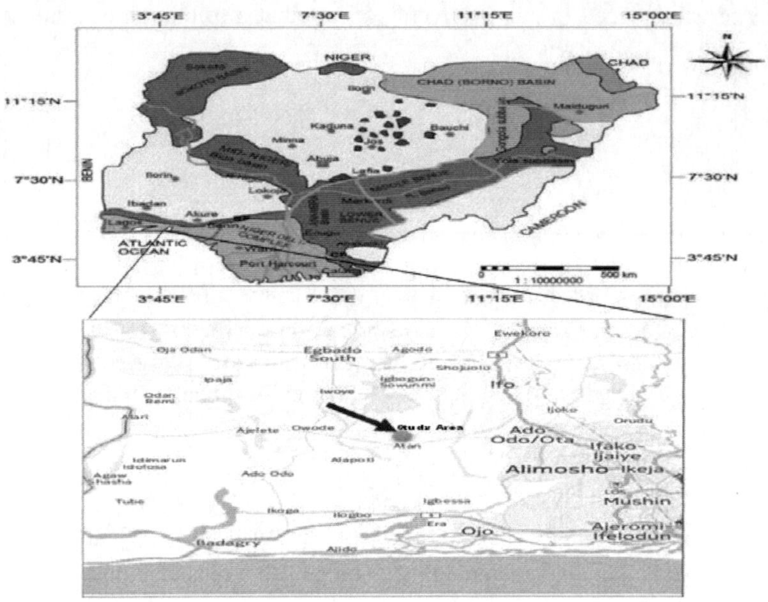

Fig. 1:(b) Map of Nigeria with arrow showing plant collection site

3.1. Optical properties of Ag/Co BNPs

Uv-vis spectra of the phytosynthesized silver hybrid at room temperature are depicted in Fig. 2. There was a noticeable colour change in the reaction from light brown to deep brown as a result of excitation of surface plasmon vibrations [24]. Appearance of surface plasmon resonance (SPR) was an evidence of nanoparticles formation; but, onset growth and nucleation were delayed till the 48th, 120th and 191st hours of the reaction time from the cluster prepared from 1.0, 2.0 and 3.0 mM precursor solutions respectively. Growth comparison in Ag/Co BNPs prepared at room temperature and 70°C is presented in Fig. 3. It is obvious that temperature increased kinetic energy of the reaction by increasing the rate of collision between metal ions (Ag^{+} and Co^{2+}) and the reducing agents (phytochemicals) present in the plant extract, thereby lowering activation energy of the reaction medium. Peak broadening indicated formation of polydispersed particles [25]. The particles were expected to be large in size due to the longer reaction time, as plant-mediated green synthesis is time dependent. Ageing of particles was also predictable.

Red shift in the SPR band (450-550 nm) is thought to be the effect of Co in the hybrid nanoparticles. This observation pointed to increase in particle size of $Ag_{core}Co_{shell}$. According to Maaz (2007) and David (2003), an increase in particle size results in shift of absorption edge to longer wavelength [26-28]. The observed two absorption peaks were in agreement with the results attained from TEM. Maximum absorption wavelength in the Ag/Co BNPs was more shifted than that of single metal Ag NPs (420-480 nm) from previous works [14, 29]. No characteristic plasmon band resonance (PBR) peak was observed in the particles synthesized in 0.5 mM precursor solution. This indicated that the reducing agent at this concentration was not significant for hybrid NPs synthesis [30]. This could also plausibly be that the working concentration was too low for the reduction of metal ion. In addition, PBR was not pronounced in the particles at 3.0 mM precursor solution, possibly because high concentration of metal salt proved to inhibit the synthesis of nanoparticles.

Luminescence properties ofAg/Co BNPs is presented in Figure 4. There was a noticeable strong local field in the nanoparticles as photoluminescence emission peaks was observed at 739 nm after excitation at 330 nm. This was plausibly due to the surface of the hybrid nanoparticles predominantly occupied by Co atom. Existence of a single emission peak in the longer wavelength band region indicated electron quantum confinement [31]. Moreover, the observed higher emission wavelength suggested stoke's shift. The sturdy peak intensity displayed by the nanohybrid was a consequence of the allowed vibrational transition in the nanocluster [32].Hence, the optical properties possess by Ag/Co BNPs is an indication for potential plasmonic applications aswideband optical modulators in the visible range of the spectrum [33], also in devices that source, detect and control light.

3.2. *Morphology of Ag/Co BNPs*

Particle size distribution and representative TEM image of the silver allied nanoparticles are presented in Figs. 5(a) and 4(b) respectively. Another representative micrograph with average size of 97.77 ± 25.53 nm is depicted in Fig. 5(c). The image (Fig. 5b) revealed silver (core) with dark colour and cobalt (shell) which corroborates the two absorption bands observed in the Uv-vis spectra. This is in agreement with previous work done by Sheny*et al.* [34]. Furthermore, this could also be as a result of Ag^{+} which got reduced first to form the core because of its higher positive electrochemical potential than cobalt [16]. Average size of the nanocluster was 202.5 ± 62.75 nm. EDX analysis confirmed formation of Ag/Co BNPs which was enriched with plant extract which acted as the capping agent (Fig. 6). Similar finding was reported by Mntungwa*et al.* [35].

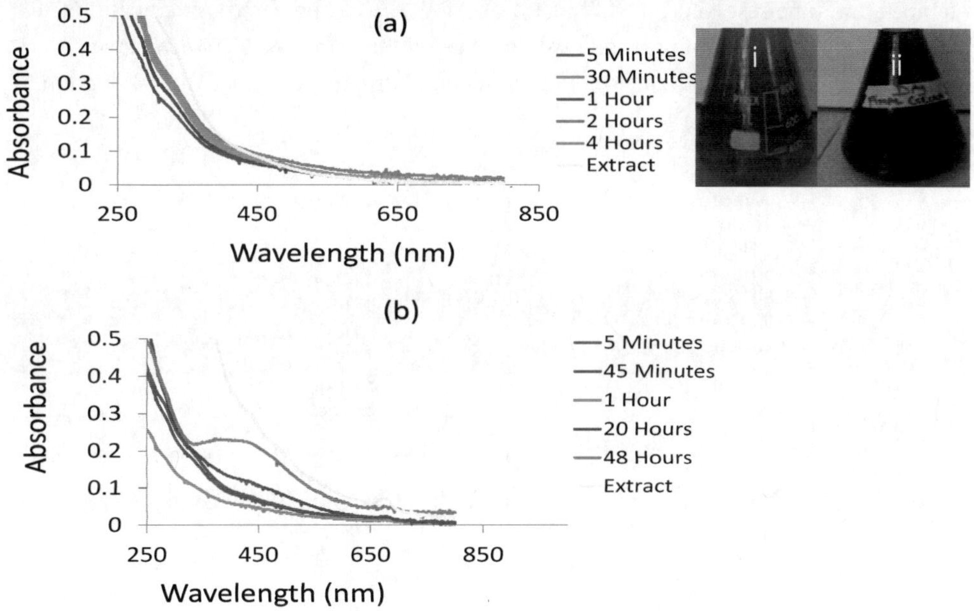

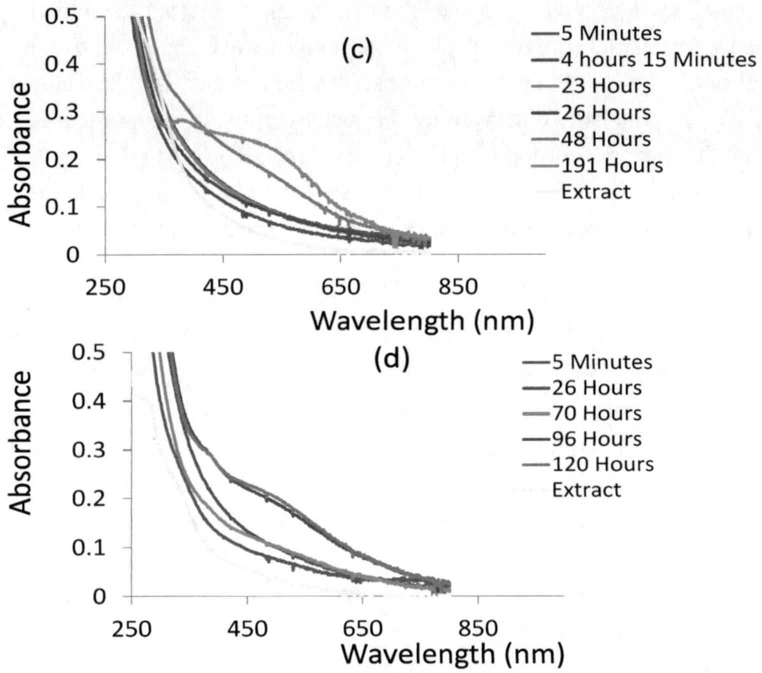

Fig. 2: Room temperature time-resolved Uv-vis spectra of Ag-Co BNPs prepared by reducing (a) 0.5 mM (b) 1.0 mM (c) 2.0 mM (d) 3.0 mM precursor solutions using the extract of *C. indica* leaves. Inset: (i) Precursor mixture with extract before reduction (ii) Final dispersion formed after reduction.

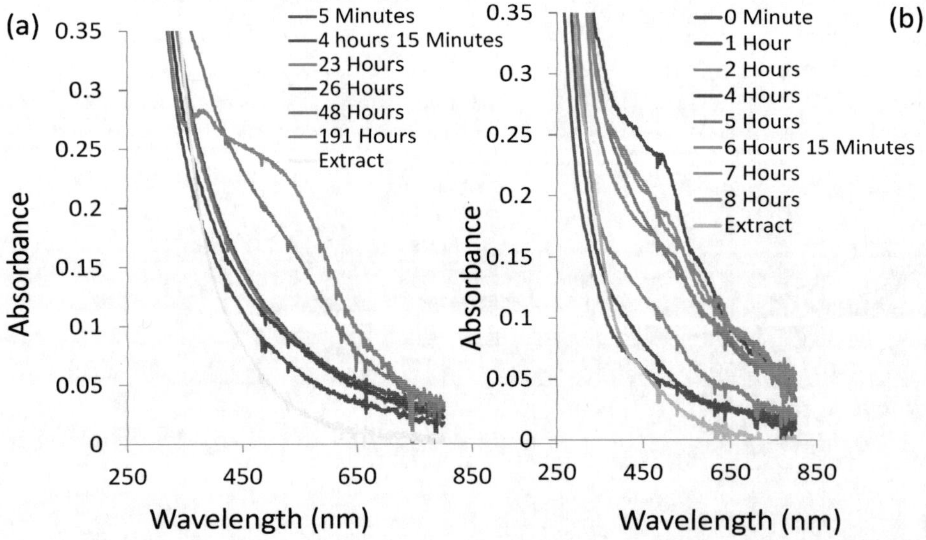

Fig. 3: Growth comparison of Ag/Co BNPs using 2.0 mM precursor (a) Room temperature, (b) 70°C

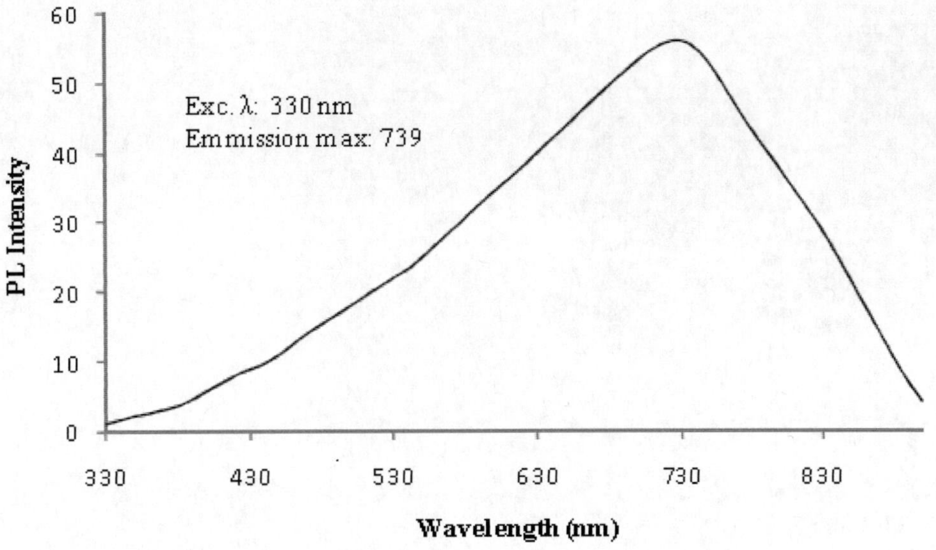

Figure 4: The PL emission spectrum of Ag/Co BNPs synthesized

3.3 FT-IR analysis

Possible functional groups present in the biosynthesized Ag/Co BNPs were identified in the FT-IR spectrum of Ag/Co BNPs (Fig. 7) which displayed bands due to O-H stretching vibrational mode at 3360 cm^{-1} which overshadowed N-H stretching in terpenoid found within this region. C-H stretching at 2938 cm^{-1} (medium absorption) was a sharp peak, C=C stretching at 1659 cm^{-1}, C=N stretching and the C-O deformation at 1557 cm^{-1} and 1065 cm^{-1} bands respectively. Glycosides, alkaloids and terpenoids were detected in our earlier study when leaves of *C. indica*were extracted in water by cold extraction [16]. These secondary metabolites were responsible for the reduction of metal ions in the precursor, capping and stability of the Ag/Co BNPs as they were formed [36]. Terpenoids and glycosides in the leaf extract were considered to be the adsorbed biochemicals on the surface of Ag/Co BNPs as spotted in the FT-IR analysis.

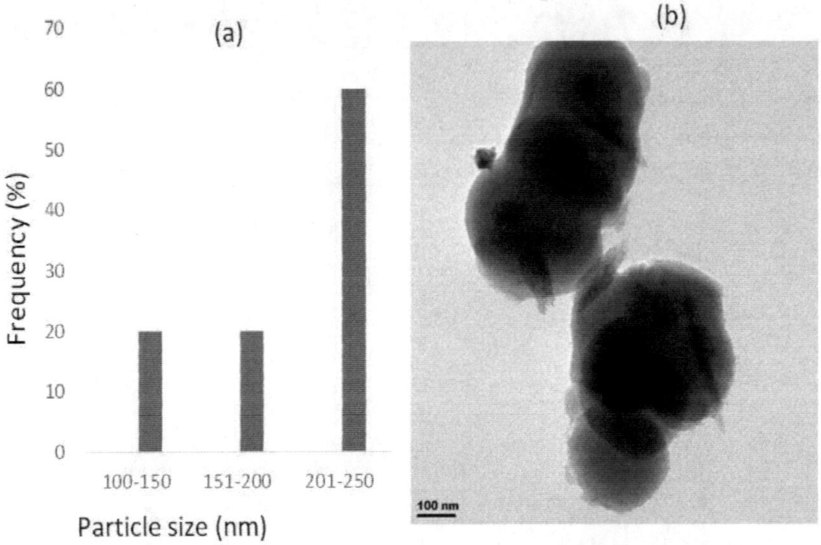

Figure 5: (a) Particle size distribution histogram of Ag/Co determined from TEM image
(b) Representative TEM image of Ag/Co BNPs

IOP Conf. Series: Earth and Environmental Science **173** (2018) 012019 doi:10.1088/1755-1315/173/1/012019

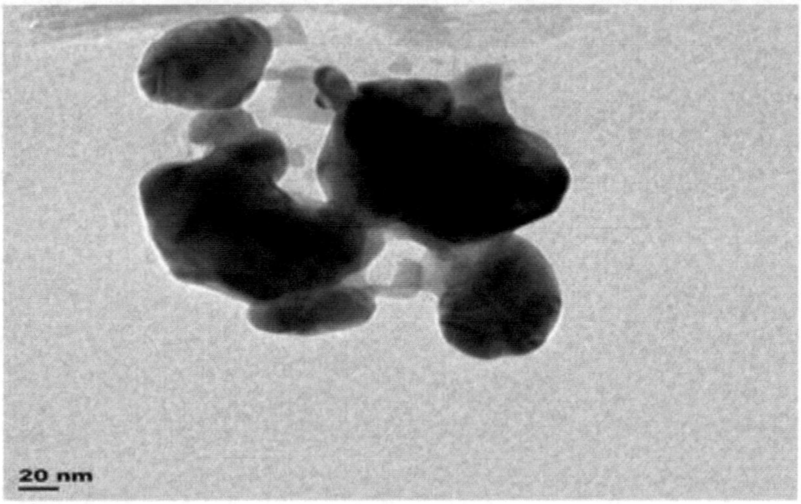

(c) Representative TEM image of Ag/Co BNPs

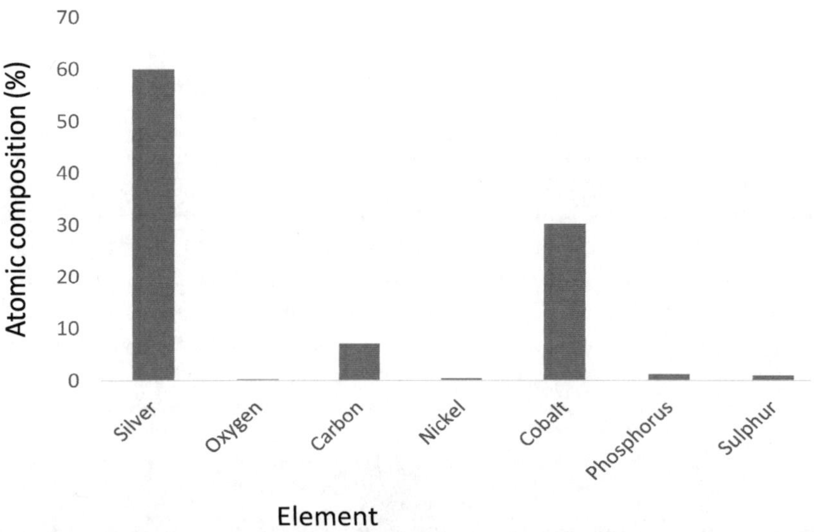

Fig. 6: EDX showing atomic composition of elements present in Ag/Co BNPs

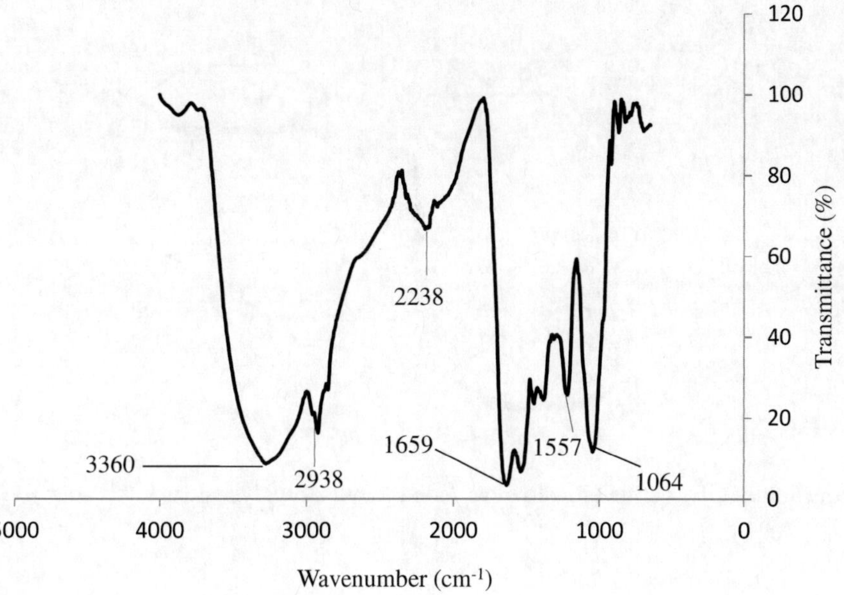

Fig. 7: FT-IR spectrum of Ag/Co BNPs

3.4. XRD analysis
XRD patterns of the synthesized Ag/Co BNPs are depicted in Fig. 8. Diffraction peaks at 2θ values = (38.15°), (44.37°), (64.57°), (77.39°) and (81.08°) corresponding to (111), (200), (220), (311) and (222) diffraction planes respectively were attributed to the crystal planes of cubic Ag structure [37]. Moreover, the spectrum also shows 2θ values = (41.78°), (44.37°), (44.37°), (52.64°) and (58.29°) equivalent to (100), (111) and (001) hcp respectively, also 52.64° and 58.29°. Core-shell structural morphology of Ag/Co are depicted in the diffraction patterns.

3.5 Mechanisms of reactions
Proposed reaction mechanisms are shown in Schemes 1 and 2 below:

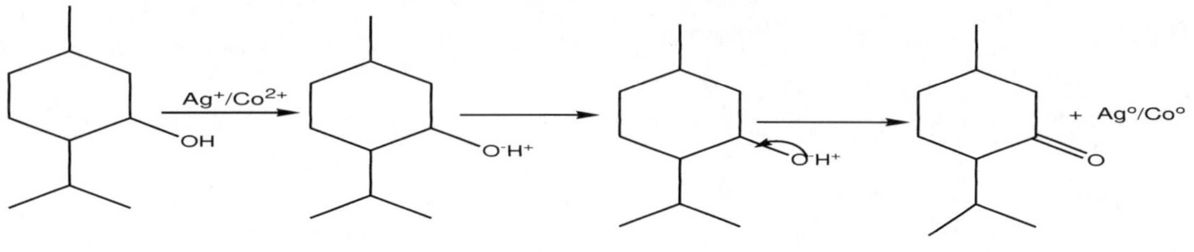

Scheme 1: Bioreduction of silver/cobalt ionsto silver/cobalt BNPs by terpenoids

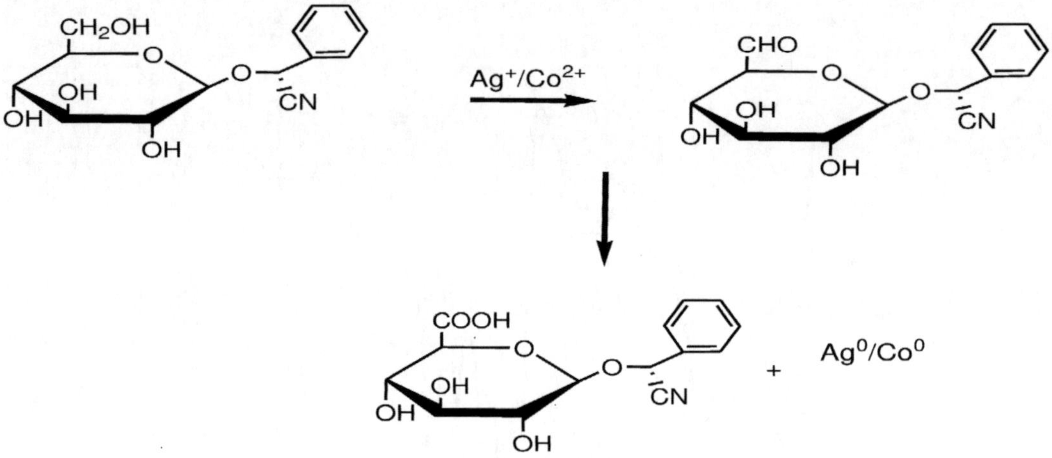

Scheme 2: Bioreduction of silver/cobalt ionsto silver/cobalt BNPs by glycosides

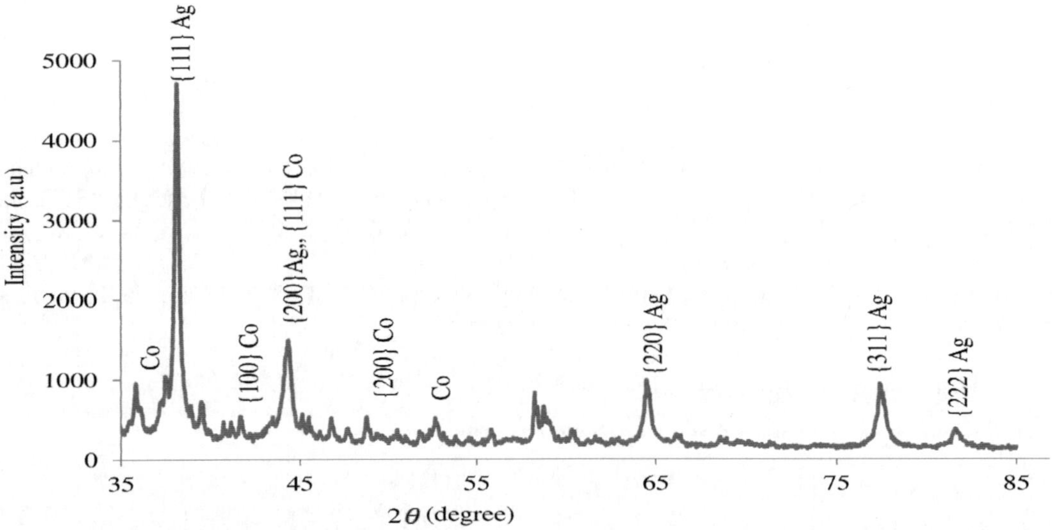

Fig. 8: XRD patterns of Ag/Co BNPs

4. Conclusion

In summary, core-shell/alloy structured Ag/Co BNPs were produced via phytoreduction process of *Canna indica* aqueous leaf extract. The nanoparticles were crystalline. The FT-IR pointed the moieties of the terpenoids and glycosides adsorbed on the nanocluster surface, which promoted capping and stability. Red shift in the SPR band was in the region 450-550 nm. Optical properties of the nanoparticles have indicated them as potential optical materials. This study can also take advantage of the surface plasmon absorption for therapeutic and diagnostic uses.

5. Acknowledgements

Authors are grateful to Mr. Olusola Rotimi of the University of Western Cape in South Africa for the FT-IR, XRD, TEM and EDX characterizations. The authors also acknowledge Covenant University.

References

[1] Shameli, K., Ahmad, M.B., Yunus, W., Ibrahim, N.A. and Darroudi, M. (2010). Synthesis and characterization of silver/talc nanocomposites using the wet chemical reduction method. *International Journal of Nanomedicine*. 5: 743–751.

[2] Kalantari, K., Ahmad, M.B., Masoumi, H.R.F., Shameli, K., Basri, M. and Khandanlou, R. (2015). Rapid and high capacity adsorption of heavy metals by Fe_3O_4/montmorillonite nanocomposite using response surface methodology: preparation, characterization, optimization, equilibrium isotherms, and adsorption kinetics study. Journal of the Taiwan Institute of Chemical Engineers, 49: 192–198.

[3] Huang, J., Cao, Y., Shao, Q., Peng, X., Guo, Z., 2017. Magnetic nanocarbon adsorbents with enhanced hexavalent chromiumremoval: morphology dependence of fibrillar vs particulate structures. Journal of Industrial and Engineering Chemistry Research, 56: 10689–10701.

[4] Noghabi, M. P., Parizadeh, M. R., Ghayour-Mobarhan, M., Taherzadeh, D., Hosseini, H. A. and Darroudi, D. (2017). Green synthesis of silver nanoparticles and investigation of their colorimetric sensing and cytotoxicity effects. *Journal of Molecular Structure*. 1146:499-503.

[5] Zhang, L., Yu, W., Han, C., Guo, J., Zhang, Q., Xie, H., Shao, Q., Sun, Z., Guo, Z., 2017. Large scaled synthesis of hetero structured electrospun TiO_2/SnO_2 nanofibers with an enhanced photocatalytic activity. *Journal of Electrochemical Society,* 164: 651–656.

[6] Darroudi, M., Khorsand, A., Muhamad, M. R., Huang, N. M. and Hakimi, M. (2012). Green synthesis of colloidal silver nanoparticles by sonochemical method.Materials Letters, 66, 1:117-120.

[7] Rostami, S., Mehdinia, A. and Jabbari, A. (2017) Seed-mediated grown silver nanoparticles as a colorimetric sensor for detection of ascorbic acid. Spectrochimica Acta Part A: Molecular and Biomolecular Spectroscopy, 180: 204-210.

[8] Izadiyan, Z., Shameli, k., Miyake, M., Hara, Mohamad, B., Kalantari, K., Taib, S. H. and a, Rasouli, E. (2018). Cytotoxicity assay of plant-mediated synthesized iron oxide nanoparticles using *Juglans regia* green husk extract. *Arabian Journal of Chemistry*. (In press).

[9] Schrofel, A., Kratošova, G., Šafarik, I., Šafarikova, M., Raška, I. and Shor, L. M. (2014). Applications of biosynthesized metallic nanoparticles – A review. *Acta Biomaterialia,* 10: 4023–4042.

[10] Khan, S.B., Ali, F., Kamal, T., Anwar, Y., Asiri, A.M., Seo, J., 2016. CuO embedded chitosan spheres as antibacterial adsorbent for dyes.*International journal of biological macromolecules,* 88: 113-119.

[11] Kamal, T., Khan, S.B., Haider, S., Alghamdi, Y.G., Asiri, A.M., 2017. Thin layer chitosan-coated cellulose filter paper as substrate for immobilization of catalytic cobalt nanoparticles. *International Journal of Biological Macromolecules,* 104: 56-62.

[12] Ismail, M., Khan, M.I., Khan, S. B., Khan, A., Akhtar, K. and Asiri, A. M. (2018). Green synthesis of plant supported Cu/Ag and Cu/Ni bimetallic nanoparticles in the reduction of nitrophenols and organic dyes for water treatment. *Journal of Molecular Liquids,* 260: 78–91.

[13] Gopinath, K., Kumaraguru, S., Bhakyaraj, K., Mohan, S., Venkatesh, K. S., Esakkirajan, M., Kaleeswarran, P., Alharbi, N. S., Kadaikunnan, S., Govindarajan, M., Benelli, G. and Arumugam, A. (2016). Green synthesis of silver, gold and silver/gold bimetallic nanoparticles using the *Gloriosa superba* leaf extract and their antibacterial and antibiofilm activities. *Microbial Pathogenesis,* 101 :1-11.

[14] Akinsiku, A. A.,Dare, E. O., Ajanaku, K. O., Adekoya, J. A., Alayande, S. O. and Adeyemi, A. O. (2016). Synthesis of silver nanoparticles by plant-mediated green method: Optical and biological properties. *Journal of Bionanoscience*. 10; 3: 171–180.

[15] Akinsiku, A. A., Dare, E. O., Ajanaku, K. O., Adekoya, J. A. and Ayo-Ajayi, J. (2018a). Green synthesized optically active organically capped silver nanoparticles using stem extract of African

cucumber (*Momordica charantia*). *Journal of Materials and Environmental Science,* 9; 3: 902-908.[16] Akinsiku, A. A., Dare, E. O., Ajanaku, K. O., Ajani, O.O., Olugbuyiro, J. A. O., Siyanbola, T. O., Elude, O. and Emetere, M. E. (2018b). Modeling and Synthesis of Ag and Ag/Ni Allied Bimetallic Nanoparticles by Green Method: Optical and Biological Properties. International Journal of Biomaterials, volume 2018, Article ID 9658080, 17 pages.

[17] Adekoya, J. A., Mlowe, S., Dare, E. O., Mesubi, M. A. and Revaprasadu, N. (2015). Synthesis and characterization of polyol stabilized Ag/Co allied nanocomposites. *Superlattices and Microstructures,* 78: 97–105.

[18] Bankole, O. M., Osifeko, O. and Nyokong, T. (2016). Enhanced nonlinear optical responses of zinc diaminopyrimidin-2-ylthio phthalocyanine conjugated to Ag_xAu_y alloy nanoparticles. *Journal of Photochemistry and Photobiology A: Chemistry*, 329, 155-166.

[19] Ke, D., Li, Y., Wang, J., Zhang, L., Wang, J., Zhao, X., Yang, S. and Han, S. (2016). Facile fabrication of poly(amidoamine) (PAMAM) dendrimers-encapsulated Ag-Co bimetallic nanoparticles for highly efficient dehydrogenation of ammonia borane. *International Journal of Hydrogen Energy,* 41: 2564-2574.

[20] Joo, Y., Ahmed, M. S., Han, H.S., Jeon, S. (2017). Preparation of electrochemically reduced graphene oxide-based silver-cobalt alloy nanocatalysts for efficient oxygen reduction reaction. *International Journal of Hydrogen Energy*, 42: 21751-21761.

[21] Sharma, G., Kumar, D., Kumar, A., Al-Muhtaseb, A.H., Pathania, D., Naushad, M., Mola, G.T., (2017). Revolution from monometallic to trimetallic nanoparticle composites, various synthesis methods and their applications: a review. Materials Science and Engineering C, 71: 1216–1230.

[22] Thakkar, K.N., Mhatre, S.S. and Parikh, R.Y. (2010). Biological synthesis of metallic nanoparticles. Nanomedicine: Nanotechnology, Biology and Medicine, 6: 257–262.

[23] Meye-Misso, N., Leonid, R., Obiang, S., De- La Croix, N.J., Privat, O. J., Felix, O. A., Engonga, O. and Clément, L. (2018). Phytochemical screening, antioxidant, anti-inflammatory and antiangiogenic activities of *Lophiraprocera* A. Chev. (Ochnaceae) medicinal plant from Gabon. *Egyptian Journal of Basic and Applied Sciences*, 5: 80–86.

[24]Krishnaraj, C., Jagan, E.G., Rajasekar, S., Selvakumar, P., Kalaichelvan, P. T. and Mohan, N.(2010). Synthesis of silver nanoparticles using *Acalypha indica* leaf extracts and its antibacterial activity against water borne pathogens. *Colloids and Surfaces B: Biointerfaces*, 76: 50-56.

[25] Link, S., Wang, Z. L. & El-Sayed, M. A. (1999). Alloy formation of gold-silvernanoparticles and the dependence of the plasmon absorption on their composition.*Journal of Physica*l *Chemistry B*, **103**, 3529.

[26] David, G.C., Wayne, K.F., Kenneth, E.G., Gerald, D.M. and Arun, M. (2003). Nanoscale thermal transport.Journal of Applied Physics, 93: 793-802.

[27] Maaz, K. et al. (2007). Synthesis and magnetic properties of cobalt ferrite ($CoFe_2O_4$) nanoparticles prepared by wet chemical route. *J. Magn. Magn Mater.*, 308

[28] Philip, D, Mishra, A.N., Bhadauria, S., Gaur, M.S., Pasricha, R. and Kushwah, B. S. (2008). *International Journal of Green*Spectrochimica Acta, Part A, 71: 80–85.

[29] Raveendran, P., Fu, J. & Wallen, S. L. (2003). Completely "Green" synthesis andstabilization of metal nanoparticles. *Journal ofthe American Chemical Society*, **125,** 46: 13940–13941.

[30] Dare, E. O, Oseghale, C. O., Labulo, H. O., Adesuji, E. T., Elemike, E. E., Onwuka, J. C., …Bamgbose, J. T. (2014). Green synthesis and growth kinetics of nanosilver under bio-diversified plant extracts influence. J*ournal* of *Nanostructure* in *Chemistry.*

[31] Rahim, A. F., Hashim, M. R. & Ali, N. K. (2011). High sensitivity of palladium on porous silicon MSM photodetector. *Physica B: Condensed Matter*, 406 4:1034-1037.

[32]Condon, E. (1928). *A theory of intensity distribution in band systems (Meeting abstract). Physical Review,* 27**:** 640-660.

[33] Vasireddi,R., Paul R. and Mitra, A.K. (2014). Green Synthesis of $Ag_{core}Cu_{shell}$ Nanoparticles: Structural and Optical Characterization. *Journal of Green Science and Technology,* 1: 1–6.

[34] Sheny, D.S., Mathew, J. and Philip, D. (2011). Phytosynthesis of Au, Ag and Au–Ag bimetallic nanoparticles using aqueous extract and dried leaf of *Anacardiumoccidentale. Spectrochimica Acta Part A,* 79: 254–262.
[35] Mntungwa, N., Pullabhotla, V. S. R. and Revaprasadu, N. (2012). Facile synthesis of organically capped CdTe nanoparticles. *Journal ofNanoscience and Nanotechnology*. 12, 3: 2640–2644.
[36] Smith, A. M., Duan, H., Rhyner, M. N., Ruan, G. and Nie, S. (2006). A systematic examination of surface coatings on the optical and chemical properties of semiconductor quantum dots. *Physical Chemistry Chemical Physics,* 8, 33: 3895-3903.
[37] Gupta, K., Jana, P. C. and Meikap, A. K. (2010). Optical and electrical transport properties of polyaniline silver nanocomposite. *Synthetic Metals*. 160, 13: 1566-1573.

Dust pollution and its potential health risk exposure to inhabitants of Covenant University and canaanland, Ogun state, Nigeria

Omeje Maxwell[1], AdewoyinOlusegun .O[1], Joel Emmanuel S.[1], OkolieSociisT.A.[2], Sam-Inyang Samuel,[1]ArijajeTheophilus .E[1]Akinpelu Akinwumi[1]

[1]Department of Physics, College of Science and Technology, Covenant University, P.M.B 1023, Ota, Ogun State, Nigeria
[2] Department of Mechanical Engineering, College of Engineering, Covenant University, Nigeria

maxwell.omeje@covenantuniversity.edu.ng

Abstract. Understanding the background concentration of heavy metals in dust samples is so much important for identifying and managing pollution. Thus, these concentrations in dust can be distributed indoors and outdoors in order of magnitudes. This study demonstrates the risk analysis of dust samples collected from selected buildings in both Covenant University and Canaanland using Atomic Absorption Spectrometer (AAS). The obtained concentrations were used to determine the Geo-accumulation, Contaminant Factor and the Pollution Loading Index (PLI). The highest geo-accumulation index and pollution loading index of 3.108 and 0.5836 due to chromium (Cr) contents in the dust samples were noted in Lecture Theatre 2 (LT2). The higher values found in LT2 may be due to the wide open of both doors and windows. Both the Geo-accumulation Index and Pollution Load Index were found to be lower than the permissible level suggested by World Health Organization (WHO) and United State Environmental Protection Agency (USEPA). Significantly, these observations from this study will provide the basis for the background concentrations and potential risks of some important heavy metals of environmental concern

1. Introduction

Dust is made up of particles in powdered form, spread on the earth's surface or spread around about by natural or mechanical forces [1]. There are two types of dusts indoor and outdoor dust but for this study we are going to focus on indoor dust. Dust found indoor is a combination of particulate matter gotten from different sources. Dust can get into buildings through different ways, which are infiltration from outside sources and internally from smoking, incense burning, building and furniture material, consumer product and occupants activities [2] [3]. Dusts on modern impervious surface have become one of the most pressing issues in modern environmental management. On one hand, exterior dusts can be easily re-suspended under definite outside dynamic condition; pollutants adsorbed on them enter human body by the pathways of respiratory inhalation and direct skin contact and cause negative health effects. Children and the elderly, whose immune systems are either underdeveloped or age-compromised as well as the inadvertent ingestion of significant quantities of dust through hand-to-mouth pathways, are more vulnerable to toxicity. Biological mechanism is due to the effects from heavy metals especially from the bioaccumulation. Set of human beings (such as infant or child) are extremely prone to environmental interference due to the growth of their organ at the early stage of development, which cause them to be more susceptible to the disorder or functional damage [4]. It has been found out that the crystalline silica dust is a major health concern, considering the lodges in the lungs of human, as such causing respiratory damage like silicosis. It also increases the potential risk of lung disease like bronchitis, tuberculosis, etc. [5]. This present study aimed at assessing the geo-accumulation and pollution load indices to the dwellers of the selected buildings under assessment in both Covenant University and Canaanland.

.

1.1 The Geological Location of the Study Area

Covenant University is in Ogun State, which falls within the Eastern Dahomey (Benin) Basin of south-western Nigerian that stretches along the continental margin of the Gulf of Guinea. Rocks in the Dahomey basin are Late Cretaceous to Early Tertiary in age. The stratigraphy of the basin has been classified into Abeokuta Group, Imo Group, Oshoshun, Ilaro and Benin Formations. The Cretaceous Abeokuta Group consists of Ise, Afowo and Araromi Formations consisting of poorly sorted ferruginized grit, siltstone and mudstone with shale-clay layers.

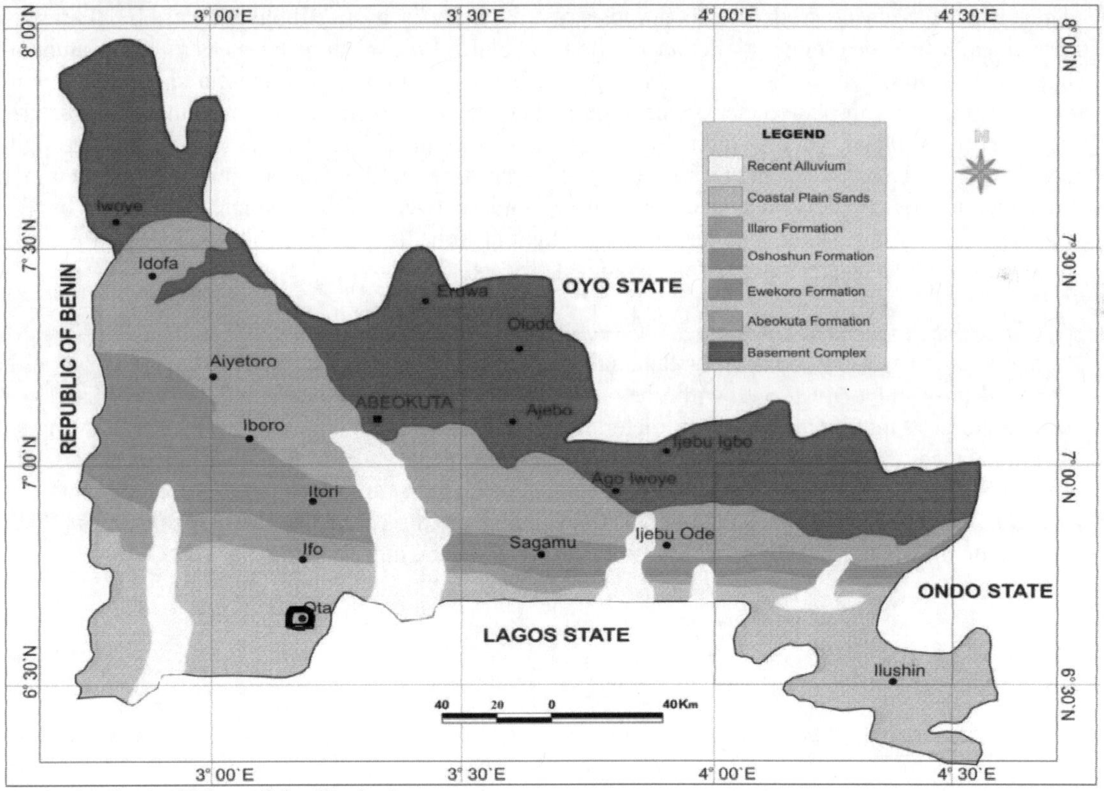

Figure1: Geological Map of the Study Area

2. Materials and Methods

Dust samples were collected from 5 different buildings in both Covenant University (Lecture Theaters One and Two, African Development Centre, Chapel) and Canaanland (Faith Tabernacle) using a manual filter membrane with wooden handle and plastic packer. Each of the filters is destroyed after use to avoid cross-contamination. This method was according to the standard of Airborne Agents through Filtration[6]

2.1. Data collection

The samples were collected thrice per location for 3 weeks for accurate analysis and sieved through 75µm mesh for homogeneity and easy digestion. About 0.2g of the samples was accurately weighed into a container per fluoroalkoxy polymer, which was then placed in a microwave pressure vessel (Ethos Plus Microwave Lactation, Milestone Inc., Shelton, CT, USA), using the standard of USEPA method 3052.[7]One replicate per digestion method was conducted per sample. The total content of heavy metals in the dust was analyzed using GF- AAS instrument connected to the intuitive WinLab32 software system comprises of the tools to analyze, report as well as achieving the measured data.[8] To calibrate the equipment, the standard solutions (panreac) of 100mgL^{-1} of all metals were used, as such, were calibrated from 10 – 100ppb.

2.1.1. Instruments.

The major instrument used for the purpose of this study is a spectrophotometer. It is simply an instrument which determines the concentration of solutes in a solution by measuring the amount of light that is absorbed by the spectrophotometer. The light intensity is measured as a function of wavelength. A spectrophotometer optically determines the absorbance or transmission of characteristic wavelengths of radiant energy (light) by a chemical species in solution. Each molecule absorbs light at certain wavelengths in a unique spectral pattern because of the number and arrangement of its characteristic functional groups such as double bonds between carbon atoms. It is based on Beer-Lamberts Law which states that the amount of light absorbed at the particular wavelength is directly proportional to the concentration of the chemical species.

Quality Control for the analysis of Heavy Metals Dust Sample

In this study, the quality control for the analysis of the dust samples using GF-AAS with model No: Perkin Elmer A Analyst 600 with the standard operation procedures (SOPs) according to the manufacturer. All other measurement meters such as TDS, pH and conductivity meters as well as the weigh balance were operated according the instructions of the SOPs to reduce analysis errors. All the equipment used in this study was calibrated before taken measurements. A calibration curve close to 1 was obtained for GF-AAS before the analysis was conducted on the bottled water samples so that the absorption of the atom of each element to be measured will be more accurate.

2.1.2. Risk assessment

2.1.3.1 Geo-accumulation Index

In this study, the risk analysis for Geo-accumulation Index (I_{geo}) of each heavy metal was calculated using the formula below:

$$I_{geo} = ln\left(\frac{C}{1.5*BV}\right) \quad 1$$

Where,

I_{geo} = Geoaccumilation Index.

C = Contaminant Concentration.

BV = Background Value

2.1.3.2 Contamination Factor

Contamination factor is gotten from the equation stated below.

$$CF = \frac{Concentration \ \ in\ the\ dust}{Background \ \ Value} \quad 2$$

Contamination is gotten from Table 1 and Background value is gotten from Table 2

2.1.3.3Pollution Load Index (PLI)

Pollution Load Index (PLI) is an index that is utilized in measuring the levels of heavy metals at the buildings for this study. The calculation was done using this equation stated below.

$$PLI = (CF_1 * CF_2 * \ldots . CF_n)^{\frac{1}{n}} \quad 3$$

Where,

CF = Contaminant Factor.

N = Number of Elements.

2.1.3.4Dermal Effect

The dermal effect was calculated using formula used is stated below:

$$D = \frac{(C*A*AF*EF*CF)}{BW} \quad 4$$

Where,

D = Exposure Dosage ($mgkg^{-1}day^{-1}$)

C = Contaminant concentration ($mgkg^{-1}$)

A = Total soil adhered (mg)

EF = Exposure Factor (unit less) = 1

CF = Conversion Factor (10^{-6} $kgmg^{-1}$)

BW = Body Weight (kg)

2.1.3.5 The Inhalation Effect

The Equation used is for inhalation effect is stated below:

$$D = \frac{(C*IR*EF)}{BW} \quad 5$$

Where,

D = Exposure Dose ($mgkg^{-1}day^{-1}$)

IOP Conf. Series: Earth and Environmental Science **173** (2018) 012020 doi:10.1088/1755-1315/173/1/012020

C = Contaminant Concentration (mgm^{-3})

IR = Intake rate (m^3day^{-1})

EF = Exposure Factor (unit less)

BW = Body Weight (kg)

3. Results and discussion

3.1. Concentrations of Measured Heavy Metals in the Various Dust Samples

The concentrations of Lead (Pb), Chromium (Cr), Cadmium (Cd), Nickel (Ni), Iron (Fe), Arsenic (As), Copper (Cu) in all the samples collected from the buildings are presented in table below.

Table 1: Concentration of Heavy Metal in the dust samples from African Leadership & Development Centre (Aldc), Chapel, Faith Tabernacle (Ft), Lecture Theatre 1 (Lt 1) and Lecture Theatre 2 (Lt 2).

Sample	Pb	Cr	Cd	Ni	Fe	As	Cu
Aldc (Week1)	0.2521	12.8209	0	9.9964	3299.703	0	10.9979
Aldc (Weel2)	0.212	12.5612	0	9.88	3468.1	0	11.6612
Aldc (Week3)	0.3675	10.4912	0	11.72	4567.32	0	11.8771
Chapel (Week1)	0.1133	2.4891	0	8.93	3561.11	0	14.2119
Chapel (Week2)	0.1173	5.8761	0	10.55	3201.001	0.0001	13.998
Chapel(Week 3)	0	5.6723	0	7.69	4001.783	0	11.0135
Ft (Week1)	0	9.3572	0	7.9915	3012.557	0.00012	11.8334
Ft 2(Week2)	0.0001	14.1772	0	10.8751	3789.121	0	12.2435
Ft (Week3)	0.0081	7.1002	0	9.9214	3904.4	0	11.7017
Lt 1 (Week 1)	0.2712	0	0	10.576	4012.871	0	10.9993
Lt 1 (Week 2)	0.3299	0	0	10.1044	4077.101	0	12.7812
Lt 1 (Week 3)	0.3761	5.1207	0	8.9955	3102.41	0	12.9711
Lt 2 (Week 1)	0	14.9991	0	9.2391	3869.98	0	10.1274
Lt 2 (Week 2)	0.0092	9.0771	0	10.0001	3901.999	0.0021	12.1145
Lt 2 (Week 3)	0	14.0921	0	9.4112	3892.095	0	10.7331

3.2. Risk Analysis

The estimated concentration of Pb in all the samples varies from $0.0001mgkg^{-1}$ to $0.3761mgkg^{-1}$ with the highest value of $0.3761mgkg^{-1}$ found in Lecture theatre 1, whereas the lowest value of $0.0001mgkg^{-1}$ reported in Faith Tabernacle as shown in Figure1. Comparing the highest value of Lead (Pb) obtained in this present study with the international standard by World Health Organization (WHO) and United States Environmental Protection Agency (USEPA) with a value of $55mgkg^{-1}$, this present study is lower by a factor of 146.24.

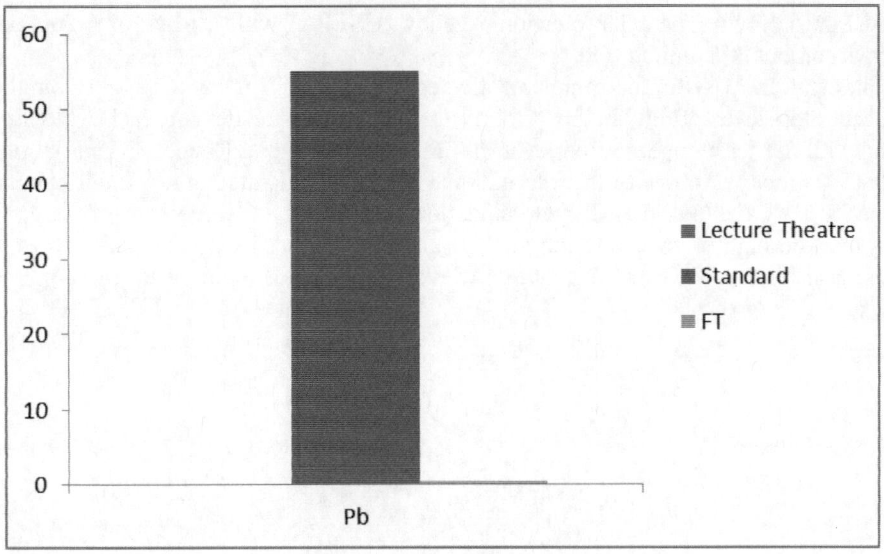

Figure 2: Comparing the Highest Values of Pb in the Dust Samples and the International Reference Standard

The concentration of Chromium (Cr) in the various samples ranges from 2.4891mgkg^{-1} to 14.9772mgkg^{-1}with the highest value of 14.9772mgkg^{-1} which is found in Lecture Theatre 2 and the lowest value of 2.4891mgkg^{-1} found in Chapel as shown in Figure 3. When comparing the highest value of Chromium (Cr) gotten in this present study with the international standard by United States Environmental Protection Agency (USEPA) and World Health Organization (WHO) with a value of 3.8mgkg^{-1}, this current study is higher by a factor 3.94 which means faith tabernacle is harmful to individual due to the chromium concentration and precautions should be taken.

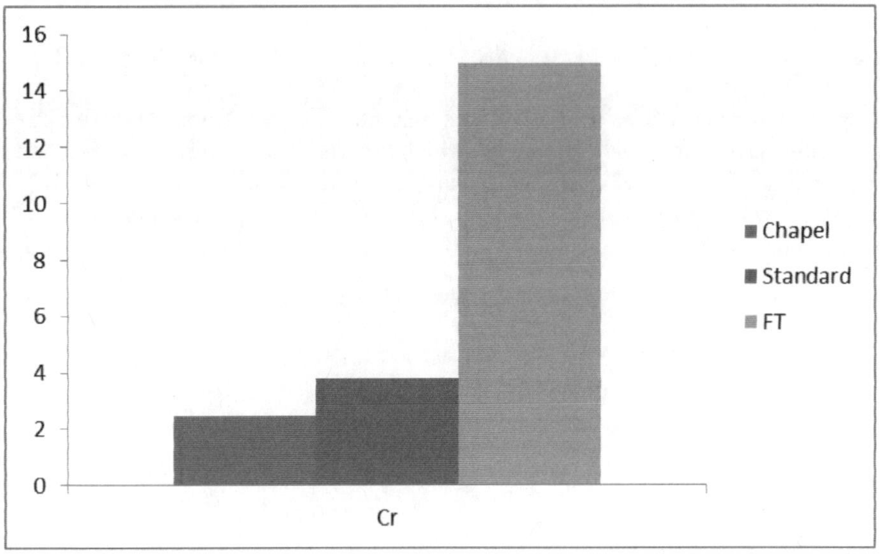

Figure 3: Comparing the Highest Values of Cr in the Dust Samples and the International Reference Standard

The concentration of Cadmium (Cd) in the samples collected from the various locations is not detected. Therefore it doesn't conflict with the international standard by World Health Organization

(WHO) and United States Environmental Protection Agency (USEPA) with a mark of 0.76mgkg^{-1} which means the environment is cadmium free.

The concentration of Nickel (Ni) in the samples collected varies from 7.69mgkg^{-1} to 11.72mgkg^{-1} with the most concentrated level of 11.72mgkg^{-1} found at the African Leadership & Development Centre (ALDC) and with the least concentrated level of 7.69mgkg^{-1} found at Chapel. When the most concentrated amount is compared to that of the international standard by United States Environmental Protection Agency (USEPA) and World Health Organization (WHO) with a value of 2.6mgkg^{-1}, this recent study shows that the highest concentration of the samples is higher by a factor of 4.51 which identifies ALDC as harmful to people to concentration levels and steps should be taken to prevent ailments to individuals.

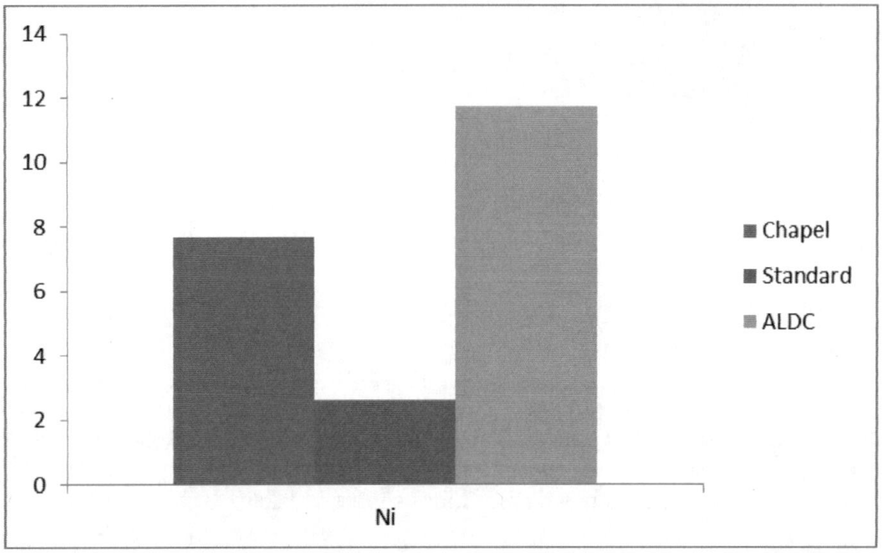

Figure 4: Comparing the Highest Values of Ni in the Dust Samples and the International Reference Standard

The concentration of Arsenic (As) in the various samples ranges from 0.0001mgkg^{-1} to 0.0012mgkg^{-1}with the highest value of 0.0021 mgkg^{-1} which is found in Lecture theatre 2 and the lowest value of 0.0001mgkg^{-1} found in Chapel. When comparing the highest value of Arsenic (As) gotten in this present study with the international standard by United States Environmental Protection Agency (USEPA) and World Health Organization (WHO) with a value of 4.5mgkg^{-1}, this current study is lower than the standard.

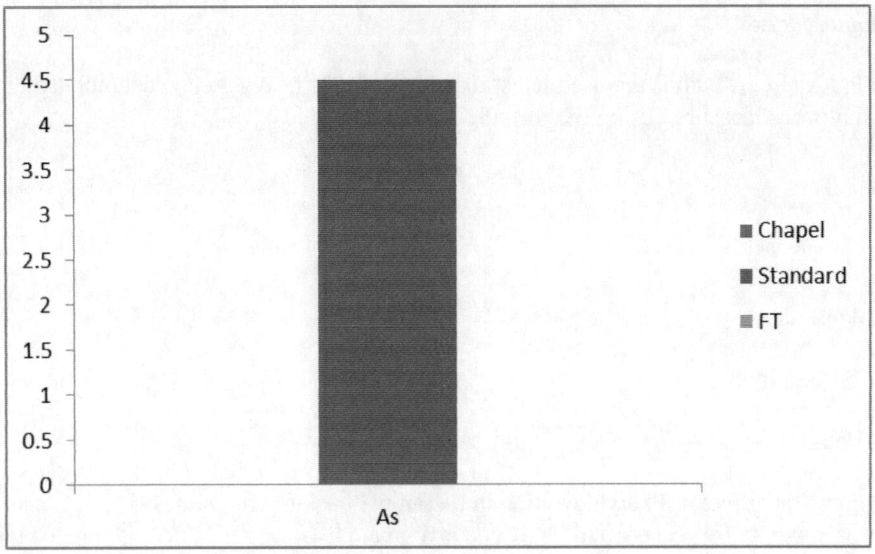

Figure 5: Comparing the Highest Values of As in the Dust Samples and the International Reference Standard

The concentration of Copper (Cu) in the samples collected varies from 10.1274mgkg^{-1} to 14.2119mgkg^{-1} with the most concentrated level of 14.2119mgkg^{-1} found at Chapel and with the least concentrated level of 7.69mgkg^{-1} found at Lecture Theatre 2. When the most concentrated amount is compared to that of the international standard by United States Environmental Protection Agency (USEPA) and World Health Organization (WHO) with a value of 3.5mgkg^{-1}, this recent study shows that the highest concentration of the samples is higher by a factor of 4.06 which identifies Chapel as harmful to people due to concentration levels and steps should be taken to prevent ailments to individuals.

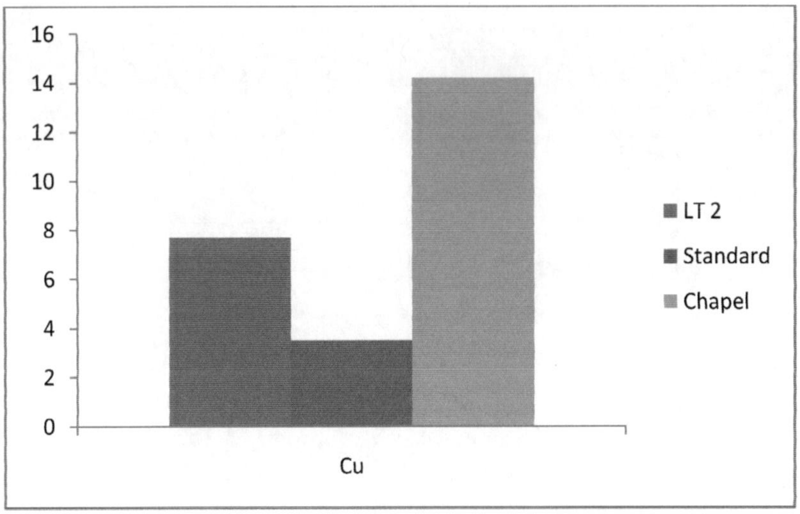

Figure 6: Comparing the Highest Values of Cu in the Dust Samples and the International Reference Standard

The pollution risk assessment methods used for this study which are the Geo-Accumulation Index (I_{geo}) and Pollution Load Index which is also known as PLI. They were adopted in this present study to assess the levels of contamination due to heavy metals originating from the dust samples.

IOP Conf. Series: Earth and Environmental Science **173** (2018) 012020 doi:10.1088/1755-1315/173/1/012020

3.3. Geo-accumulation Index

Geo-accumulation Index (I_{geo}) which is the used to estimate the level of heavy metal accumulated or contained in the sample was measured using the formula below suggested by[9]

$$I_{geo} = \ln\left(\frac{C}{1.5 * BV}\right) \quad 6$$

Where,

I_{geo} = Geoaccumilation Index.

C = Contaminant Concentration.

BV = Background Value.

The 1.5 is to bring down the effect of likely deviation in thesample background values. The I_{geo} for each heavy metal was computed and classified [10] "uncontaminated" ($I_{geo} \leq 0$); "uncontaminated to moderately contaminated" ($0<I_{geo}\leq 1$); "moderately contaminated" ($1<I_{geo}\leq 2$); "moderately to heavily contaminated" ($2<I_{geo}\leq 3$); "heavily contaminated" ($3 <I_{geo} \leq 4$); "heavily to extremely contaminated" ($4 <I_{geo} \leq 5$); "extremely contaminated" ($I_{geo} \geq 5$)

Note: Contaminant Concentration can be gotten from Table 1

Table 2: This represents the background value of chromium, Nickel, Iron and Copper.

Background Value			
Cr	**Ni**	**Fe**	**Cu**
0.42	68	47600	45

Table 3: This shows the geo-accumulation index of Chromium, Nickel, Iron and Copper.

	Geo-Accumulation Index of Heavy Metals			
Sample ID	**Cr**	**Ni**	**Fe**	**Cu**
ALDC (Week 1)	3.013112	-2.32275	-3.07447	-1.81442
ALDC (Week 2)	2.992648	-2.33446	-3.02469	-1.75586
ALDC (Week 3)	2.812572	-2.16368	-2.74937	-1.73752
Chapel (Week 1)	1.373957	-2.43556	-2.99823	-1.55805
Chapel (Week 2)	2.232929	-2.26885	-3.10483	-1.57321
Chapel (Week 3)	2.19763	-2.58505	-2.88156	-1.81301
FT (Week 1)	2.698182	-2.54659	-3.16551	-1.7412
FT (Week 2)	3.068565	-2.2385	-2.93616	-1.70713
FT (Week 3)	2.422158	-2.33028	-2.90619	-1.75239
LT 1 (Week 1)	0	-2.26639	-2.87879	-1.8143
LT 1 (Week 2)	0	-2.312	-2.86291	-1.66415
LT 1 (Week 3)	2.095327	-2.42825	-3.13612	-1.6494
LT 2 (Week 1)	2.955167	-2.40153	-2.91505	-1.89688
LT 2 (Week 2)	2.66779	-2.32238	-2.90681	-1.71772
LT 2 (Week 3)	3.10765	-2.38307	-2.90935	-1.8388

This present study of the Geo-Accumulation Index shows that Chromium (Ct) was found to be lower at Chapel (week 1) with a value of 1.373957, whereas he highest value was noted in Lecture theatre 2 (Week 2) with a value of 3.10765as shown in Table1. With this value closer to the recommended value, it implies that Lecture Theatre may be contaminated if not properly swept daily

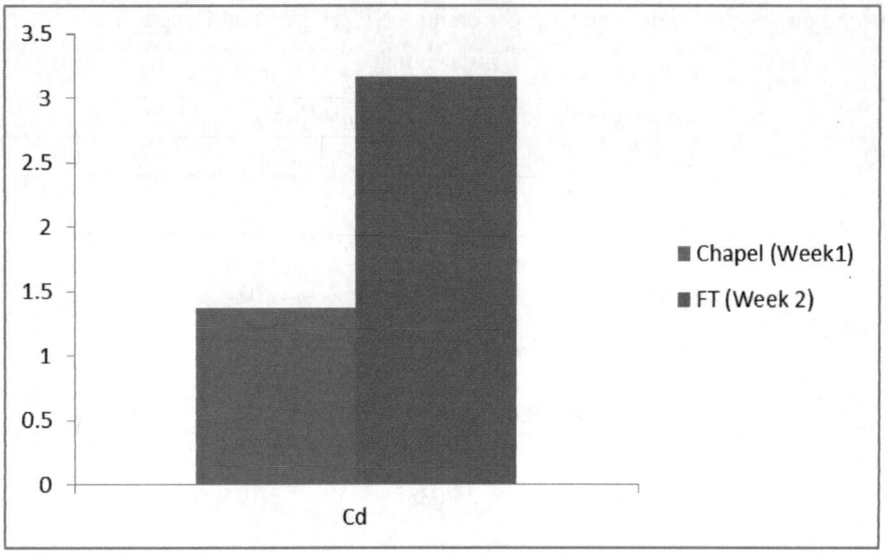

Figure 7: Comparing the Highest Values of Geo-Accumulation Index Due to Cr in the Dust Samples and the International Reference Standard

Chromium (Cr) is found to be lower at Chapel (week 3) with a value of -2.58505 and the highest value was noted at ALDC (Week 3) with a value of -2.16368 as shown in Table 2. Geoaccumulation due to Nickel (Ni) was found to be lower at Faith Tabernacle (Week 1) with a value of -3.16551 and the highest was noted at ALDC (Week 3) of about -2.74937 as shown in Table 2For Copper (Cu), it is lower at Lecture Theatre 2 (Week 1) with a value of -1.89688 and the highest value was found at Lecture Theatre 1 (Week 2) with a value of -1.66415 as shown in Table 2 which means no building is contaminated by copper. With these values, lower than the recommended values indicate that none of the buildings iscontaminated.

3.4. Contamination Factor

Contamination factor is gotten from the equation stated below.

$$CF = \frac{Concentrationinthedust}{BackgroundValue} \quad 7$$

Concentration is gotten from Table 1 and Background value is gotten from Table 2.

IOP Conf. Series: Earth and Environmental Science **173** (2018) 012020 doi:10.1088/1755-1315/173/1/012020

Table 4: This shows the conservation factors of Chromium, Nickel, Iron and Copper.

	Contamination Factor			
Sample ID	**Cr**	**Ni**	**Fe**	**Cu**
ALDC (Week 1)	30.52595	0.147006	0.069322	0.244398
ALDC (Week 2)	29.90762	0.145294	0.072859	0.259138
ALDC (Week 3)	24.97905	0.172353	0.095952	0.263936
Chapel (Week 1)	5.926429	0.131324	0.074813	0.31582
Chapel (Week 2)	13.99071	0.155147	0.067248	0.311067
Chapel (Week 3)	13.50548	0.113088	0.084071	0.244744
FT (Week 1)	22.27905	0.117522	0.063289	0.262964
FT (Week 2)	35.66	0.159928	0.079603	0.272078
FT (Week 3)	16.90524	0.145903	0.082025	0.260038
LT 1 (Week 1)	0	0.155529	0.084304	0.244429
LT 1 (Week 2)	0	0.148594	0.085653	0.284027
LT 1 (Week 3)	12.19214	0.132287	0.065177	0.288247
LT 2 (Week 1)	28.80738	0.135869	0.081302	0.225053
LT 2 (Week 2)	21.61214	0.14706	0.081975	0.269211
LT 2 (Week 3)	37.55262	0.1384	0.081767	0.238513

3.5. Pollution Load Index

Pollution Load Index (PLI), is an integrated pollution index was also utilized in measuring the levels of pollution of heavy metals at the selected buildings. The calculation was done using this equation stated below according to [11][12].

$$\boldsymbol{PLI = (CF_1 * CF_2 * \ldots.CF_n)^{\frac{1}{n}}} \qquad 8$$

Where,

CF = Contaminant Factor.

N = Number of Elements.

The result of PLI = 0 means "background concentration", 0 < PLI < 1 means "unpolluted",

1 < PLI < 2 indicates "unpolluted to moderately polluted", 2 < PLI < 3 means "moderately polluted", 3 < PLI < 4 signifies "moderately to highly polluted", 4 < PLI < 5 depicts "highly polluted" whereas PLI > 5 corresponds to "very highly polluted".

Note: Concentration factor can be found in the table above and Background value is gotten from Table 2.

IOP Conf. Series: Earth and Environmental Science **173** (2018) 012020 doi:10.1088/1755-1315/173/1/012020

Table 5: This Table represents the values for the pollution load index.

	Pollution Load Index (PLI)
ALDC (Week 1)	0.525100322
ALDC (Week 2)	0.535194105
ALDC (Week 3)	0.574627945
Chapel (Week 1)	0.368246326
Chapel (Week 2)	0.461613968
Chapel (Week 3)	0.421038388
FT (Week 1)	0.456888624
FT (Week 2)	0.512833268
FT (Week 3)	0.478924759
LT 1 (Week 1)	0
LT 1 (Week 2)	0
LT 1 (Week 3)	0.417218198
LT 2 (Week 1)	0.517312461
LT 2 (Week 2)	0.514625869
LT 2 (Week 3)	0.583577407

In this recent study, Pollution Load Index (PLI) is found to be lower in Chapel (Week 3) with a value of 0.421038388 and higher at Lecture theatre2 (Week 2) with a value of 0.583577407as shown in Table 4. In contrast with the international standard, all of the selected buildings indicate unpolluted with the pollution loading index reference values.

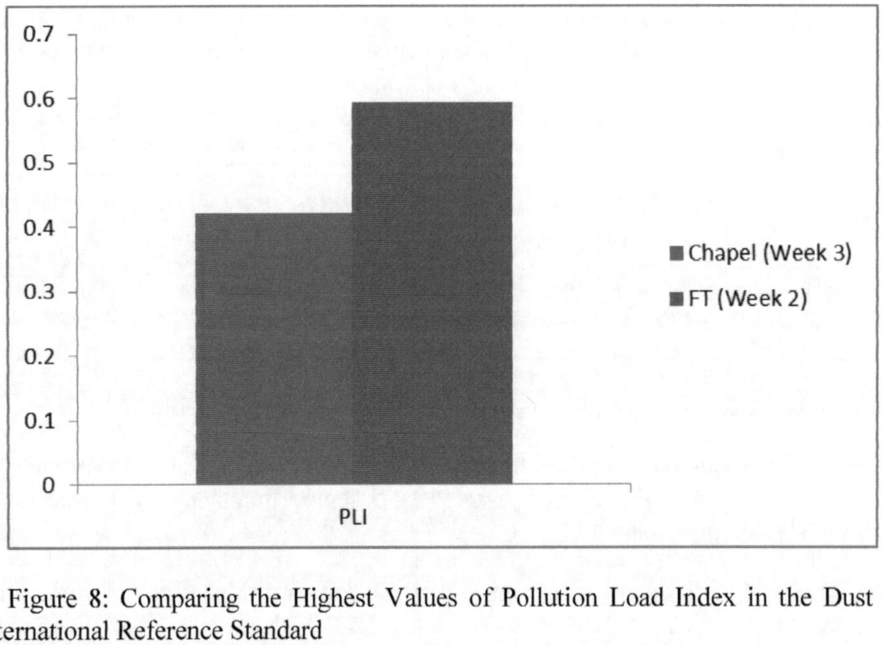

Figure 8: Comparing the Highest Values of Pollution Load Index in the Dust Samples and the International Reference Standard

IOP Conf. Series: Earth and Environmental Science **173** (2018) 012020 doi:10.1088/1755-1315/173/1/012020

4 Conclusion

This present study aimed at monitoring the impact of distribution of heavy metal contamination in surface dust samples in selected buildings to understand the exposure and health assessments due to fine particle size dust to dwellers. The result revealed that the geoaccumulation index and pollution loading index for all the areas under study do not exceed the pollution limit level suggested by United State Environmental Agency (USEPA).This present study will serve as the pollution control/management to promote the health of the people living in both urban and highly industrialized areas. There is real urgency for Nigerian government to look into creating a separate agency solely for monitoring dust pollution. This study also suggests the need for awareness to dangers of dust to be brought to the attention of people. Further studies can be carried out using more enhanced methods and instruments to detail the contaminant levels as well as risks on inhabitants.

Acknowledgement

The authors would like to thank the management of Covenant University for supporting this project as well as the student that actualized this work.

References

[1] Adekola, F.A. and Dosumu, O.O. (2001). Heavy metal determination in household dust from Ilorin City, Nigeria. J. Nigeria Society for Experimental Biology, 3 (1), 217 – 221

[2] Al-Rahji, M.A. and Seaward M.R.D. (1996). Metal level in indoor and outdoor dust in Riyadh, Saudi Arabia. Environmental International, 22 (3), 315 – 324

[3] Turner, A and Simmonds, L. (2006) Elemental concentration and metal bioaccessibility in UK household dust. Science of the Total Environment, 371, 74–81.

[4] Yaaqub, R.R., Davies, T.D., Jickells, T.D., Miller, J.M. (1991) Trace elements in daily collected aerosols at a site in south-eastern England. Atmospheric Environment, 25, 985 – 996.

[5] Last, J.M. (1998). Public Health and Human Ecology, (2ndEdition) pp153200.McGraw-Hill Medical Publishing Prentice – Hall Int. Edition, Canada

[6] Technical Data Bulletin, 2018. Respiratory Protection for Airborne Exposures to Biohazards pp 4, #174

[7] A. Agazzi, C. Pirola, Fundamentals, methods and future trends of environmental microwave sample preparation. Microchemical Journal Volume 67, Issues 1–3, December 2000, Pages 337-341

[8] C. Blake and B. Bourqui, Determination of Lead and Cadmium in Food Products by Graphite Furnace Atomic Absorption Spectroscopy. Vol. 19(6), November/December 1998

[9] Lu, X.W.; Zhang, X.L.; Li, L.Y.; Chen, H. Assessment of metals pollution and health risk in dust from nursery schools in Xi'an, China. Environ. Res. 2014, 128, 27–34.[CrossRef][PubMed][CrossRef]

[10]Wang,X.;Huang,Z.;Su,M.;Li,S.;Wang,Z.;Zhao,S.;Zhang,Q.Characteristicsofreferenceandbackground values of soils in Hetao area. Rock Miner. Anal. 2007, 26, 287–292

[11] Sajid, I.; Mohammad, W.; Muhammad, T.; Mohammad, A.; Muhammad, M.C. Elemental contamination in urban parks of Rawalpindi/Islamabad-a source identification and pollution level assessment study. Environ. Monit. Assess. 2012, 184, 5497–5510.

[12] Lu, X.W.; Li, L.Y.; Wang, L.J.; Lei, K.; Huang, J.; Zhai, Y. Contamination assessment of mercury andarsenic in roadway dust from Baoji, China. Atmos. Environ. 2009, 43, 2489–2496.

Potential Health Risks of Heavy Metal Contents in Bottled Water from Lagos State and Its Environs, Nigeria

Omeje Maxwell[1], AdewoyinOlusegun .O[1], Joel Emmanuel S.[1], OkolieSociis T.A.[2],AyoweOmorotemu Efemena[1]Akinpelu Akinwumi[1], ArijajeTheophilus E.

[1]Department of Physics, College of Science and Technology, Covenant University, P.M.B 1023, Ota, Ogun State, Nigeria
[2] Department of Mechanical Engineering, College of Engineering, Covenant University, Nigeria

maxwell.omeje@covenantuniversity.edu.ng

Abstract. The concentrations of lead (Pb), Cadmium (Cd), Chromium (Cr), Nickel (Ni), B and Fe in twenty (20) different brands of bottled water samples were investigated to ascertain the risk exposure to consumers using Atomic Absorption Spectrometry (AAS). The concentrations of the heavy metals analyzed varied from bottled water to bottled water samples. The BE bottled water sample was found to contain the least concentration of Pb with a value of 0.0232 mg/l. The risk of chronic daily intake (CDI) was determined based on the United State Environmental Protection Agency (USEPA) model for health risk. It was observed that Fe, Mg and Ca reported in all the bottled water samples. The estimated Chronic Daily Intake (CDI) of different metals from the water samples is found to be in order of magnitude of Pb> Fe>Cd>Ni. The highest CDI found in VA, SO, SO, LAT water samples are higher than the International Reference Dose Level according to WHO and USEPA respectively. This study suggest that some bottled water factories should be sited in zones that are safer from these heavy metals if Reverse Osmosis machine for proper removal of these trace elements from the raw water is not in use for water treatment.

Keywords: Water; AAS; Heavy Metals; Health Risk;

1. Introduction

Water is an indispensible part of the human environment and the source of portable water is an essential priority for the public health. Almost 70 % of the body system constitutes water which plays a vital role in biochemical processing. It will be difficult for human to survive without water.

Heavy metals are metals with high densities or high atomic weight such as Chromium (Cr), Cadmium (Cd)Arsenic (As), Mercury (Hg), Nickel (Ni), and Lead (Pb) are regarded as heavy metalsof health concern. This heavy metals pose health risk if found in water and lead to several health problems to the human populace. These heavy metals are known to be the elements with a special weight of about 4-5 times as much as that of water [1, 2].These are metals with atomic weight of over 40 and do not break down easily once they are in the environment. Coal mining and smelting release arsenic, herbicides, pesticides, fungicides which contaminate both the surface and groundwater system

IOP Conf. Series: Earth and Environmental Science **173** (2018) 012021 doi:10.1088/1755-1315/173/1/012021

in an environment[3,4]. If arsenic that if found in pesticide is washed from the surface to the plant by rainfall, it could be a source of contaminant to the stream, river and groundwater

The major source of arsenic exposure to human to Arsenic-containing water associate with diabetes, hepatitis, cancer, and cardiovascular disease in [5-8].An ecologic standardized mortality ratio analysis showed that exposure to low-to-moderate levels of arsenic in drinking water was associated with increased mortality rates for CVD in both men and women [9]. Hi level of zinc in water may cause nausea, abdominal pain and vomiting [9]. In addition, it could lead to lethargy, anemia, and dizziness [9]. Lead in water according to studies shows those adverse effects on the functioning of the nervous system as well as hypertension, impaired thyroid function and preterm birth [9].

Presence of these heavy metals in bottled water, which is considered to be one of the purest forms of commercially sold water in Nigeria, would be hazardous to the health of the general public if found above the recommended level. This study is aimed at investigating the concentration of heavy metals and the potential chronic daily intake to the public that rely on the bottled water.

1.1Geology of the Areas where the Water are Produced

1.1.1 Lagos.

Lagos State is situated in the south-western piece of the Nigerian Federation. On the North and East, it is limited by Ogun State. In the West it shares boundaries with the Republic of Benin. Behind its southern outskirts lies the Atlantic Ocean. Lagos is described by a wet tropical climate with mean yearly rainfall over 1800mm. There are two main seasons, specifically; the rainy season and dry season, which normally last from April to October and October to March separately. It experiences an average temperature of 270C. The vegetation cover is dominated by swamp forest, wetlands and tropical swamp forest. Water is the most significant topographical feature in Lagos State. Water and wetlands cover over 40% of the total land area that is within the state and an additional 12% is prone to seasonal flooding.

The geology of Lagos State is majorly sedimentary of tertiary sediments and quaternary sediments. Where tertiary sediments are unconsolidated sandstones, grits with mudstone band and sand with layers of clay, and quaternary sediments are recent deltaic sands, mangrove swamps and alluvium near the coast. The state is located on sedimentary rock mainly of sand and alluvium.

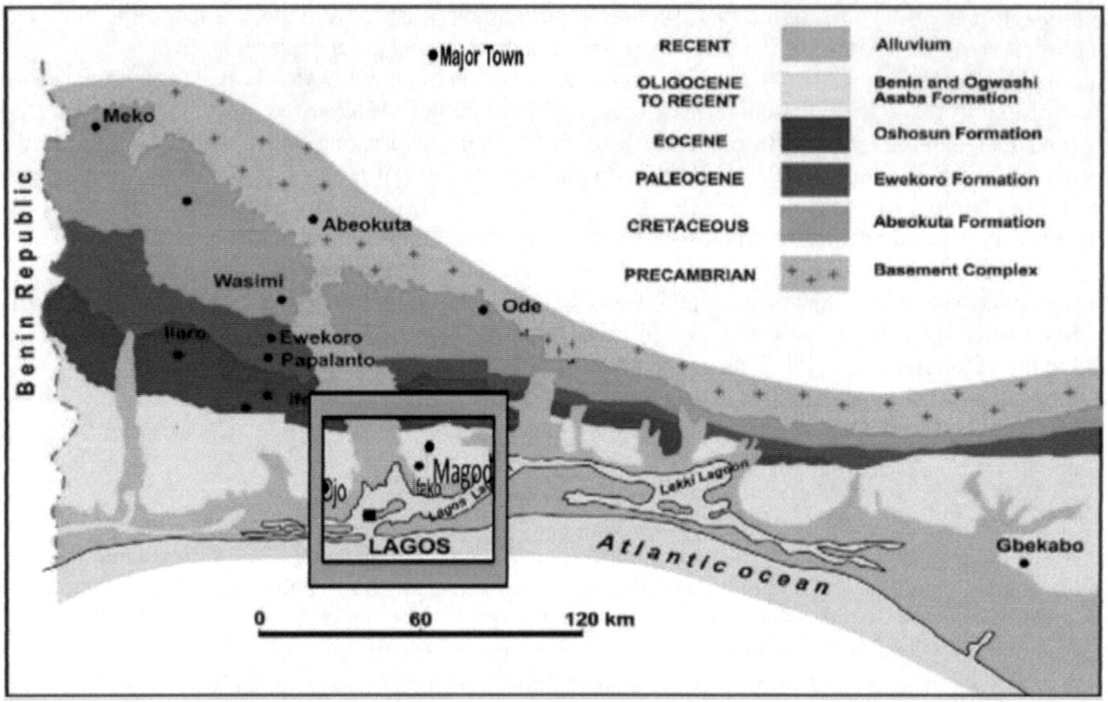

Fig 1. The geologic map of Lagos showing in a Square Box

2. Materials and Methods for Bottled Water Analysis of Heavy Metals

20 different brands of bottled water samples purchased in different shops in and around Lagos were analyzed for this present study. Each was put in 250 ml beaker for testing of pH, total dissolved solid (TDS) and conductivity. These parameters were measured directly before other analysis. The TDS, temperature and conductivity were analyzed using HACH Sension meter (TM5). The pH was analyzed using LAMOTTE Tracer Pochesester meter. Also, heavy metal contents were measured in the water samples using Perkin Elmer A Analyst 600 Graphite Furnace Atomic Absorption Spectrometer (GF-AAS) instrument connected to the intuitive WinLab32 software system comprises of the tools to analyze, report as well as achieving the measured data.

2.1 Quality Control for the analysis of Heavy Metals in the Water Samples

The quality control for the measured water samples was carried out using GF-AAS (Perkin Elmer A Analyst 600) with a standard operation procedures suggested by the manufacturer. All other measurement meters such as TDS, pH and conductivity meters as well as the weigh balance were operated according the instructions of the SOPs to reduce analysis errors. All the equipment used in this study was calibrated before taken measurements. A calibration curve close to 1 was obtained for GF-AAS before the analysis was conducted on the bottled water samples so that the absorption of the atom of each element to be measured will be more accurate.
Where the Chronic Daily Intake of each heavy metal is measured with the formula below

Exposure doses from ingestion of water can be calculated as follows

$$D = (C * IR * EF) / BW$$

Where
D= exposure dose (mg/kg/day)
C= contaminant concentration
IR= intake of contaminated water (L/day)
EF= exposure factor (unitless)

BW= body weight (kg)qw
Default Sinking Water Intake Rates
2 L/day – adult
Note: L/day – liters per day and BW (body weight) for Adults= 70kg

3.Results and Discussion

3.1. Water Quality Parameters

3.1.1. The pH Levels in All the Measured Water Samples

The pH values measured in this present study, which varies between 6.6 and 8.8 is in the range of acceptable value. Comparing this value with Turkish bottled water which ranges from 5.5 to 8.5, it can be observed that ours are within the range. The result was similar with the study by [11] in Lower Saxony, Germany who found that tap water was slightly basic and the mean pH value was 7.83. The standard pH for quality water made for drinking is between 7.2 to 8.5. All the water samplesanalyzed fall within this recommended level of quality water except VA water sample that is increasing to acidic range.

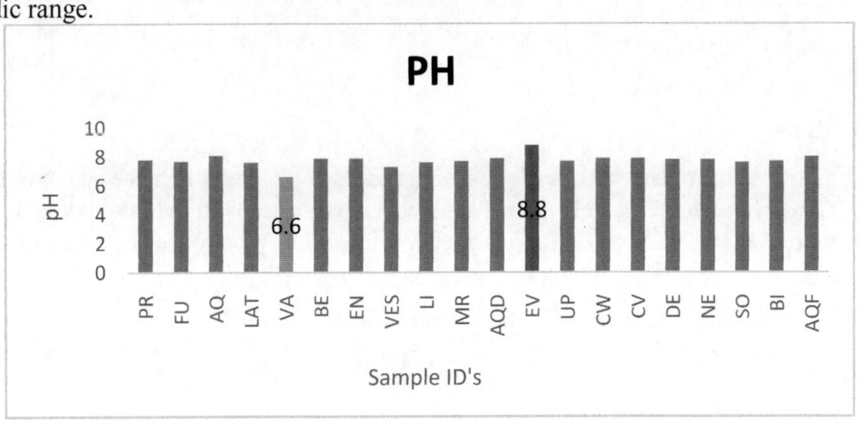

Fig. 2. The Plot of pH in measured water samples against the Sample ID

3.1.2 The Total Dissolved Solid (TDS) in the Measured Water Samples

The total dissolved solids (TDS) are shown in Fig. 2. The TDS which describes the inorganic salts and small amounts of organic matter present in solution in water. The more the TDS the less quality the water is whereas the less the TDS the purer the water is. It also indicates that the suspended solids that may enhance the pollution level in the water samples varies from 3 mgL^{-1} to 288 mgL^{-1}. The highest value of 288 mgL^{-1} noted in BE water sample, whereas the lowest value of 3 mgL^{-1} was noted in MR water sample. According to NAFDAC, the recommended level of 100 mgL^{-1} is the safe level for our local community whereas 500 mgL^{-1} is the permissible level by WHO, 2006. TDS value values of the measured samples reported very low, as such, do not exceed the limit value [12] and far lower than the 1000 mgL-1 by [13]. Till date, no reliable data on possible health implication of TDS associated with drinking water [12]

IOP Conf. Series: Earth and Environmental Science **173** (2018) 012021 doi:10.1088/1755-1315/173/1/012021

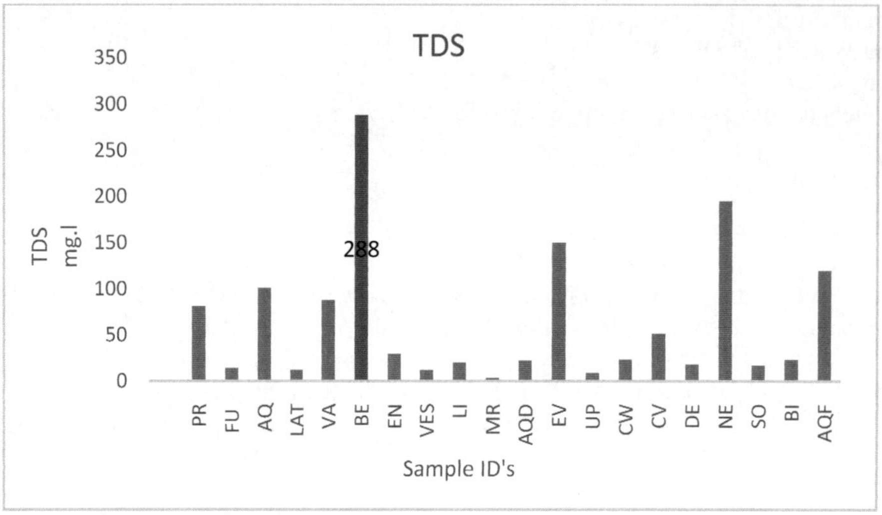

Fig. 3. The TDS against the Sample ID

3.2 Heavy Metal Analysis

Table 1.presents the variations in concentrations of the heavy metals in different brands of the bottled water samples. The details of individual metals were discussed and compare with the international reference levels recommended by United State Environmental Protection Agency are shown below.

Table 1. Concentrations of the Measured Heavy metals in the different Bottled Water

Bottled water samples	Units	Cd	Cr	Ca	Mg	Caco3	Pb	Ar	Ni	Fe
PR	mg/l	0	0	17.1	0	17.1	0	0	0	0.012
FU	mg/l	0.0332	0	34.08	0.12	34.2	0	0	0.0059	0.0213
AQ	mg/l	0	0	16.97	0.13	17.1	0	0	0.0064	0.0182
LAT	mg/l	0	0	16.98	0.17	17.1	0	0	0.0124	0.0226
VA	mg/l	0	0	16.9	0.2	17.1	0.3481	0	0.0069	0.0046
BE	mg/l	0.021	0	17.01	0.09	17.1	0	0	0.0073	0.0065
EN	mg/l	0	0	0	0	0	0.0232	0	0	0
VES	mg/l	0	0	16.95	0.15	17.1	0	0	0	0.0082
LI	mg/l	0	0	34.02	0.18	34.2	0	0	0.0001	0.0158
MR	mg/l	0	0	16.96	0.14	17.1	0.172	0	0	0.0105
AQD	mg/l	0	0	16.92	0.18	17.1	0	0	0.0022	0.0243
EV	mg/l	0	0	16.94	0.16	17.1	0	0	0.0004	0.0641
UP	mg/l	0.0221	0	16.92	0.18	17.1	0	0	0.0003	0.0408

CW	mg/l	0	0	16.96	0.14	17.1	0	0	0	0.0329
CV	mg/l	0	0	51.1	0.2	51.3	0.1412	0	0	0.0323
DE	mg/l	0.022	0	17.02	0.08	17.1	0.0311	0	0	0.046
NE	mg/l	0	0	34.04	0.16	34.2	0	0	0	0.0571
SO	mg/l	0.0721	0	34	0.2	34.2	0	0	0	0.0715
BI	mg/l	0.0031	0	16.92	0.18	17.1	0	0	0	0.0528
AQF	mg/l	0.0021	0	16.95	0.15	17.1	0	0	0	0.0609

3.2.1The Concentration of Lead (Pb) in the Water Samples

The concentrations of lead (Pb) measured in the selected bottled water samples show slight variations as presented in Figure 3. The highest value of 0.3481mgL^{-1} was found in VA and the lowest value of 0.0232 mgL^{-1} was found in EN water sample. It can be observed that PR, FU, AQ, LAT, BE, VES, LI AQD, EV, UP, CW, NE, SO BI and AQF do not have any Pb contaminant. A slight relationship could be observed between pH and Pb in Figure 3 in VA water sample. It may be that the increase in the water acidity increases the concentration of Pb in water. The concentration of Pb reported in VA with the same sample recoding more acidic value. Comparing the highest value of 0.3481 mgL^{-1} obtained from this preset study with the value reported by [12] for treated sewage water in France, with a value of 0.034 mgL^{-1}, it can be observed that this present study is distinctly higher. Figure4 shows the comparison of this present study with the International Standard.

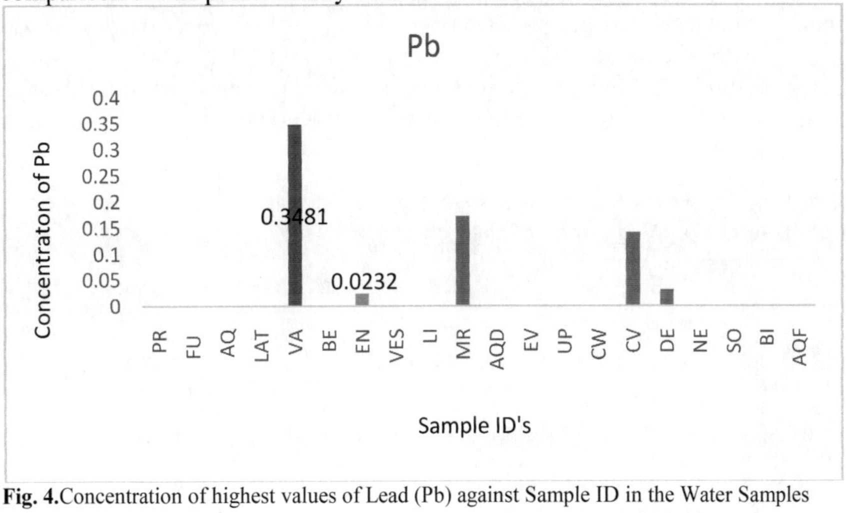

Fig. 4. Concentration of highest values of Lead (Pb) against Sample ID in the Water Samples

IOP Conf. Series: Earth and Environmental Science **173** (2018) 012021 doi:10.1088/1755-1315/173/1/012021

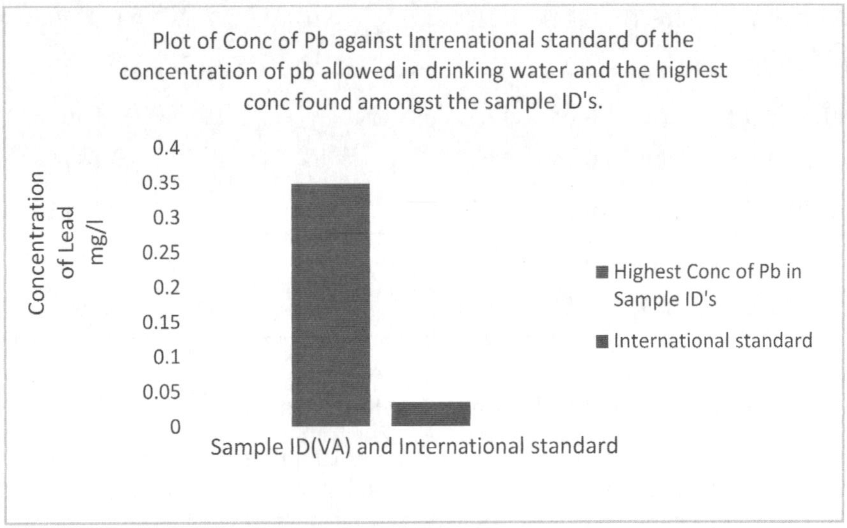

Fig 5: Concentration of highest values of lead against international standards

3.2.2The Concentration of Cadmium (Cd) in the Water Samples

The concentration of Cd in the water samples varies from 0.0021 to 0.072 mgL^{-1} with the highest value of 0.072mgL^{-1} noted in SO water sample as presented in Figure5. The lowest value of 0.0021mgL^{-1} was found in AQF bottled water sample. It can be observed that there may be no relationship between the concentration of Cd and pH/TDS level in the water samples. Comparing the highest value of 0.072 mgL^{-1} obtained from this preset study with the value reported by [12] for treated sewage water in France, with a value of 0.011mgL^{-1}, it can be observed that this present study is slightly higher. Comparing the highest value of this present study with[12] standard of 0.003 and 0.005mgL^{-1}, this present study is distinctly higher. . Figurre6. shows the comparison of this present study with the International Standard.

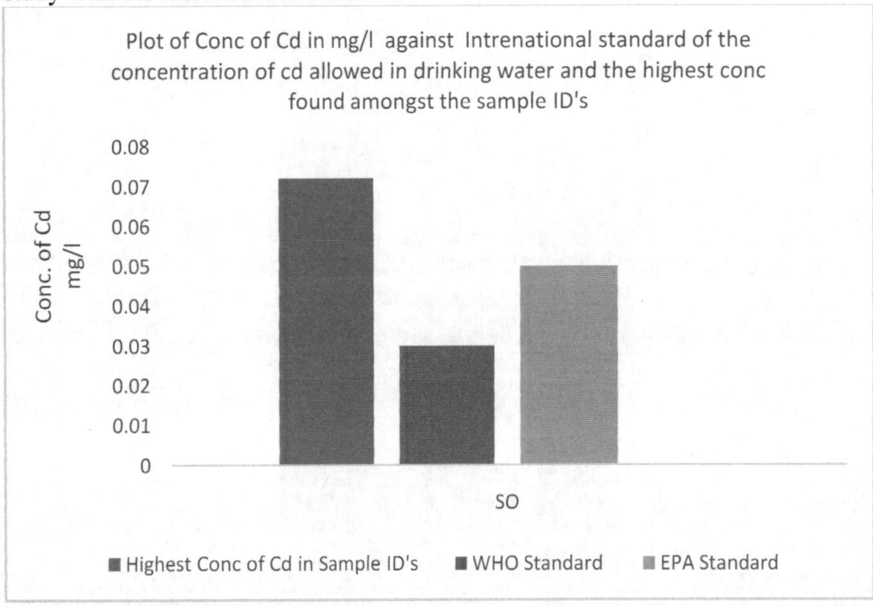

Fig. 6: Concentration of Cadmium (Cd) against highest concentration of cadmium, comparing with the international standards.

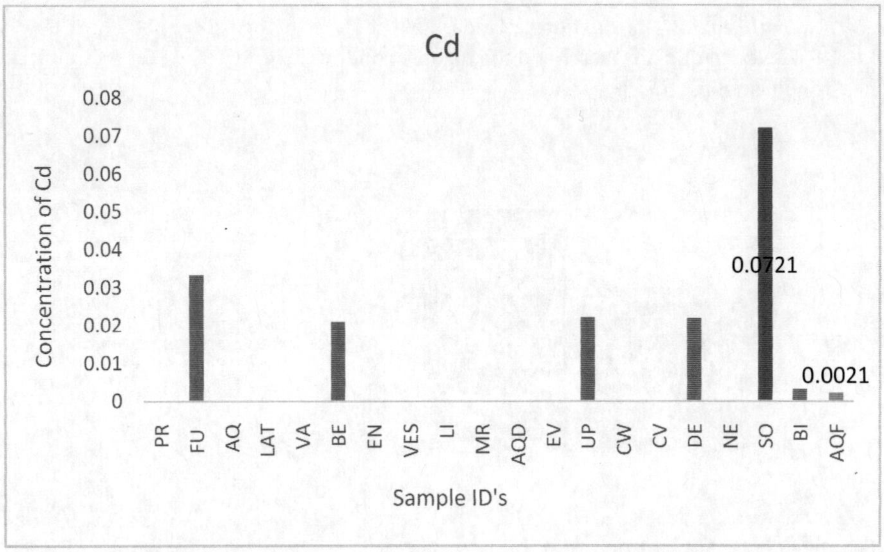

Fig.7: Concentration of highest values of Cadmium (Cd) against Sample ID in the Water

3.2.3The Concentration of Nickel (Ni) in the Water Samples

The concentration value noted for Ni all the samples ranges between 0.0003 mgL^{-1} (LI water sample) to 0.0124mgL^{-1} (LAT bottled water sample) is presented in Figure7. It can be observed that there is no relationship between the concentration of Ni and pH/TDS levels in the water samples. Comparing the highest value of 0.0124 $mgL^{-1}$from this present study with [12] standard of 0.02 mgL^{-1} (Provisional guideline value (this value is used for constituents for which there is some evidence of a potential hazard) and 0.1 mgL^{-1}, this present study is slightly lower. Fig.8 shows the comparison of this present study with the International Standard.

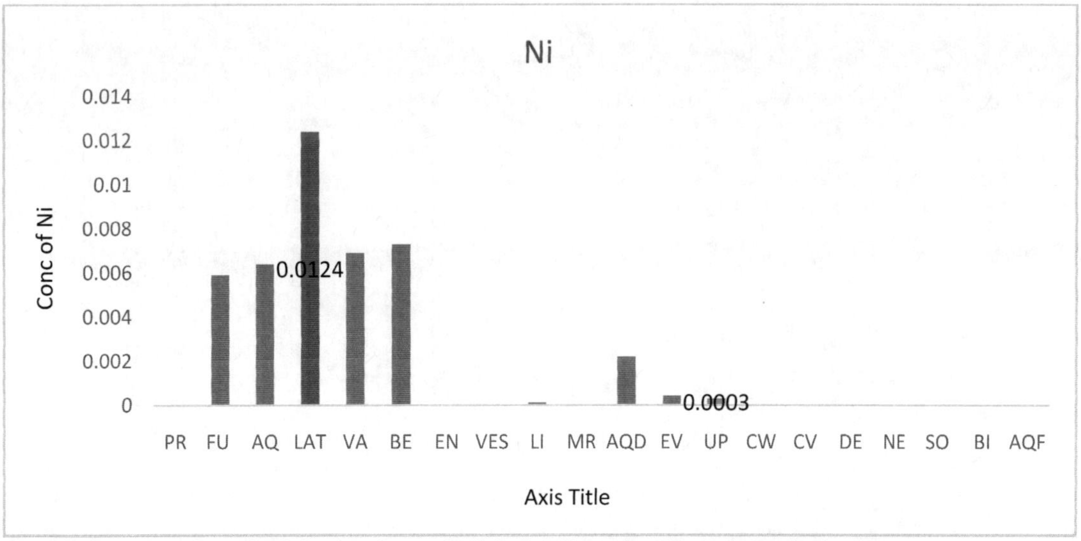

Fig. 8: Concentration of Highest values of Nickel (Ni) against Sample ID in the Water

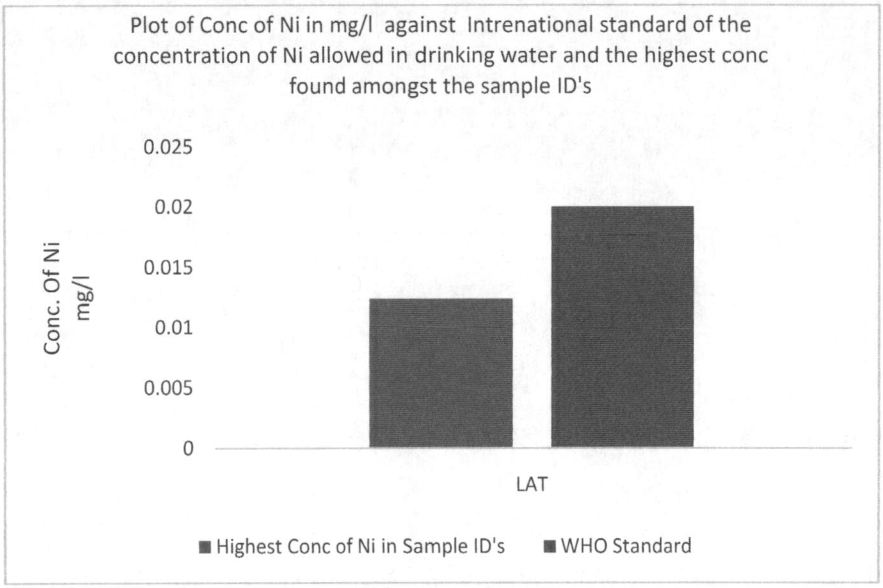

Fig. 9: Concentration of Highest values of Nickel (Ni) against international standard.

3.2.4The Concentration of Iron (Fe) in the Water Samples

The results show that the concentration of Fe in the water samples varies from 0.0046 to 0.0715 mgL^{-1} with the highest value of 0.0715mgL^{-1} found in SO water sample whereas the lowest value of 0.0065mgL^{-1} was found in VA bottled water sample as presented in Fiure.9. It can be observed that there may be no relationship between the concentration of Cd and pH/TDS level in the water samples. Comparing the highest value of this present study with ref[12 & 13] standard of 0.3 mgL^{-1} (Secondary maximum contaminant level (SMCL, which are [not enforceable] guidelines established by the USEPA for use in evaluating esthetic properties in water), this present study is distinctly lower. . Figure 10. shows the comparison of this present study with the International Standard.

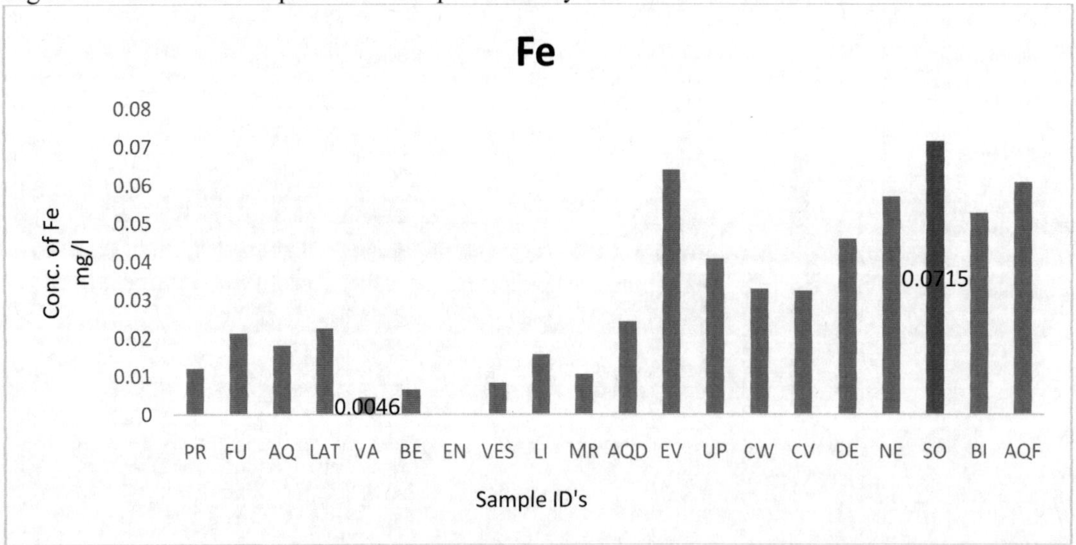

Figure 10: Concentration of Highest values of Iron (Fe) against Sample ID in the Water

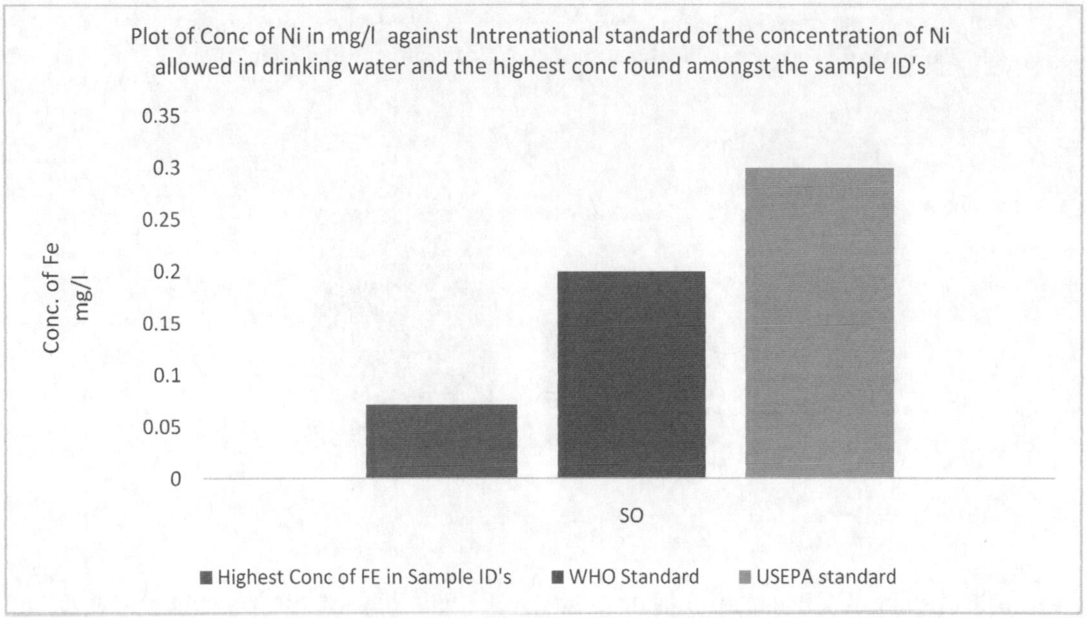

Fig. 11: Concentration of Iron (Fe) against Sample ID with the highest concentration of Fe in the Water Samples and the international standards.

3.2.5.The Concentration of Calcium (Ca) in the Water Samples

The concentration of Ca in the water samples ranges from 16.9 to 51.1 mgL^{-1} with the highestvalue of 51.1mgL^{-1} found in CV water sample whereas the lowest value of 16.9mgL^{-1} was found in VA bottled water sample as shown in Figure 12. It is noted that there no relationship between the concentration of Ca and pH/TDS level in the water samples. Comparing the highest value obtained for this present study with the Turkish Legislation, 1979 for bottled water standard of 100mgL^{-1}, this present study is distinctly lower by a factor of 1.9.Figure 13. shows the comparison of this present study with the International Standard.

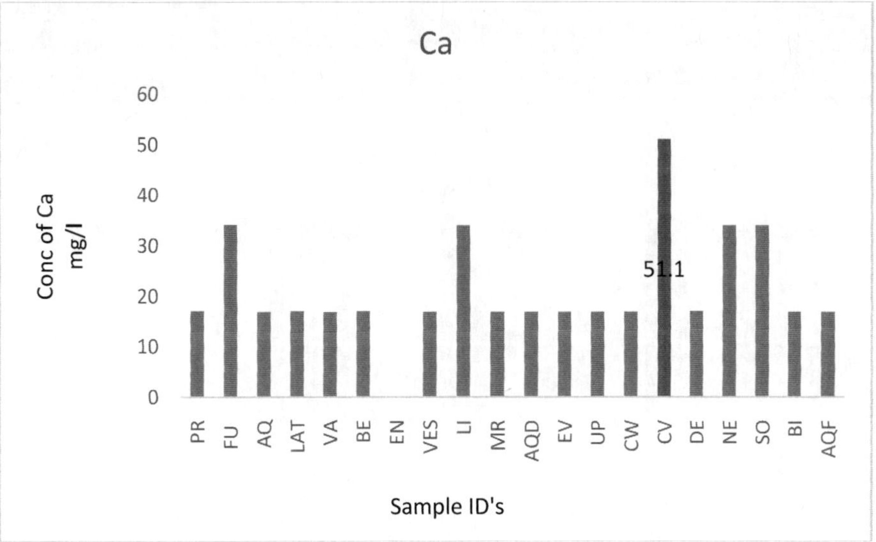

Fig 12: Concentration of highest values of Calcium (Ca) against Sample ID in the Water Samples

IOP Conf. Series: Earth and Environmental Science **173** (2018) 012021 doi:10.1088/1755-1315/173/1/012021

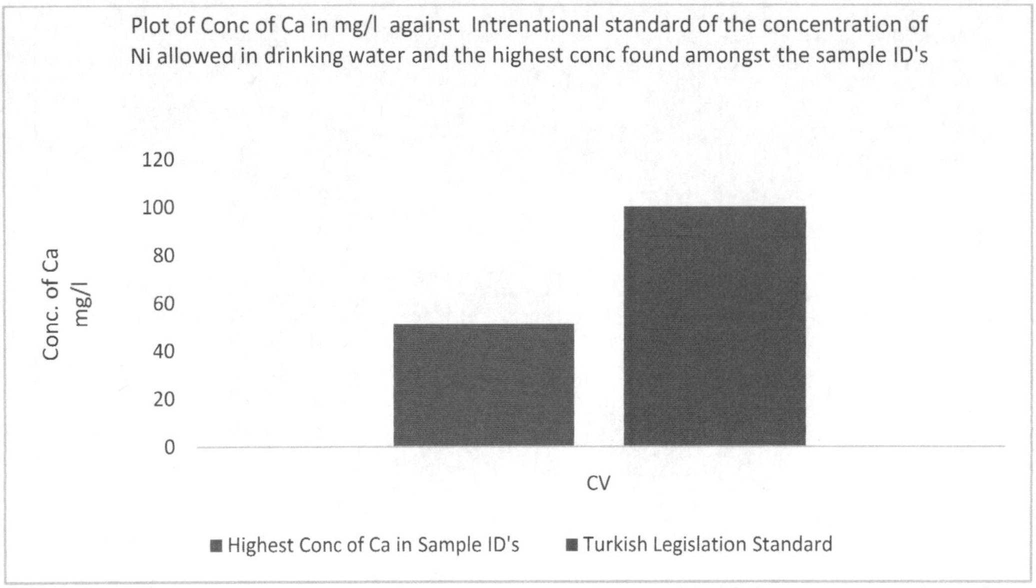

Fig 13: Concentration of Calcium (Ca) against Sample ID with highest concentration of Ca in the Water Samples and the international standard.

3.2.6 The Concentration of Magnesium (Mg) in the Water Samples

The results show that the concentration of Mg in the water samples varies from 0.08 to 0.2 mgL^{-1} with the highest value of $0.2mgL^{-1}$ found in CV and VA and SO water samples respectively, whereas the lowest value of $0.08mgL^{-1}$ was found in DE bottled water sample as presented in Figure14. It can be observed that there may be no relationship between the concentration of Cd and pH/TDS level in the water samples. Comparing the highest value obtained for this present study with the Turkish Legislation, 1979 for bottled water standard of $50mgL^{-1}$, this present study is far lower. Figure15.shows the comparison of this present study with the International Standard.

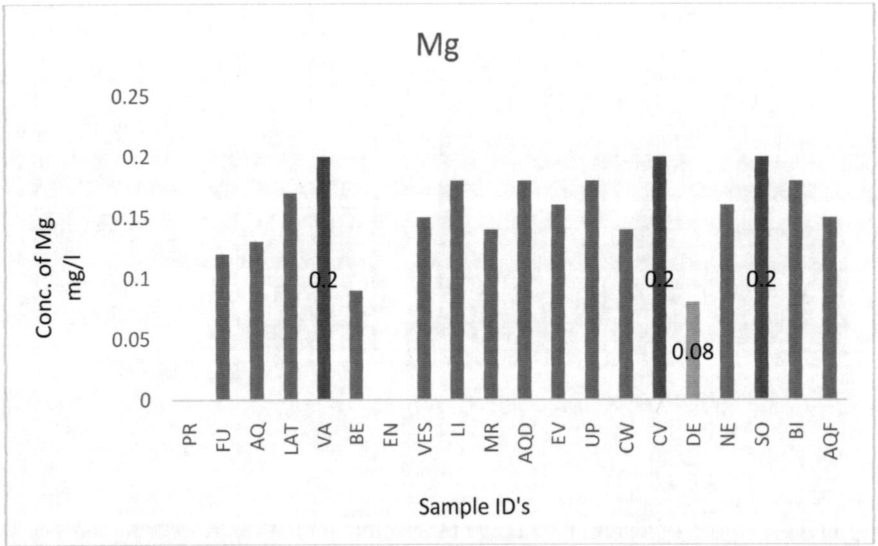

Fig. 14: The Concentration of highest Magnesium (Mg) against Sample ID in the Water Samples

IOP Conf. Series: Earth and Environmental Science 173 (2018) 012021 doi:10.1088/1755-1315/173/1/012021

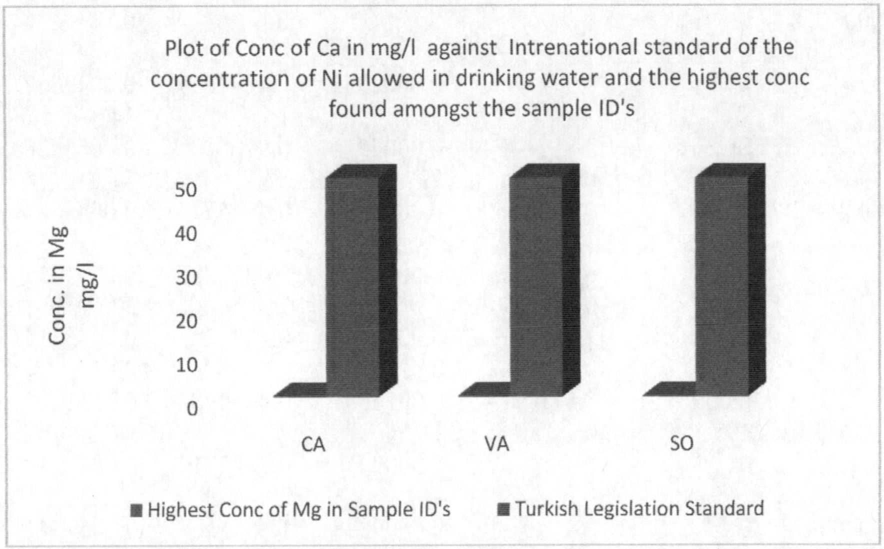

Fig. 15: The Concentration of Magnesium (Mg) against Sample ID's with the highest concentration of Mg in the Water Samples and international standard.

3.3 RISK ANALYSIS OF HEAVY METALS IN THE BOTTLED WATER SAMPLES

3.3.1 Chronic Daily Intake of Heavy Metals in the Selected Bottled Water Sample

Table 2: Showing the CDI results of cadmium in the different bottle water samples.

Dose Intake	**Body weight**	**Pb (mg/kg/d)**	**Cadmium (mg/kg/d)**	**NI (mg/kg/d)**	**Fe (mg/kg/d)**	**Ca (mg/kg/d)**	**Cd (mg/kg/d)**
DI (L)	**BW**	**CDI (Pb)**	**CDI ()**	**CDI (NI)**	**CDI (Fe)**	**CDI (Ca)**	**CDI (cd)**
2	70	0	0	0	0.000342857	0.488571429	0
2	70	0	0.000948571	0.000168571	0.000608571	0.973714286	0.003428571
2	70	0	0	0.000182857	0.00052	0.484857143	0.003714286
2	70	0	0	0.000354286	0.000645714	0.485142857	0.004857143
2	70	0.009945714	0	0.000197143	0.000131429	0.482857143	0.005714286
2	70	0	0.0006	0.000208571	0.000185714	0.486	0.002571429

2	70	0.0006628 57	0	0	0	0	0
2	70	0	0	0	0.0002342 86	0.4842857 14	0.0042857 14
2	70	0	0	2.85714E-06	0.0004514 29	0.972	0.0051428 57
2	70	0.0049142 86	0	0	0.0003	0.4845714 29	0.004
2	70	0	0	6.28571E-05	0.0006942 86	0.4834285 71	0.0051428 57
2	70	0	0	1.14286E-05	0.0018314 29	0.484	0.0045714 29
2	70	0	0.0006314 29	8.57143E-06	0.0011657 14	0.4834285 71	0.0051428 57
2	70	0	0	0	0.00094	0.4845714 29	0.004
2	70	0.0040342 86	0	0	0.0009228 57	1.46	0.0057142 86
2	70	0.0008885 71	0.0006285 71	0	0.0013142 86	0.4862857 14	0.0022857 14
2	70	0	0	0	0.0016314 29	0.9725714 29	0.0045714 29
2	70	0	0.00206	0	0.0020428 57	0.9714285 71	0.0057142 86
2	70	0	8.85714E-05	0	0.0015085 71	0.4834285 71	0.0051428 57
2	70	0	0.00006	0	0.00174	0.4842857 14	0.0042857 14

The chronic daily intake (CDI) of Cd between the ages of 1- 12 and using the risk model suggested by [14], the Equation 1 used for this risk estimation is shown section 2.2. For the CDI estimated for adult in this present study, it varies from .000631 to 0.00206 mg/l. The highest value of 0.00206 mg/l reported in SO bottled water sample whereas the lowest value of 0.000631 noted in UP bottled water sample. It was observed that PR, AQ, LAT, VA, EN, VES, LI, MR, AQD, EV, CW, CV and NE Bottled water samples are free from Cd contamination.

The CDI due to lead (Pb) accumulation from different bottled water samples were calculated using USEPA risk model presented in Equation 1. The CDI varies from water sample to water sample with the highest value of 0.00995 mg/l recorded in VA bottled water sample and lower value of 0.000663 mg/l was found in EN bottled water sample.

The Nickel (Ni) risk exposure from the angle of CDI was calculated using the USEPA risk model published 1998 as presented in Equation 1. The Ni calculated varies from 0.0000857 to 0.000354 mg/l with the highest value recorded in LAT bottled water sample and the lowest value of 0.0000857 mg/l reported in UP bottled water sample.

The CDI calculated according to [14] for (Fe) indicates that the highest exposure from Iron (Fe) intake was higher in SO bottled water sample with a value of 0.00204 mg/l. The lowest value of 0.000186 mg/l was noted in BE bottled water sample.

The calcium (Ca) risk exposure from the angle of CDI was calculated using the USEPA risk model published 1998 as presented in Equation 2.1. The Ca calculated varied from 0.4834 to 0.9737 mg/l with the highest value recorded in FU bottled water sample and the lowest value of 0.4834 mg/l reported in BI, UP, AQD bottled water sample.

The Magnesium (Mg) risk exposure from the angle of CDI was calculated using [14]as presented in Equation 1. The Mg calculated varied from 0.002571 to 0.005714 mg/l with the highest values recorded in SO, CV, VA bottled water sample and the lowest value of 0.002571 mg/l reported in BE, bottled water sample. All the samples are contaminated with Mg except PR and EN bottled water samples.

4.0 Conclusion

From the level of concentrations and the potential health risks of Pb>Fe>Cd>Ni from the bottled water samples measured has the mean values of 0.001022mg/kg/day for adult while Iron(Fe) has mean values of 0.0086051mg/kg/day for adults and Cadmium(Cd) has mean values of 0.0002508 mg/kg/day for adult while NI has mean values of 5.9187E-05mg/kg/day for adults. This shows that the water samples from VA, SO, SO, LAT are considerably contaminated with these heavy metals. However, the pollution of these metals is in order ofPb>Fe>Cd>Ni respectively. Thus, these bottled water samples from VA, SO, SO and LAT were polluted significantly. The health risk due to chronic daily intake were found to be comparatively light in some bottled water samples such as EN, while some have shown to pose risks due to moderate level of heavy metals found in them. Within the scope of this study, a deep investigation of heavy metals as well as the health risks on the public is suggested as well as the bio-monitoring the level in blood and urine as a follow-up work.

Acknowledgment

The authors thank Covenant University for their financial support.

References

[1] Duruibe, J., Ogwuegbu, M., andEgwurugwu, J. "Heavy metal pollution and human Biotoxiceffects,"(2007).*International Journal of Physical Sciences*, 60 (2): 112-118.
[2] Raikwar, M. K., Kumar, P., Singh, M., and Singh, A. T"oxic effect of heavy metals in livestock health. *Veterinary world*,"(2008).**1**: 28-30.
[3] Dogaru, D., Zobrist, J., Balteanu, D., Popescu, C., Sima, M., Amini, M., and Yang, H. "Community perception of water quality in a mining-affected area: A case study for the Certej catchment in the Apuseni Mountains in Romania.*Environmental management*,"(2009).**43**1131-1145.
[4]ATSDR.Toxicological profile for arsenic. 2007. Available online: http://www.atsdr.cdc.gov/toxprofiles/tp.asp?id=22&tid=3 (
[5] Hopenhayn, C. "Arsenic in drinking water: impact on human health. Elements"**2006**, 2, 103–107.
[6] Smith, A. H., Hopemhayn-Rich, C., Bates, M. N., *et al* "Cancers risks from arsenic in drinking water. *Environmental Health Perspectives*,"**97**: 259–267.
[7] Ferreccio, C., Gonzalez, C., Solari, J., *et al*. "Lung cancer and arsenic concentrationsin drinking water in Chile. *Epidemiology*,"(2000). **11**: 673–679.
[8] Rahman, M. M., Mordal, D. *et al*. "Status of groundwater arsenic contamination in all 17 blocks of Nadia district in the state of West Bengal, India," a 23-year study report. *Journal of Hydrology*, (2014) **518**: 363-372.
[9] Caussy, D. (2003). Lessons from case studies of metals:investigating exposure bioavailability, and risk .*Ecotoxicology and Enviromental Safety* , **47** (2): 45-51.
[10] WHO (2011): Guide for drinking-water quality, fourth edition. World Health Organization, Geneva.
[11] World Health Organization, (1996).Guidelines for drinking-water quality.vol 2, health criteria and other supporting information. World Health Organization.
[12] WHO, (2006).Guidelines for drinking-water quality: First addendum to volume 1, Recommendations.World Health Organization.
[13] USEPA, (2012). Integrated Exposure Uptake Bio-Kinetic model for Lead in Children (IEUBK),

US Environmental Protection Agency.
[14] USEPA Environmental Protection Agency (EPA) (1999) Cancer risk coefficients for nvironmental exposure to radionuclides. United State Environmental ProtectionAgency. *Federal Guidance Report* No -13(EPA. **402** pp-99-101).

Tree Species for bio-monitoring and Green Belt Design: A Case Study of Ota Industrial Estate, Nigeria

Winifred U Anake[1*], Jacinta E Eimanehi[1], Hotom S Riman[2] and Conrad A Omonhinmin[3]

[1]Analytical and Environmental Chemistry Unit, Department of Chemistry, Covenant University, Km 10, Idiroko Road, Ota, Ogun State, Nigeria.
[2]Environmental Chemistry and Pollution Control Unit, Department of Chemistry, University of Ibadan, Ibadan, Oyo State, Nigeria.
[3]Applied Biology & Biotechnology Unit, Department of Biological Science, Covenant University, Km 10, Idiroko Road, Ota, Ogun State, Nigeria.

E-mail: winifred.anake@covenantuniversity.edu.ng

Abstract. Bio-chemical and physiological parameters of plants have been employed in the screening of suitable bio - monitors via the estimation of air pollution tolerance index (APTI) and anticipated performance index (API). In the present study, five plant species, *Elaesis guineensis, Mangifera indica, Terminate catappa, Musa spp and Araucaria heterophylla* within high concentrated industrial areas of Ota industrial estate were evaluated based on these two indices. APTI for all plant species ranged from 3.43 to 10.1 signifying their bio-monitoring status. Out of five species, *Terminate catappa* was identified as the most sensitive. Following API classification, *Mangifera indica* was graded as a good performer while *Elaesis guineensis* and *Terminate catappa* were grouped as moderate performers for green belt development. Further evaluation of screened trees for high grade tolerant species and carbon sequestration potential is recommended.

1. Introduction

Ambient air pollutants such as particulate matter and gaseous mixture released into the atmosphere from industrial chimneys, stacks, open burning of solid waste and mobile sources like cars have dramatically increased in Ota industrial estate, Nigeria [1] – [3]. The pollutants concentrations depend on the characteristics of the source of pollution such as influence area, meteorological and topographical conditions [4].

Different plant species have shown varied potentials in reducing air pollution levels by providing sizeable leaf area for air pollutant impression, absorption and accumulation [5], [6]. The impact of air pollution on the plants can be evaluated from a combination of Air Pollution Tolerance Index (APTI) and Anticipated Performance Index (API). APTI is a tool used for the identification and classification of plants into tolerant and sensitive group based on variation in the biochemical parameters namely relative water content, pH of leaf extract, total chlorophyll content and ascorbic acid content. It has been used by several researchers for the selection of air pollution tolerant species in industrial areas [7], [8]. Based on the response of plant towards a particular stress, they can be classified as sensitive or tolerant [9].

The anticipated performance index value gives a good and sensible reason to classify different plant species for the development of green belt, reforestation and afforestation. It takes into consideration the air pollution tolerance index as well as the biological and socio-economical parameters of plant. API is very useful in the selection of plants that can perform dual functions of improving air quality and providing appealing and recreational value [10], [11].

Hence, the objectives of the present study were to investigate the potentials of trees within the high concentrated industrial areas of Ota industrial estate, Ogun state, Nigeria for (a) monitoring environmental pollution, and (b) green belt design.

2. Materials and Methods

2.1. Sample collection
Plant species namely *Elaesis guineensis, Mangifera indica, Terminate catappa, Musa spp* and *Araucaria heterophylla* were collected in triplicates between 7 and 11am for a period of 3 months (January to March, 2018) around Federated mill, gas plant, horticulture garden and Alluminium tower in Ota industrial estate, Ogun State, Nigeria. The collected samples were wrapped in a foil paper, preserved in ice chest box and transported immediately to the laboratory for identification. Afterwards, samples were washed with distilled water, air dried, ground and refrigerated.

2.3. Determination of Leaf Bio-chemical Parameters
In order to determine the bio-chemical parameters of plant species, four (4) parameters were evaluated, they are; pH of leaf extract, total chlorophyll content, ascorbic acid content and relative water content of the leaf. For pH of leaf extract, 4 g of fresh leaves was extracted in 40 mL distilled water, filtered and measured using a glass electrode pH [12]. The relative water content was determined according to [12], [13]. Analysis of total chlorophyll content was done by extracting 1 g of ground fresh leaves with 10 mL acetone for 15 minutes using a centrifuge at 2500 rmp for 3 minutes and the absorbance was taken at 643nm, 645nm and 663nm with ultraviolet spectrophotometer (Genesys). Further calculations were done according to Arnon equation [12], [13]. For ascorbic acid determination, 1g leaves was extracted with 4 mL oxalic acid-EDTA solution, 1 mL orthophosphoric acid, 2 mL of 5% sulphuric acid, 4 mL of ammonium molybdate and 3 mL of water. After 15 minutes, the absorbance of the filtered extract was taken at 730 nm with a spectrophotometer [13], [14].

2.4. *Determination of Air Pollution Tolerance Index (APTI)*
The method of calculating the APTI was proposed by [15]

$$APTI = A(T + P) + R/10 \tag{1}$$

where: A=Ascorbic acid content mg/g; T = Total chlorophyll content mg/g; P = pH of leaf extract and R= Relative water content of leaf (%).

2.5. *Determination of Anticipated Performance Index*
The APTI value combined with some biological and socio economic characters such as plant habit, canopy structure, type of plant and economic value as shown in Table 1 was used in evaluating the API of

various plants. Table 2 indicates the various categories of plant species based on their API score [9], [16]. Anticipated performance index (API) was calculated as follows:

$$API = \frac{No.of\ "+"\ obtained}{Total\ No.\ of\ "+"} \times 100 \quad (2)$$

Table 1. Grade distribution of plant species with respect to APTI, biological parameters and socioeconomic importance.

Character	Character State	Pattern of assessment	Grading allotted
Tolerance	APTI	9.0–12.0	+
		12.1–15.0	++
		15.1–18.0	+++
		18.1–21.0	++++
		21.1–24.0	+++++
Biological	Plant habit	Small	-
		Medium	+
		Large	++
	Canopy structure	Sparse/irregular/globular	-
		Spreading crown/open/semi-dense	+
		Spreading dense	++
	Type of plant	Deciduous	-
		Evergreen	+
Laminar structure	Size	Small	-
		Medium	+
		Large	++
	Texture	Smooth	-
		Coriaceous	+
	Hardness	Delineate	-
		Hardy	+
Socio economic	Economic value	< 3 uses	-
		3 or more uses	+
		5 or more uses	++

Table 2. Anticipated performance index (API) of plant species.

Grade	Scores (%)	Assessment category
0	< 30	Not recommended
1	31-40	Very poor
2	41-50	Poor
3	51-60	Moderate
4	61-70	Good
5	71-80	Very good
6	81-90	Excellent
7	91-100	Best

3. Results and Discussion

3.1 *Bio-chemical Analysis of Plants*

The results of the bio-chemical parameters for various plants are presented in Table 3. The RWC ranged from 91.5 to 25.1 as observed in *Musa spp* and *Terminate catappa* respectively. High RWC in plants regulate the physiological performance of plants that are exposed to high concentration of atmospheric pollutants and favours tolerance to pollution [14], [17]. Maximum and minimum TC content in this study was observed in *Musa spp* (1.13 mg/g) from horticulture farm and Elaesis guineensis (0.356 mg/g), located near Federated mill surrounded by several industries. A reduction in the TCC content in plants leaves is a function of blockage of stomatal opening/ closing by pollutants, drought, heavy metals accumulation in the soil etc. [18], [19]. Hence, increased in pollution levels lowers the chlorophyll content [20]. The low and high TTC contents in plant leaves exhibits sensitive and tolerance ability of plants in polluted environments [21]. pH of plant species showed the highest and lowest potential of hydrogen ion concentration in leaf extract of *Araucaria heterophylla* (6.60) and *Musa spp* (5.47) respectively. A lower pH value shows good correlation with sensitivity to air pollution and also reduces photosynthesis process in plants.Plants with pH lower than 7 are more susceptible while those with pH around 7 are more tolerant to pollution. Earlier findings indicate that a reduction of ascorbic acid content in plants is pH dependent [22]–[25]. As such, change in leaf extract pH can influence the stomatal sensitivity due to air pollution [26] – [28]. In addition, *Araucaria heterophylla* (1.58) and *Elaesis guineensi* (0.578) recorded the highest and lowest ascorbic acid contents respectively. Ascorbic acid is a strong reductant and it activates many bio-chemical and physiological activities of the cell such as cell wall synthesis and cell division. It is generally higher in tolerant plant species due to its stress reducing factor [29], [30].

Table 3. Biochemical parameters and air pollution tolerance indices (APTI) of the studied plant species.

Sample code	Taxon	Common name	A(mg/g)	P	T(mg/g)	R(%)	APTI	Rating
Federated Mill	*Elaesis guineensis*	Oil Palm	0.578	6.00	0.356	89.4	9.31	Sensitive
	Mangiferia indica	Mango	1.40	5.91	0.599	66.6	7.57	Sensitive
By Gas plant	*Terminate catappa*	Almond	1.43	5.73	0.691	25.1	3.43	Sensitive
Honda/Horticulture	*Musa spp*	Plantain	1.36	5.47	1.13	91.5	10.1	Sensitive
Aluminum Tower	*Araucaria heterophylla*	Christmas Tree	1.58	6.60	0.549	28.9	4.02	Sensitive

3.2.*Assessment of APTI and API of tree species for green belts*

Table 3 shows the APTI values for the five investigated tree species within the Ota industrial estate. APTI, provides information on plants based on their sensitivity and tolerance to air pollutants using the bio-chemical parameters and APTI grading. APTI of screened plant species ranged from 3.43 (*Terminate catappa*) to 10.1 (*Musa. spp*). Based on APTI, plant species have been categorized into three gradations. Plants with APTI value 1 to11 are termed sensitive, 12 to 16 are known as intermediate and those with APTI $\geq$ 17 classified as tolerant [10], [14], [31]. Present finding suggested that *Elaesis guineensis, Mangifera indica, Terminate catappa, Musa spp* and *Araucaria heterophylla* can be used for biomonitoring of air pollution since their APTI value occurred in the ranged of sensitive species (APTI=1 to 11).

The API score categories presented in Table 1, were used in allotting scores for different selected plants as shown in Table 4 and 5. Plant species found under API category of excellent, very good, good and moderate performers can be recommended for cultivation as green belts species [16], [18], [32]. In the present study, according to table 4 and 5, tree species *Mangifera indica* (API = 4) was screened as a good performer, *Elaesis guineensis* and *Terminate catappa* (API = 3) as moderate performers while *Musa spp L.* and *Araucaria heterophylla (API =2)* were evaluated as poor performers. The API assessment of the tree species showed that *Mangifera indica, Elaesis guineensis* and *Terminate catappa* could be cultivated as ideal species for green belts, in the study area, to alleviate pollution [17].

Table 4. Evaluation of plant species based on their APTI values, biological parameters and socioeconomic importance.

Taxon	APTI	PH	CS	TT	LT	LS	EI	Hardiness	Grade allotted	
									Total plus (+)	Scoring %
Elaesis guineensis	+	++	+	+	-	+	++	+	9	56.2
Mangifera indica	-	++	++	+	+	++	++	+	11	68.8
Terminate catappa	-	++	++	+	+	-	++	+	9	56.2
Musa spp L.	+	+	+	+	-	++	+	_	8	50
Araucaria heterophylla	+	++	+	+	+	-	+	+	8	50

APTI: Air pollution tolerance index; PH: Plant habit; CS: Canopy structure; TT: Type of tree; LS: Laminar size; LT: Laminar texture; H: Hardiness and EI: Economic importance

Table 5. Anticipated Performance Index (API) value of the studied plant species.

Taxon	Grade allotted		API value	Assessment
	Total plus (+)	Percentage %		
Elaesis guineensis	9	56.2	3	Moderate
Mangifera indica	11	68.8	4	Good
Terminate catappa	9	56.2	3	Moderate
Musa spp L.	8	50	2	Poor
Araucaria heterophylla	8	50	2	Poor

4. Conclusions

Evaluation of anticipated performance index of plants is an important index for screening appropriate tree species. APTI gradation highlighted *Terminate catappa* as the most sensitive among other five species. *Mangifera indica, Elaesis guineensis* and *Terminate catappa* were found suitable for greenbelt design in Ota industrial estate.

5. Acknowledgment

The authors gratefully thank the management of Covenant University for financial support in processing the paper.

6. References

[1] Etim E U2012 Estimation of pollution load from an industrial estate, South-western Nigeria *Afr J Environ Sci and Technol***6**125–9

[2] Anake W U, Ana G R E E, Williams A B, Fred-Ahmadu1 O H and Benson N U 2017 Chemical Speciation and Health Risk Assessment of Fine Particulate Bound Trace Metals Emitted from Ota Industrial Estate, Nigeria *Earth& Environ Science***68**1–7

[3] Anake W U, Benson N U and Ana, GREE 2016 Characterization of airborne fine particulate matter ($PM_{2.5}$) and its air quality implications in Ogun State, Nigeria*Proc. 3rdCovenant University Int. Conf. on African Development Issues*543–7

[4] PahlavaniP, Sheikhain Hand Bigdeli B 2017 Assessment of an air pollution monitoring network to generate urban air pollution maps using Shannon information index, fuzzy overlay and Dmpster-shafer theory, A case study: Tehran, Iran *Atmospheric Environment***167**254–69

[5] Rai PK 2016 Impacts of particulate matter pollution on plants: Implications for environmental biomonitoring *Ecotoxicology and Environmental Safety***129**120–36

[6] Rai PK and Panda LS 2014 Dust capturing potential and air pollution tolerance index (APTI) of some roadside tree vegetation in Aizawl, Mizoram, India: An Indo- Burma hotspot region *Air Qual. Atmos. Health*793–101

[7] Ogunkunle C O, Suleiman L B, Oyedeji S, Awotoye O O, Fatoba P O 2015 Assessing the air pollution tolerance index and anticipated the performance index of some tree species for biomonitoring environmental health *Agroforest Syst***89**447–54

[8] Achakzai K, Khalid S, Adrees M and Bibi A 2017 Air pollution tolerance index of plants around brick kilns in Rawalpindi, Pakistan *J. Environ Manage***190**252–58

[9] Patel D and Kumar J I 2018 An evaluation of air pollution tolerance index and anticipated performance index of some selected plant species *Open Journal of Air Pollution*71–13

[10] Sahu H and Sahu S K 2015 Air Pollution Tolerance Index (APTI), Anticipated Performance Index (API), Carbon Sequestration and Dust Collection Potential of Indian Tree Species *International Journal of Emerging Research in Management &Technology***4***37–40*

[11] Rai, PK 2016b Impacts of particulate matter pollution on plants: Implications for environmental biomonitoring *Ecotoxicology and Environmental Safety***129** 120–136

[12] Arnon D I, 1949 Copper enzymes in isolated chloroplasts: Polyphenol oxidise in Beta `vulgaris *Plant Physiology***2**1–15

[13] Zhang, PQ, Liu, Y, Chen, X, Yang, Z, Zhu, MH and Li, Y 2016, Pollution resistance assessment of existing landscape plants on Beijing streets based on air pollution tolerance index method *Ecotoxicology and Environmental Safety***132**212–23

[14] Bharti, KS, Trivedi, A and Kumar, N 2017, Air pollution tolerance index of plants growing near

an industrial siteUrban Climate. http://dx.doi.org/10.1016/j.uclim.2017.10.007

[15] Singh S K and Rao D N 1983 Evaluation of plants for their Tolerance to Air PollutionProc. *of the symposium on AirPollution Control* 218–24

[16] Pathak V, Tripathi B D, Mishra V K 2011Evaluation of Anticipated Performance Index of some tree species for green belt development to mitigate traffic generated noise. *Urban Forestry & Urban Greening***10**61–6

[17] Tsega, Y C and Devi-Prasad, A G 2014 Variation in air pollution tolerance index and anticipated performance index of roadside plants in Mysore *India. J Exp Biol***35**185–90

[18] Pandey A K, Pandey M, Mishra A, Tiwary S M and Tripathi B D 2015 Air pollution tolerance index and anticipated performance index of some plant species for development of urban forest *Urban Forestry & Urban Greening***14**866–71

[19] Socha, A N, Kandziora-Ciupa M, Trzesicki M and Barczyk, G 2017 Air pollution tolerance index and heavy metal bioaccumulation in selected plant species from urban biotopes *Chemosphere***183**471– 82

[20] BakiyarajR and AyyappanD 2014 Air Pollution tolerance index of some terrestrial plant around an industrial area*International Journal of Modern Research and Reviews***2**1–7

[21] Balasubraman,A,Prasath,CN,Gobalakrishan, K and RadhakrishanS 2018 Air pollution tolerance index and anticipated performance index assessment in trees species of Coimbatore urban city, Tamil nadu, India *International Journal of Environmental and Climate change***8**27–38

[22] Subramani S and Devaanandan S 2015 Application of air pollution tolerance index in assessing the air quality*International journal of pharmacy and pharmaceutical sciences*7216–21

[23] Meerabai G, Venkata R C and Rasheed M 2012 Effect of industrial pollutants on physiology of Cajanus cajan (l.)-Fabaceae. *International Journal Environmental Science***2**1889–94

[24] Shalom N C, Opeyemi C E, Franklyn N I, Solomon O R and Jacob O P 2018 Phylogenetic Relationship and Genetic Variation among Thaumatococcus daniellii and Megaphrynium macrostachyum Ecotypes in Southwest Nigeria *Asian J. Plant Sci.* **17**27–36

[25] Iweala, EEJ, Uhegbu, FO, Adesanoye, OA 2013 Biochemical effects of leaf extracts of Gongronema latifolium and selenium supplementation in alloxan induced diabetic rats*Journal of Pharmacognosy and Phytotherapy***5**91–7

[26] EneteI C and OgbonnaC E 2012 Evaluation of air pollution tolerance index (APTI) of some selected ornamental shrubs in Enugu city, Nigeria *Journal of Environmental Science, Toxicology and Food Technology***1**22–5

[27] Ogboru,R O, Agboje,I, Mangodo,C and OkolieL P 2015 Effects of Air Pollution on Plants Around Okpai Gas Plant Area, Ndokwa- East Local Government Area, Delta State, Nigeria *International Journal of Applied Research and Technology*470 –5

[28] Sharma,M, Panwar,N, Arora,P, Luhach,J and ChaudhryS 2013 Analysis of biological factors for determination of air pollution tolerance index of selected plants in Yamuna Nagar India. *J. Environ. Biol***34**509–14

[29] Tripathi,A K and GautamM 2007 Biochemical parameters of plants as indicators of air pollution *J. Environ. Biol***28**127–32

[30] Rathore, DS, Kain, T, Gothalkar, P 2018 A study of air pollution status by estimation of APTI of certain plant species around Pratapnagar Circle in Udaipur CityIJAEB**11**33 – 8

[31] Padmavathi, P, Cherukuri, J and Reddy, MA 2013 Impact of air pollution on crops in the vicinity of a power plant: a case study *Int J Eng Res Technol***2**3641–51

[32] Sarala, T D and Sabitha, M A 2011 Variation in air pollution tolerance index and anticipated performance near a sugar factory: implication for landscape-plant selection for industrial areas *Journal Resource Biology*7494–502

Construction of an in-situ smart device that measures some basic environmental factors for agricultural monitoring

T.A. Adagunodo, J.J. Ajigbotosho, L.N. Obafemi, M.R. Usikalu, S.A. Akinwumi, W.A. Ayara

Department of Physics, Covenant University, Ota, Ogun state, Nigeria

theophilus.adagunodo@covenantuniversity.edu.ng

Abstract. Most of the rural agricultural and extension officers have no access to the daily variations in meteorological data that affect the growth and yield of crops. This is as a result of little or no access to the data that could be used to predict the best agricultural practices for optimum output in the rural environments. In this work, a portable and user friendly smart Automated Weather Station (AWS) was constructed. The device that is programmed with Arduino UNO is capable of measuring the temperature, relative humidity and moisture content of the soil. The performance evaluation of the device was observed in the laboratory and on-site (Covenant University farm) both on the sunny and rainy days respectively. It was revealed that there is positive linear relationship between the volume of water added to the soil sample and the soil moisture content in the laboratory measurements. The on-site measurements showed that the temperature is inversely proportional to the relative humidity and soil moisture content. This device is efficient in taking the short-term essential parameters that could improve the yield of agricultural products in the rural environments if properly utilized.

Keywords: Agricultural monitoring, Relative humidity, Moisture content, temperature, In-situ smart AWS,

1. Introduction

The impact of solar heat radiation on climate diurnal variations cannot be underrated [1]. This effect causes variations in the moisture content, relative humidity and soil temperature, which are the chief environmental factors that sustain the plant growth. The contributions of soil temperature to the soil properties are not limited to plant growth alone, it also contributes to the seed germination, physical and chemical constituents of soil, solute diffusion, absorption of water and nutrition, as well as a medium of exchange between the atmosphere and the topsoil [2]. Crops respond to soil temperature differently and as such have different impact on plant growth [1].

The plant-soil interactions are capable to influence the soil moisture variations. Soil moisture has many favourable advantages to plant growth. It takes part in the biochemical processes that occur in the plant. Water serves as carrier of nutrients within the plant, it regulates the internal temperature of the plant, and maintains the turgidity of the fruits and leaves of the plants [3]. A high level of soil moisture needs to be maintained if the ecological systems would retain more water in the subsurface [4].

IOP Conf. Series: Earth and Environmental Science **173** (2018) 012023 doi:10.1088/1755-1315/173/1/012023

Humidity can simply be defined as the amount of water in the atmosphere. Relative humidity is one of the factors that can affect the growth of plants. Different species of plants have different humidity ranges in which they can germinate well. Humidity affects plant growth because it affects when plants open or close their stomata. The stomata are responsible for plant's transpiration.

There are sophisticated Automated Weather Station (AWS) that are capable to measure some weather parameters in all seasons. Apart from the fact that this AWS is cumbersome, which is only capable of ex-situ measurements, it is considered too expensive in a developing nation like Nigeria. Few previous works have been recorded in the applications of weather parameters to monitoring of agricultural products [5-8], most of the works from SW Nigeria is based on the impacts of weather parameters on loss of signal in communication environment, which have been reviewed in Refs. [9-13]. However, this work was designed to construct a portable AWS that is capable to measure the three (3) basic weather parameters (temperature, soil moisture and relative humidity), which are vital for sustainability of crops. This device is capable to take remote measurements while the data are retrieved through the aid of cloud technology without staying on the field with the portable device.

2. Materials and Methods

In order to construct a device that monitors soil moisture, humidity and temperature, there are some essential components that are required. These are an Arduino, a soil moisture sensor probe, a humidity sensor, a Liquid Crystal Display (LCD), a temperature sensor and WIFI adapter among others. The moisture sensor probe is a two probe sensor made of pure nickel. The nickel is chosen because of its strength against corrosivity and its fair conductive properties when buried in the subsurface. The temperature and humidity sensor that was used in this study was a DHT11 sensor, which uses a thermistor to measure temperature. The Arduino UNO was used for coding the software section of the device, and then the LCD that displays the values of the three parameters to be measured by the device. A simple circuit diagram for this setup is presented in Fig.1.

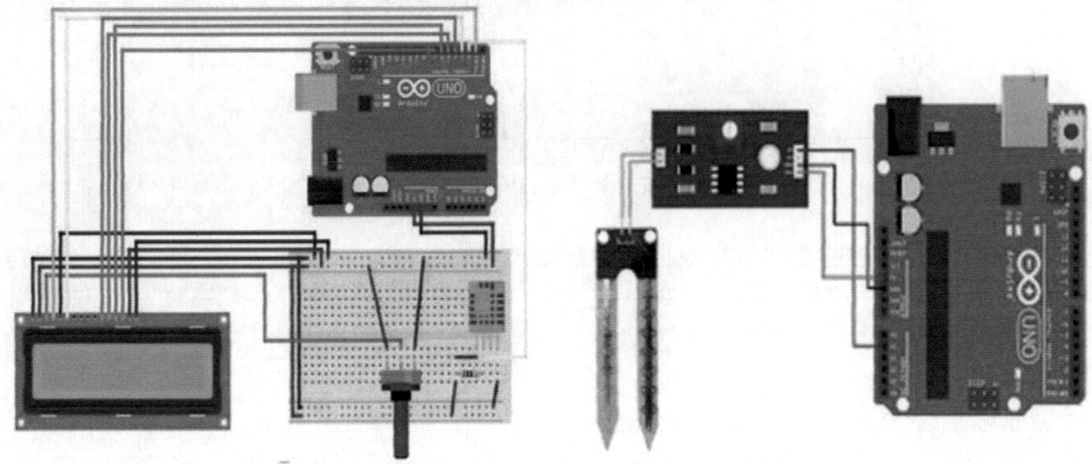

Fig. 1. A simple circuit diagram for the smart device.

The soil moisture sensor has two probes that are used to measure the moisture content of the soil. The two probes measure the soil moisture by allowing electric current to go through the soil, which gives the resistance value. The DHT11 consists of a temperature and humidity sensor, which uses the thermistor. A thermistor is a kind of variable resistor that measures temperature by measuring the change in resistance of the soil. For the measurement of humidity, the component that senses humidity has two electrodes, with a moisture holding substrate in between them. The conductivity of the holding substrate changes as humidity varies. These changes in resistance are processed and measured by an integrated circuit and then it will be read by the processor. The coded microcontroller then processes the information and sends it to the LCD.

3. Results and Discussion

After the construction of the device, it was tested in the laboratory for performance evaluation and later taken to the Covenant University (CU) farm field measurements. The laboratory setup of the device was shown in Fig. 2, while the laboratory measurements of temperature, humidity and soil moisture were presented in Table 1. The measurements in air and with soil samples (dry and wet) were presented. The result of measurements in air and that of dry soil sample showed no value for the moisture content. However, a known volume of water ranging from 2.0 to 10.0 ml was added consecutively to the dry soil sample. It was revealed that there is positive linear relationship between the volume of water added to the soil sample and the soil moisture content. This shows the effectiveness of the sensors.

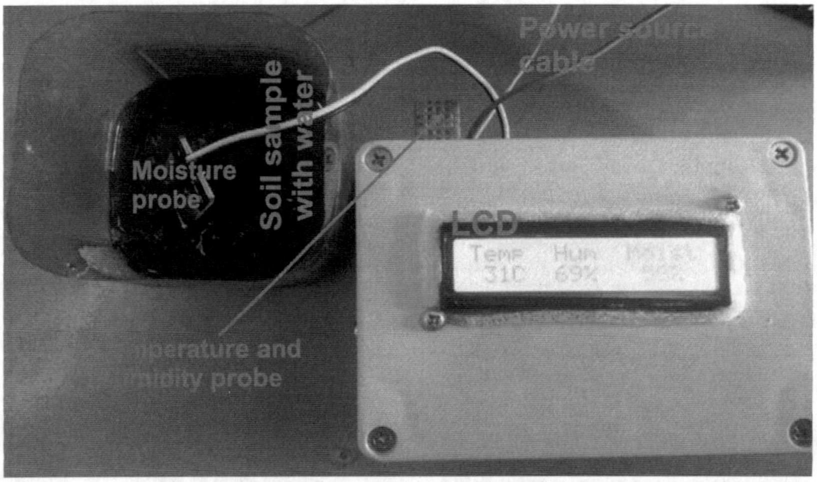

Fig. 2. Laboratory setup of the smart device

Table 1. Laboratory measurement from the smart device

Sample	Temperature (°C)	Humidity (%)	Moisture content (%)
Measurement in air	32.00	72.00	0
Dry soil sample	32.00	73.00	0
Soil sample with water (2.0 ml)	32.00	71.00	5.00
Soil sample with water (4.0 ml)	32.00	71.00	10.00
Soil sample with water (6.0 ml)	32.00	71.00	26.00
Soil sample with water (8.0 ml)	32.00	70.00	56.00
Soil sample with water (10.0 ml)	31.00	70.00	63.00

The field measurements taken at the CU farm recorded the data for a sunny and rainy days respectively. The results showed varied responses of the device to different weather conditions. The spatial variations of each data for the two periods were presented in Figs. 3 to 8 respectively.

Fig. 3 shows the spatial variation of humidity on a rainy day in CU farm. It was revealed that the southeastern part of the map has the lowest value which increases upwardly towards the northwestern and southwestern regions of the study area. There are two points on the map with clustered contour lines. These points depict the highest humidity values. Fig. 4 shows the spatial variation of moisture on a rainy day in CU farm. The south-east region of the map has the lowest soil moisture content, which increases towards the western part of the CU farm. The two points on the map with clustered contour lines depicts the zones with high soil moisture contents. Fig. 4 is similar to the relative humidity map (Fig. 3), because the two parameters are directly proportional to each other. Fig. 5 shows the spatial variation of temperature on a rainy day in CU farm. It was revealed that the western part of the map has the lowest temperature values, which increases downwardly towards the southeastern region of the map. The two points on the map with clustered contour lines showed the minimum values for temperature. The spatial variation on CU farm showed that temperature is inversely proportional to relative humidity and soil moisture.

Fig. 6 shows the spatial variation of humidity on a sunny day in CU farm. It was revealed that the eastern part of the map has the lowest humidity content, which increases towards the western region of the map. Fig. 7 shows the spatial variation of moisture content on a sunny day in CU farm. It was shown that the eastern part of the map has the lowest moisture content, which increases towards the western region of the map. Fig. 8 shows the spatial variation of temperature on a sunny day in CU farm. This map showed that the temperature decreases in NE – SW direction.

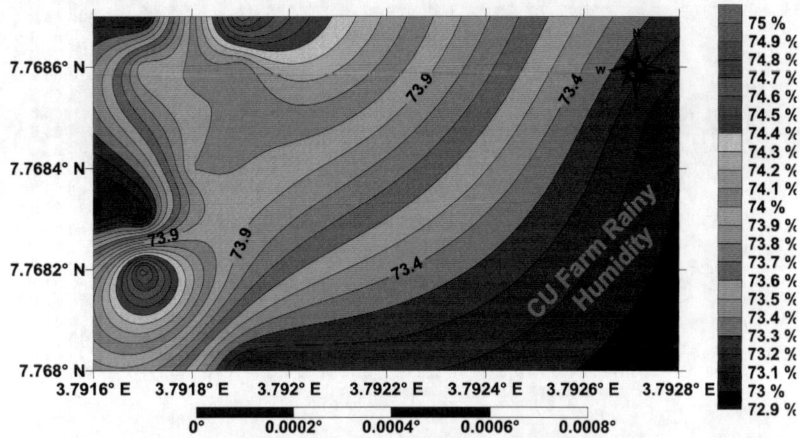

Fig. 3. Variation of humidity in CU farm during the rainy period.

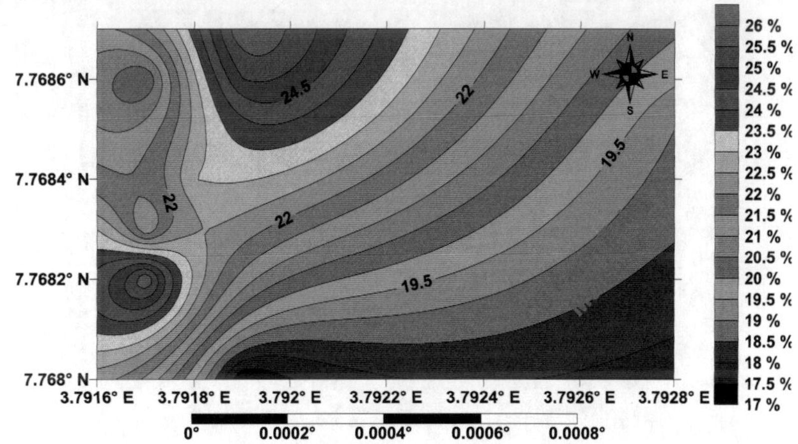

Fig. 4. Variation of soil moisture in CU farm during the rainy period.

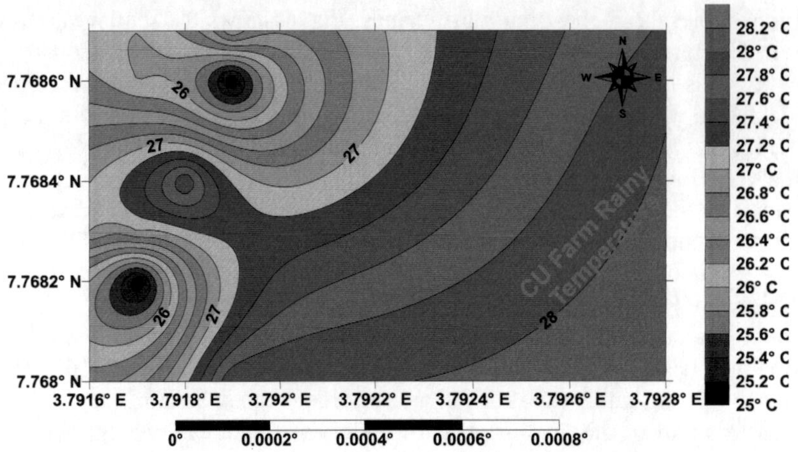

Fig. 5. Variation of temperature in CU farm during the rainy period.

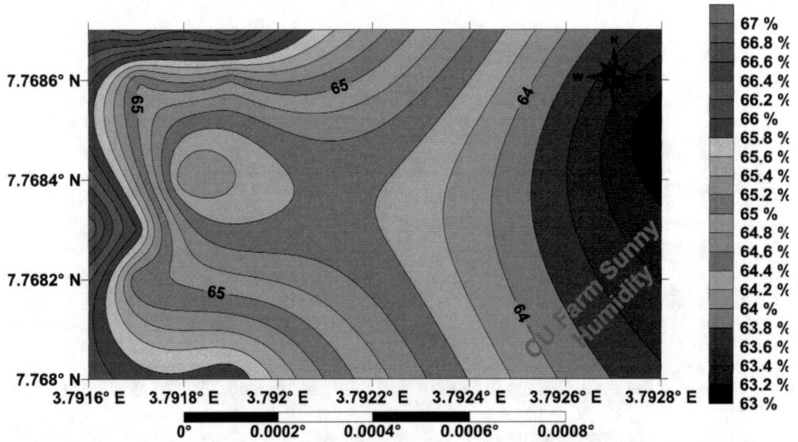

Fig. 6. Variation of humidity in CU farm during the sunny period.

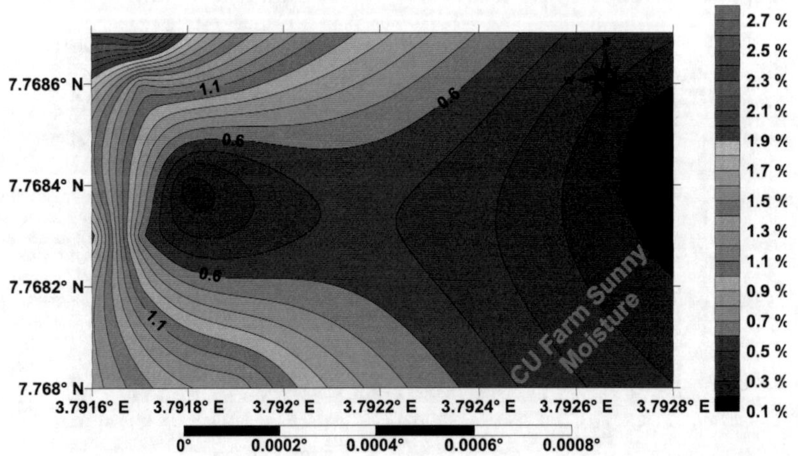

Fig. 7. Variation of soil moisture in CU farm during the sunny period.

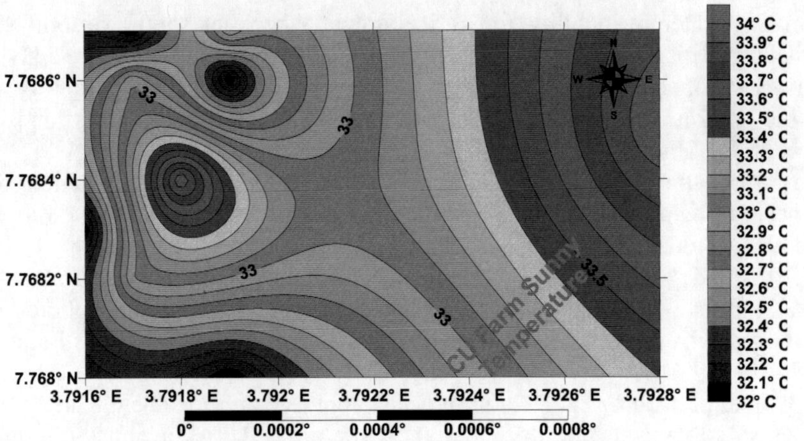

Fig. 8. Variation of temperature in CU farm during the sunny period.

4. Conclusion

In this study a smart device that is capable of measuring relative humidity, soil moisture content and temperature was constructed. The performance evaluation of the device was tested both in the laboratory and on the field. This device is capable to measure some basic parameters that could be used to predict crop growths and yield for sustainable agricultural practices especially in Nigeria. The device is portable, user friendly and quite easy to handle. It can be operated by any farmer. The moisture of the soil can be measured up to the root region of the crop. Remote access to the data through WIFI connection range is possible in this portable device, which would also assist the agricultural rural and extension workers to perform their role effectively in remote villages.

Acknowledgement

We appreciate the management of Covenant University, Nigeria for the publication support received from them.

References

[1] Sriboon W., Tuntiwaranuruk U., Sanoamuang N. (2017). Hourly soil temperature and moisture content variations within a concrete pipe container for planting lime trees in Eastern Thailand. Case studies in Thermal Engineering, **10**: 192–198.

[2] Banimahd S.A., Zand-Parsa S.H. (2013). Simulation of evaporation, coupled liquid water, water vapour and heat transport through the soil medium. Gric. Water Manag., **130**: 168–177.

[3] Ma Y., Qu L., Wang W., Yang X., Lei T. (2016). Measuring soil water content through volume/mass replaconcrete using a constant volume container. Geoderma., **271**: 42–49.

[4] Zheng H. Gao J., Teng Y., Feng C., Tain M. (2015). Temporal variations in soil moisture for three typical vegetation types in inner Mongolia, Northern China. Plos One, **10**(3): e0118964.doi:10.1371/journal.pone.0118964.

[5] Alabi A.A., Akinyemi O.D., Olowofela J.A., Salako F.K., Ajiboye G.A., Olurin O.T. (2017). Temporal variation of ground temperature at depths 2cm to 200cm in an experimental field in Abeokuta, South-western, Nigeria. Arab. J. Geosci., **10**: 424. Doi:10.1007/s12517-017-3201-z.

[6] Ufoegbune G.C., Bello N.J., Dada O.F., Eruola A.O., Makinde A.A. and Amori A.A. (2012).

Estimating water availability for agriculture in Abeokuta south western Nigeria. Global Journal of Science Frontier Research Agriculture and Veterinary Sciences, **12**(9): 13–24.

[7] Adoghe A.U., Popoola S.I., Chukwuedo O.M., Airoboman A.E., Atayero A.A. (2017). Smart weather station for rural agriculture using meteorological sensors and solar energy. Proceedings of the World Congress on Engineering, **1**: 323–326.

[8] Emetere M.E., Omotosho T.V., Kayode O. (2010). On the determination of agricultural prospects using remote sensing and field technique. AIP Conference Proceedings **1705**, 020039. https://doi.org/10.1063/1.4940287.

[9] Afolabi, O.O., Adagunodo, T.A., Ayorinde, T.T., Rabiu, A.B., Akinwumi, S.A., Usikalu, M.R. (2017). Impact of Magnetic Activity on Occurrence of Ionospheric Amplitude Scintillation over Lagos, Nigeria. IEEE RADIO 2017 at Protea Hotel by Marriott Sea Point, Cape Town, South Africa. DOI: 10.23919/RADIO.2017.8242229.http://ieeexplore.ieee.org/document/8242229

[10] Alagbe G.A., Afolabi O.O., Adagunodo T.A., Rabiu A.B., Akinwumi S.A., Omotosho T.V. (2017). Study of Ionospheric Amplitude Scintillation during Geomagnetic Activities of 2012 at Low Latitude Region. Journal of Informatics and Mathematical Sciences, **9**(2): 251 – 256.

[11] Adagunodo, T.A., Akinwumi, S.A., Omotosho, T.V., Akinyemi, M.L. 2017. Estimation of Specific Attenuation of Radio Signal in Southwest Nigeria. IEEE RADIO 2017 at Protea Hotel by Marriott Sea Point, Cape Town, South Africa. DOI:10.23919/RADIO.2017.8242225.http://ieeexplore.ieee.org/document/8242225.

[12] Akinwumi, S.A., Omotosho, T.V., Usikalu, M.R., Ometan, O.O., Adewusi, M.O., Adagunodo, T.A. (2018). Study of Oxygen and Water Vapour Attenuation in West Africa. Interciencia Journal, **43**(1): 180 – 191.

[13] Omotosho T.V., Mandeep J.S., Abdullah M. (2013). Cloud attenuation studies of the six major climatic zones of Africa for Ka and V satellite system design. Annals of Geophysics, **56**(5): A0568. doi: 10.4401/ag-6342.

Integrated geophysical methods for subsurface characterisation and health hazard assessment in parts of southwestern basement Nigeria

Omowumi Ademila

Department of Earth Sciences, Adekunle Ajasin University, Akungba-Akokoa, Nigeria

omowumi.ademila@aaua.edu.ng

Abstract. Aeromagnetic and radiometric data of Ikole Sheet 245 were interpreted for structural mapping, rock characterization and radionuclides hazard indices assessment. The data acquired were processed, filtered and enhanced to improve signal to noise ratio of rocks magnetisation. The Total Magnetic Intensity (TMI) map revealed high and low intensity values between 65 and 180 nT and -150 to -65 nT respectively. The residual map revealed clearer the undistorted signatures of the area geology and the interpretive boundaries of the underlying geology. Reduction to Magnetic Equator (RTE) map showed NE-SW regional lineament (R-R') that diagonally divides the study area into approximately two equal halves; a central ridge between two depressions trending in NW-SE direction within Akoko complex and N–S faults of F1-F'1, F2-F'2 and F3-F'3 within Ikole axis. The upward continued maps (1 *km*, 2 *km* and 4 *km*) revealed the attitude of the basement rocks and structures with depth, some of the identified structures are deep-seated beyond 1.6 km downward, and as well revealed a general trend of NW-SE for the basement rocks. The radioelement maps revealed varying radioelement concentrations. Areas with high concentrations implied that the rocks are crystalline, undeformed and rich in feldspar with U-Th bearing minerals majorly within Ikole complex. Other regions with low concentrations on the other hand depicted varying geologic framework compositions. The maps also revealed uranium deposit in Ikole and mineralized structures such as veins, joints, fractures and dykes. The Ternary map showed that the two depressions (A) and deep-seated faults (F-F') contain low and high radioactive rich bearing-minerals respectively. The spectral depth analysis result showed that the shallow and deeper magnetic sources have an average depth estimate of 20 m and 3.60 km respectively. The Absorbed dose and Annual effective dose equivalent (AEDE) of the study area are below the world standard limit for radiological hazard. Hence, the study areas are mineralized and have several structures that retain important radioactive minerals with no hazardous risk.

Keywords: Aeromagnetic, Aeroradiometric, Structural mapping, Radionuclides hazard indices, Magnetic sources

1. Introduction

Human existence and development has relied on natural resources obtained from both near surface and subsurface. Magnetic geophysical survey exploits the considerable differences in the magnetic properties of minerals with the ultimate objective of characterizing the earth's subsurface. Magnetic survey method determines the subsurface spatial distribution of rock magnetization properties, (or susceptibility and remanence) which cause small changes in the earth's magnetic field strength and direction. Subsurface characteristics of interest to earth scientists include distribution and structure of rock types, grain size and rock strength, porosity, permeability and allowing decision making in assessing the risks and costs from subsurface information [1]. The physical principles of aeromagnetic methods are based on taking measurements of the ambient magnetic susceptibility of the surface geology and using the data to determine the distribution of magnetic minerals and changes in lithology [1]. High resolution aeromagnetic method has proved useful in mapping the basement structures and rapid selection of targeted areas for mineral exploration in Basement Complex of Southwestern Nigeria [2]. According to [3], the principles for the application of aeromagnetic surveys to geological mapping include the unique qualities of aeromagnetic data to comprise responses from magnetic minerals only, magnetic minerals are widely distributed in all terrains and the data are unaffected by non-magnetic cover, including vegetation, water and surficial deposits. The ability to accurately map a single geological parameter in a uniform manner over a very large area makes aeromagnetics highly suited to structural and lithological mapping over a wide range of scales [3].

Radiometric method involves the measurement of naturally occurring radioactive elements that exist in rock forming minerals and soil profiles. [4] indicated that airborne radiometric survey likewise is used to measure variations in the radioactive mineral composition in order to delineate lateral lithological changes. The decay products of radioactive elements such as uranium, thorium and isotopes of potassium are important in geologic mapping and mineral exploration [4]. Radioactive elements occur naturally in the crystals of particular minerals. Since the energy of gamma rays is related to the source radioactive element, they can be used to measure the abundance of those elements in an area. So by measuring the energy of gamma rays being emitted in an area, we can infer the presence of particular minerals in the earth's surface. The gamma rays are detected by a spectrometer, which counts the number of times each gamma ray of particular energy intersects it [5; 6].

Radiometric survey is one of the most cost-effective and rapid techniques for geochemical mapping of the radioactive elements. Nowadays, the method is mainly applied for geological mapping and exploration of other types of economic minerals; geochemical and environmental monitoring such as mapping of radioactive contamination from fallout of nuclear accidents and plumes from power plants; allow the interpretation of regional features over large areas, and applicable in several fields of science [7; 8; 9]. They may be used to estimate and assess the terrestrial radiation dose permissible to human population and to identify areas of potential natural radiation hazard. The airborne datasets can provide detailed information about the characteristics of the soil and its parent geological materials, including surface texture, weathering, leaching, soil depth, moisture and clay mineralogy [10]. The ground survey provides opportunity to have direct access to take direct measurement on different rock types and to enhance signal to noise ratio than when measurements are taken few meters above the ground level. Gamma-ray have proved valuable in mapping the stratigraphy and structure of non-magnetic rock units, such as sedimentary and low-grade metasedimentary sequences, and for recognizing individual phases and zoning within units mapped out on the basis of differing K, Th and U concentrations. One of the most important recent applications of gamma-ray spectrometry is in the field of regolith-landform, soil, and land resource mapping, where it has been used to map geomorphic activity, regolith types, land degradation, soil types and soil loss [4].

Variation of magnetic and radiometric mineral composition in rocks as a result of the different lithological setting and mineralization processes like hydrothermal mineralization now form an integral part of most mineral exploration programs and form the basis of modern geological mapping most especially in basement terrain. High resolution gamma ray spectrometry, flown simultaneously with magnetics, is a cost effective mapping tool which provides a three-element- K, Th, U chemical map of the ground surface [11]. Gamma-ray data have proved complementary to magnetic data in many terrains, enabling mapping of bedrock and regolith in vegetated regions with an effective penetration, unlike remotely sensed data such as Landsat Thematic Mapper and SPOT Image, whose effective penetration is limited to the top few microns and masked by vegetation cover [12].

Increase failures of borehole development and other engineering projects such as roads, dams and bridges have prompted this study as a result of unpredictable depth to bedrock usually encountered in basement terrains that is controlled by various geological factors. Hence, this study is designed to have detailed information about the characteristics of the subsurface towards decision making in predicting target areas for detailed exploration, environmental monitoring and estimate and assess the terrestrial radiation dose permissible to human.

2. Location and Geologic Setting of the Study Area

Ikole Sheet 245 is used as the study area for this research work. It lies within the Southwestern Basement Complex of Nigeria and lies within latitudes 7° 30' N and 8° 00' N (830000 – 885000 mN) and longitudes 5° 30' E and 6° 00' E (773500 – 830000 mE) of the Minna Zone 31 N (Figures 1 and 2). The towns in the study area include Ikole and other towns (Ekiti State) in the Northern to Western part; Ikare, Arigidi, Ajowa, and Akunu among others in Akoko area (Ondo State) at the Southern part, and Ayere to some parts of Kabba (Kogi State) situated within a small area at the Northeastern part. The areas are accessible mainly by roads and footpaths.

The Basement Complex of Southwestern Nigeria is located in a triangular portion of the Nigerian basement, an extension of the Dahomeyide shield of the West African Craton. Rocks of the region include Migmatized-Gneiss Complex (MGC) that is characterized by grey foliated gneiss, ultramafic rocks and felsic component comprised of pegmatite, aplite and granitic rocks [13]. The MGC in Southwestern Nigeria is affected by three major geotectonic events ranging from Early Proterozoic of 2000 Ma to Pan African events of ~600 Ma [14; 15; 16]. The rocks of the basement have been affected by medium pressure Barrovian metamorphism [16; 17]. The attitudes of tectonic structures in the Nigerian Basement have been documented in terms of orientation and magma-induced veins and dykes such as quartz veins and pegmatites [17]. Deformation of the Nigerian basement complex occurred in two phases, a ductile phase, which is responsible for the formation of planar structures (foliations) and a brittle phase resulting in jointing and fractures, many of which have been filled with quartzo-feldspathic veins, dolerite dykes, pegmatite and aplitic veins and dykes [17].

The major lithological units of the study area include the migmatites, granite gneiss, charnockite, granite, and other felsic and mafic intrusives. The basement rocks show great variations in grain size and in mineral composition. The rocks are predominantly quartz gneisses and schists (Ikole axis) consisting essentially of quartz with small amounts of white micaceous minerals. The rocks of the basement complex in the area have been subjected to intense regional metamorphism in which shearing stress was the dominant control resulting in the absence of minerals of high metamorphic grade. Magmatisation is widespread throughout the area reflected by rapid alternation of granite, biotite gneiss and biotite schist which grade into one another. Granite gneiss in the study area is of two types; the biotite rich gneiss and the banded gneiss. The biotite rich gneiss is fine to medium grained that show strong foliation trending westwards and is usually dark in colour. The banded gneiss show parallel alignment and alteration. Selective granitisation has resulted in biotite-rich layers in the gneisses being converted into porphyritic granite, while the leucocratic bands have been converted to aplitic granite. Minor folds are very common in the gneiss and schist of the study area [18; 19], the basic geological structure of Southwestern Nigeria is a complementary anticliriorium and synclinorium with northwards plunging axes.

IOP Conf. Series: Earth and Environmental Science **173** (2018) 012024 doi:10.1088/1755-1315/173/1/012024

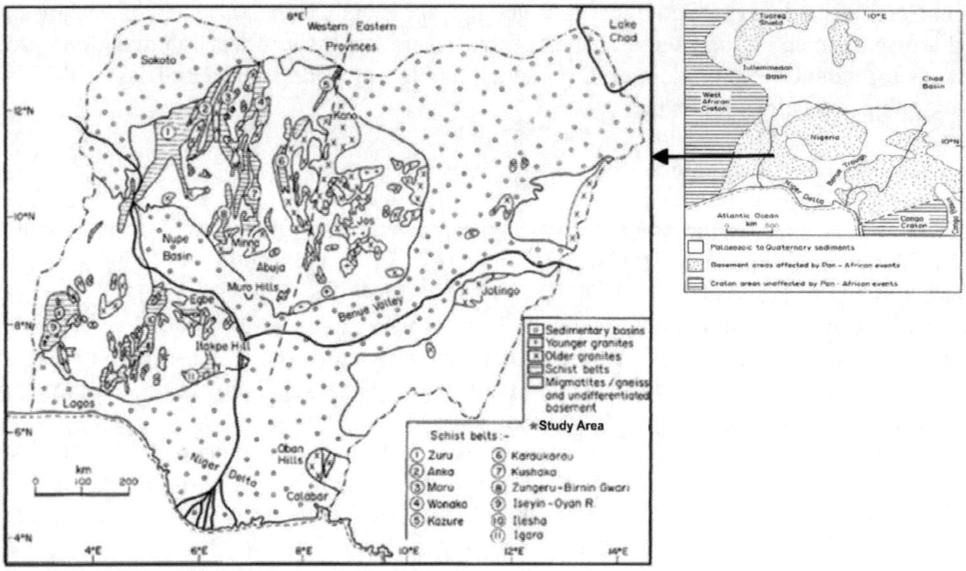

Figure 1. Regional geological setting of Nigeria (modified after [14]).

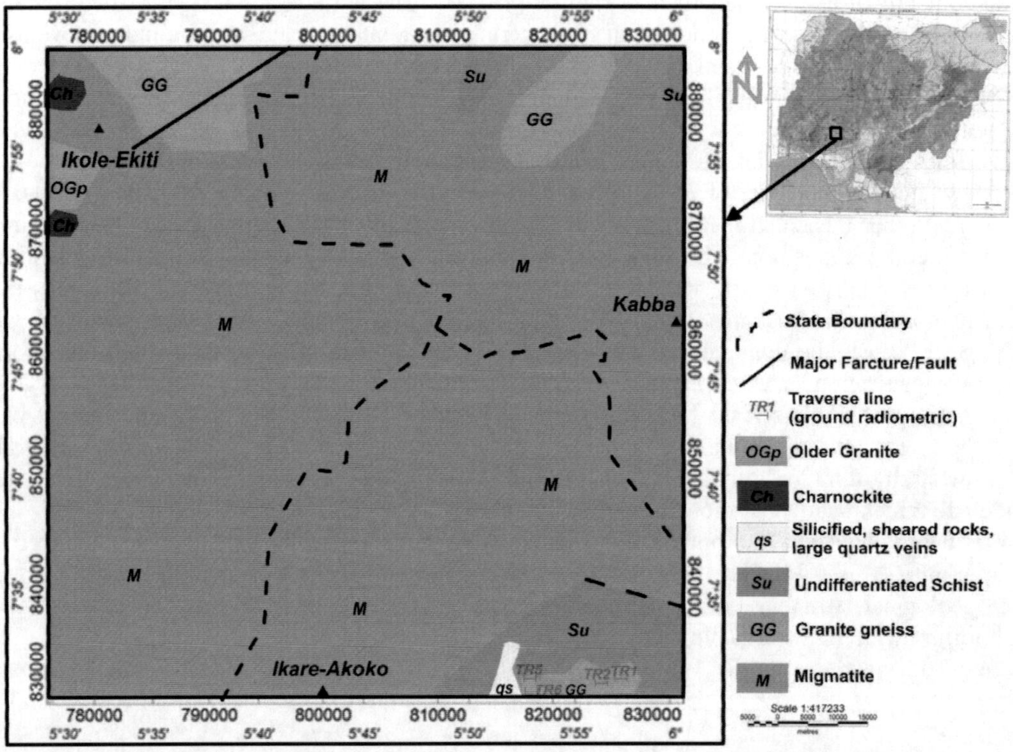

Figure 2. Geological Map of the Study Area showing the various litho-structures and ground radiometric survey traverses (modified after [20]).

3. Materials and Methods

Aero-geophysical data (total magnetic intensity and radiometric) covering Ikole Sheet 245 (1:100 000) were acquired from Nigeria Geological Survey Agency (NGSA), while the ground radioelements data were acquired within Akoko area using Gamma-Ray Spectrometer along eight (8) traverses (TR) taken on outcrops of granite-gneiss, charnockite, granite and grey-gneiss using a spread length of 100 *m* with spacing of 5 *m* between station positions. These aero-geophysical were processed using Oasis Montaj^{TM} Software and Microsoft Excel Package.
The aeromagnetic TMI data (Ikole Sheet 245) within latitudes 7° 30' N and 8° 00' N (830000 – 885000 mN) and longitudes 5° 30' E and 6° 00' E (773500 – 830000 mE) of the Minna Zone 31 N is that of the total field and is in a gridded form. This method fits minimum curvature curves (which is the smoothest possible surface that would fit the given data values) using method described by [21]. The aeromagnetic data filtering and reductions commenced with the removal of the Near Surface Noise (NSN) caused by metallic materials, fences, cables (both buried and surface) and others using the Butterworth filter. The derived data after the removal of the NSN was reduced to the magnetic equator using the Reduction to Equator filter (RTE) with declination and inclination of -1.541° and -11.119° respectively, in order to centre the peaks of anomalies directly over their sources. Thereafter, the Regional Field associated with the measured magnetic data was produced to get the Residual data, by subtracting the derived Regional field data from the observed data (RTE). Other filtering process such as Upward Continuation to depth of 1, 2 and 4 km to see the attenuation of the high wave number anomalies and the accentuation of anomalies from deep-seated features was applied. Finally, the Residual map was divided into four Quadrants (1, 2, 3 and 4) and were windowed out to determine respective radially averaged power spectrum (depth analysis) used for depth to top of magnetic sources estimation.
The airborne radiometric gridded data from Nigerian Geological Survey Agency (NGSA), which have been corrected for aircraft noise and height attenuation, cosmic and background radiations removal, as well as stripping corrections derived from calibration data based on protocols described in [8] and [22 and 23] were imported into Oasis Montaj^{TM} Software where appropriate filters were used to enhance signals attributed to important subsurface signatures and natural radioactivity. A Ternary map was also produced from same software after respective radioelements had been normalized (to an equal-area histogram) to annul the effect of potassium being the most crustal abundant radioactive elements from dominating and overriding other radioelements. [8] and USGS standard colour notation of red representing % K, green for eTh and blue for eU were used. Blue was used for eU because blue tends to reduce the poorest signal-to-noise ratio and accommodate the erratic nature of eU that would have been obvious to the eyes if other colours were to have been used.
The mean values of the ground radiometric data presented in Table 1 from eight (8) traverses were used to compute the mean values for Absorbed Dose Rate (D), Annual Effective Dose Equivalent (AEDE), and Internal and External Hazard Indices using standard equations. The results were displayed as bar plots for pictorial representations and compared with worldwide potential hazards index standard, in order to determine if rock materials sourced within the selected area is safe for people inhabiting the study area and for other construction purposes without posing any significant radiological health hazard.
The gamma radiation doses can be estimated by employing the convenient formula [24].

$$D = (0.462\ C_U + 0.621\ C_{Th} + 0.0417\ C_K)\ nGyh^{-1} \quad (1)$$

where D is the absorbed dose rate in $nGyh^{-1}$, C_U, C_{Th} and C_K represent the activity concentrations of ^{238}U, ^{232}Th and ^{40}K respectively. It is assumed that the contribution from other radionuclides, such as ^{137}Cs, ^{235}U, ^{87}Rb, ^{90}Sr, ^{138}La, ^{147}Sm and ^{176}Lu to the total dose rate are negligible. UNSCEAR reported that the world average absorbed gamma dose rate mean is 55 nGy/h.
The annual effective dose equivalent (AEDE) received outdoor by a member of the public is calculated from the absorbed dose rate by applying dose conversion factor of 0.7 Sv/Gy and occupancy factor for outdoor and indoor was 0.2 and 0.8 respectively [25].

$$\text{AEDE (Outdoor)} (\mu Sv/y) = D\ (nGyh^{-1}) \times T \times Q \times 0.2 \times 10^{-3} \quad (2)$$

where D is the absorbed dose rate in air, Q is the conversion factor of 0.7 Sv/Gy, which converts the absorbed dose rate in air to human effective dose received and T is the time for one year, i.e. 8760 hrs. The external (H_{ex}) and internal (H_{in}) radiation hazard indices were calculated respectively due to natural radionuclides of ^{238}U, ^{232}Th and ^{40}K. For the radiation hazard indices to be negligible, both the external (H_{ex}) and internal (H_{in}) radiation hazards must be less than unity [26]. Internal exposure to radon is very hazardous which can lead to respiratory diseases like asthma [27]. Natural radionuclide in soil, sediment and rocks produce an external radiation field to which all humans are exposed. Therefore, the H_{ex} and H_{in} indices were evaluated by using equations 4 and 5 respectively [28].

$$H_{ex} = C_U/370 + C_{Th}/259 + C_K/4810 \leq 1 \quad (4)$$

$$H_{in} = C_U/185 + C_{Th}/259 + C_K/4810 \leq 1 \quad (5)$$

where C_U, C_{Th} and C_K are the radioactivity concentrations of ^{238}U, ^{232}Th and ^{40}K in $BqKg^{-1}$ respectively. The elemental concentrations of Uranium-238 (in ppm), Thorium-232 (in ppm) and Potassium (in %) can be calculated from measured activity concentrations of ^{238}U, ^{232}Th and ^{40}K in $Bqkg^{-1}$ using the conversion factors recommended by the [29] as follows:

$$1\ ppm = 10^{-4}\ \% \quad (6)$$

$$1\%\ K = 313\ Bq/kg\ of\ ^{40}K \quad (7)$$

$$1\ ppm\ U = 12.35\ Bq/kg\ of\ ^{238}U \quad (8)$$

$$1\ ppm\ Th = 4.06\ Bq/kg\ of\ ^{232}Th \quad (9)$$

4. Results and Discussion

Generally, the maps are depicted as high – moderate – low magnetic intensity across the whole area. It is evident from the various maps produced that there are different rock types; infilled geologic materials from different degrees of weathering and movement of fluid along faults lines; anomaly trends and closures; structural features with depth from one location to another that brought about the separation of the study area into different complexes.

From the TMI map (Fig. 3), several deductions were made from the range of magnetic intensity variations for respective rock types and other geologic features in the study area. The map shows magnetic intensity of highs (55 – 180 nT) classified as undeformed crystalline basement rocks; magnetically quiet zones in greenish colour as zones marked to fall within the undeformed and zone of alteration (yellowish); and low to very low magnetic intensity (-150 to -40 nT) designated as highly susceptible materials, infilled geologic materials and fluid. The amplitude of a magnetic anomaly is directly proportional to magnetization which depends on magnetic susceptibility of the rocks; the high and low magnetic susceptible areas on the TMI maps are depicted as low and high magnetic intensity values respectively, and a low/negative magnetic peak values represent typical anomalous signatures [11; 30; 31]. The total magnetic intensity varies from one complex to the other due to the mineral contents of rocks and level of geological processes such as fractures, weathering, etc.

The colour shaded Reduction to the Magnetic Equator (RTE) (Fig.4) map shows rocks of low intensities caused by high degree of deformation and weathering, classified as migmatite and other minor rock types with large numbers of intrusive rocks. Also, a NE-SW trending regional lineament (R-R') that diagonally divides the study area into approximately two equal halves and a central ridge

between two depressions trending in NW-SE direction within Akoko complex are revealed on this map. The RTE map clearly reveal the N–S faulting system in Ikole complex to be a normal faults of F1-F'1, F2-F'2 and F3-F'3, contributing to the identified low magnetic intensities in this area.
The Residual map (Fig. 5) gives a clearer and undistorted signatures of the study area geology from which the interpretive/inferred boundaries on the geological map (Fig. 2) are being proper delineated relative to their positions, based on different complexes, variations in the compositions of the crystal lattices framework, degree of weathering and metamorphism, separating the study area into three complexes namely; Ikole complex at the northwestern to southwestern part, Akoko complex from the southwestern to northeastern part and Kabba complex as a small complex on the northeastern part of the map. The map also reveals the presence of faults, lineaments and other structures.

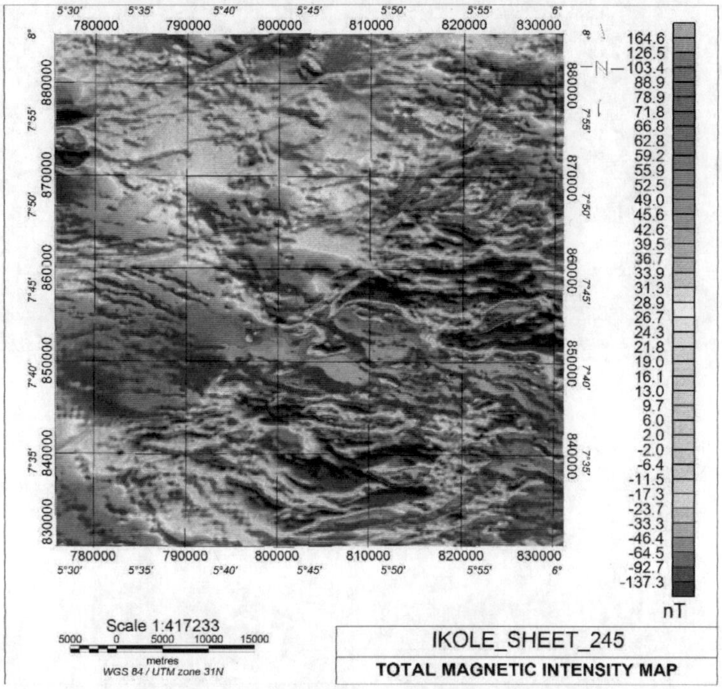

Figure 3. Total Magnetic Intensity (TMI) Map

IOP Conf. Series: Earth and Environmental Science **173** (2018) 012024 doi:10.1088/1755-1315/173/1/012024

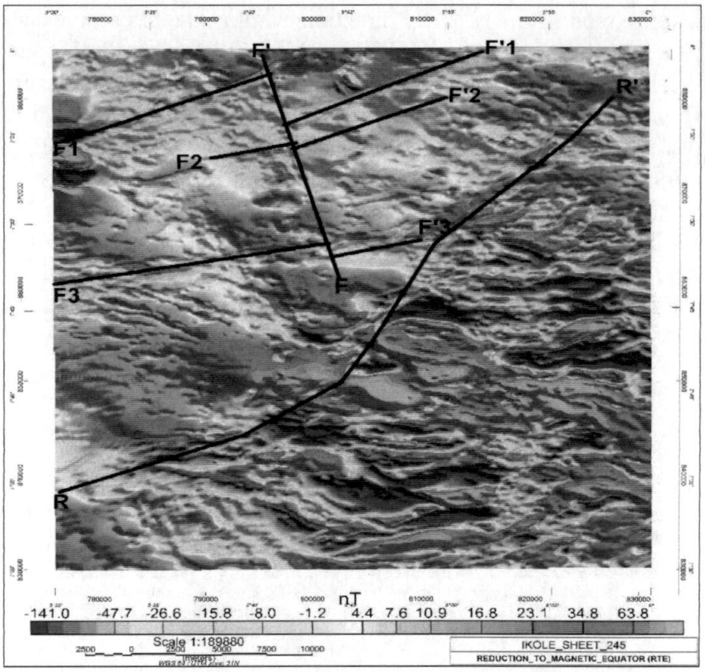

Figure 4. Colour Shaded Reduction to Magnetic Equator (RTE).

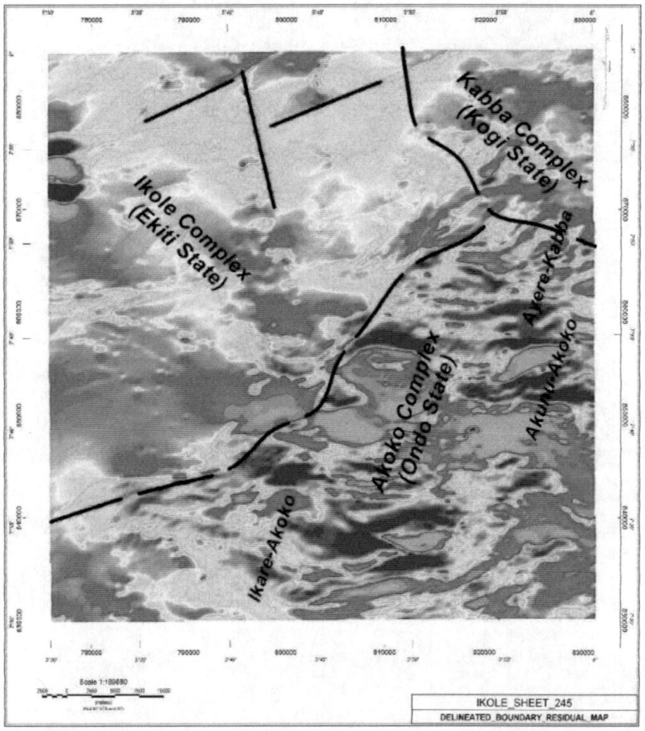

Figure 5. Residual and Delineated Boundaries Map of the Study area.

Figures 6a, b and c show the upward continued maps to 1 *km*, 2 *km* and 4 *km* respectively, in order to accentuate the responses from the basement rocks/deep-seated features that are useful in the interpretation of magnetic anomaly fields over areas containing many near surface magnetic sources such as dykes and other intrusions. The maps reduce ambiguities in the earlier identified litho-structures with depth. The upward continuation maps reveal a regional NW-SE trend for the rocks in the study area. The mapped faults within the Ikole complex, the two large depressions formed on both sides of the NW-SE trending ridge within Akoko complex and the NE-SW regional lineament are deep-seated because they are seen to be far beyond 1 *km* (Fig. 6a). The identified faults appear to be less than 2 *km* in depth (Fig. 6b), while other features are seen beyond this depth, even the depressions are far beyond 4 *km* (Fig. 6c). It was inferred that the intense metamorphism that brought about high fracturing density and degree of weathering in the area are pronounced in rocks around Akoko and the structures associated with these processes also deep-seated. It is also established that the deep sources could be attributed to basement and intra-basement features like lineaments/faults, dykes and veins.

Figures 7 show the eU (Fig. 7a), eTh (Fig. 7b) and K (Fig. 7c) concentrations maps respectively. The eU concentration map (Fig. 7a) depicts moderately high to very high eU concentration ranging from 4.5 – 13.0 *ppm* mostly where the faults were identified around Ikole and as low to very low ranging from -1.5 to 3.0 *ppm* mostly around Akoko and Kabba areas. The eTh concentration map (Fig. 7b) evinces somewhat similar radioactivity patterns and variations as observed in some sections of the eU map. High eTh concentration ranged from 25.0 – 70.0 *ppm*, with dominance around the fault zones in Ikole and some areas in Akoko, while the low eTh concentration ranged from 8.5 – 16.0 *ppm* and very low (< 8.5 *ppm)* dominating Kabba and some of the rocks around Akoko areas. The % K concentration map (Fig. 7c) on the other hand shows high % K radiation ranging from 2.0 – 4.0 % over a wide area extent around Ikole, especially where the faults occurred and within some areas around the southern part, while very low % K (< 0.7 %) radiations are seen around the central, western and eastern part of the study area.

The areas with moderately high to very high radioelements concentrations could be described as undeformed rocks with uranium, thorium (such as thorite, zircon) and potassium rich bearing-minerals. The low areas are considered to be infilled geologic materials or fluids within the host rocks caused by pronounced geological processes such as weathering, fracturing and also that the veins, joints, lineaments and faults may not contain radioactive rich minerals that could have given off high radioactivity. On the eU concentration map, high radioactive emission rate around and along the N-S trending faults around Ikole area confirms the presence of a deep-seated uranium deposit associated with other radioactive elements. These maps also reveal the strike directions of the rocks, generally in NW-SE direction with fewer rocks trending NE-SW and E-W. Interestingly, radioactive responses of various rock types in the study area have helped in lithological differentiations; mapping of the faults and regional lineament features; rate of weathering, and identify areas that contain radioactive rich bearing-minerals.

The Ternary map for the three (3) radioelements concentrations (Fig. 8) is produced for proper summary and to show the dominance of respective radioelements contributions in various geologic materials. In areas where we have black (denoted as A) and white (denoted as B) colourations indicate that the three radioelements concentrations are low and high respectively, while other colourations (denoted as C, D and E) may suggest variations in radioelements concentrations or level of radiation by one or two radioactive elements. The two depressions on both side of the ridge (delineated in green polygonal line) around Akoko reveal accumulated geologic materials of low radioactivity, while the ridge contains minerals rich in U, Th and K. The Ternary map also reveals NE-SW regional lineament trend (R-R') separating the complexes, deep-seated faults (F-F') rich in radioactive minerals with high spontaneous emission and various geological boundaries.

The radially averaged power spectrum plots (Fig. 10) were produced from four quadrants (Fig. 9) to infer the total depth estimate to top of magnetic sources that produced the observed anomalies in the study area using spectral analysis. The gradients of the layers were calculated at two different points as shallower and deeper sources based on the wavelength of magnetic sources. The gradients are: Quadrant 1 (0.5 and 20.0), Quadrant 2 (0.47 and 5.5), Quadrant 3 (0.25 and -20.0) and Quadrant 4 (1.75 and 45.0) with approximate depth estimate. The calculated depths are: Quadrant 1 (40 m and 1.6 km), Quadrant 2 (37 m and 440 m), Quadrant 3 (20 m and 1.6 km) and Quadrant 4 (140 m and 3.6 km) respectively. The total depth estimates to top of magnetic sources for shallower and deeper sources are 20 *m* and 3.60 *km* respectively.

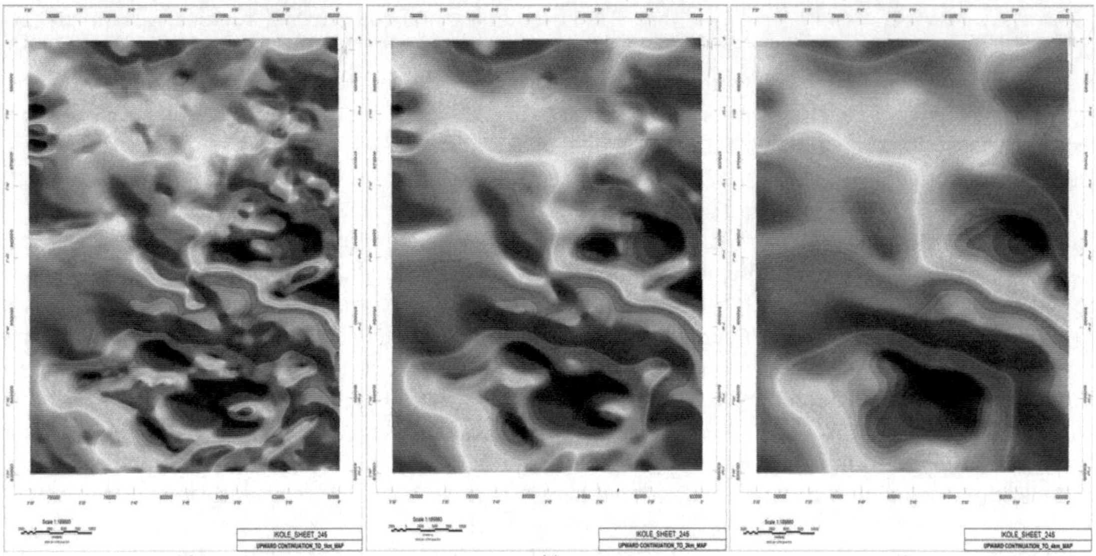

(a) (b) (c)

Figures 6. Upward continuation maps for (a) 1 km, (b)2 km, and (c) 4 km

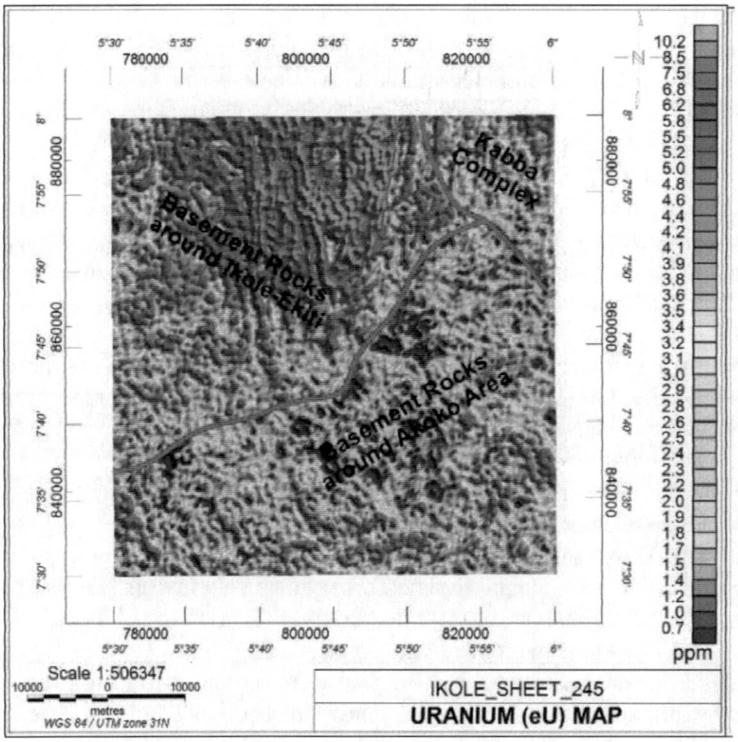

(a)

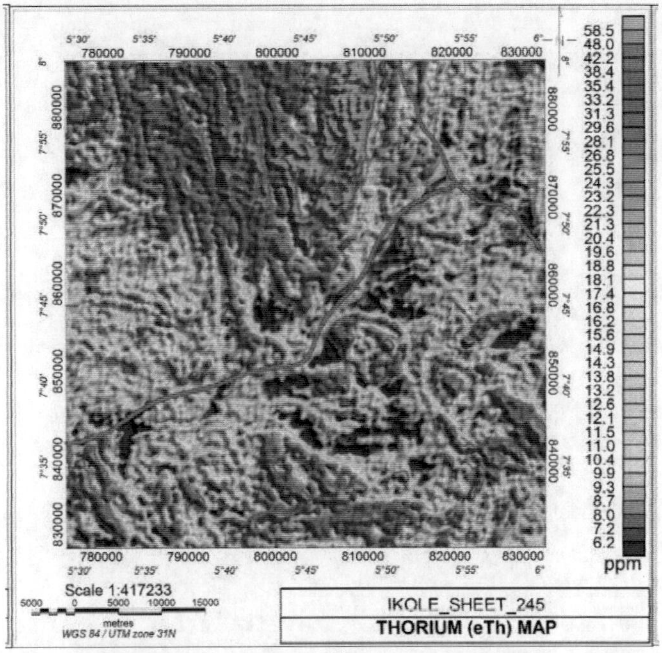

(b)

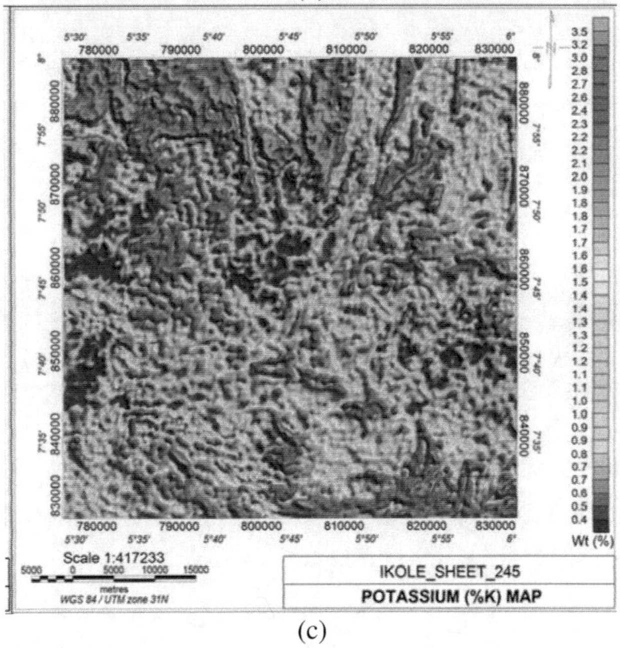

(c)

Figures 7: Radioactive Concentration Maps of Uranium (a), Thorium (b), and Potassium (c)

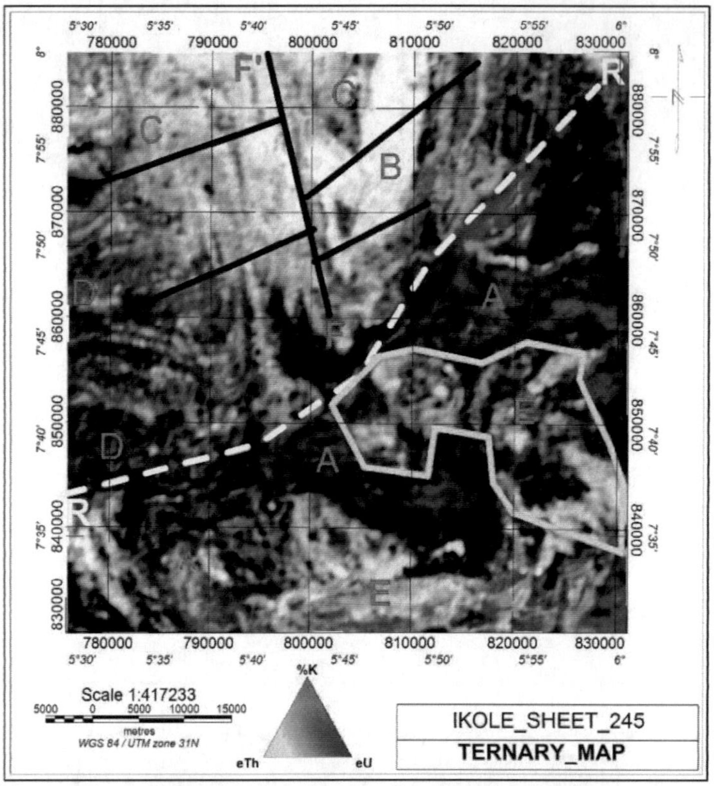

Figure 8. Ternary Map of Study Area showing the distribution of radioelements concentrations.

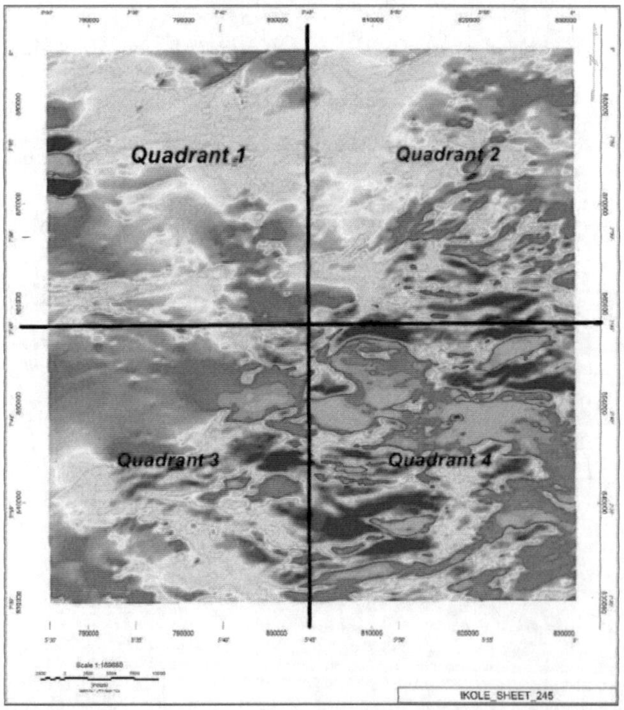

Figure 9. TMI showing quadrants selected for Radial Average Power Spectrum.

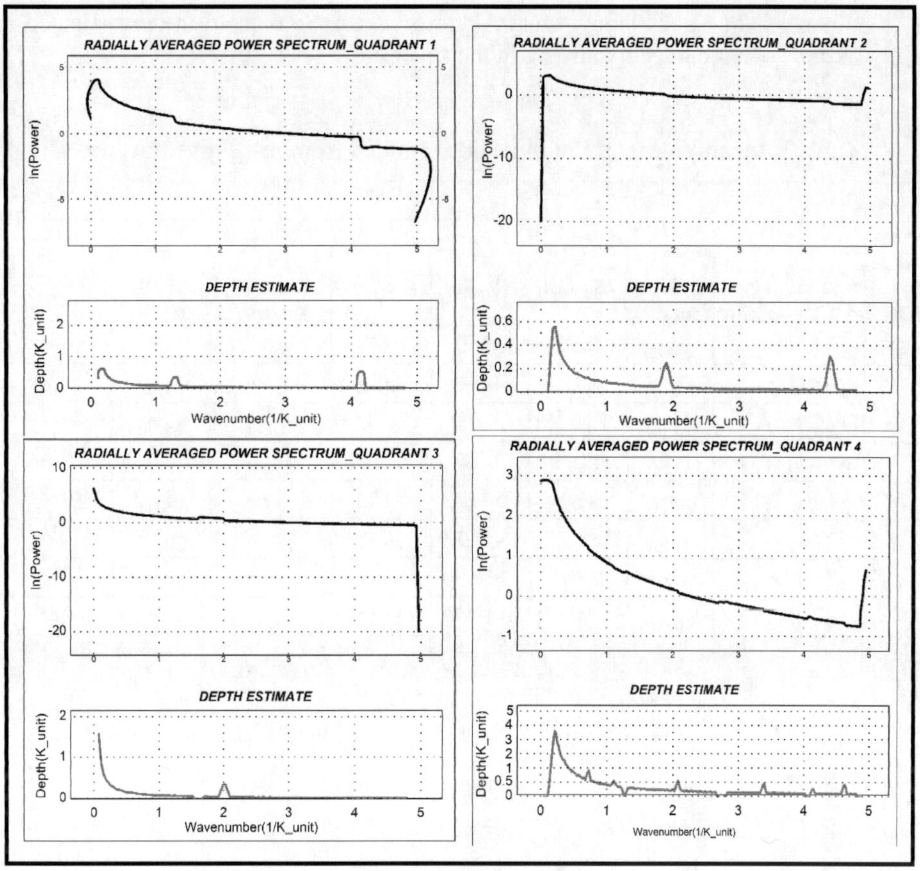

Figure 10. Radial Average Power Spectrum Plots of selected quadrants for depth estimation.

The results of the Total Mean values for eU, eTh, % K, Dose rate, Absorbed and Effective Dose Rates for ground radiometric survey are shown in Table 1. The mean values were computed from field data and those calculated using standard equations through Microsoft Excel. From Table 1 and Figure 11, variations in radioelements from the ground survey were seen to vary from one lithology to another i.e. from traverses (TRs) 1 to 8. TR 4 has the highest eU, eTh and % K; TR 5 has the lowest % K; while, TR3 with the lowest eU and eTh. TR4 responses could be used to infer that the rocks are crystalline, undeformed and rich in feldspar, U - and – Th – bearing minerals. TRs 3, 5, 6, 7 and 8 depict varying geologic framework compositions, degree of metamorphism and weathering in various rocks. From the ground radiometric survey, it is evident that the rocks around Akoko area of this study have undergone intense deformation than other parts. Its worthwhile recording the gamma-ray exposure/dose levels obtained from the survey data to deduce some certain parameters for the hazard indices. The mean ground-level dose rates obtained for different traverses varied approximately from 96.5 to 130 *(nGy/h)*.

Figures 12 and 13 show the pictorial representation plots for the mean Absorbed Dose Rate (D) and Annual Effective Dose Equivalent (AEDE) for outdoor/external human index radiological hazard, with TR 1 having highest values, TR 3 has the lowest of all, while other traverses have nearly same values. The highest mean Absorbed Dose (16.1 *nGy/h*) in the study area is below the background cosmic rays value of 55 nGy/hr (Nano Gray per hour) [24]. Also, the highest mean Annual Effective Dose Equivalent (AEDE) value of about 19.75 *µSv/y* (micro Sievert per year) (*i.e.*0.01975 *mSv/y)* is far below the worldwide effective dose of 2.4 *mSv/y* [24], and below the minimum permissible limit of 1 *mSv/y* recommended by [32]. The health hazard impact assessment for both external and internal radiations to human reveal values below unity Therefore, radiations from the investigated areas are radiologically hazard risk free to people living in the study area.

Table 1: Mean Values of Radionuclides emitted from respective traverses

LOCATION	MEAN									
	K (%)	eU (ppm)	eTh (ppm)	K *(Bq/Kg)*	U *(Bq/Kg)*	Th *(Bq/Kg)*	Absorbed Dose Rate (D) *(nGy/h)*	AEDE *(μSv/y)*	H_{ex}	H_{in}
TR1	3.78	3.2	23.95	1183.14	39.52	97.24	16.51	20.25	0.728	0.835
TR2	2.45	3.94	21.85	766.85	48.66	88.71	15.50	19.00	0.633	0.765
TR3	3.26	2.57	15.02	1020.38	31.74	60.98	10.65	13.06	0.533	0.619
TR4	3.82	6.85	30.05	1195.66	84.60	122.00	21.99	26.96	0.948	1.176
TR5	2.07	4.07	20.50	647.91	50.27	83.23	14.70	18.03	0.592	0.727
TR6	2.21	3.37	23.74	691.73	41.62	96.38	16.39	20.10	0.628	0.741
TR7	2.39	3.13	20.97	748.07	38.66	85.14	14.57	17.87	0.599	0.693
TR8	2.48	3.34	20.48	776.24	41.25	83.15	14.36	17.62	0.593	0.705

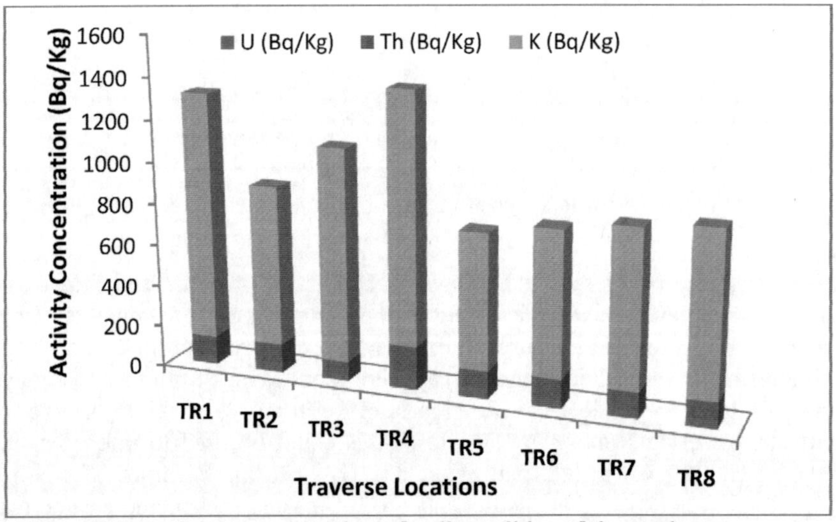

Figure 11. Activity concentration of radionuclides of the study area

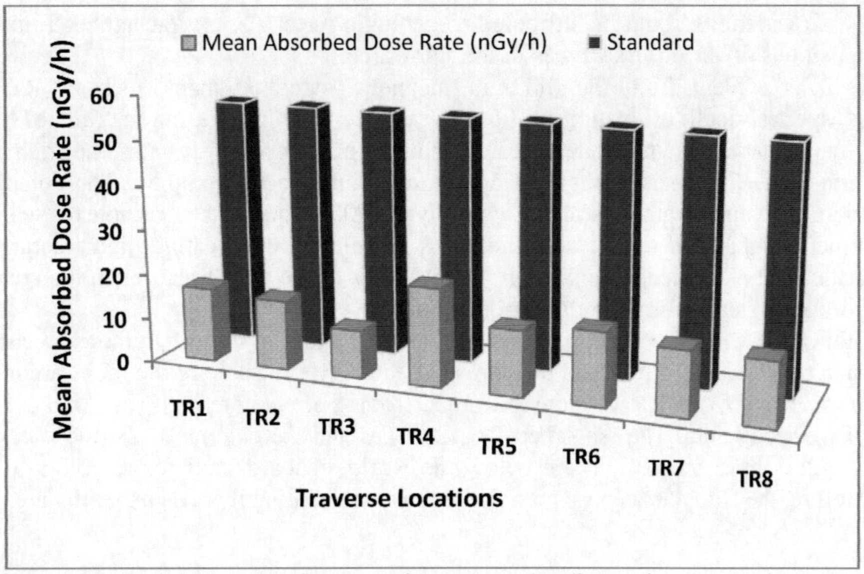

Figure 12. Mean absorbed dose rate (nGy/h) from ground radiometric survey within the Akoko Complex of the study area compared with the [24].

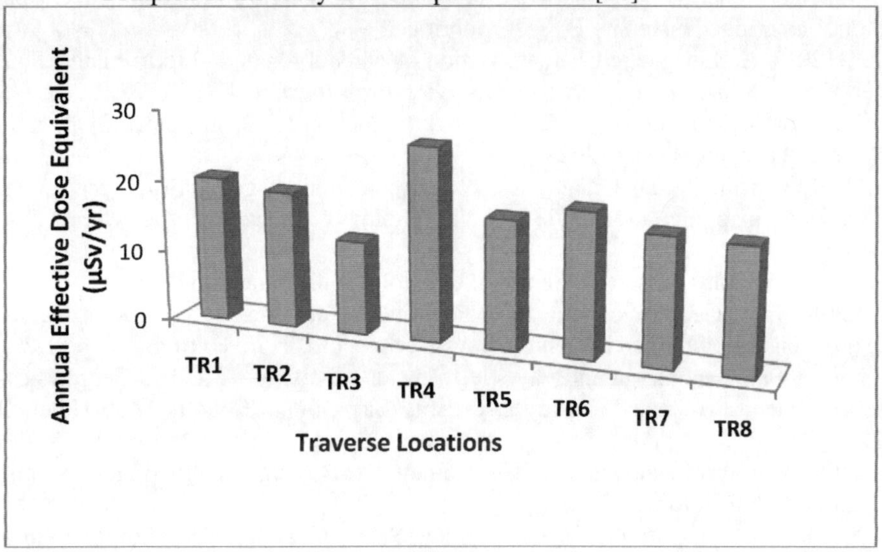

Figure 13. Annual Effective Dose Equivalent (AEDE) (outdoor) from ground radiometric survey within the Akoko Complex of the study area.

5.0 Conclusion

High resolution airborne interpretation of Ikole Sheet 245 have shown the robustness of using integration methods that is further complemented by ground survey in characterizing magnetic intensities; radioelements concentrations and distributions of eU, eTh and % K; effective mapping of lithologies and geological structures such as fractures, lineaments, faults and dykes with their trends; depth to magnetic sources; radiological impact risk assessments and amount of dose rate permissible to human.

The maps produced for this study revealed some rocks/materials with different range of magnetic intensities and susceptibilities, radioactivity, some infilled geologic materials based on amplitude variations, structural trends of approximately N–Sand a regional lineament trending NE–SW

IOP Conf. Series: Earth and Environmental Science **173** (2018) 012024 doi:10.1088/1755-1315/173/1/012024

directions. These variations could be attributed to geologic materials, crystal lattices framework and mineral compositions of various rock types in the study area.
Interestingly, it is evident from the different magnetic reduction maps such as RTE, upward continuations, spectral depth estimation of 3.6 km, radioelements concentrations (eU, eTh and % K) and Ternary maps, that the faults are deep-seated and host uranium deposit with other rich radioactive bearing-minerals. The lineament is regional and extends into Kabba complex. The complexes have been highly deformed and weathered, most especially the Akoko and Kabba complexes which showed intense deformations that had restructured the rocks by folding and creating high fracturing density with both smaller and elongated lineaments in them. Points of the deep fractures in the area would be important hydrogeologically for groundwater development.
In addition, the Ikole Complex comprises rocks that have granitic crystal framework compositions making the area to be less deformed and the veins and faults within some of the rocks are mineralized. The geology of Ikole Complex is quite different from their counterparts in Akoko and Kabba complexes of the study area. The Absorbed Dose, AEDE and the external hazard (H_{ex}) and internal hazard (H_{in}) indices have values that are below the world standard limit for radiological hazard in humans. Therefore, the study areas pose no risk to humans inhabiting there or in nearby areas.

References

[1] Telford, W. M., Geldart, L. P., Sheriff, R. E. and Keys, D. A. 1990. Applied Geophysics. 2nd edition. Cambridge University Press, Cambridge.
[2] Ademila, O. 2017. Aeromagnetic Characterisation of parts of Ondo and Ekiti States, Southwestern Nigeria. *Environmental and Earth Sciences Research Journal*, 4(3): 66-75.
[3] Kearey, P., Brooks, M. and Hill, I. 2002. An Introduction to Geophysical Exploration, 3rd ed. Oxford: Blackwell Science. 262.
[4] Wilford, J., Bierwirth, P. and Craig, M. 1997. Application of gamma-ray spectrometry in soil or regolith mapping and applied geomorphology. *Journal of Australian Geology and Geophysics, 17*(2): 201-216.
[5] Faure, G.1977. Principles of Isotope Geology, UK, John Willey and Sons Inc.
[6] Milsom, J.2003. Field Geophysics: The geological field guide series, John Milsom University College, London. Published by John Wiley and Sons Ltd. Third edition. 51-70.
[7] Grasty, R. L. 1998. Airborne gamma-ray spectrometer surveying. Airborne Geophysics, I.A.G.E. Weekend course, 16th and 17th May at Galway Bay Golf & Country Club Hotel, Oranmore, Ireland.
[8] IAEA, 1991. Airborne gamma ray spectrometer surveying, International Atomic Energy Agency, Technical Report Series, No. 323: 97.
[9] IAEA, 2007. International Atomic Energy Agency. Safety Glossary, Terminology used in Nuclear Safety and Radiation Protection, 2007 Edition.
[10] Bierwirth, P. N., 1997. The use of airborne gamma-emission data for detecting soil properties. Proceedings of the Third International Airborne Remote SensingConference and Exhibition. Copenhagen, Denmark.
[11] Parasnis, D. S. 1996. Principles of Applied Geophysics, 5th ed. Chapman Hall, London. 456.
[12] Jaques, A. L., Wellman, P., Whitaker, A. and Wyborn, D. 1997. High resolution geophysics in modern geological mapping. *AGSO Journal of Australian Geology and Geophysics*, 17(2): 159-173.
[13] Rahaman, M. A. 1981. Recent Advances in the Study of the Basement Complex of Nigeria. First Symposium on the Precambrian Geology of Nigeria, Summary.
[14] Woakes, M., Ajibade, C. A., and Rahaman, M. A. 1987. Some metallogenic features of the Nigerian Basement. *Journal of Africa Science*, 5: 655-664.
[15] Ajibade, A. C. and Fitches W. R. 1988. The Nigerian Precambrian and the Pan –African Orogeny, Precambrian Geology of Nigeria. 45-53.

[16] Oyinloye, A. O. 2011. Geology and Geotectonic Setting of the Basement Complex Rocks in Southwestern Nigeria: Implications on Provenance and Evolution. *Earth and Environmental Sciences*, 98 – 117. ISBN: 978-953-307-468-9.

[17] Rahaman, M. A., Emofurieta, W. O. and Cean-Vachette, M. 1983. The potassic granite of Igbetti area. Further evidence of the polycyclic evolution of the Pan-African belt in Southwestern Nigeria. *Journal of Precambrian Resources*, 22: 75-92.

[18] Jones, H. and Hockey, R. 1964. Geology of parts of southwestern Nigeria. *Geological Survey of Nigeria Bulletin,* No. 31.

[19] Rahaman, M. A., 1989. Review of the basement geology of southwestern Nigeria. In: Kogbe, C.A., (ed) Geology of Nigeria, Rock View (Nig.) Limited, Jos, Nigeria. 39-56.

[20] Nigerian Geological Survey Agency (NGSA), 2009. Geological map of Nigeria. Map prepared by Nigerian Geological Survey Agency.31, Shetima Mangono Crescent Utako District, Garki, Abuja.

[21] Briggs I. C. 1974. Machine contouring using minimum curvature. *Geophysics*, 39: 39-48.

[22] Grasty, R. L. and Minty, B. R. S. 1995a. A guide to the technical specifications for airborne gamma ray surveys. Australian Geological Survey Organization, Record.

[23] Grasty, R. L. and Minty, B. R. S. 1995b. The standardization of airborne gamma-ray surveys in Australia. *Exploration Geophysics*, 26: 276-283.

[24] UNSCEAR, 2000. United Nations Scientific Committee on the Effects of Atomic Radiation Sources, effects and risks of ionizing radiation. Report to the general assembly, annexes B: Exposure from natural radiation sources. New York. 678-679.

[25] Veiga, R. G., Sanches, N., Anjos, R. M., Macario, K., Bastos, J., Iguatemy, M., Aguitar, J. G., Santos, M. A., Mosquera, B., Carvaiho, C., Baptista, M. and Umisedo, N. K. 2006. Measurement of natural radioactivity in Brazilian beach sands. *Radiation Measurements,* 41: 189-196.

[26] Beretka, J. and Mathew, P. J. 1985. Natural radioactivity of Australia building materials industrial wastes and by-products. *Health Physics*, 48: 87-95. https://doi.org/10.1097/00004032-198501000-00007.

[27] Tufail, M., Akhar, N., Jaried, S. A. and Hamid, T. 2007. Natural radiation hazard in building bricks fabrication from soils of two districts of Pakistan. *Radiological Protection*, 27: 481-492. https://doi.org/10.1088/0952-4746/27/4/009

[28] Ramasamy, V., Suresh, G., Meenakshisundaram, V. and Gajendran, V. 2009. Evaluation of natural radionuclide content in river sedi-ments asnd excess lifetime cancer risk due to gamma radioactivity. *Research Journal of Environmental and Earth Sciences*, 1(1): 6-10.

[29] IAEA, 2003. International Atomic Energy Agency (IAEA) guidelines for radioelement mapping using gamma ray spectrometry data, Vienna.

[30] Gunn, P. J., Maidment, D. and Milligan, P. R. 1997a. Interpreting magnetic data in areas of limited outcrop. *AGSO Journal of Australian Geology and Geophysics*, 17(2): 175-185.

[31] Fieberg, F. C. 2002. Ground magnetic investigations for gold prospecting in Southwestern Nigeria. Presentation given at the 62nd Meeting of the German Geophysical Society, Hannover.

[32] IAEA, 2007. International Atomic Energy Agency. Safety Glossary, Terminology used in Nuclear Safety and Radiation Protection- 2007 Edition.

IOP Conf. Series: Earth and Environmental Science **173** (2018) 012025 doi:10.1088/1755-1315/173/1/012025

Prediction of transmissivity of aquifer from geoelectric data using artificial neural network

Oluseun Adetola SANUADE[1]*, Kehinde David OYEYEMI[2], Joel Olayide AMOSUN[3], Julius O. FATOBA[3], Olaide S. HAMMED[4]

[1]Geosciences Department, King Fahd University of Petroleum & Minerals, Dhahran, Saudi Arabia
[2]Department of Industrial Physics, Covenant University, Ota, Ogun State, Nigeria
[3]Department of Geophysics, Federal University Oye-Ekiti, Ekiti State, Nigeria
[4]Department of Physics, Federal University Oye-Ekiti, Ekiti State, Nigeria.

*sheunsky@gmail.com

Abstract. We have used artificial neural network (ANN) for the prediction of transmissivity (T) from geoelectric parameter in an aquifer of a typical basement complex, Southwestern, Nigeria. The study area is composed of migmatite-gneiss, charnockite and granite gneiss. A geophysical investigation involving Vertical Electrical Sounding (VES) was carried out in the study area. Twenty VES were acquired using Campus Ohmegaresistivity meter. VES curves were interpreted quantitatively by using the technique of partial curve matching and with 1D forward modeling by means of computer using *WinResist*software. The interpreted data were used to determine transverse resistance, T_R. Transmissivity were measured in the boreholes drilled at the 20 VES locations. T_R and measured T were subjected to ANN analysis using *MATLAB 2017a software* in order to predict T of the aquifer. Root mean square error (RMSE) was used to test the performances of our model. The results show that T_R range from 133.00 - 381.20 Ωm^2, T also vary from 0.4 to 4.1 m^2/day. However, ANN model was able to predict T values with coefficient of correlation (R) values of 0.97, 1.00, 0.99 and 0.94 for training, test, validation and all network models respectively. RMSE value for the ANN model was found to be 0.085 which implies high performance of our model. A linear relationship was suggested for the ANN analysis to predict T. It can therefore be concluded that with ANN model, it is possible to predict T of aquifer in the study area where geoelectric data such as T_R is known and T values unknown.
Keywords:Aquifer, transmissivity. ANN, geoelectric, prediction, MATLAB.

1. Introduction

Hydrogeological parameters including transmissivity, hydraulic conductivity, recharge, and so on, can be determined in an aquifer by means of classical hydrogeological techniques such as pumping tests, grain size analysis, tracer techniques. Most of these techniques are expensive. However geophysical methods such as electrical, seismic, gravity, electromagnetic have been used for groundwater investigation to determine the geometry of aquifer, subsurface electrical conductivity, depth to basement and so on [1]. These methods, when compared with hydrogeological methods, are cost effective, relatively fast and non-destructive. There are established relationships between different geophysical and hydrogeological parameters. For instance, a relationship exists between electrical resistivity (ρ) and transmissivity (T). This relationship is made possible because both parameters are associated with the structure of pore spaces and heterogeneity [2-3]. Therefore, the combination of aquifer parameter estimated from borehole and electrical resistivity parameter estimated from surface electrical resistivity measurement could be a useful tool of estimating hydrogeological properties and potentials of aquifers.

Several studies have been done to know the relationship between hydrogeological and geophysical parameters [4-7]. [8] calculated the transmissivity of an aquifer using Dar Zarrouck parameters by taking into consideration two important laws which are Darcy's and Ohm's laws. The authors made use of the correlation that exists between the two parameters. [9] also reviewed the relationship that exist between transverse resistance and transmissivity by means of geostatistical analysis. [10] determined the correlation between aquifer resistivity that were normalized and hydrogeological parameter and normalized transverse resistance and transmissivity of aquifer in a weathered granite geological environment that are overlain by sand dunes in Jalore, northwestern part of India.

However, most of the previous studies on the relationship between hydrogeological and geophysical parameters used correlation studies and geostatistical approaches. The objective of this research, therefore, is to predict the transmissivity of aquifer from geoelectric data (transverse resistance) in a typical basement complex environment using artificial neural network (ANN).

The unique attribute of ANN is that it is data driven, which provides a unique opportunity to predict transmissivity effectively from transverse resistance as opposed to previous studies that finds a relationship between hydrogeological parameters and Dar Zarrouck parameters using other approaches. This will reduce non-uniqueness of parameter estimation and lead to improved transmissivity estimation.

The study area is in southwest area of Nigeria (Fig. 1). It lies within longitude 748600 – 749900 and latitude 850100 – 850600. Fig 2 shows a simplified geological map of the study area. The area is situated in the Precambrian Basement Complex where Migmatite-Gneiss, Charnockite and Granite Gneiss are the rock types in the area.

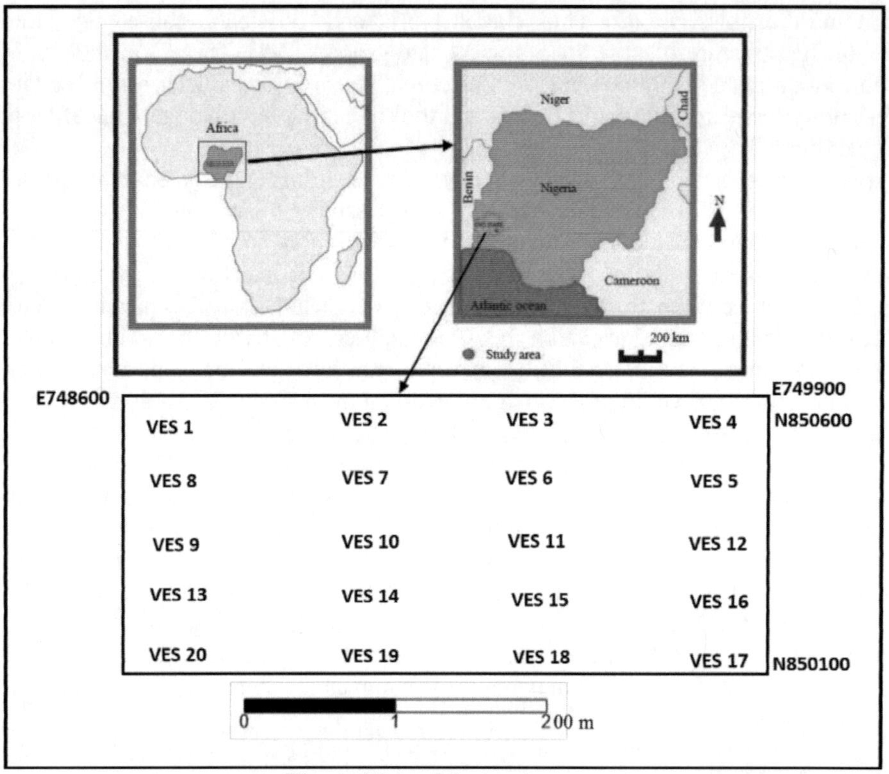

Fig. 1: Map of the study area

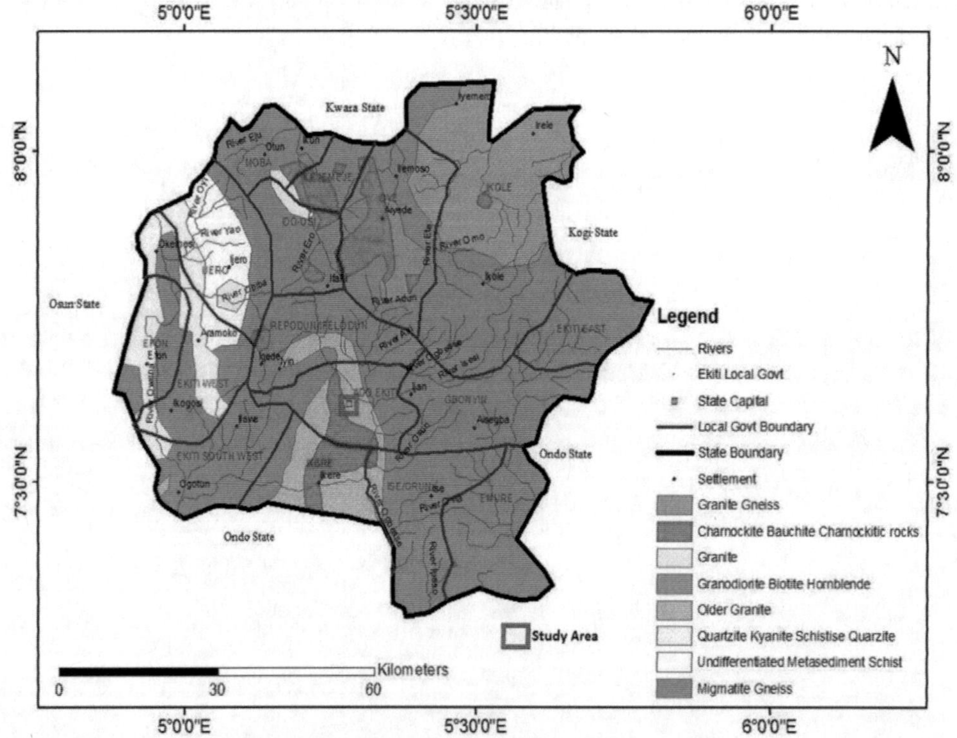

Fig. 2: Geological map of the study area [11]

2. Data acquisition and processing

2.1 Geophysical investigations

Geophysical investigations involving the Vertical Electrical Sounding (VES) was carried out to assess the groundwater potentials in the study area. Twenty (20) VES were acquired using Campus Ohmegaresistivity meter. The Schlumberger configuration, having current electrode spacing (AB/2) that varies from 1 to 100 m, was used during the data acquisition. VES curves were interpreted quantitatively using the technique of partial curve matching and 1D forward modeling using computer with the *WinResist* version 1.0 software. The processed and interpreted data were then used to determine aquifer resistivity, aquifer thickness and Dar Zarrouck parameter (transverse resistance (T_R) for the aquifer using equation 1 [12].

$$T_R = h.\rho \tag{1}$$

Where h and ρ are the thickness and resistivity of the aquifer respectively

2.1 Hydrogeological investigations

Twenty (20) boreholes were drilled to depth of about 35 m at the VES points. Pumping test was carried out at the 20 VES points. The test was done for about 3 to 6 hours using constant discharge rate method. Transmissivity (T) of aquifer was determined from the pumping test [13]. The data obtained from both geoelectrical (Transverse resistance) and hydrogeological data (transmissivity) were subjected to ANN analysis for the prediction of transmissivity of the aquifer from the geoelectric data.

2.3 Artificial neural network (ANN)

ANN is an artificial intelligence method, which mimics the human brain's behaviour by acquiring knowledge through a learning process [14-15]. ANN is capable of modeling both non-linear and linear systems without assuming implicit functions as it is in the unlike most conventional statistical approaches. ANN has been utilized in several areas of science and engineering for classification, regression analysis, time-series forecasting, clustering and dynamic system modelling [16-17]. In this research, the multi layered perception (MLP) supervised feed-forward Levenberg-Marquardt based backpropagation algorithm (LMA) was used to predict transmissivity in the aquifer. The network consist of three layers (Fig.3): (i) input layer (transmissivity), which accepts and presents the input data to the network through a set of neurons (ii) hidden layer, which runs a set of algorithms to compute non-linear transformation of the input data (iii) output layer, which computes desired result through a set of linear output neurons in an iterative process. The output (T) depends on the weighted sum (w) of input variable (T) plus additional term called bias (b) to ensure numerical stability [16]. *MATLAB* software was used to implement the procedures. The procedures [15] is expressed as shown in equations 2–3:

$$p_j(n) = f\left(U_j(n)\right) = f(\sum_{i=0}^{n_i} \omega_{ji} x_i(n) + b_j)\ for\ j = 1,2,3,\dots\dots\dots.n_j \tag{2}$$

$$q_k(n) = f\left(U_k(n)\right) = f(\sum_{j=0}^{n_j} \omega_{kj} p_j(n) + b_k)\ for\ k = 1,2,3,\dots\dots\dots.n_k \tag{3}$$

where $x_i(n)$ = input to neuron i, $p_j(n)$ is the parameter estimated by neuron j of the hidden layer, $q_k(n)$ is the output calculated by the neuron k, ω_{ji} and ω_{kj} are parameters that control connection strength between input neuron i and hidden neuron j and j and k respectively. We used a non-linear hyperbolic tangent sigmoid transfer function (equation 4) and a linear transfer function in both the hidden and output layers.

$$f(U_j) = \frac{1 - e^{-2U_j}}{1 + e^{-2U_j}} \tag{4}$$

whereU_j ε [-∞ ∞] and f (U_j) is bounded on (-1 1).

We used trial and error method of ANN for the development of our models for the estimation of the hidden layer's neutrons. We increased the numbers of neurons at each stage of analysis in order to optimize the ANN model. Back-propagation method is used for the iteration of the network learning process so as to ensure the error between the input and output data is minimized. The gradient vector of the error surface is then calculated and the networks are updated by as described in equations 5 and 6 [14].

$$\Delta\omega(n+1) = -\eta\frac{\partial f}{\partial\omega}|\omega = \omega(n) \quad (5)$$

$$\omega(n+1) = \omega(n) - \eta\frac{\partial f}{\partial\omega}|\omega = \omega(n) \quad (6)$$

where ω = strength that connects nodes of input and hidden layer and hidden and output layer nodes, n = epoch that means the amount of learning iterations, η = rate of learning which determines the number of change to connection weights of the network, Δω = increment in weight and f = sigmoid activation function.

Levenberg-Marquardt Algorithm (LMA) used in the network is a kind of backpropagation algorithm, which is a combination of two minimization techniques, namely, Gauss-Newton and gradient descent algorithms. This quality makes it sophisticated and powerful. In LMA, weights are calculated by using equation 7 [18].

$$\Delta\omega(n+1) = (J_k^T J_k + \mu I)^{-1} J_k^T e \quad (7)$$

I is identity unit matrix, k = the output nodes and J = Jacobian of input errors.

The parameter μ is adjusted automatically at each iteration of Jacobian matrix computation. When μ is small, LMA turns to Gauss-Newton algorithm, and when it is very large, it transform to gradient descent algorithm. To prevent the process of learning being trapped in local minimal as a result of assigning random weights, we trained our ANN network for a minimum of 20 times and we chose the best model based on the goodness of fit. The process of training were terminated automatically when the sum of mean square error (MSE) is found within a predefined tolerance error. The neural network structure was employed to construct the ANN model as given in equation 8 and the structures of the ANN model are also provided in Table 1.

$$T = function\ (T_R) \quad (8)$$

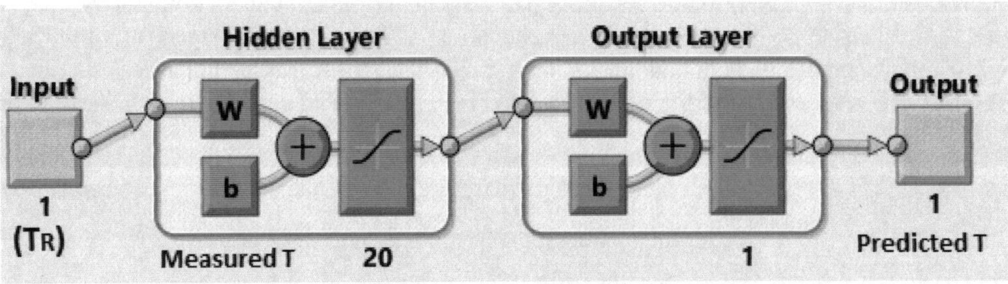

Fig. 3: Artificial neural network architecture

Table 1: The structures of the ANN model

Input neuron number	Hidden neuron number	Output neuron number	Type of network	Transfer function	Training parameter	Algorithm used for training
1	20	1	Feed-forward back-propagation	Tansig	Learning rate epochs 114	LM (Levenberg-Marquardt

3. Data interpretation

3.1 Geoelectric interpretation

The interpreted electrical resistivity survey shows that the study area is made up of 3 to 4 geoelectric layers. The dominant curve is the QH curve (Fig. 4). The typical curve types are given in Fig 5. The lithology is generally consist of top soil, laterite/clay, weathered layer and fresh basement rock.

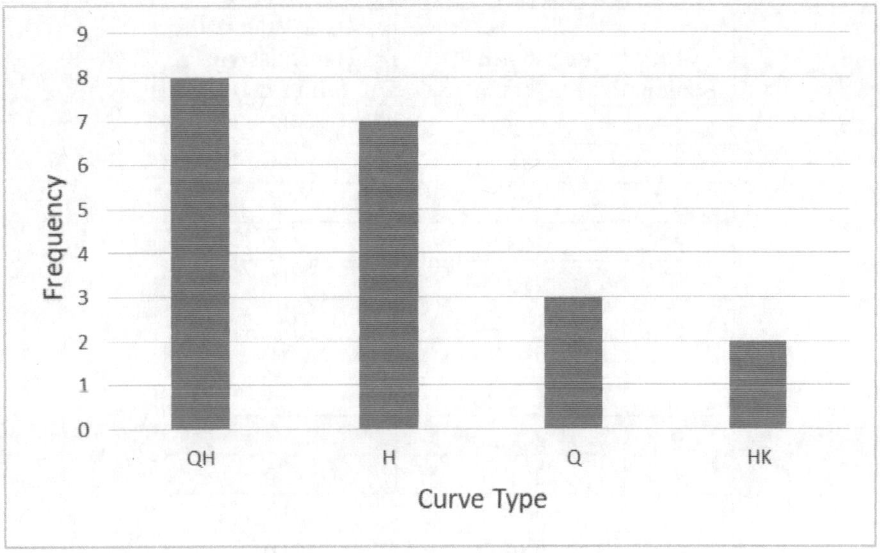

Fig. 4: The frequency of occurrence of the various curve types.

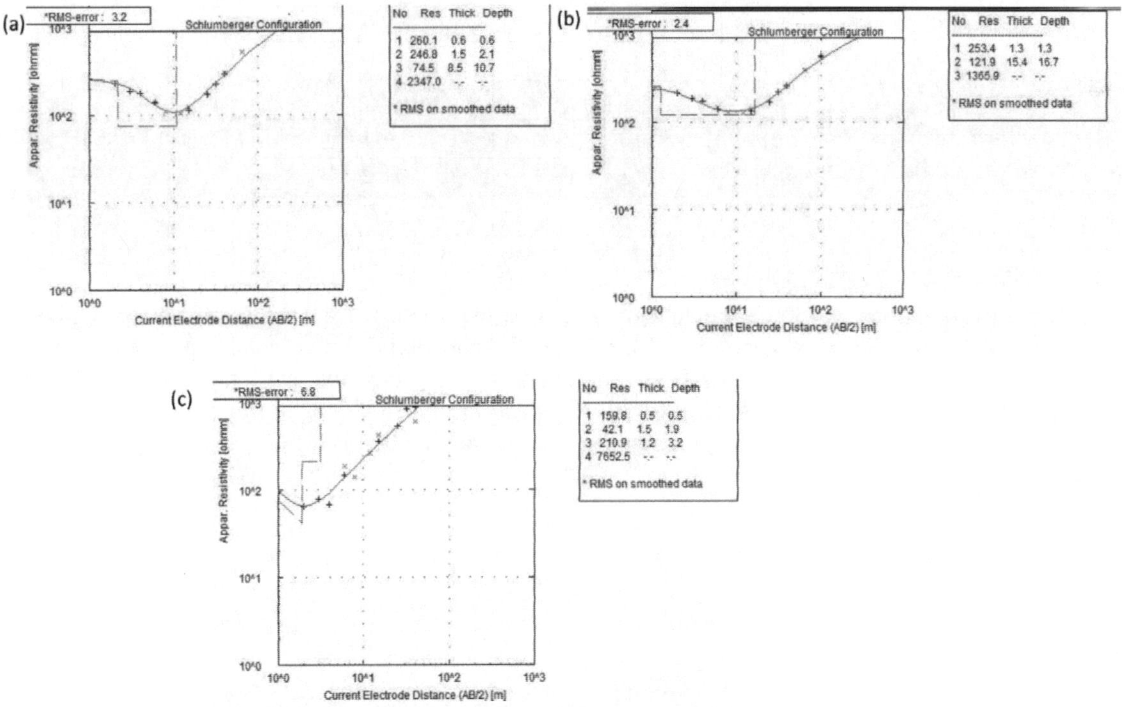

Fig. 5: Typical VES curve types in the study area (a) QH-type (b) H-type (c) HK-type

*3.1.1 Transverse Resistance.*The transverse resistance (T_R) is a parameter that describes the areas with high potentials for groundwater exploration [19]. T_R for each VES stations is presented in Table 2 as well as the measured transmissivity (T). High values of T_R indicates that the area have high potential for groundwater exploration and vice versa. T_R values vary from 133.00 to 381.20 Ωm^2 across the 20 VES whereas T range from 0.4 to 4.1 m^2/day. The highest T_R value was observed at VES station 3.

Table 2: Calculated Transverse resistance

VES Station	Transverse Resistance (T_R) (Ohm.m^2)	Measured Transmissivity (m^2/day)
1	202.10	0.8
2	276.15	0.6
3	381.20	1.5
4	168.60	0.8
5	218.90	0.5
6	133.20	0.4
7	261.50	3.9
8	172.20	4.1
9	242.20	2.2
10	261.10	3.5
11	202.10	0.8
12	276.15	0.6
13	281.00	1.5
14	136.00	4.2
15	218.90	0.5
16	130.20	0.4
17	261.50	3.9
18	172.20	4.1
19	242.20	1.2
20	261.10	2.2

3.2 Artificial neural network (ANN)

The regression model for ANN implementation is shown in Figs 6-9. Fig 6 shows the regression model for training data. The training data compute gradient, estimate and update appropriate weights and biases to capture the pattern/non-linear correlation between the input and output data. The degree of correlation of predicted transmissivity with ANN and measured values of transmissivity was found to be high for the training data with coefficient of correlation (R) that is very close to unity (R = 0.97).

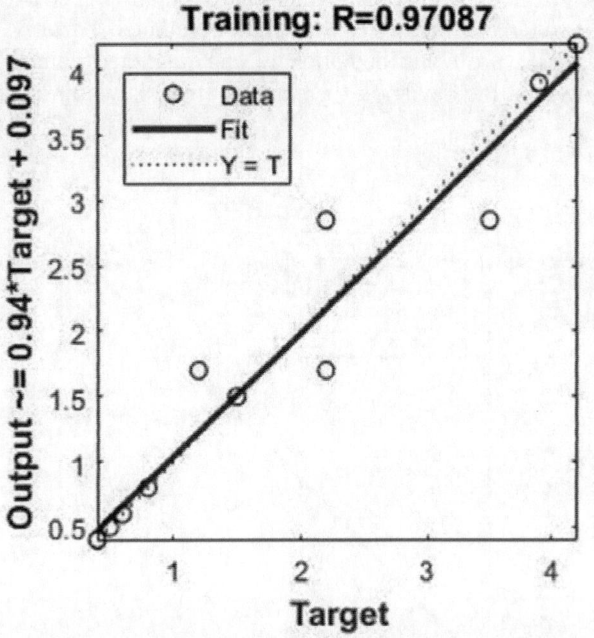

Fig. 6: Regression model for the training network

Fig 7 shows the regression model for validation. This model computes input and output parameters using a subset of the input data. It is an indication of checking the performance of prediction of the model [20-21]. The result of this model gives the correlation coefficient between the predicted and actual transmissivity to be perfect with R of 0.99. The capability of ANN to simulate transmissivity of aquifer was proven on the testing data set.

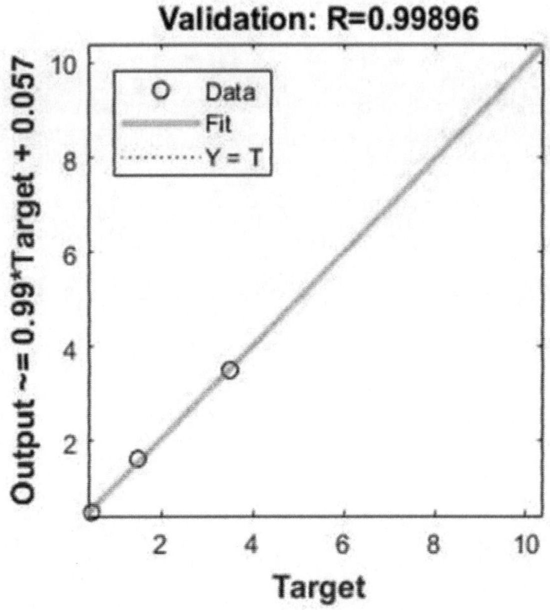

Fig. 7: Regression model for the validation network

Also Fig 8 shows the plot of testing data which has correlation coefficient of 1.0. We therefore compared the values of measured transmissivity with that of predicted T from ANN analysis as seen in Table 3. From the table, there are some deviation of the predicted values of T from the values measured on the field. However, the deviation (percentage error) is within an acceptable limit of less than 5 % [21].

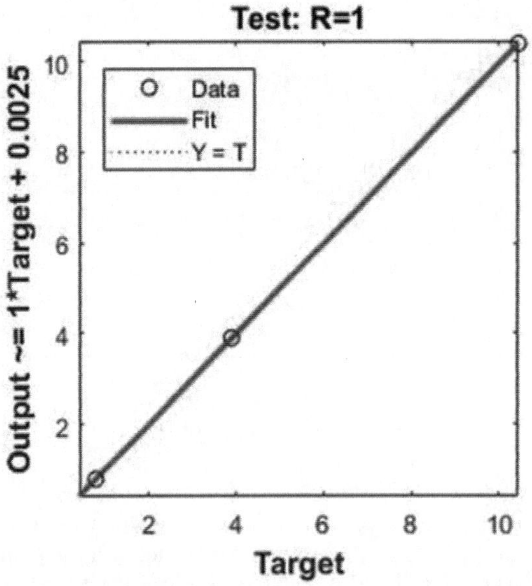

Fig. 8: Regression model for the testing network

Table 3: Measured and Predicted transmissivity

VES Station	Measured Transmissivity (m^2/day)	Predicted Transmissivity (m^2/day)	Error (%)
1	0.80	0.79	+1.25
2	0.60	0.60	0.00
3	1.50	1.51	-0.67
4	0.80	0.80	0.00
5	0.50	0.50	0.00
6	0.40	0.39	+2.5
7	3.90	3.89	+0.26
8	4.10	4.10	0.00
9	2.20	2.20	0.00
10	3.50	3.50	0.00
11	0.80	0.79	+1.25
12	0.60	0.60	0.00
13	1.50	1.50	0.00
14	4.20	4.22	-0.48
15	0.50	0.51	-2.00
16	0.40	0.40	0.00
17	3.90	3.90	0.00
18	4.10	4.10	0.00
19	1.20	1.20	0.00
20	2.20	2.21	-0.45

Moreover, the 'all' network phase of the model combined training, testing and validation plots as seen in Fig 9. The network phase shows a very good relationship between the measured and predicted T with coefficient of correlation (R) of 0.94.

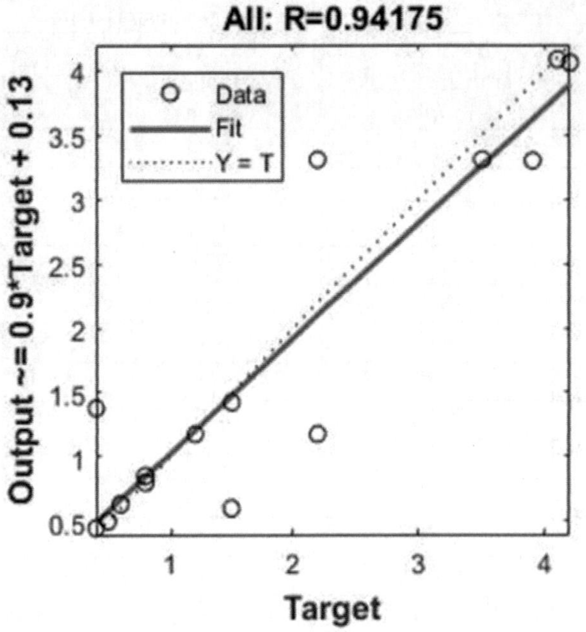

Fig. 9: Regression model for network phase

The plot of the neural network model generated is given in Fig. 10. This plot shows the relationship between the measured transmissivity and the predicted transmissivity using the ANN model developed in this study. The figure shows a very good correlation between the values of our predicted and measured T ($R^2 = 1.00$).

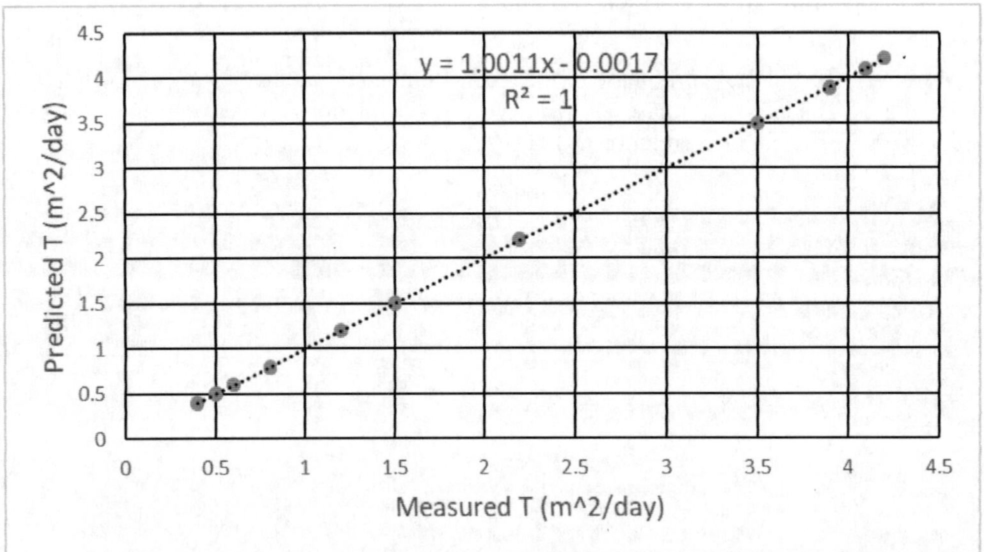

Fig. 10: Plot of the predicted transmissivity versus measured transmissivity

We also estimated root mean square error (RMSE) to examine the effectiveness of the predictive ANN model (equation 9).

$$RMSE = \sqrt{\frac{\Sigma(y - y')^2}{N}} \quad (9)$$

where y represents the measured parameter, y′ represents the predicted parameters and N is the total data points which is 20 in this study.
The implication of this statistical index is that a good model should have low values of RMSE. However, the calculated RMSE for our model shows a value of 0.085. The low value of the calculated RMSE indicated high prediction performances for our mode

4. Conclusions
In this study, ANN have been used to predict T from geoelectric data in a basement complex area of Southwest, Nigeria. The measurement of the goodness of the fit statistics of our ANN model reveals that the ANN model generated in the study area are reliable in predicting T. The determination of correlation coefficient between measured and predicted T values was 1.0. Also, the RMSE value for the ANN model was found to be 0.085 which indicate high performances of our model. A linear relationship was suggested for the ANN analysis to predict T. This would lead to reduction in cost, effort and time to carry out standard laboratory procedure or pumping test to determine T of aquifer from several discrete points along several transects. However, it is worthy to note that the predicted equation derived by the authors are valid only for understudied aquifer with similar characteristics or geology as the study area. Further studies are needed for the verification of these relationships.

Acknowledgments
The authors thank KFUPM for the provision of valuable software.

References
[1] Chandra S, Kumar D, Ahmed S, Perrin J, Dewandel B (2008)"Contribution of geophysical methods in exploration and assessment of groundwater in hard rock aquifers". The 3rd International Conference on Water Resources and Arid Environments and the 1st Arab Water Forum.
[2] Mažac O, Kelly WE, Landa I (1985) A hydrogeological model for relations between electrical and hydraulic properties of aquifers. J. Hydrol. 79: 1-19.
[3] Soupios PM, Kouli M, Vallianatos F, Vafidis A, Stavroulakis G (2007) Estimation of aquifer hydraulic parameters from surficial geophysical methods: a case study of Keritis Basin in Chania (Crete-Greece). Journal of Hydrology 338: 122-131. doi:10.1016/j.jhydrol.2007.02.028.
[4] Sri Niwas, Lima OAL (2003) Aquifer parameter estimation from surface resistivity data. Ground Water 41(1): 95-99.
[5] Chand R, Chandra S, Rao VA, Singh VS, Jain SC (2004) Estimation of natural recharge and its dependency on subsurface geoelectric parameters. J. Hydrol. 299, 67-83.
[6] Singh KP (2005) Nonlinear estimation of aquifer parameters from surficial resistivity measurements. Hydol. Earth Syst. Sci. Discuss. 2: 917-938
[7] Chandra S (2006) Contribution of geophysical properties in estimating hydrogeological parameters of an aquifer. PhD, BHU, Varanasi.
[8] Sri Niwas, Singhal DC (1981) Estimation of aquifer transmissivity from Dar-Zarroukparameters in porous media. Journal of Hydrology 50: 393-399. doi:10.1016/0022-1694(81)90082-2.
[9] Ahmed S, Marsily GD, Talbot A (1988) Combined use of hydraulic and electrical properties of an aquifer in a geostatiscal estimation of transmissivity. Ground Water 26(1): 78-86.
[10] Yadav GS, Abolfazli H (1998) Geoelectrical soundings and their relationship to hydraulic

parameters in semi-arid regions of Jalore, Northwestern India. Jour. Appl..Geophys. 39: 35-51.

[11] Ademilua, O. L. (1997) A Geoelectric and geologic evaluation of groundwater potential of Ekiti and Ondo States, Southwestern, Nigeria. Unpublished M.Sc. Thesis, Dept. of Geology, Obafemi Awolowo University, Ile-Ife, Nigeria. pp. 1- 67

[12] Zohdy AAR, Eaton GP, Mabey DR (1974) Application of surface geophysics to Ground-Water Investigation. Techniques of Water Resources Investigation, US Geological Survey.

[13] Marechal JC, Dewandel B, Subrahmanyam K (2004) Use of hydraulic test at different scales to characterize fracture network properties in the weathered-fractured layer of hard rock aquifer. Water Resour. Res. 40 (W11508): 1-17.

[14] Haykin SS (2009) Neural networks and learning machines. Pearson Education Upper Saddle River, vol. 3.

[15] Singh UK, Tiwari R, Singh S (2013) Neural network modeling and prediction of resistivity structures using VES Schlumberger data over a geothermal area. Computers & Geosciences 52: 246–257.

[16] Atkinson PM, Tatnall A (1997) Introduction neural networks in remote sensing. International Journal of Remote Sensing 18(4): 699–709.

[17] Chantasut N, Charoenjit C, Tanprasert C (2004) Predictive mining of rainfall predictions using artificial neural networks for Chao Phraya River. 4th International Conference of the Asian Federation of Information Technology in Agriculture and the 2nd World Congress on Computers in Agriculture and Natural Resources, Bangkok, Thailand, August, pp. 9–12.

[18] Ranganathan A (2004) Thelevenberg-marquardt algorithm. Tutorial on LM algorithm 11(1): 101–110.

[19] Worthington PF (1977) Geophysical investigation of groundwater resources on the Kalibari Basin. Geophysics 42 (4): 838-849.

[20] Gokceoglu C, Zorlu K. (2004) A fuzzy model to predict the uniaxial compressive strength and the modulus of elasticity of a problematic rock. Engineering Applications of Artificial Intelligence 17(1): 61–72.

[21] SanuadeOA,AdesinaRB, AmosunJO, FajanaAO, Olaseeni OG (2017) Using artificial neural network to predict dry density of soil from thermal conductivity. RMZ – M&G 64:.1–12.

Monitoring of seepages around dams using geophysical methods: a brief review

O.T. Kayode[1], A.M. Odukoya[2], T.A. Adagunodo[1], A.A. Adeniji[3]

[1]Department of Physics, Geophysics Unit, Covenant University, Ota, Ogun State, Nigeria
[2]Department of Geosciences, University of Lagos, Lagos State, Nigeria
[3]Department of Physics and Solar Energy, Bowen University, Iwo, Osun State, Nigeria

solajesreign@yahoo.com; olusola.kayode@covenantuniversity.edu.ng; taadagunodo@yahoo.com

Abstract. The slow escape of a liquid or gas through a porous material or small openings is called seepage. It is a process of seeping in soil engineering whereby water in soils move. Seepages often pose a grave problem in building foundations and also a common problem in earth dams (Dams are structures built to retain water/fluild) due to abnormal or excessive leakage. Seepage through or around dams have been responsible for most dam failures. Dam failures are usually catastrophic with many fatalities and causing the destruction of infrastructure and properties. Therefore, monitoring of seepage through and around dams becomes a necessity in other to maintain dam's stability. The main field of application of this investigation is subjected to site characterization and foundation quality assurance. This paper reviews the geophysical techniques which have been considered for the monitoring and control of seepages around dams.

1. Introduction

The incessant incidence of dam failure is becoming an important topic in the past few decades. The improper knowledge about the subsurface geological material, inadequate information about the soil, poor dam foundation design and also seepages around dams have been attributed to the cause of this failure (Lukman et al., 2011). In other to prevent the significant potential hazards that always accompany such failures there is an urgent need for seepage studies (Close et al., 2002). Hence, a need to discuss the importance of geophysical techniques for the thorough mapping of dams, which is a vital factor in civil engineering as well as hydrogeological sectors.

Geophysical methods have contributed immensely to the monitoring of seepages around dams and to various aspects of human endeavours (Adagunodo et al. 2013a; Adagunodo et al. 2014) and more attention should be given to its intrinsic applications in environmental studies. Rapid and cost effective subsurface information/geology of large area are made possible by the applications of geophysical methods (Telford et al., 1976). Generally, dam and reservoirs are basically constructed for irrigation, water supply, hydro-electric power generation, flood control or some combination (Greager and Hinds, 2002). Many advanced countries like USA and China, make use of dams and reservoirs in such that the development of China's hydropower and its regulation of water resources were as a result of the construction of dam and reservoir projects (Wang et al., 2012).

Major earth dam failures are due to seepages, although if it is within the design limit the stability is not threatened but excessive seepage will eventually lead to dam failure because seepage flow renders dam vulnerable to failure during their staged construction (Wang et al. 2012). The combined ERT results and geotechnical monitoring data are good indicators of likely mechanism for abnormal seepages (Lin et al., 2013). Most of the available seepage monitoring systems are insensitive to small changes in the seepage flow. However, the non-intrusive geophysical techniques have been on high demand to facilitate early detection or diagnosis of anomalous seepage occurences (Voronkov et al., 2004).

However, the relevance of geophysical methods in the investigation of seepage in earth-dam have been established in several works. Among this is the electrical resistivity tomography (ERT) method which have been progressively applied for seepage investigations, dams and buildings stability control (Okko et al. 1994; Abuzeid 1994; Panthulu et al. 2001; Karastathis et al. 2002; Turkmen et al. 2002; Oh et al. 2003; Lim et al. 2004; Sjodahl et al. 2005; Song et al. 2005; Kim et al. 2007 and Cho and Yeom 2007; Adagunodo et al. 2013b; 2015; 2017; Akinwumi 2014; Kayode and Akinwumi, 2014; Adewoyin et al. 2017). Also, Chin-ping et al., (2013) applied ERT to the investigation of seepage in earth dams in Taiwan. This particular dam was reconstructed to increase its water level and two-dimensional (2D) ERT was deployed at the study area to investigate several abnormal leaks that was observed on the downstream portion of the earth-dam. For time-lapse measurements, periodic measurements were collected on the downstream side. ERT results were combined with geotechnical data to indicate clearly the seemly mechanism of the abnormal seepage.

Seepage anomalies can be further enhanced in its capabilities by time-lapse measurement (Johansson and Dahlin 1996; Sjödahl et al. 2010). Also, Ani and Arewa (2013) used 2D electrical resistivity tomography to image suspected seepage channels in an earth dam in Zaria, Nigeria. This method was used to diagnose the variations in the volume of water content in the dam whether it is as a result of an anomalous seepage at the subsurface or a mere seasonal effect. Various zones of relatively uniform resistivity values were mapped and identified from the interpreted acquired data. The results revealed that the rather low resistivity zones within the bedrock represent probable pathways for seepages. Hence, geophysical method was successfully used to detect and map the seepage pathways within the bedrock of the earth dam (Ani and Arewa, 2013; Alemaw et al. 2016).

Moreso, Walid Al-fares (2011; 2014) carried out survey involving various geophysical methods such as the superficial electromagnetic (EM), electrical sounding and ERT to identify and discern the cause of the water losses problem in the Afamia B dam located at Al- Ghab basin in Syria. He noted that the ERT technique was more effective and convenient to detect water infiltration problems and leakages through the bedrocks. The integrated results of the methods revealed some geological structures that have adverse effect which may eventually lead to leakages in the dam foundation. The vertical infiltration in the dam was later detected as the principal cause for the seepage problem.

Also, Oladapo et al. (2013) carried out a feasibility study for constructing a dam on Gurara dam phase II using seismic refraction method. Off-end/reverse and split spread shootings were the seismic field techniques employed on the field. The study site is on the crystalline basement complex rocks of North Central Nigeria. The results showed significant seepage features within the bedrock, that is, on the western flank of the proposed dam axis. Significant fractured bedrock with probability for seepage was delineated on the eastern flank of the dam. The overall result revealed the competence and suitability of the materials investigated within the dam.

The present day seepage pattern and hydrogeological mechanisms that influence seepages were carried out in Raymond, California (Burke et al. 2011) using the self-potential and direct current resistivity surveys. The background knowledge of the hydrogeology and the information of a typical subsurface structures from resistivity surveys aided the interpretations as several flow scenarios were investigated. Evidence of seepage was discovered from the self-potential data interpretation of the dam which is consistent with past observations (Sjodahl et al. 2009). When properly integrated geophysical surveys, well data and numerical modeling can provide better understanding of seepages at sites. (Burke et al. 2011). In view of the above, this article utilizes the performance of ERT and other related geophysical methods to detect the abnormal seepage at earth dams and to further understand the possible ways of monitoring abnormal seepage in earth-dams.

2. Causes of seepage

So many factors have been researched to lead to the cause of seepage in dams. They include; open scams, joint in rocks, unconsolidated rocks in the foundation, poorly compacted soil, rat holes, digging of drains, deep rooted trees, cracks, frost action, shrinkage cracking in the environmental soil, uprooted trees, earthquakes; excessive uplift pressure and trapped groundwater and insufficient structural drainage (Greager et al. 2002; FERM, 2006; Flores-Berrones et al. 2011; Lukman et al 2011; Omofunmi et al. 2017).

Seepage is peculiar to all earth dams because water percolates slowly through dams and foundations. If seepage forces are large enough they may lead to eroded soil which may weaken the soil and eventually result in landslides. Seepage may result in environmental failure if not properly monitored (Auvinet et al 2010).

Other recognizable signs of seepage according to Auvinet et al. (2010) and Omofunmi et al. (2017) are

i. Seasonal temperature variation inside the dam.
ii. Rapid water level decrease below the level expected with normal use.
iii. Accumulated portion of water in areas around the dam site.
iv. Reduced growth rate in vegetation around the dam compared with other places far away.
v. Change of vegetation colour with distorted growth.

2.1 Effects of seepages on earth dams

Dams and embankments are most vulnerable to seepage during water conditions. Piping, excessive internal pressure or saturation, solutioning of soluble rocks such as gypsum, limestone and rock-salt, internal erosion, excessive uplift and heave or blowout, are some of the common seepage problems which could lead to dam failures (Alberro, 2006; Omofunmi et al. 2017).

2.1.1 Piping

Piping is a kind of seepage failure mode that occurs under the influence of seepage force in soil materials. The poor soils formed channels which transport materials downstream. These channels (hollow) grow wider as a result of infiltration of materials downstream and if not quickly checked it may eventually lead to subsidence of the dam. (Wang et al. 2017; Omofunmi et al. 2017).

2.1.2 Internal erosion

Internal erosion is another common way that water can damage embankment dams. This is the movement of water in cracks or other defects along borderlines of soil and bedrock (FERC, 2006; Fell et al. 2003; Flores-Berrones et al. 2011). Failures of internal erosion are commonly found in areas with slicken slides, careless construction, cracks and horizontal piping (FERC, 2006).

2.1.3 Solutioning

Dam foundation on soluble rocks that dissolve in groundwater can cause seepage problems. Rainfall dissolves soluble rocks in zones above the water table and below by the groundwater movement. Extreme caution should be taken on dams to be constructed on materials prone to solutioning such as gypsum, rock salts and limestone rocks (FERC, 2006; Flores-Berrones et al., (2011). The dissolved salts if not checked may gradually lead to the loss of strength of subsurface formations.

2.1.4 Internal pressures and saturations
Seepage normally takes place at the foundation soil of the dam especially when water is stored in the dam. Pore pressure, reduced shear strength are some of the adverse effects of seepages (FERC, 2006). The dam becomes so attenuated to withstand the water pressure as a result of continuous saturation and slumping processes and this may eventually leads to dam failure (FERC, 2006).

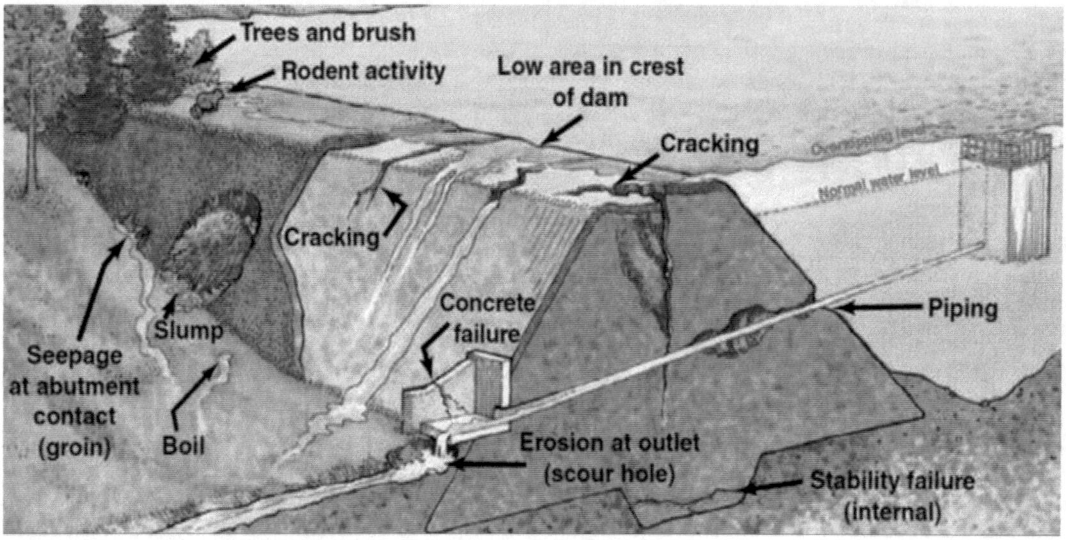

Figure 1: Effects of seepages in dams. Courtesy: the constructor

2.1.5 Blowout, uplift or heave
This is one of the symptoms of internal pressure and saturation. It occurs when there is an excessive uplift pressure, that is, the previous layer apply an excessive force on an overlying confining layer at the seepage foundation. Dam failure occurs when the pore pressure on the underlying confining layer surpasses the overburden pressure of the overlying soils. This scenario eventually results into a blowout, commonly forming a boil as it breaks through or breaches the confining layer (FERC, 2006).

3. Methodology and Applications

3.1 Electrical Resistivity Tomography (ERT)
Electrical resistivity method is non-intrusive geophysical method whereby profiles of the length of the proposed dam of investigation were spread in up and downstream sides, parallel and close to the dam embankment. Resistivity profiles have been successfully used to detect seepage paths in dams (Panthulu, 2001). The intrinsic electrical resistance of materials and the resistance contrasts between them is basically what the geophysical resistivity methods rely on. The resistivity of earth materials is determined by injecting electrical current into the ground and measuring the resulting potential difference. Self -potential (SP) or Induced polarization (IP) method is another proven method for monitoring seepage in the subsurface (Panthulu, 2001). Electric potential generated from fluid flow can be recorded on the surface survey lines. Hence, a comparison of SP/IP surveys from different reservoir levels can actually reveal seepage flow path. However, a combination of SP/IP and resistivity profiles can facilitate the detection and delineation of seepage path in dams (Brosten et al. 2005).

3.2 Electromagnetic method (EM)
This is also a non-intrusive geophysical method commonly used for seepage monitoring in dams. Low or high conductivities produced from air filled subsurface features and water or clay-filled features respectively are good indicators for seepage pathways using EM (Boston et al. 2005). EM can be used

selectively to correlate survey results to known and existing seeps or groundwater seepage sources. Understanding of site hydrogeology of the study area is necessary when using EM so that the presence of power lines, buried cables, weak electrical conductors and clay layers in the soils can be delineated. Changes in water conductivity due to changing ion concentration and other related changes in groundwater should also be considered when using this method (Montgomery and Kofoed 2001).

3.3 Seismic refraction method

This geophysical method has also been used for seepage monitoring in dams but rarely used because of its high cost implication. The method has possibility to give an image of the subsurface and map lateral and vertical variations in the subsurface geology of the required site under investigation (Mustafa et al. 2012). A combination of seismic refraction and resistivity methods will give robust information of the seepage pathways in and around dams.

4. Conclusion

Monitoring of seepages around embankment dams is very important in civil engineering and hydro-geophysical structures. Generally, the effects of seepage on dams are functions of soil water retention capability which is a vital factor in engineering geology. Geological mapping and field investigations for dam projects should delineate soil properties and rock formations that could cause seepage failures. An in depth study of the geological and geotechnical properties of these materials should be carried out. A combination of electrical resistivity and other tools such as self-potential/Induced polarization have proven to give quantity and quality interpretation of seepage flow. Performing resistivity and self-potential profiles along several lines across the abutment permits a quasi-three dimensional view of the seepage pattern to be obtained (Brosten et al. 2005). Lastly, geophysicists and the civil engineers should work hand in hand in other to build and maintain dam structures to avoid the damage and waste of resources that may eventually result in the project if done otherwise.

Appreciation

We appreciate the management of Covenant University, Nigeria for publication support received from them.

References

Abuzeid N. (1994). Investigation of channel seepage areas at the existing Kaffrein dam site Jordan using electrical-resistivity measurements. *Journal of Applied Geophysics*, 32: 163-175.

Adewoyin O. O., Joshua E. O., Akinyemi M. L., Omeje M and Joel E.S (2017). Application of 2D electrical resistivity imaging and cone penetration test to assess harzardous effect of near surface water on foundations in Lagos Nigeria, Journal of Physics-Conference series (1) 852.

Adagunodo T. A., Adeniji A. A., Erinle A. V., Akinwumi S. A., Adewoyin O. O., Joel E. S., Kayode O. T. (2017). Geophysical Investigation into the Integrity of a Reclaimed Open Dumpsite for Civil Engineering Purpose. Interciencia Journal, 42(11): 324 – 339.

Adagunodo T. A., Sunmonu L. A. and Adeniji A. A. (2015). Effect of Dynamic Pattern of the Saprolitic Zone and its Basement on Building Stability: a Case Study of a High-Rise Building in Ogbomoso. *Journal of Applied Physical Science International*, 3(3): 106–115.

Adagunodo T.A., Sunmonu L.A. and Oladejo O.P. (2014). Electromagnetic Investigation into the cause(s) of Road Failure along Takie-Ikoyi Road, Ogbomoso. *International Journal of Business and Applied Sciences*, 1(1), 78–85.

Adagunodo T.A., Sunmonu L.A., Oladejo O.P. and Ojoawo I.A. (2013a). Vertical Electrical Sounding to Determine Fracture Distribution at Adumasun Area, Oniye, Southwestern Nigeria. *IOSR Journal of Applied Geology and Geophysics*, 1(3), 10–22.

Adagunodo T. A., Sunmonu L. A., Ojoawo A., Oladejo O. P. and Olafisoye E. R. (2013b). The Hydro Geophysical Investigation of Oyo State Industrial Estate Ogbomosho, Southwestern Nigeria Using Vertical Electrical Soundings. *Research Journal of Applied Sciences, Engineering and Technology*, 5(5), 1816–1829.

Akinwumi, I.I. (2014) "Soil Modification by the Application of Steel Slag", Periodica Polytechnica Civil Engineering, 58(4), 371–377.

Alemaw B. F, Keaitse E.O and Chaoka T.R (2016). Management of water supply reservoirs under uncertainties in arid and urbanized environment, Journal of Water Resources and Protection, 8, 990 – 1009.

Alberro J. (2006) Effect of transient flows on the behaviour of earth structures, 8th Nabor Carrillo Lecture, Mexican society of geotechnical engineering SMIG, Tuxtla Gutierrez, Chiapas Mexico

Ani D. C and Arewa J. O (2013). Electrical Resistivity Imaging of suspected seepage channels in an Earthen Dam in Zaria, North – Western Nigeria. Open Journal of Applied Sciences, 3: 145 – 154.

Auvinet G. and Lopez-Acosta N. P (2010). Rapid drawdown condition in submerged slopes. A presentation at the Institute of Engineering and the Mexican Society of Geotechnical Engineering (SMIG). 1, 67189.

Brosten, T.R., J.L. Llopis, and J.R. Kelley (2005). "Using Geophysics to Assess the Condition of Small Embankment Dams," ERDC/GSL TR-05-17. U.S. Army Corps of Engineers Research and Development Center, Vicksburg, Mississippi.

Burke J., Minsley L., Bethany L., Burton1., Scott I. and Michael H. P. (2011). Hydrogeophysical Investigations at Hidden Dam, Raymond, California. Journal of Environmental and Engineering geophysics, 16 (4): 145–164.

Close ME, Stanton GJ, Pang L (2002). Use of rhodamine WT with XAD7 resin for determining groundwater flow paths. Journal of Hydrogeology, 10: 368 - 376.

Chin-Ping Lin, Yin-Chun Hung, Zen-Hung Yu, and Po-Lin Wu (2013). Investigation of abnormal seepages in an earth dam using resistivity tomography. *Journal of Geo-Engineering,* 8 (2): 61 – 70.

Cho I. K., and Yeom J. Y (2007). Crossline resistivity tomography for the delineation of anomalous seepage pathways in an embankment dam. *Geophysics,* 72, 31 - 38.

Fell R., Wan C. and Foster M (2003) Time for development of internal erosion and piping in embankment dams. *Journal of Geotechnical and Geoenvironmental Engineering,*129, 307.

Federal Energy Regulatory Commission (2006). Training aids for dams safety evaluation of seepage. 18 - 27.

Flores-Berrones R., Ramirez Reynaga M and Macari E. J (2011). Internal erosion and rehabilitation of an earth dam. *Journal of Geotechnical and Geoenvironmental Engineering,* 137 (2): 150 - 160.

Greager V and Hinds N (2002). Engineering of dams John Wiley and sons, New York.

Johansson S. and Dahlin T. (1996). Seepage monitoring in an earth embankment dam repeated resistivity measurements. *European Journal of Environmental and Engineering geophysics,* 1: 229 – 247.

Lukman S, Otun S. A, Adie D. B, Ismail A and Oke I. A (2011). A brief assessment of a dam and its failure prevention. *Journal of Failure Analysis and Prevention*, 11, 97 – 109.

Lim H. D., Kim K. S., Kim J. H., Kwon H. S and., Oh B, H (2004). Leakage detection of earth dam using geophysical methods. International Commission on Large Dams, 72th Annual Meeting, 212 - 224.

Lin C., Hung Y., Yu H., and Wu P (2013). Investigation of abnormal seepages in an earth dam using resistivity tomography, *Journal of GeoEngineering*, 8, (2): 61- 70.

Karastathis, V. K., Karmis, P. N., Drakatos, G., and Stavrakakis, G. (2002). Geophysical methods contributing to the testing of concrete dams, application at the Marathon Dam. *Journal of Applied Geophysics*, 50, 247 - 260.

Kayode, O.T.; Akinwumi, I.I. (2017). Residual soils derived from charnockite and migmatite as road pavement layer materials. Journal of Materials and Environmental Science, 8(2), 657–665.

Kim J. H., Yi M. J., Song Y., Seol S. J. and Kim K. S (2007). Application of geophysical methods to the safety analysis of an earth dam. *Journal of Environmental and Engineering Geophysics,* 12, 221 - 235.

Montgomery, J.R. and V.O. Kofoed (2001). "Mapping, tracking, and monitoring flow paths through earthen dams," in: Dam Safety 2001, the 18th annual conference of the Association of State Dam Safety Officials, held September 9–13, 2001, in Snowbird, Utah: Lexington, Kentucky: Association of State Dam Safety Officials.

Mustafa S.R, Ibrahim E. H, Elawadi E, Metwaly M, Agami N.A (2012). Seismic refraction and resistivity imaging for assessment of groundwater seepage under a dam site, southwest of Saudi Arabia. *International Journal of the Physical Sciences*, 7 (48): 6230 – 6239.

Oh Y. C., Jeong H. S., Lee Y. K. and Shon H. (2003). Safety evaluation of rock-fill dam by seismic (MASW) and resistivity method. Proceedings of the 16th Annual Symposium on the Application of Geophysics to Engineering and Environmental Problems, 1377 - 1386.

Okko O., Hassinen P. and Korkealaakso, J. (1994). Location of Leakage Paths Below Earth Dams by Geophysical Techniques: Proceedings of the 13th International Conference on Soil Mechanics and Foundation Engineering (ICSMFE) New Delhi, 1349 – 1352.

Omofunmi E. O; Kolo J. G, Oladipo A. S, Diabana P. D and Ojo A. S. (2017). A review on effects and control of seepage through earth-fill dams, *Current Journal of Applied Science and Technology* 22 (5): 1 – 11.

Oladapo M. I., Adeoye O., Oluwakemi O., Adebobuyi F. S., Badejo O. and Ifarajimi W. (2013). Seismic refraction study of Gurara dam phase II, northwestern Nigeria. *Journal of Geology and Mining Research,* 5 (11): 239 – 249.

Panthulu, T.V., Krishnaiah C. and Shirke, J. M (2001). Detection of seepage paths in earth dams using self potential and electrical resistivity methods. *Engineering Geology*, 59, 281–295.

Sjodahl P., Dahlin, T. and Johansson, S (2005). Using resistivity measurements for dam safety evaluation at Enemossen tailings dam in southern Sweden. *Environmental Geology,* 49, 267 - 273.

Sjodahl P, Dahlin T and Johansson S (2009). Embankment dam seepage evaluation from resisitivity monitoring data, *Near Surface Geophysics,* 7 (5, 6): 463 – 474.

Sjödahl, P., Dahlin, T., and Johansson, S. (2010). Using the resistivity method for leakage detection in a blind test at the Røssvation embankment dam test facility in Norway. *Bulletin of Engineering Geology and the Environment*, 69, 643 - 658.

Song SH, Song YH, Kwon BD (2005). Application of hydrogeological and geophysical methods to delineate leakage pathways in an earth fill dam. *Exploration Geophysics,* 36: 92-96.

Telford W. M., Geldart L. P., Sheriff R. E. and Keys D. A (1976). Applied Geophysics.

Cambridge University Press.

Turkmen S, Ozguler E, Taga H and Karaogullarindan T (2002). Seepage problems in the karstic limestone foundation of the Kalecik dam south Turkey, *Engineering Geology*, 63 (3): 247 – 257.

Voronkov, O. K., Kagan, A. A., Krivonogova, N. F., Glagovsky, V. B., and Prokopovich, V. S. (2004). Geophysical methods and identification of embankment dam parameters. Proceedings of the 2nd International Conference on Geotechnical and Geophysical Site Characterization, 593 - 599.

Walid A. (2011). Contribution of the geophysical methods in characterizing the water leakage in Afamia B dam, Syria.*Journal of Applied Geophysics,* 75 (3): 464 – 471.

Walid A. (2014). Application of Electrical Resistivity Tomography Technique for Characterizing Leakage Problem in Abu Baara Earth Dam, Syria. *International Journal of Geophysics,* 1 – 9.

Wang Q. G., Du Y. H., Su Y. and Che K. Q. (2012). The environmental impact post-assessment of dam and reservoir projects: a review. Procedia Environmental Sciences, 13: 1439 – 1443.

Wang Y, Li C, Zhou X and Wei X (2017). Seepage piping evolution characteristics in Bim soil –an environmental study, *Water,* 9 (458): 1 – 18.

Soil characterization using satellite remote sensing in southwestern Nigeria: Implications for precision agriculture

A P Aizebeokhai[1], U N Okenwa[1], K D Oyeyemi[1], O T Kayode[1] and G A Adeyemi[2]

[1]Applied Geophysics Unit, Department of Physics, Covenant University, Nigeria
[2]Department of Civil Engineering, Covenant University, Nigeria

ahzegbobor.aizebeokhai@covenantuniversity.edu.ng; uchenna.okenwa@stu.cu.edu.ng

Abstract. Timely and reliable soil information with respect to their nature extent and spatial distribution is very essential for the optimal utilization of available natural resources in sustained bases. The technology advances in the field of remote sensing; Geographical Information System (GIS) have augmented the efficiency of soil survey. The use of advanced computer technologies with database can be used in decision making, risk assessment and environmental modelling. Precisely this project aims at the application of high spatial resolution satellite data (LANDAT 7) for assessing soil properties which include soil pH, soil texture and soil drainage. This study was carried out in South Western Nigeria in order to map out some soil characteristics with the use of remotely sensed data to assess their variability within the area in order to improve precision agriculture within the area.

Keywords: Remote sensing, GIS, Soil characterization, Spatial distribution, Southwestern Nigeria.

1. Introduction

The growth of plant is very essential to human as it serves a lot of purpose such as the supply of food for mankind, provision of oxygen for humans, animals and the environment, protection of the soil from erosion, protection of vegetation from desertification and reduction of CO_2 present in the atmosphere. Soil, being the topmost layer of the surface of the crust, has numerous characteristic which influence its property. These characteristics such as texture, class, acidity, alkalinity, hydrology, moisture, porosity, resistivity and consistency has relative effect on their individual property such as aeration, resistance to pH change, drainage rates susceptibility to water erosion. Soil survey is a prognostic study of soil as geographic bodies; however it governs the special relations of the set of soil properties that are perceived in nature. Soil data and interpretation are very important in precision agriculture research effort. Interpretation of soil data has been essential to the development of better management practices that makes agricultural productivity effective. The interpretations of soil data are can be applied to soil with desired characteristics for specific experiments and uses.

Precise and accurate spatial characterization of soil information is critical for the modelling of environment, assessment of risk and making of decision. Several works have been done using geophysical and remotely sensed data to solve problems involving environmental contamination, groundwater, agriculture (salinity) and geotechnical engineering [1-9]. Data obtained from remote sensing is being used as a secondary source of information in digital soil mapping is being presented as a feasible substitute to the traditional form of mapping methods this is because remote sensing is cost effective and less time consuming. Digital mapping method is defined as formation and population of soil spatial data using the observational methods within laboratory within either non-spatial or spatial inference technique of soil [10] [11] [12]. This research is aimed at soil characterization in some parts of the southwestern Nigeria using remote sensing technique with a view to making recommendations as to the level of soil fertility in the area of study towards improving crop yields and agricultural productivity. The study area covers the six states in the western Nigeria (Figure 1).

Figure 1: Location map of the study area

2. Methodology

The main datasets used for this project were Sheets 1 to 8 existing hardcopy soil map of Nigeria, each of these sheet were at a visible scale of 1:650000 acquired from the Federal Department of Agricultural Land Resource (FDALR) Abuja, and four Sheets of political map of Nigeria acquired from Office of the Surveyor General of federation of Abuja Nigeria. Leica Geosystem ERDAS IMAGINE 9.1 and ESRI ArcGIS 9.3 were both use to carry out GIS operation. The data was first studied with a view to have optimal quality control of the data. Then they were scanned and spatially referenced to identifying their weakness. The scanned data were spatially referenced to WGS (1984) UTM geographic coordinate system. In GIS interface, the resulting raster images were subset, checked and vectorized on the screen. A non-spatial data base was created consisting soil parameters such as drainage, geology and pH values were generated.

3. Results and Discussion

3.1. Soil pH

Soil pH is defined as the measure of soil acidity or alkalinity and it affects the soil nutrient; the range of soil pH determines how best the nutrients in the soil and the choice of the best crop to plant in the soil. Soil pH distribution in the study area is presented in Figure 2. In Lagos, there is high dominance of acidity that ranges between 4.5 to 5.0 pH units. In Ogun state it is observed that the soil pH ranges between the acidic and alkaline. In the western part of Ogun state towards central the soil pH of 6.6 to 9.1 pH unit is of high dominance, while in the eastern part Ogun state the soil pH of 5.1 to 5.5 pH unit dominates the region, in the western part of Ogun state (especially Ota) has acidic pH unit that ranges from 3.9 to 4.5 pH unit. Considering Ondo state, the result shows that the soil pH ranges from 3.9 to 6.6 pH unit, the acidic soil pH ranges between 3.9 to 5.0 pH unit with high dominance of acidity in the northward, southward and eastern part of Ondo. It was also observed that the eastern part of the soil ranges between 6.5 to 6.6 pH unit and it is of high prominence. Ekiti state has the soil pH ranging from 3.9 to 5.6. It was observed that soil pH of value 5.1 pH unit is of high dominance while soil pH of 3.9 to

4.2 is scanty. Osun generally ranges from 3.4 to 9.1 pH unit, soil pH of 3.9 to 4.5 is of high dominance in the west while in the east the soil pH value is 5.1, and there is an even distribution of soil acidity and alkalinity. Oyo on the other hand generally ranges from 4.5 to 9.1 pH unit.

Soil pH influences the measure of chemicals and nutrients that can be dissolved in soil water, and in this way the measure of nutrients available to crops. Some nutrients are more obtainable in acidic soil while others are more obtainable in the alkaline soil. The increase of strongly acidic soils (that is less than 5 pH units) can result in poor crop productivity as a result of the following factors: manganese toxicity, aluminum toxicity, deficiency in calcium and magnesium, low levels of vital plant nutrients such as molybdenum and phosphorus.

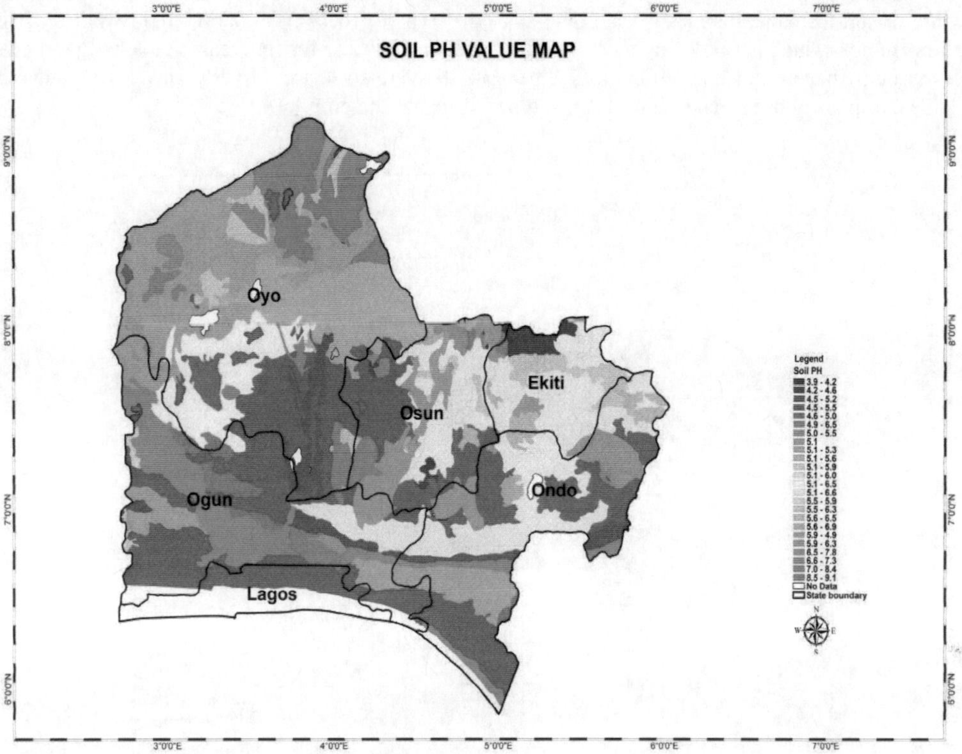

Figure 2: Soil pH map of southwestern Nigeria

3.2. Soil Texture

Soil texture is defined as the potion of sad, silt and clay size particle that make up the mineral fraction. Soil texture distribution in the study area is presented in Figure 3. The soil in Lagos has a texture of sandy clay. In Oyo, it is observed the texture sandy clay is of high dominance followed by the sandy loam, it is also observed that the texture of soil northward is mostly loamy sand and the texture of soil southward is clay loam. Considering Ogun, it is observed that the texture of sandy clay is dominating followed by the sandy loam and then the loamy sand. In Ekiti state, the texture loamy sand is predominant and then there are some traces of clay loam, sandy and concretionary clay. Ondo has an even distribution among the soil texture sand clay, sandy loam and loamy sand. In Osun, the texture loamy sand is prominent; the soil texture on the west is shown to be sandy clay while the south is composed of texture sandy loam and sandy. South western Nigeria has soil texture mostly composed of sandy loam, sandy clay, loamy sand and clay loam this indicates that it is composed of both sand and clay but there's more percentage of sand in the soil, sandy soil have larger particles than clay and very unrestricted draining due to the air space in between particles; they are easy to work on and till but they dry out very fast and usually have poor productivity as the nutrients are basically washed away. Clayey soils are poised of minute mineral particles that shrink the air space in the soil; and cause the soil to absorb water and become easily compacted. Clayey soils are usually very fertile but need some enhancement to reveal some nutrients.

Sandy loam are predominated by sand particles, but contain sufficient clay and sediment to deliver some structure and fertility, they have the capability of draining water quickly and cannot retain significant amount of water and nutrients. Growing of plants in this soil type would necessitate regular irrigation and fertilization than soils with a greater clay concentration and sediments. Sandy loam soils are often poor in specific nutrients and may require additional fertilization to support healthy crop productivity. Loamy sand comprises of more than 25 percent of medium particles, less than 25 percent of the two coarsest particle types and less than 50 percent of fine or very fine particles. Clay loam is a mixture of soil that contains mostly clay than any other type of rocks or minerals; the particles of clay are very minor, which is one of its most significant characteristics; for this reason, loams that contain excessive clay have a tendency to be heavy, because they are so dense. Usually clay loam contains a good deal of plant nutrients and supports most types of plants and crop.

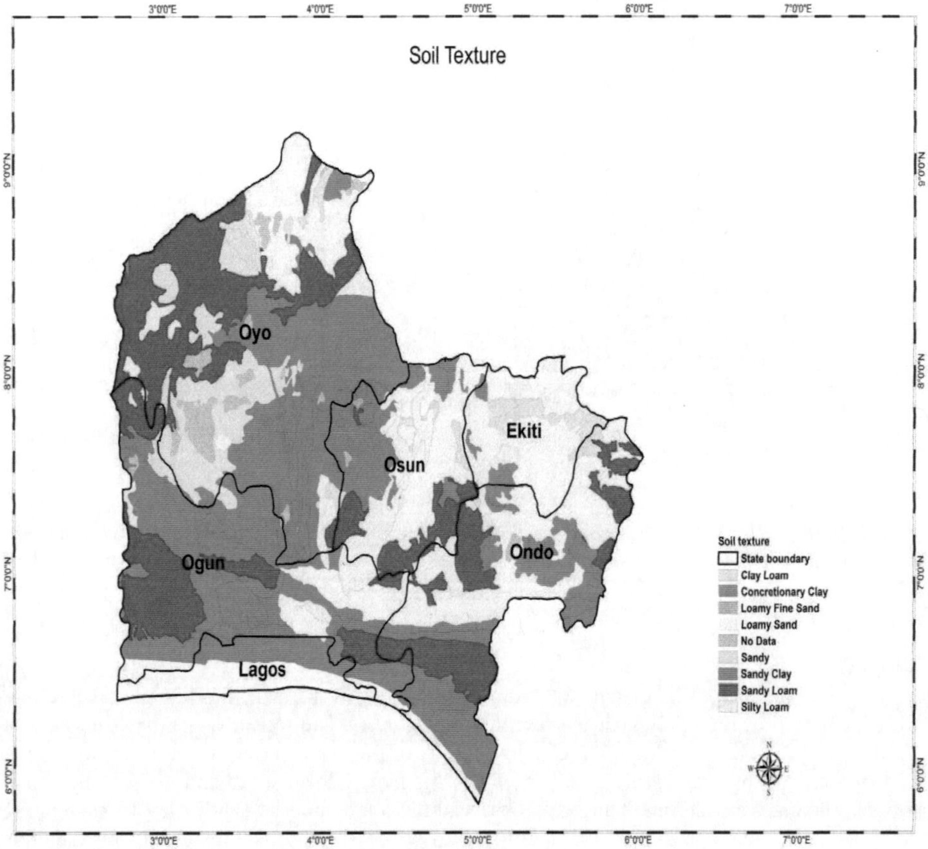

Figure 3: Soil texture map of southwestern Nigeria

3.3. Drainage Pattern

The soil in Lagos is observed to be well drained; majority of the soil in Ogun is shown to vary from well drained to somewhat poorly drained, some areas varies from moderately to poorly drained to very poorly drained. In Ekiti state, the soil varies from well drained to moderately to poorly drained soil. In Osun, the majority of the soil is well drained. In Ondo state, the soil varies from well drained to somewhat poorly drained and then to poorly drained. Well-drained soil is most prominent. In Oyo state, it is observed that the soil in the east varies from well drained to very poorly drained soil; in the west, the soil is shown to be moderately well drained. Soil drainage distribution in the study area is presented in Figure 4. In the south western, Nigeria the soil drainage varies from well drained to poorly drained

IOP Conf. Series: Earth and Environmental Science **173** (2018) 012027 doi:10.1088/1755-1315/173/1/012027

soil. One of the major physical properties of soil which is important to crop production is drainage through the rooting zone. This characteristic greatly influences aeration in the rooting zone, and the degree of aeration greatly influences several important biochemical reactions of economic importance to crop production.

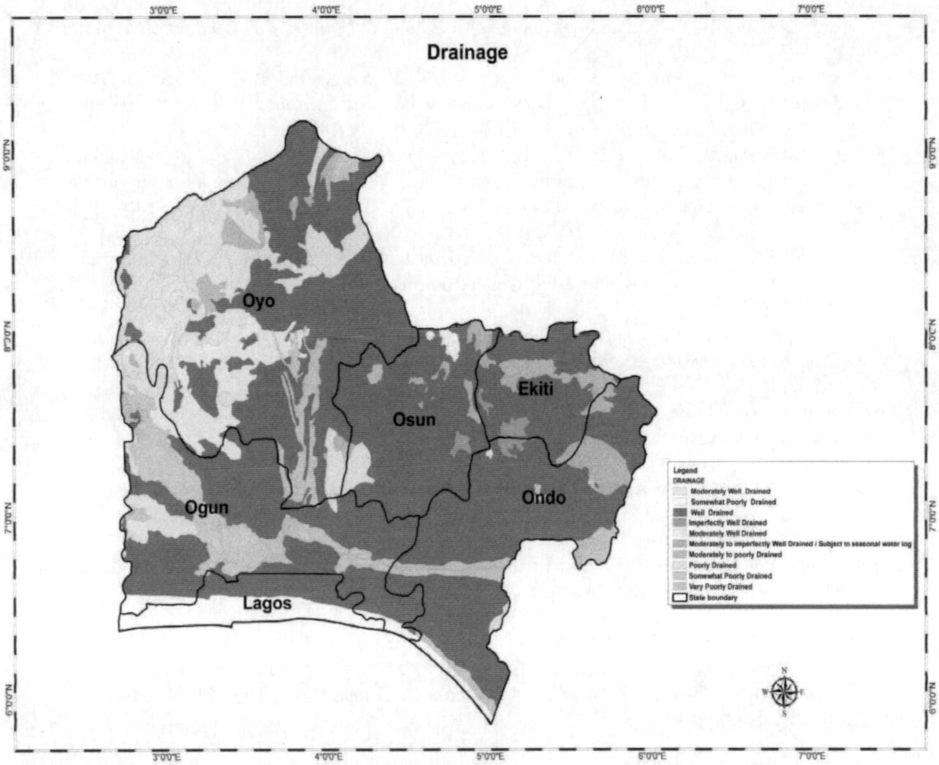

Figure 4: Soil drainage map of southwestern Nigeria.

4. Conclusion

Accurate and detailed spatial information of soil is important for modelling environment, risk assessment and decision making. This study explored the use of remote sensed data and geographic information system to map the spatial distribution of the soil properties such as soil pH, soil texture and soil drainage in southwestern Nigeria towards efficient precision farming and improved crop productivity. It is recommended that crops such as oranges, peaches, and pears are suitably for planting in Lagos state. In Ogun, it is recommended that crops such as bean, cabbage, and lettuce and water melon can be planted. In Ondo as potatoes is suitable for planting, in Ekiti crops like black berry and blue berries are recommend. It is also recommended that high resolution spatial information technique using remote sensing for digital soil mapping is more efficient than the traditional soil mapping technique which involves ground based study and laboratory analysis. This is because it is less time consuming and inexpensive.

Acknowledgements
The authors wish to appreciate the Covenant University Centre for Research, Innovation and Discovery for the conference support sponsorship.

References

[1] Aizebeokhai A P and Oyeyemi K D 2014 The use of the multiple gradient array for geoelectrical resistivity and induced polarization imaging: *Journal of Applied Geophysics* **111** 364–375 doi:10.1016/j.jappgeo.2014.10.023.

[2] Aizebeokhai A P Oyeyemi K D and Kayode O T 2015 Assement of soil petrophysical parameters using electrical resistivity tomography (ERT) and induced polarization techniques. *Research Journal of Applied Sciences* **10(9)** 479-485.

[3] Oyeyemi K D Aizebeokhai A P and Oladunjoye M A 2015 Integrated Geophysical and Geochemical investigation of saline water intrusion in a coastal alluvial terrain, Southwestern Nigeria. *International Journal of Applied Environmental Sciences* **10(4)** 275–1288.

[4] Aizebeokhai A P Oyeyemi K D and Joel E L 2016a Groundwater potential assessment in a sedimentary terrain southwestern Nigeria. *Arabian Journal of Geoscience* **9** 110–117. doi.org/10.1007/s12517-016-2524-5

[5] Aizebeokhai A P Oyeyemi K D Joel E L 2016b Electrical resistivity and induced polarization imaging for groundwater exploration. SEG International Exposition and Annual Meeting, Texas.

[6] Oyeyemi K D and Olofinnade O M 2016 Geoelectrical –Geotechnical studies for near surface characterization, case history: Lagos, SW Nigeria. *Electronic Journal of Geotechnical Engineering* **21(10)** pp 3735–3750.

[7] Aizebeokhai A P and Oyeyemi K D 2017 Geoelectrical characterization of basement aquifers: the case of Iberekodo, southwestern Nigeria. *Hydrogeology Journal* 1–14 doi.org/10.1007/s10040-017-1679-9.

[8] Aizebeokhai A P Oyeyemi K D Noiki F R, Etete B I Arere A U E Eyo U J and Ogbuehi VC 2017 Geoelectrical resistivity data sets for characterization and aquifer delineation in Iyesi, southwestern Nigeria. *Data in Brief* **15** 828–832 doi:10.1016/j.dib.2017.10.057.

[9] Oyeyemi, K D Aizebeokhai A P Adagunodo T A Olofinnade O M Sanuade A O Olaojo A A (2016). Subsoil characterization using geoelectrical and geotechnical investigations: Implications for foundation studies. *International Journal of Civil Engineering and Technology* **10(8)** pp 302-314.

[10] McBratney A B Mendonca Santos M L and Minasny B 2003. On digital soil mapping. *Geoderma*, **117(1-2)** pp 3-52.

[11] Carré F McBratney A B Mayr T Montanarella L 2007 Digital soil assessments: Beyond DSM, *Geoderma*, **142(1-2)** pp 69-79.

[12] Lagacherie P McBratney A B Voltz M 2007 Digital Soil Mapping: An Introductory Perspective, 1 (ed.), Elsevier, Amsterdam.

Estimation of aquifer hydraulic parameters from surficial geophysical methods: a case study of Ota, Southwestern Nigeria

K D Oyeyemi[1], A P Aizebeokhai[1], J M Ndambuki[2] O A Sanuade[3], O M Olofinnade[4], T A Adagunodo[1], A A Olaojo[5] and G A Adeyemi[4]

[1]Applied Geophysics Unit, Department of Physics, Covenant University, Nigeria
[2]Department of Civil Engineering, Tshwane University of Technology, Pretoria, South Africa
[3]Department of Geoscience, King Fahd University of Petroleum and Minerals, Saudi Arabia
[4]Department of Civil Engineering, Covenant University, Nigeria
[5]Department of Earth Sciences, Ajayi Crowther University, Nigeria

kdoyeyemi@yahoo.com, kehinde.oyeyemi@covenantuniversity.edu.ng

Abstract. Geoelectrical resistivity surveys have been carried out using Schlumberger configuration within the Iyana Iyesi area of Ota, Ogun state. The aim of this research was to experimentally estimate the hydrogeophysical parameters of an aquifer (porosity, transmissivity, hydraulic conductivity and permeability) which have been completed successfully. Since drilling of boreholes specifically to compute the hydraulic parameters is relatively expensive, estimation of the parameters from vertical electrical soundings is considered a reliable alternative. The results showed that the study area has majorly low value of overburden materials serving as the protective capacity to the aquifers that are characteristically high in porosities and transmissivities. This low protective capacity denote the high vulnerability of the aquifer system to the influx of surface-based contaminants. The aquifer systems within the study area possess significantly high storativity property based on their high porosity and transmissivity.

Keywords: Electrical resistivity, Groundwater exploration, Hydraulic parameters, Sedimentary terrain, Aquifers porosity, Aquifer transmissivity

1. Introduction

Adequate availability and sufficient access to good water for domestic and industrial use is a critical issue worldwide. The groundwater is a dependable source of portable water supply for both home-use and agricultural activities (irrigation). This source is characteristically cost effective with high quality water supply in major urban centers in most developing countries. Considering the importance of groundwater, the quantitative aquifer characterization has become important in addressing some hydrogeological issues such as low yields and productivity. In order to estimate hydro-geophysical parameters of the subsurface aquifer, geophysical surveys need to be carried out. Geophysical methods provide information about the subsurface over a variety of spatial resolution and they depend on the characteristic physical properties of rocks. These techniques are relatively cheap, labour intensive and sensitive to different properties. There are many methods used in geophysical surveys for hydro-geophysical investigations but the most commonly used method is the geoelectrical resistivity method which can either be conducted in electrical profiling or vertical electrical sounding modes, a combination of which gives 2D, 3D and 4D time-lapse surveys. Several works have been carried out on the evaluation, exploration and exploitation of groundwater within the crystalline basement and sedimentary terrains of Nigeria [1-15].

This research focuses on estimating hydro-geophysical parameters such as transmissivity, hydraulic conductivity, porosity and permeability from surface electrical resistivity measurements. Surficial geoelectrical resistivity survey involves the injection of electrical current into the ground using electrodes in order to understand the lateral changes and vertical cross sections of the natural hydro-geologic setting. This method of geophysical investigation is also useful to delineate a localized buried objects and monitor the presence and mobility of contaminants in the subsoil and groundwater.

2. Methodology

2.1. *Location and the geology of the study area*

The study area is Iyana Iyesi, Ota situated within the Ado–Odo LGA of Ogun state, southwestern Nigeria. The state lies approximately between latitude 6.2°N and 7.8°N. The area is characterized by a gentle slope with a low lying area, and the two major climatic seasons are dry and season. Ota and its environs fall in the eastern portion of the Dahomey (or Benin) basin of Southwestern Nigeria extending from the continental margin of the Gulf of Guinea (Fig. 1). The stratigraphy of the Dahomey basin consists of the Abeokuta Group which is subdivided into Ise, Afowo and Araromi Formations [17]. The Abeokuta Group lies directly above the base of the basement complex; this in turn is overlain by Ewekoro Formation and others [18] [19].

2.2. *Data acquisition and processing*

Vertical electrical sounding (VES) survey was carried out using Schlumberger array at 12 different locations in the area (Fig. 2). The overburden materials are not too thick to necessitate the use of large current electrode spacing for deeper penetration, therefore the maximum AB/2 used was 320m on a logarithmic scale which began at a distance of 1.0m. ABEM Terrameter was used for the data acquisition. An initial spacing was chosen, and the current electrodes were moved outward while the potential electrodes were maintained at fixed points. At some points where the AB/2 became large enough, an increase of the potential electrode spacing was needed. The resistivity readings at every VES point were automatically displayed on the digital screen and saved. VES field curves may have subtle inflections which require the interpreter to make decisions as to how real or how significant such features are. Often a noisy field curve is smoothed to produce a graph which can then be modelled more easily. The field data smoothed and corrected as necessary, were plotted on a log-log graph and interpreted on a set of standard master curves. The estimated geoelectric parameters from the partial curve matching process were adopted as initial model parameters for computer-based iterations on an iteration software (WIN RESIST) which was used to iterate the sounding curves of (VES 1-12) to obtain the final model geoelectric layers parameters.

2.3. *Hydraulic parameters estimation*

The similar expressions for both electrical flow through a conductive medium and the fluid flow are described below based on the Ohm's law and Darcy's law.

$$J = -\sigma * dV / dr \quad (1)$$

$$Q = -K * \frac{dh}{dr}; K = k * y / \mu \quad (2)$$

where in equation (1) J, σ and V represent current density, electrical conductivity (reciprocal of electrical resistivity ρ) and the electric potential at a point with distance r. In equation (2), Q, h, k, μ and y are the flow rate, hydraulic head, permeability, viscosity and specific weight. The physical relationship between rock properties was first attempted by Archie [20] on a clay free sands.

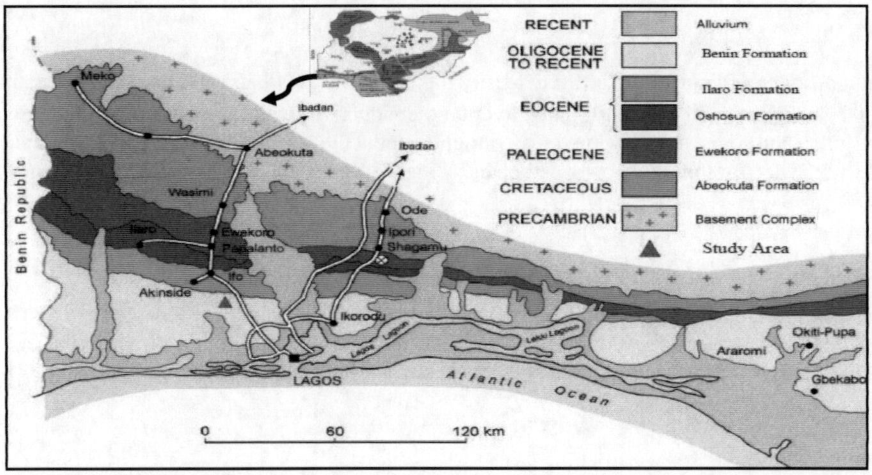

Fig.1: Geological sketch map of Nigeria showing the major geological components: basement, younger granites, and sedimentary basin (after [16])

Fig. 2: Displaying the location of the VES points within the study area using google map.

$$F_f = {\rho_r}/{\rho_w} \quad (3)$$

$$\rho_r = a * \rho_w \phi - mS_w - n \quad (4)$$

where Ff , ρr, ρw, ϕ, a and Sw are the formation factor, bulk and fluid resistivities, porosity, tortuosity (a) and water saturation, the cementation and saturation exponents are m and n respectively. In this research, the computation of hydraulic conductivity was carried out using the Kozeny – Carman-Bear equation as presented in equation (5).

$$k = \left(\frac{\delta_w g}{\mu}\right) \bullet \left(\frac{d^2}{180}\right) \bullet \left[\frac{\varphi^3}{(1-\varphi)^2}\right] \quad (5)$$

where d and δ_W are the grain size and fluid density (taken to be 1000kg/m^3). The μ is the dynamic viscosity (0.0014 kg/ms) [21]. The estimated hydraulic conductivity values are in m/sec using Eq.5. The properties of an aquifer are not governed by hydraulic conductivity (K) alone, but by the parameter, transmissivity (T). The transmissivity was estimated using Darcy's law for groundwater which was defined as:

$$T = K * h \qquad (6)$$

where K and h are the hydraulic conductivity and thickness. Water samples were gotten from the different VES locations and were tested with an apparatus called the conductivity meter which measured the electrical conductivity of the samples.

3. Results and Discussion

The results of the VES data inversion is presented in Figs (3 and 4). The number of the geoelectrical layers range from six to eight with the topsoil having depths range of 0.9 – 1.7 m, while the resistivity of this substratum ranges 49.4 – 295.8 ohm-m. Different delineated lithologies which are the Top Soil, Sandy Soil, Lateritic Clay, Confining Bed, Sand (aquifer) and Shale/Clay. The main aquifer which is fairly coarse sand unit is interpreted as the seventh geoelectrical layer (Table 1) with thickness ranging from 11.7 m to 13.1 m, and average bottom depth of about 60.0 meters. The estimated hydraulic parameters such as hydraulic conductivity, porosity, formation factor, transmissivity and permeability parameters of the rock (Table 2). Water samples were gotten at each location and tested for the hydraulic conductivity values. The thickness and resistivity were gotten from the inverse model while the formation factor and porosity were determined using Archie's law. Table 2 shows that the transmissivity of each VES is relatively high and porosity is also high. Since aquifers are characterized by its ability to store, transmit and retain water. Thus, the possible aquifer present in the study area is highly transmissible and porous.

4. Conclusion

Vertical electrical resistivity surveys have been carried out using Schlumberger configuration in Iyana-Iyesi area of Ota, Ogun state. The aim of this research was to experimentally estimate the hydrogeophysical parameters of an aquifer (porosity, transmissivity, hydraulic conductivity and permeability) which have been completed successfully. High porosity and transmissivity of the aquifer system in the area prove them to be highly viable and productive. Since drilling of wells specifically to compute the hydraulic parameters is relatively expensive, estimation of the parameters from vertical electrical soundings is considered a reliable alternative. Although this research has illustrated the power of hydrogeophysical methods for improving the resolution and understanding the subsurface properties, they are often still limited in their ability to inform about parameters that may be most relevant at the larger scales where water resources or environmental contaminants are managed. Based on our results, it has been concluded that the vertical electrical sounding surveys and analysis of water samples provide cost effective methodologies to estimate the hydraulic parameters of subsurface aquifers other than the expensive pumping test method. Therefore, with improved delineation of aquifer geometry and accurate estimation of hydraulic parameters for a viable aquifer system from surface-based geoelectrical resistivity surveys, sustainable groundwater resource exploitation and management can be planned for effectively.

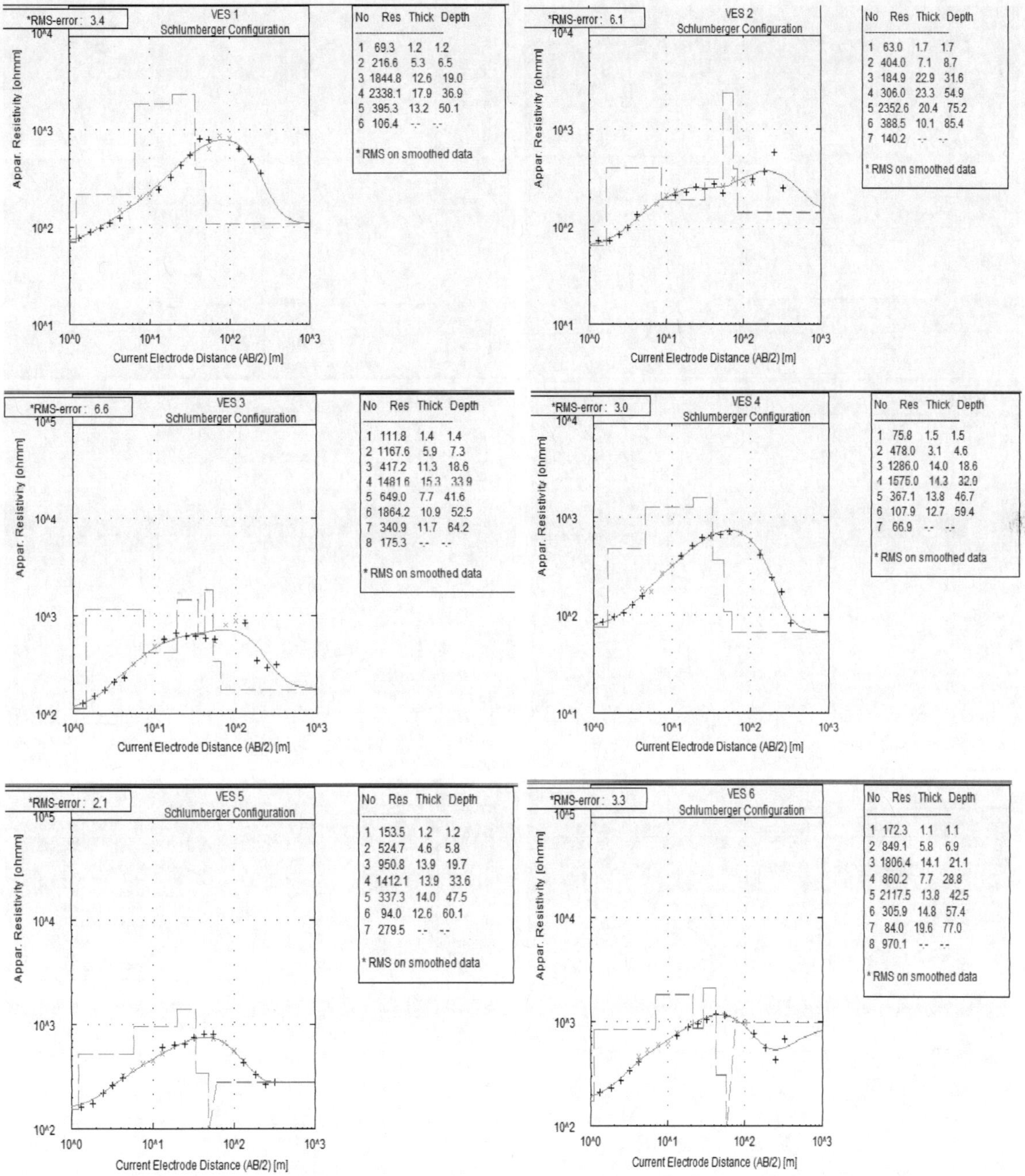

Fig. 3: Sounding curves interpretation for VES (1-6).

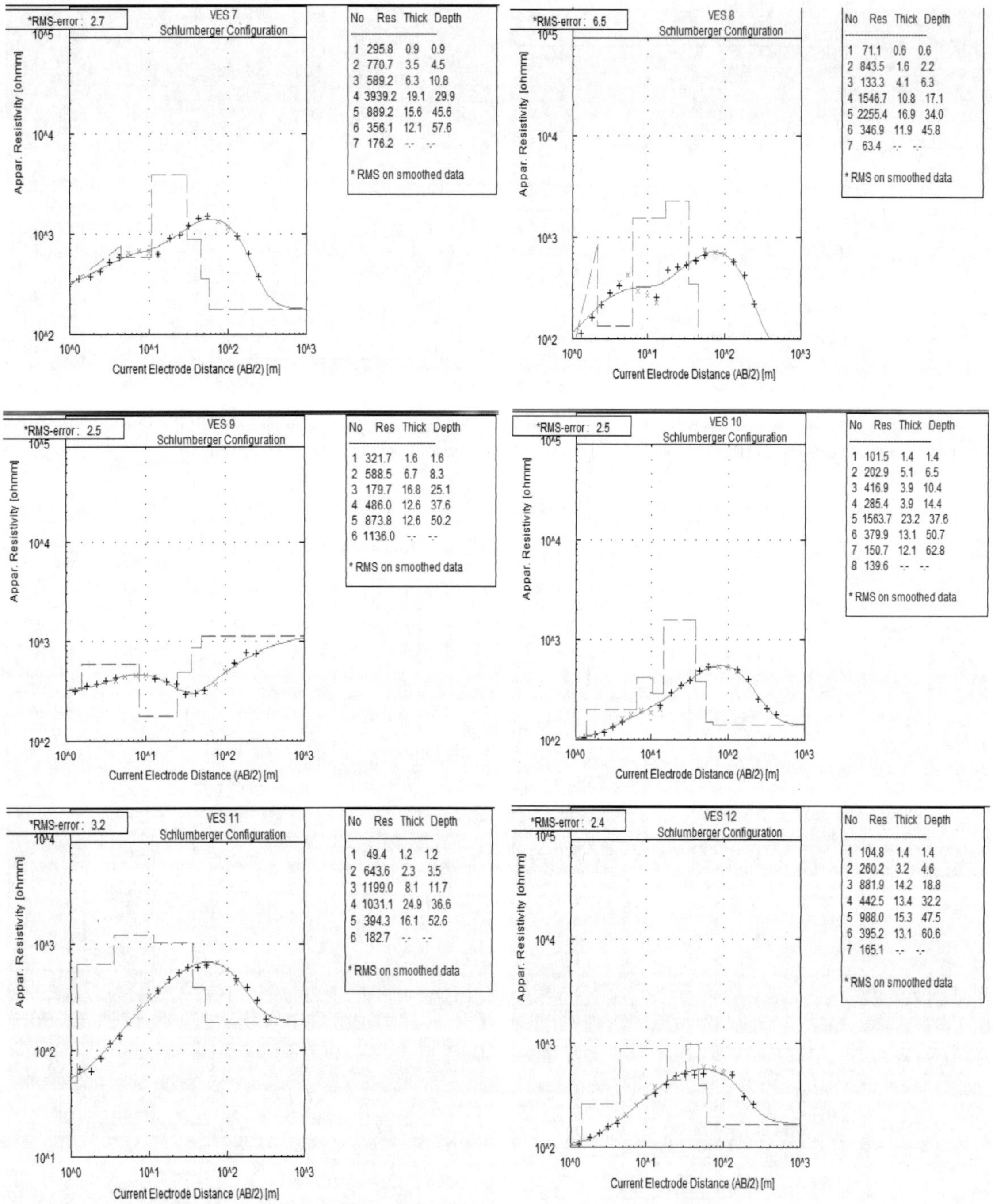

Fig. 4: Sounding curves interpretation for VES (7-12).

IOP Conf. Series: Earth and Environmental Science **173** (2018) 012028 doi:10.1088/1755-1315/173/1/012028

Table 1: Table showing geoelectrical parameters for the delineated layers from the VES data

VES		Layer 1	Layer 2		Layer 3	Layer 4	Layer 5	Layer 6	Layer 7	Layer 8
	Probable Lithology	Topsoil	Sandy Clay		Lateritic clay			Confining Clayey Sand	Sand (Main Aquifer)	Shale/Clay
1	Resistivity	69.3	216.6			1844.8	2338.1	395.3	106.4	
	Thickness	1.2	5.3			12.6	17.9	13.2		
	Depth	1.2	6.5			19.0	36.9	50.1		
2	Resistivity	69.0	404.0		184.9	306.0	2352.6	140.2		
	Thickness	1.7	7.1		22.9	23.3	20.4			
	Depth	1.7	8.7		31.6	54.9	75.2			
3	Resistivity	111.8	1167.6	417.2	1481.6	649.0		1861.2	340.9	175.3
	Thickness	1.4	5.9	11.3	15.3	7.7		10.9	11.7	
	Depth	1.4	7.3	18.6	33.9	41.6		52.5	64.2	
4	Resistivity	75.8	478.0		1286		1574	367.1	107.9	66.9
	Thickness	1.5	3.1		14		14.3	13.8	12.7	
	Depth	1.5	4.6		18.6		32.9	46.7	59.4	
5	Resistivity	153.5	524.7		950.8		1412.1	337.7	94.0	279.5
	Thickness	1.2	4.6		13.9		13.9	14.0	12.6	
	Depth	1.2	5.8		19.7		33.6	47.5	60.1	
6	Resistivity	172.3	849.1		1806.4	860.2	2117.5	305.1	84.0	970.1
	Thickness	1.1	5.8		14.1	7.7	13.8	14.8	19.6	
	Depth	1.1	6.9		21.1	28.8	42.5	57.4	77.0	
7	Resistivity	295.8	770.7		589.2	3939.2		889.2	356.1	176.2
	Thickness	0.9	3.5		6.3	19.1		15.6	12.1	
	Depth	0.9	4.5		10.8	29.9		45.6	57.6	
8	Resistivity	71.1	843.5		133.3	1546.7		2255.4	346.9	63.4
	Thickness	0.6	1.6		4.1	10.8		16.9	11.9	
	Depth	0.6	2.2		6.3	17.1		34.0	45.8	
9	Resistivity	321.7			588.5	179.7		486	873.8	1136.0
	Thickness	1.6			6.7	16.8		12.6	12.6	
	Depth	1.6			8.3	25.1		37.6	50.2	
10	Resistivity	101.5	202.9		416.9	285.4	1563.7	379.9	150.7	139.6
	Thickness	1.4	5.1		3.9	3.9	23.2	13.1	12.1	
	Depth	1.4	6.5		10.4	14.4	37.6	50.7	62.8	
11	Resistivity	49.4	643.6			1199	1031.1	394.3		182.7
	Thickness	1.2	2.3			8.1	24.9	16.1		
	Depth	1.2	3.5			11.7	36.6	52.6		
12	Resistivity	104.8	260.2		643.6	881.9	442.5	988.0	395.2	
	Thickness	1.4	3.2		2.3	14.2	13.4	15.3	13.1	
	Depth	1.4	4.6		3.5	18.8	32.2	47.5	60.6	

Table 2: Table showing aquifer geologic properties for VES data

VES NO.	THICKNESS (m)	BULK RESISTIVITY (Ωm)	AQUIFER RESISTIVITY (Ωm)	HYDRAULIC CONDUCTIVITY (m/sec)	FORMATION FACTOR	TRANSMISSIVITY (m^2/s)	POROSITY	PERMEABILTY (m^2) x 10^7
1	10.04	106.4	45.45	2.375	2.3410	23.845	0.52	3.39314
2	14.25	140.2	71.43	4.998	1.9627	71.2215	0.595	7.14029
3	9.17	175.3	76.92	2.649	2.2790	24.29133	0.531	3.78454
4	12.7	107.9	83.33	66.236	1.2949	841.197	0.8197	94.6229
5	12.6	94	83.33	381.234	1.1280	4803.548	0.912	544.62
6	19.6	84	62.5	49.82	1.344	976.425	0.80	71.168
7	9.58	176.2	100	8.459	1.7620	81.03722	0.647	12.0848
8	7.65	63.4	55.55	304.574	1.141	2329.991	0.903	435.106
9	12.6	873.8	58.82	0.009913	14.86	0.1249	0.125	0.014161
10	12.1	150.7	55.55	1.353	2.7129	16.3713	0.464	1.9333
11	10.52	182.7	55.55	0.692	3.2890	7.27984	0.4	0.988444
12	10.1	165.1	55.55	0.983	2.9721	9.9283	0.433	1.40402

Acknowledgements
The authors wish to appreciate the Covenant University Centre for Research, Innovation and Discovery for the conference support sponsorship.

References

[1] Edet A E (1990) Applications of Photogeologic and Electromagnetic Techniques to Groundwater Exploration in Northwestern Nigeria. Journal of African Earth Sciences **11(3–4)** pp 321–328.
[2] Olorunfemi M O and Fasuyi S A 1993 Aquifer types and geoelectric/hydrogeologic characteristics of part of central basement terrain of Nigeria (Niger State). Journal of Africa Earth Science **16(3)** pp 309–317.
[3] Edet A E and Okereke C S 1997 Assessment of hydrogeological conditions in basement aquifers of the Precambrian Oban Massif, southeastern Nigeria. Journal of Applied Geophysics **36** pp 195–204
[4] Olayinka, A I and Weller A 1997 The inversion of geoelectrical data for hydrogeological applications in crystalline basement areas of Nigeria. *Journal of Applied Geophysics* **37 (2)** pp 103–105.
[5] Adepelumi A Ako B and Ajayi T 2001 Groundwater contamination in the basement-complex areaof Ile-Ife, southwestern Nigeria: A case study using the electrical-resistivity geophysical method *Hydrogeology Journal* **9(6)** pp 611– 622.
[6] Ehinola O A Opoola A O and Adesokan H A 2006 Empirical analysis of electromagnetic profiles for groundwater prospecting in rural areas of Ibadan, southwestern Nigeria. *Hydrogeology Journal* **14(4)** pp 613–624.

[7] Aizebeokhai A P and Oyeyemi K D 2014 The use of the multiple gradient array for geoelectrical resistivity and induced polarization imaging. *Journal of Applied Geophysics* **111** pp 364–375 doi:10.1016/j.jappgeo.2014.10.023.

[8] Aizebeokhai A P Oyeyemi K D and Kayode O T 2015 Assement of soil petrophysical parameters using electrical resistivity tomography (ERT) and induced polarization techniques. *Research Journal of Applied Sciences* **10(9)** pp 479-485.

[9] Oyeyemi K D Aizebeokhai A P and Oladunjoye M A 2015 Integrated Geophysical and Geochemical investigation of saline water intrusion in a coastal alluvial terrain, Southwestern Nigeria. *International Journal of Applied Environmental Sciences* **10(4)** pp 275–1288.

[10] Aizebeokhai A P Oyeyemi K D and Joel E L 2016a Groundwater potential assessment in a sedimentary terrain southwestern Nigeria. *Arabian Journal of Geoscience* **9** pp 110–117. doi.org/10.1007/s12517-016-2524-5

[11] Aizebeokhai A P Oyeyemi K D and Joel E L 2016b Electrical resistivity and induced polarization imaging for groundwater exploration. SEG International Exposition and Annual Meeting, Texas.

[12] Oyeyemi K D and Olofinnade O M 2016 Geoelectrical –Geotechnical studies for near surface characterization, case history: Lagos, SW Nigeria. *Electronic Journal of Geotechnical Engineering* **21(10)** pp 3735–3750.

[13] Aizebeokhai A P Oyeyemi K D Noiki F R Etete B I Arere A U E Eyo U J and Ogbuehi V C 2017 Geoelectrical resistivity data sets for characterization and aquifer delineation in Iyesi, southwestern Nigeria. *Data in Brief*, **15** pp 828–832 doi:10.1016/j.dib.2017.10.057.

[14] Aizebeokhai A P and Oyeyemi K D 2017 Geoelectrical characterization of basement aquifers: the case of Iberekodo, southwestern Nigeria. *Hydrogeology Journal* 1–14 doi.org/10.1007/s10040-017-1679-9.

[15] Oyeyemi, K D Aizebeokhai A P Adagunodo, T A Olofinnade O M Sanuade O A and Olaojo A A 2016 Subsoil characterization using geoelectrical and geotechnical investigations: Implications for foundation studies. *International Journal of Civil Engineering and Technology* **10(8)** pp 302-314.

[16] Obaje N G 2009 Geology and mineral resources of Nigeria. In: Brooklyn SB, Bonn HJN, Gottingen J. R., Graz K. S. (eds) Lecture notes in earth sciences. Springer, Berlin, pp 305-324.

[17] Omatsola M E and Adegoke S O 1981 Tectonic Evolution and Cretaceous Stratigraphy of the Dahomey basin. Nigerian Journal of Mining and Geology **18 (1)** pp 130-137.

[18] Kogbe C A 1989 Geology of Nigeria. Rockview (Nig.) Ltd, Nigeria, pp 104-136.

[19] Adegoke S O Dessauvagie T F J and Whitman A J 1970 Macrofauna of Ewekoro Formation (Paleocene) of southwest Nigeria. *African Geology,* University of Ibadan pp 269-276.

[20] Archie G E 1942 The electrical resistivity log as an aid in determining some reservoir parameters. *Transactions of the American Institute of Mining, Metallurgy, and Petroleum Engineering* **146** pp 54-62.

[21] Fetter C W 1994 Applied Hydrogeology, 3rd ed. Macmillan College Publishing, Inc., New York, 616 p.

IOP Conf. Series: Earth and Environmental Science **173** (2018) 012029 doi:10.1088/1755-1315/173/1/012029

The use of geological-based geophysical surveys for groundwater distribution in crystalline basement terrain, SW Nigeria

K D Oyeyemi[1], A P Aizebeokhai[1], O A Sanuade[2], J M Ndambuki[3], O M Olofinnade[4], A A Olaojo[5] and T A Adagunodo[1]

[1]Applied Geophysics Unit, Department of Physics, Covenant University, Nigeria
[2]Department of Geoscience, King Fahd University of Petroleum and Minerals, Saudi Arabia
[3]Department of Civil Engineering, Tshwane University of Technology, Pretoria, South Africa
[4]Department of Civil Engineering, Covenant University, Nigeria
[5]Department of Earth Sciences, Ajayi Crowther University, Nigeria

kdoyeyemi@yahoo.com, kehinde.oyeyemi@covenantuniversity.edu.ng

Abstract. This research involves the subsurface geological characterization for groundwater potential assessment within the campus of the Polytechnic of Ibadan, southwestern Nigeria. The study is directed towards groundwater resources exploration, development and management in the campus. Five 2D resistivity imaging traverses were conducted using Wenner array in addition to five VES surveys using Schlumberger array that provide layering information and geoelectrical parameters. Three geologic layers delineated from the 2D resistivity inversion models include predominantly clayey sand/ sandy clay top soil (overburden), partly weathered or fractured basement and fresh basement. Their inverse model resistivity values ranges 6.68 – 98.6 Ωm, 68.0 – 929 Ωm and $\geq$ 2252 Ωm with bottom depths ranges 3.8 – 6.4 m and 6.4 – 10 m respectively. 1D model inversion from VES results also delineate three lithologies classifying both topsoil and some part of the partly weathered basement as overburden with resistivity and thickness range 483 – 1746.9 Ωm, 1.1 – 1.8 m; partly weathered or fractured basement 60.3 – 93.5 Ωm, 8.4 -12.9 m and fresh basement 984.6 – 2078.9 Ωm. The saturated portion of the partly weathered or fractured basement at depth will favour groundwater exploration and development in this area, while the relatively shallow overburden thickness would serve as the protective layer and recharge for the fractures.

Keywords: Geoelectrical resistivity, Basement terrain, Groundwater exploration, Subsurface characterization, Basement aquifer

1. Introduction

The heavy reliance on groundwater as a source of affordable water for both industrial and domestic use throughout the world demands that the water occurrence within the subsurface is of significant quantity and high quality. Subsurface geological characterizations using surficial geoelectrical resistivity technique are sufficient to address variety of problems related hydrological investigations in complex geological terrains such as crystalline basement. Several works have been carried out on the assessment, abstraction, development and management of groundwater within the hard rock terrain of Nigeria [1-14]. In this research, Electrical resistivity imaging (ERI) and vertical electrical soundings (VES) were combined for subsurface characterization as part of preliminary studies for groundwater resource evaluation and management in a basement complex, southwestern Nigeria.

2. Methodology

2.1. Geologic setting of the study area

The study site is located within the Polytechnic Ibadan Campus, Ibadan southwestern Nigeria. Most parts of Ibadan and other neighbouring towns within the southwestern Nigeria are underlain by gneiss-migmatites complex and the metasedimens underlie [15]. The gneiss-migmatite complexes within this region include minor quartizites and calc-silicate bearing units. The major rock types present in the study area are predominantly granites and schists of metasedimentary scores (such as mica schist, quartizite and quartz-schist), biotite, biotite-hornblende gneiss, banded gneiss and granite gneiss, and migmatite gneiss (Figs 1a and b). The minor rock types within the area include augen-gneiss pegmatite and amphibolite. The gneisses that are rich in mafic minerals may undergo weathering, resulting to rigoliths and clayey soils overlying the coarse grained granitic components soil with more varying texture and lower clay contents [17].

2.2. Data acquisition and processing

The data acquisition were done manually using Omega resistivity meter in the investigation site within the Polytechnic of Ibadan campus, Ibadan southwestern Nigeria (Fig. 2). Five resistivity soundings were carried out along the five lines with maximum half-current electrode spread (AB/2) of 75.0 m using Schlumberger array, which is enough for the anticipated depth of investigation (DOI). Likewise, five 2D electrical resistivity profile lines in the West-East and North-South direction were conducted with each 2D resistivity profiles being 110 m length (Fig. 2). Wenner array with least electrode spacing of 5.0 m was adopted for the surveys reaching 5 data level of 5 for each profiles. The Wenner configuration was adopted for the 2D ERT survey due to its easy application, strongest signal strength, and greater sensitivity to vertical resistivity variation or structures lying horizontally with respect to the country rocks [18].

The processing of the acquired soundings (VES) data were done by plotting the apparent resistivity against half–current electrode spacing (AB/2) on a log–log graph sheets. The estimated geoelectric parameters (resistivities and thicknesses) from the partial curve matching were later adopted as the initial inputs for several iteration on a Win-Resist software program to get final model parameters for subsurface geoelectric layers. Also, apparent resistivity data observed for the 2D electrical resistivity tomography (ERT) were processed and inverted using RES2DINV inversion code [19] [20].

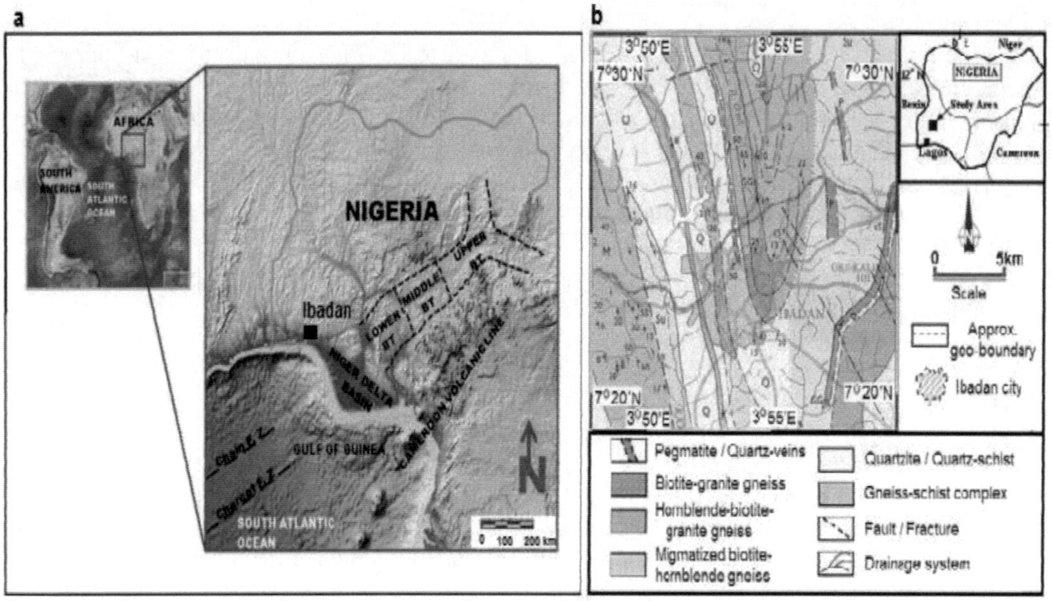

Fig. 1: Map showing the Nigeria relief and adjoining areas with extracted and modified Ibadan, using ETOPO1 global relief model [16]. **b.** Geological map of Ibadan (After NGSA, 2009).

Fig. 2: Satellite imagery map of the study site in Ibadan indicating the 2D resistivity profile lines (blue) and VES points (yellow arrows)

3. Results and Discussion

3.1. Vertical electrical sounding

The derived geoelectric parameters from several iterations of the resistivity vertical electrical soundings are shown in Figs. (3 and 4) and Table 1. Three geoelectrical layers were identified from the iterated curves. Large consistency of the geoelectric parameters were observed in the VES curves.

Based on the local geology of the area and the information available from the boreholes and hand-dug wells, the delineated subsurface strata are: overburden layer (mainly sandy clay units) with model apparent resistivity ranging between 1746.3 (Ωm) and 483.3 (Ωm) and thickness range 1.1 – 1.8 metres; intercalation of partly weathered and fracture basements with model resistivity range of 60.3 – 93.5 (Ωm) as well as thickness range of 8.4 – 12.4 metres; fresh basement with model resistivity ranging from 784.6 to 2078.9 (Ωm). The second layer is interpreted as the water-saturated aquiferous zone in the area of study. The fresh portion of the basement is inferred to be the lowermost infinitely thick layer with very high resistivity values. The outcomes of the VES surveys serve as great input to better understanding the hydrogeological setting in the study area. The water saturated fractured and weathered basement should be the target for sustainable water supply in the study area.

3.2. *Interpretation of 2D resistivity models*

The 2D inverse models of the apparent resistivity distribution within the subsurface with depths obtained from the smoothness constrained inversion are presented in Figs. 5 and 6. The inversion models reveal a range of both relatively low resistivity and high resistivity zones within the area of study. The 2D inverse models resistivity values range from about 9.18 – 10133 Ωm, 6.68 – 40441 Ωm, 17.7 – 7219 Ωm, 15.1 – 20382 Ωm, and 10.1 – 19610 Ωm for traverses 1 – 5 respectively. The variations in inverse models resistivity values 2D geoelectrical imaging to the tune of tens of thousands Ohm-metre higher than those of the 1D VES points denote the greater sensitivity and resolution of 2D electrical resistivity imaging [21].The geoelectrical parameters for the three identified geoelectric layers such as the inverse resistivity models values and thicknesses obtained from the soundings agree reasonably with those obtained from the 2D resistivity imaging.

Table 1: Interpreted geoelectric layered parameters from the VES (1-5)

Layer		VES 1	VES 2	VES 3	VES 4	VES 5	Lithology
1	Resistivity (Ωm)	483.3	739.7	723.3	973.9	1746.9	Overburden (Sandy Clay)
	Thickness (m)	1.8	1.4	1.2	1.1	1.1	
	Bottom Depth (m)	1.8	1.4	1.2	1.1	1.1	
2	Resistivity (Ωm)	60.3	84.6	83.2	93.5	71.3	Weathered /Fractured basement
	Thickness (m)	8.9	12.9	10.5	11.4	8.4	
	Bottom Depth (m)	10.8	14.3	11.7	12.5	9.4	
3	Resistivity (Ωm)	784.6	1928.4	909.1	2078.9	1532	Fresh Basement

Depth-to-base of the geoelectric layers reveals quick assessment of their lateral thickening.

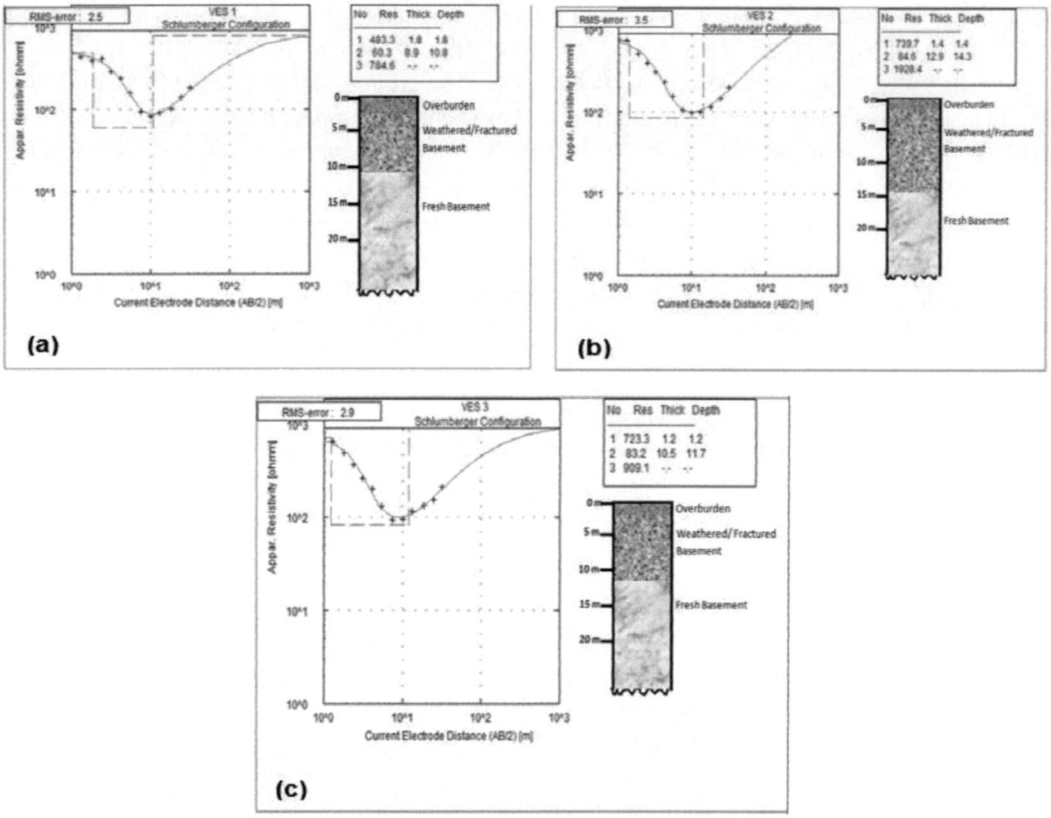

Fig. 3: VES curves and geoelectrical parameters obtained for (a) VES 1, (b) VES 2 and (c) VES 3.

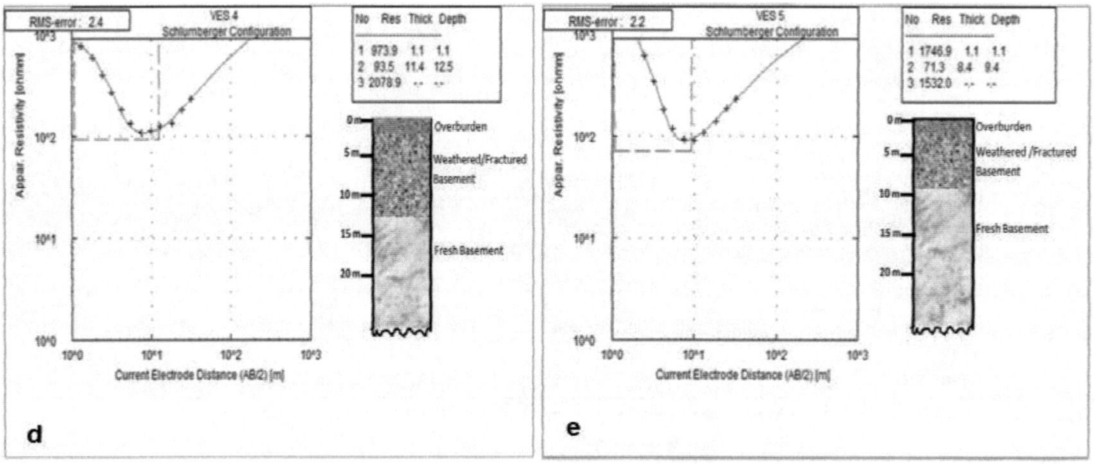

Fig. 4: VES curves and geoelectrical parameters for (d) VES 4 and (e) VES 5.

IOP Conf. Series: Earth and Environmental Science **173** (2018) 012029 doi:10.1088/1755-1315/173/1/012029

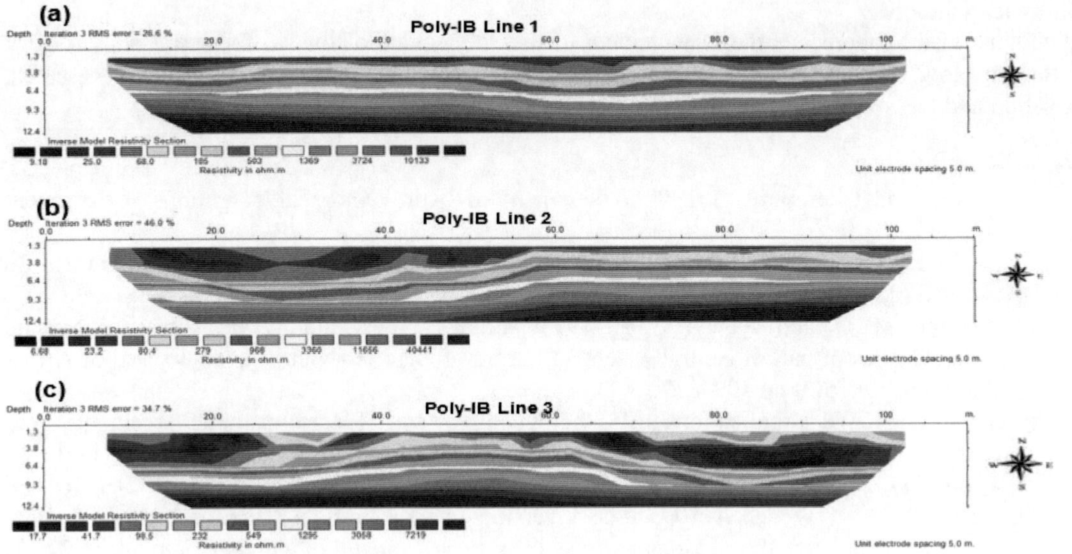

Fig. 5: 2D inverse resistivity model for Poly-IB profiles 1, 2 and 3.

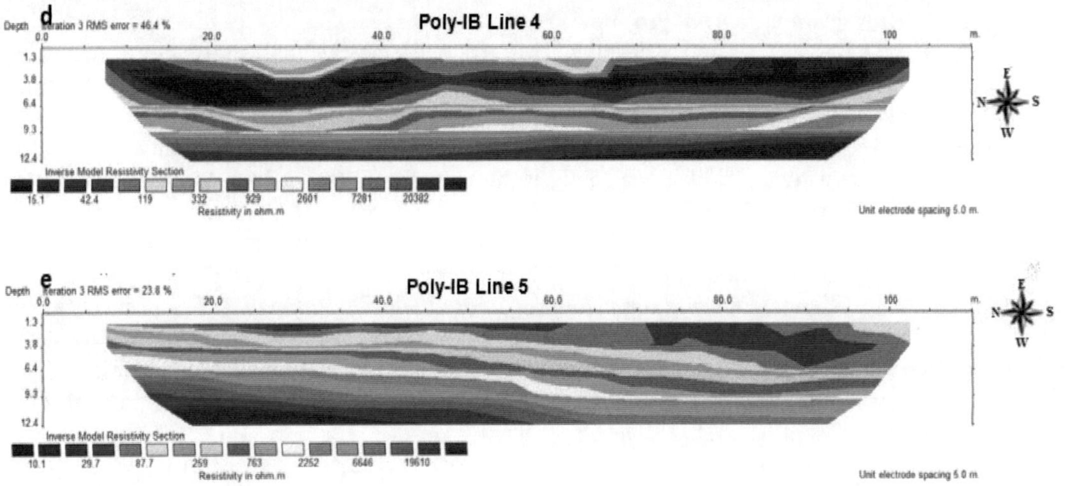

Fig. 6 2D inverse model resistivity sections for Poly-IB profiles 4 and 5.

4. Conclusion

Geoelectrical resistivity surveys involving vertical electrical soundings and ERT have been used to evaluate the groundwater potential in a crystalline basement terrain and to delineate target locations for sitting boreholes in the study area. The analyses of the soundings resistivity models and the 2D inversion models clearly shows three geoelectric layers; top layer, which is inferred to be sandy clayey/clay; water-saturated weathered and fractured basement with clayey materials serving as the aquifer unit, and fresh basement. The electrical soundings delineate higher bottom depth for the water-saturated aquiferous unit, however with little or low lateral resolution. Lateral heterogeneity of clay content within this layer was evident in the 2D electrical resistivity imaging. The rigoliths (weathered basement) unit within the area with bottom depth up to 10 metres from 2D imaging but about 14.1 metres from the VES results. Thus, groundwater exploration and development in the area should target the weathered and fractured basement layer.

Acknowledgements
The authors wish to appreciate the management of the Polytechnic of Ibadan for permission to carry out this research. We are equally very grateful to the Covenant University Centre for Research, Innovation and Discovery for the conference support sponsorship.

References

[1] Edet A E and Okereke C S 1997 Assessment of hydrogeological conditions in basement aquifers of the Precambrian Oban Massif, southeastern Nigeria. Journal of Applied Geophysics **36** pp 195–204

[2] Olorunfemi M O and Fasuyi S A 1993 Aquifer types and geoelectric/hydrogeologic characteristics of part of central basement terrain of Nigeria (Niger State). Journal of Africa Earth Science **16(3)** pp 309–317.

[3] Edet A E (1990) Applications of Photogeologic and Electromagnetic Techniques to Groundwater Exploration in Northwestern Nigeria. Journal of African Earth Sciences **11(3–4)** pp 321–328.

[4] Olayinka, A I and Weller A 1997 The inversion of geoelectrical data for hydrogeological applications in crystalline basement areas of Nigeria. *Journal of Applied Geophysics* **37 (2)** pp 103–105.

[5] Adepelumi A Ako B and Ajayi T 2001 Groundwater contamination in the basement-complex areaof Ile-Ife, southwestern Nigeria: A case study using the electrical-resistivity geophysical method *Hydrogeology Journal* **9(6)** pp 611– 622.

[6] Ehinola O A Opoola A O and Adesokan H A 2006 Empirical analysis of electromagnetic profiles for groundwater prospecting in rural areas of Ibadan, southwestern Nigeria. *Hydrogeology Journal* **14(4)** pp 613–624.

[7] Aizebeokhai A P and Oyeyemi K D 2014 The use of the multiple gradient array for geoelectrical resistivity and induced polarization imaging. *Journal of Applied Geophysics* **111** pp 364–375 doi:10.1016/j.jappgeo.2014.10.023.

[8] Oyeyemi K D Aizebeokhai A P and Oladunjoye M A 2015 Integrated Geophysical and Geochemical investigation of saline water intrusion in a coastal alluvial terrain, Southwestern Nigeria. *International Journal of Applied Environmental Sciences* **10(4)** pp 275–1288.

[9] Aizebeokhai A P Oyeyemi K D and Joel E L 2016a Groundwater potential assessment in a sedimentary terrain southwestern Nigeria. *Arabian Journal of Geoscience* **9** pp 110–117. doi.org/10.1007/s12517-016-2524-5

[10] Aizebeokhai A P Oyeyemi K D and Joel E L 2016b Electrical resistivity and induced polarization imaging for groundwater exploration. SEG International Exposition and Annual Meeting, Texas.

[11] Oyeyemi K D and Olofinnade O M 2016 Geoelectrical –Geotechnical studies for near surface characterization, case history: Lagos, SW Nigeria. *Electronic Journal of Geotechnical Engineering* **21(10)** pp. 3735–3750.

[12] Aizebeokhai A P Oyeyemi K D Noiki F R Etete B I Arere A U E Eyo U J and Ogbuehi V C 2017 Geoelectrical resistivity data sets for characterization and aquifer delineation in Iyesi, southwestern Nigeria. *Data in Brief*, **15** pp 828–832 doi:10.1016/j.dib.2017.10.057.

[13] Aizebeokhai A P and Oyeyemi K D 2017 Geoelectrical characterization of basement aquifers: the case of Iberekodo, southwestern Nigeria. *Hydrogeology Journal* 1–14 doi.org/10.1007/s10040-017-1679-9.

[14] Oyeyemi, K D Aizebeokhai A P Adagunodo, T A Olofinnade O M Sanuade O A and Olaojo A A 2016 Subsoil characterization using geoelectrical and geotechnical investigations: Implications for foundation studies. *International Journal of Civil Engineering and Technology* **10(8)** pp 302-314.

[15] Oyinloye O A 2011 Geology and geotectonic setting of the basement complex rocks in South Western Nigeria: implications on provenance and evolution. In: Earth and Environmental Sciences, I. A. Dar IA (Ed.), chapter **5**, *InTech*, Rijeka, Croatia
[16] Amante C and Eakins B W 2009 ETOPO1 1 arc-minute global relief model: procedures, data sources and analysis. NOAA Technical Memorandum NESDIS NGDC – **24**, 19p
[17] Oyawoye M O 1970 The Basement Complex of Nigeria, In: T. F. J. Dessauvagie, and Whiteman (Eds), African Geology. Ibadan University Press. Nigeria.
[18] Dahlin T and Zhou B 2004 A numerical comparison of 2-D resistivity imaging with 10 electrode arrays. *Geophysical Prospecting* **52** pp 379–398.
[19] Griffiths D H and Barker R D 1993 Two dimensional resistivity imaging and modelling in areas of complex geology. *Journal of Applied Geophysics* **29** pp 211–226.
[20] Loke M H and Barker R D 1996 Rapid least-squares inversion of apparent resistivity pseudosections by a quasi-Newton method. Geophysical Prospecting **44** pp 131–152.
[21] Loke MH (2001) Electrical Imaging Surveys for Environmental and Engineering Studies: A Practical Guide to 2D and 3D Surveys, (Available atwww.geoelectrical.com), 62 pp.

Investigation into the types of fracture and viable depth to substratum of a housing estate using geophysical techniques

T.A. Adagunodo[1], L.A. Sunmonu[2], A.V. Erinle[2], M.A. Adabanija[3], K.D. Oyeyemi[1], O.T. Kayode[1]

[1]Department of Physics, Covenant University, Ota, Nigeria
[2]Department of Pure and Applied Physics, Ladoke Akintola University of Technology, Ogbomoso, Nigeria
[3]Department of Earth Sciences, Ladoke Akintola University of Technology, Ogbomoso, Nigeria

taadagunodo@yahoo.com; erinleayo@yahoo.com; taadagunodo@gmail.com

Abstract. Mapping of the configuration of bedrock is paramount in civil engineering and hydrogeological settings. In civil engineering, areas that are extensively fractured (either shallow or deep) are adjudged as weak zones whereby these provinces are regarded as copious zones for groundwater exploration. Very Low Frequency Electromagnetic (VLF-EM) and ground magnetic survey were carried out in Olupona Housing Estate along seven traverses with interstation spacing of 20 m. For the two methods, traverses 1 to 3 were taken in North-South azimuth which covered distance 300 m while traverses 4 to 7 were acquired along East-West azimuth which covered distance 400 m. VLF-EM results revealed Type-1 fracture along traverse 1, traverses 2 to 4 showed no fracture but a fairly competent bedrock. Type-2 fractures were revealed on traverses 5 to 7 while Type-3 fracture was shown on traverse 5 alone. Ground magnetic results depicted that the depths to biotite or biotite muscovite granite are relatively deep. The mineral rocks with very thin, intermediate, and very thick bodies have their depths range as 14.7 m, 11.0 m and 8.8 m respectively. On the average, the depths to these mineral rocks from the surface varied from 6.5 to 19.6 m. The trends of the identified fractured zones are NE-SW and NW-SE directions. It is concluded that in order to avoid building collapse in future, construction of high-rise buildings is unadvisable in the study area. If the identified fractured zones are properly maximized for groundwater exploration, the fractured zones can be optimized such that reservoir(s) can be constructed so as to supply water to the houses for domestic usage.

Keywords: Fracture distribution; weak zones; ground magnetic; overburden estimates; VLF-EM.

IOP Conf. Series: Earth and Environmental Science **173** (2018) 012030 doi:10.1088/1755-1315/173/1/012030

1. Introduction

Fracture in basement terrain is a break in crystalline basement rock as a result of tectonic forces or magma intrusions into the parent rock [1]. Determining its distribution and type for hydrogeological and civil engineering purposes is paramount. Its identification in civil engineering study helps to establish the precise foundation type and appropriate depth to which the foundation of a building must be laid. In civil engineering study, areas with deeper and wider fracture are considered as weak zones. These zones would be interpreted as the promising zones for groundwater exploration. In hard rock terrain, groundwater yield is greatly dependant on the size of fracture and its interconnectivity because it is located in fractures and cracks of the local rock. However, slightly fractured and shallow fractured zones are still considered as competent zones for civil engineering purposes but the users need to seek expert's advice before any structures would be erected [1].

Geophysical methods had been proven as the economical complement to expensive drilling of test wells in fluid studies and to determine the location and orientation of fractured zones in the subsurface. They can be integrated with geologic, borehole and hydrogeological techniques to enhance well sitting or as 'stand alone' method for fracture detection. In some situations, buildings that are expected to be an abode of peace to the people have become sources of great concern to the owners and occupiers as a result of the persistent failure in these structures [2]. This is what happens when the geophysical methods (especially a pre-foundation study) that map the subsurface systems are neglected.

The developments of human beings have solely been dependent on subsurface resources, either those obtained from near surface/shallow depth (for civil engineering purposes) or those obtained from deeper depth (for mineral exploration or crude oil). Today, good source of water for domestic and industrial purposes also come from the subsurface aquifers, structures that interface safely with these shallow regions are built and human wastes are being deposited also to this subsurface [3].

In conjunction with all these factors, subsurface characteristics that attract geoscientists include location, distribution, structure and depth of rock types, porosity, permeability, strength and grain size distribution of rocks in the subsurface etc. The earth's intrinsic intricacy could make it unfeasible to deduce these attributes through direct observation. As a result of this, they are often inferred through the fundamental physical properties such as electrical conductivity, density, acoustic impedance and magnetic susceptibility [4].

Ground magnetic study is used to map a localized zone in order to understand and interpret its geology. Ground magnetic technique entails measuring the magnetic components' amplitude at discrete point along the distributed traverses that would cover the entire survey area. In ground magnetic method, the three (3) components being measured are vertical, horizontal and total components. One or two of these components have been used to map out fractures, faults, magnetic basement's depth, and other geological features. The magnetic method has been found useful in location of buried magnetite ore bodies due to its high magnetic susceptibility [3, 5].

It is more efficient to use VLF-EM method in the near surface fracture delineation than estimation of overburden thickness, because their anomalies are influenced by the overburden's conductivity than the thickness [6]. VLF-EM method exhibits very high penetration depth in crystalline (hard rock) terrain due to its high resistivity. VLF is capable to delineate fractures in lateral direction effectively compared to electrical resistivity sounding and to also characterize aquifer structures in a complex region [1].

This study is aimed to determine the fracture type in the study area which would be useful for civil engineering purposes and perhaps groundwater exploration. The outcome of this study shall complement the previous study carried out by Sunmonu *et al.* [7] in the study area in order to affirm the type of buildings to erect on the site and possible locations for groundwater exploration. In this study, VLF-EM and ground magnetic techniques were espoused. VLF-EM was considered because it is the best technique for reconnaissance survey that is capable to map geological structures while ground magnetic approach was utilized to map the magnetic anomalies in the subsurface.

Responses of VLF-EM method to subsurface variations have been studied by Paal [8], Babu *et al.* [9], Gnaneshwar *et al.* [10], Adagunodo *et al.* [11] and Olafisoye *et al.* [12]. Magnetic applications have been extensively researched on by Adagunodo and Sunmonu [13], Adagunodo *et al.* [14], [15] and Sunmonu et al. [16]. However, integration of VLF-EM and ground magnetic methods have been used by Ogungbemi and Oladapo [6] to investigate on the hydrogeophysical characterization of Ijapo Housing Estate, Akure and Babu *et al.* [9] to map basement fractures in Raigarh, India. However, other geophysical techniques have been found applicable in the interpretation of subsurface structures. Few of these techniques can be found in Refs. [17-25].

2. Geographical and geological settings of the study area

The study area is the proposed location for Olupona Housing Estate which falls within Latitude 07° 6.4430' – 07° 6.6600' north and Longitude 04° 2.0017' – 04° 2.1733' along Iwo–Ibadan road, Osun state, southwestern Nigeria. Accessibility of the study area is through major and minor road networks that link Housing Estate with Olupona village. The area under investigation belongs to the tropical rainforest region, which is experienced in most of southwest, Nigeria. There are two distinct seasons in the study area; the wet and the dry seasons. Wet season occurs between the months of March through October while dry season runs from November through February. These two seasons define the climate of the area under investigation [7].

The study area is underlain by the Precambrian to Cambrian basement Complex of Osun state, southwestern Nigeria which is subdivided into Pan African (Older) Granitoids, metasediments/metavolcanic series, and migmatite gneiss complex. Olupona, falls within Pan African (Older) Granitoids with biotite and biotite muscovite granite being the major rock groups in the vicinity the study (Fig. 1).

3. Materials and Method

ABEM-WADI, a VLF equipment was employed fro EM data acquisition while the magnetic technique employed Proton Precision Magnetometer for its data acquisition. Seven (7) traverses were occupied for VLF-EM survey with inter-station spacing of 20 m. Ground magnetic data were acquired on the same traverse as that of VLF-EM employing the same inter-station spacing. Traverses 1 to 3 were taken in North-South azimuth which covered the total length of 300 m while traverses 4 to 7 were acquired in the East-West azimuth which covered the total length of 400 m.

The theory and application that governs VLF-EM has been reported by several authors, among these researchers are Babu *et al.* [9], Gnaneshwar *et al.* [10] and Adagunodo *et al.* [11]. As reported by Gnaneshwar *et al.* [10] that, VLF-EM is one of the existing passive methods which operates on the 'Principle of EM Induction'. This technique utilizes radiation from radio transmitter (Tx), a military-based Tx at about forty-two (42) locations globally which are employed for navigation. The Tx operates within primary EM field varying from 15 to 30 kHz. These radio transmitters produce plane EM waves that could induce secondary eddy currents, particularly in electrically conductive elongated two-dimensional targets. Despite the range of frequency generated is very low for radio transmission, it is still higher than those used in standard low-frequency electromagnetic methods (1 to 3 kHz). Paal [8] observed that radio waves at very low frequencies could be used to prospect for conductive mineral deposits. Following this breakthrough, the world very low frequency transmitters have been found useful globally as electromagnetic sources for near-surface geologic mapping Babu *et al.* [9]. As Gnaneshwar *et al.* [10] have reported, "very low frequency method generally yields considerable electromagnetic anomalies, even over poor conductors such as sheared contacts, fracture zones, and faults". Therefore, this method has been so favoured and become a very renowned tool for the swift mapping of the geological structures in near-surface.

VLF-EM data were processed by downloading the filtered real and imaginary components from the Abem Wadi VLF-EM equipment. WADI VLF-EM detects the ratio (in percentage) between the vertical and the horizontal components, which is indicative of the degree of in homogeneity of the subsurface. This instrument measures the field strength and the phase displacement around the fracture zone. The filtered real and the filtered imaginary data were plotted against the station separations in order to determine the fracture types in the study area. Ground magnetic data were drift corrected. The residual data were further enhanced using Total Horizontal Derivative.

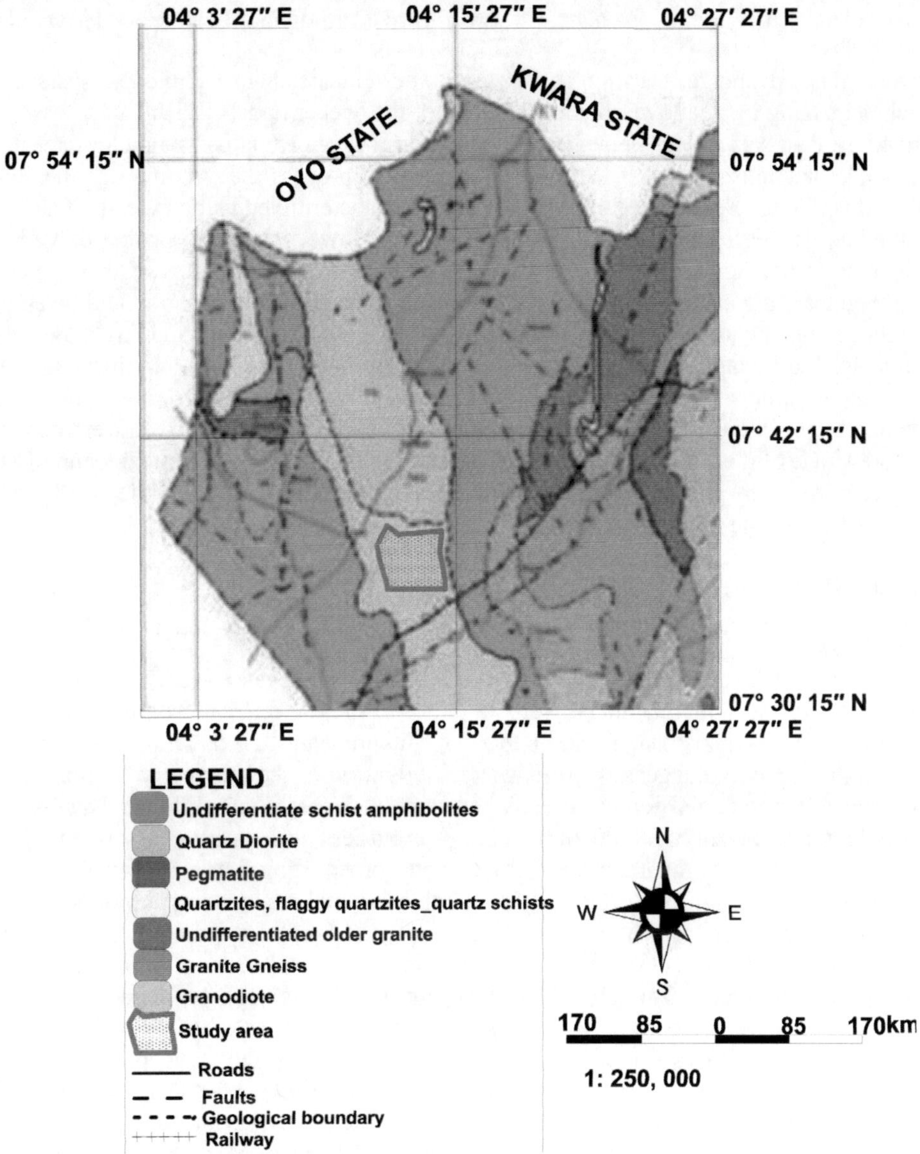

Fig. 1. The Geological domains in some parts of Osun State showing the study area

Total derivative is the maximum gradient of the magnetization. One of the applications of magnetic data to geological interpretation is the delineation of the magnetized structures' edges. Vertical and horizontal derivatives are regularly employed to intensity the magnetic data's details. The total horizontal derivative and analytic signal are two effective tools that are used to detect the edges of magnetized structures [26]. A commonly used edge detection filter is the Total Horizontal Derivative (THDR) and it is given by Cordell and Grauch [27] as:

$$\text{THDR} = \sqrt{\left(\frac{\partial T}{\partial x}\right)^2 + \left(\frac{\partial T}{\partial y}\right)^2} \quad (1)$$

where T is the magnetic field, $\frac{\partial T}{\partial x}$ and $\frac{\partial T}{\partial y}$ are the two orthogonal horizontal derivatives of the magnetic field. The advantage of using THDR over other enhancement techniques is that it is more effective when imaging for shallower bodies than that of the deeper bodies [26].

The enhanced data was finally presented as map and profiles of relative magnetic intensity values against station separations. Surfer 11 Software was used to process the 2-D map of the study area. However, the overburden thicknesses of the study area were determined using Peter-half slope method [28]. The method has recently been used by Ojo *et al.* [3]. However, the integration of VLF-EM and ground magnetic techniques used in Olupona Housing Estate enabled the subsurface characterization for fracture type mapping and depth to bedrock estimates. This characterization would be essential for civil engineering and hydrogeological purposes in the study area. Automated Euler deconvolution software was used to juxtapose the linear features that have been revealed through other techniques. This was done in order to enhance the production of geomagnetic section of the study area. Geomagnetic sections of the study area were presented along seven profiles. These were deduced from the weighted average of the results from the geology and the integrated geophysical techniques used in the study area. The map, profiles and sections generated were used to conclude on the concealed geological structures in the study area.

4. Results and Discussion

4.1. VLF-EM results

The results of VLF-EM are presented as field plot for all the seven traverses (Figs. 2a to 2g). In Fig. 2a, the filtered real and imaginary were in-phase with filtered real having higher amplitude compared to filtered imaginary that is closer to zero. This indicates the material contents in fracture zone are of high resistivity suggesting fresh water. Furthermore, the symmetry of filtered real and imaginary suggests the fractured zone is directly beneath 300 m that is non-dipping. Therefore, Type-1 fracture (single fractured zone at 300 m without overburden) is experienced towards the end of traverse 1. Thus, the point is favourable for groundwater prospecting but the absence of overburden may be a draw-back. However, the location is not suitable for civil engineering purposes.

Type-2 fracture (another form of fracture) occurred at horizontal distance of about 60 m along profile 5 (Fig. 2e). Type-2 fracture was also observed at horizontal distance of about 145 m along profile 6 (Fig. 2f); and about 20 m and 175 m along profile 7 (Fig. 2g). However, the fractures are dipping (not directly or vertically beneath the points of observations) as exemplified by asymmetry of the peaks of their amplitudes. Directional drilling will be required to explore such fractures successfully. At horizontal distance of about 245 m along profile 5 (Fig. 2e), there was a double fractured zone with overburden (Type-3 fracture) as exemplified by behaviour of both filtered real and filtered imaginary such that when the field passes through the overburden, the amplitude of filtered imaginary is reduced and its phase somewhat displaced. The presence of overburden is evidence of weathering suggesting double fracture zone with overburden and has a better prospect for groundwater exploration compared to single fracture without overburden.

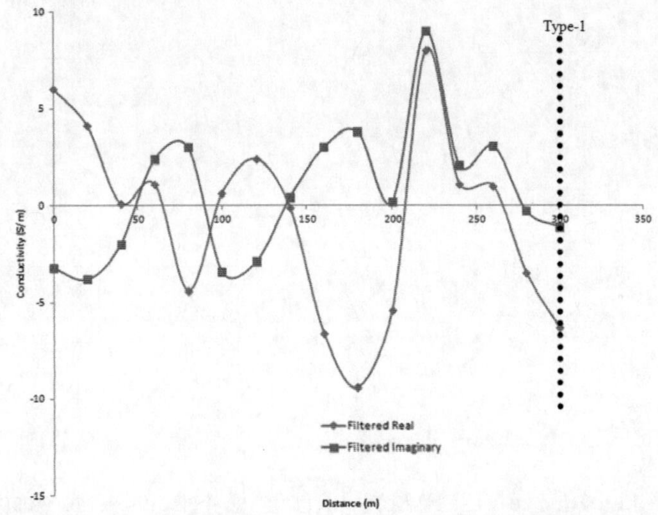

Fig. 2a. The VLF-EM plot of Traverse 1 (from north to south)

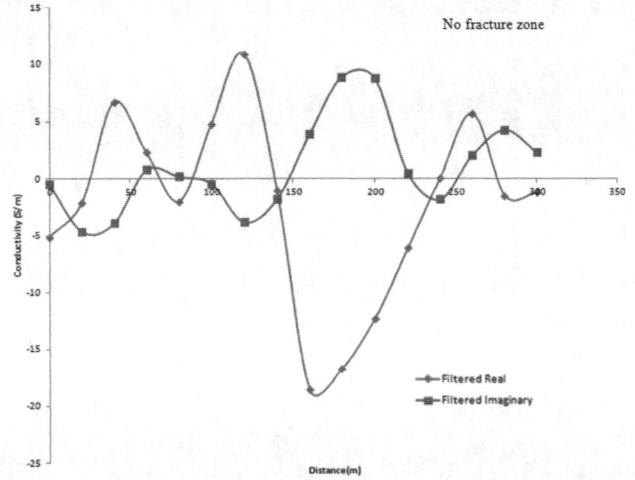

Fig. 2b. The VLF-EM plot of Traverse 2 (from north to south)

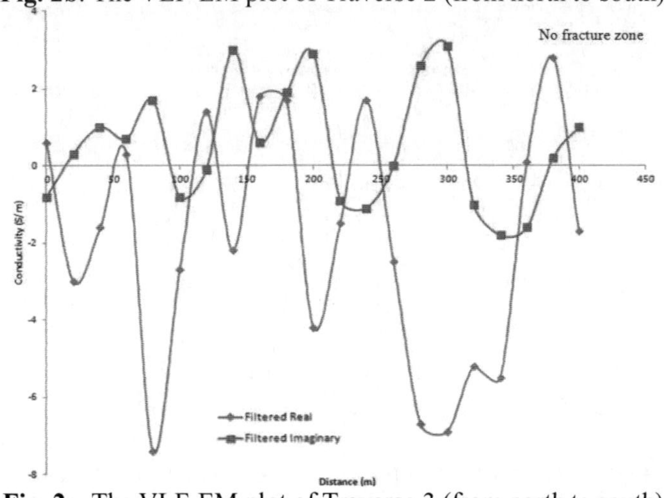

Fig. 2c. The VLF-EM plot of Traverse 3 (from north to south)

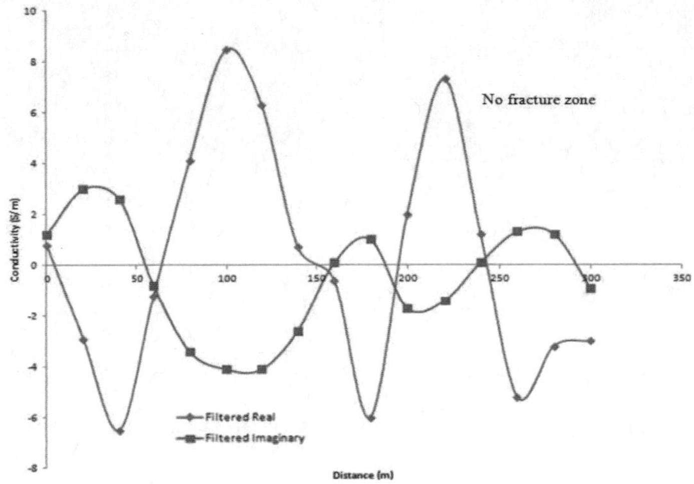

Fig. 2d. The VLF-EM plot of Traverse 4 (from east to west)

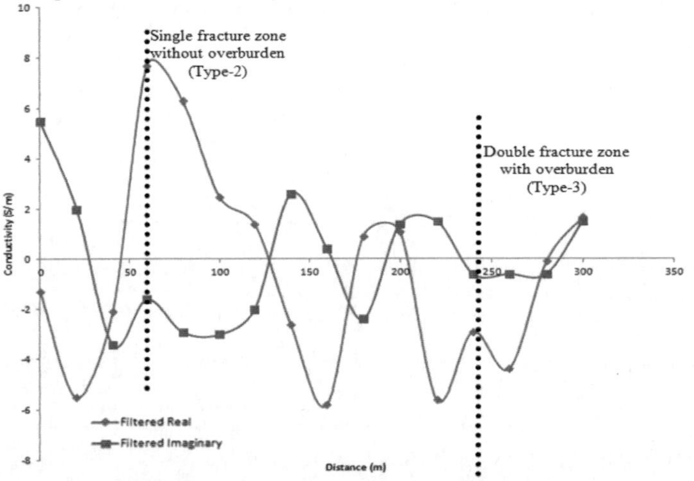

Fig. 2e. The VLF-EM plot of Traverse 5 (from east to west)

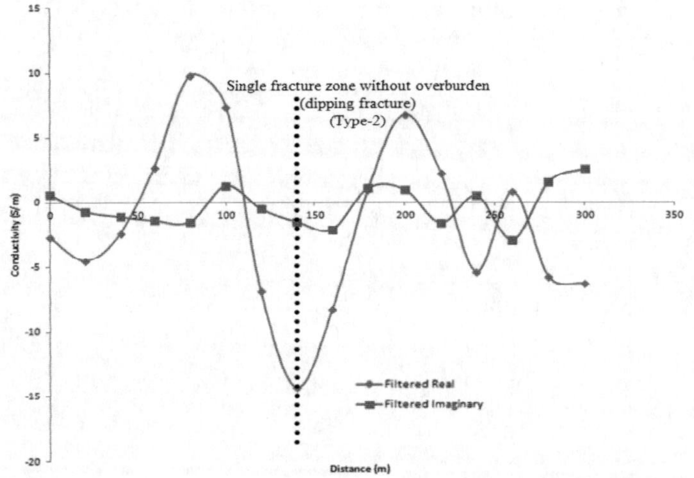

Fig. 2f. The VLF-EM plot of Traverse 6 (from east to west)

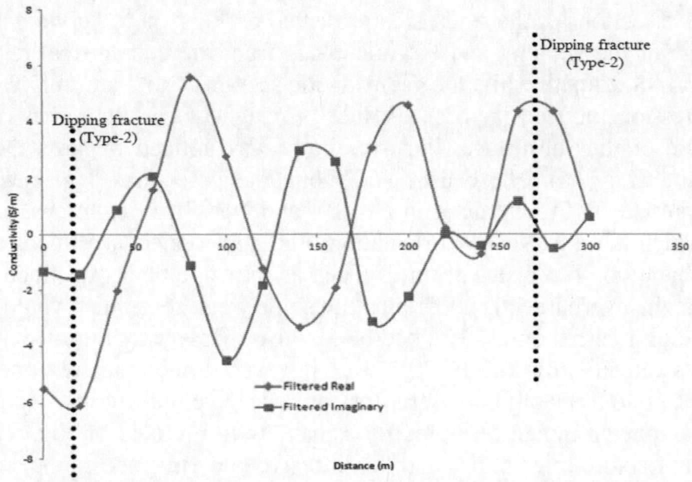

Fig. 2g. The VLF-EM plot of Traverse 7 (from east to west)

Along traverse 4 (Fig. 2d) four distinct zones of positive peaks (high conductivity) were observed on this model. The filtered real and the filtered imaginary are out of phase all through suggesting the zone is resistive, which is an indication of competent zone for civil engineering activities. It is evident from the model section that the subsurface beneath this traverse is fairly competent. The same trend was also experienced on traverses 2 and 3 (Figs. 2b and 2c) of the study area which depicts no fractured zone with fairly competent signatures for civil engineering purposes.

4.2 Ground Magnetic Results

The 2D contour map produced in the study area is presented in Fig. 3. A visual inspection of the contour map showed widely spaced contour lines in the study area apart from northeastern and southeastern tips which showed different orientations from the generalized contour arrangement in the study area. The wide spacing of the contour lines depict that the depth to biotite or biotite muscovite granite in these locations is relatively deep (i.e. overlain with thick overburden). Closely spaced contour lines at northeastern tip depict a shallow depth to bedrock. However, the linear subparallel orientation of contour lines at the northeastern and southeastern tips suggest shallow subsurface planar features such as faults or localized fractured zones passing through part of the study area.

The magnetic data were presented as seven profiles (Figs. 4a to 4g). The advantage of profile presentation over grid-based presentation is that geological details are being revealed easily on profiles compared to grid-based presentations. Spikes were experienced across the profiles which are probably due to the presence of magnetic bodies in the subsurface. The magnetic highs represent the presence of mineral rocks in the subsurface while the magnetic lows are due to the presence of planar features such as fracture, fault or contact between two rocks. These magnetic high zones are good for civil engineering activities while the magnetic low zones are probable zones for hydrogeological purposes if explored properly.

The magnetic peaks on each profile were used to estimate depth to magnetic sources. The estimated depths derived from traverses 1 to 7 were summarized on Table 1. From Table 1, mineral rocks with very thin, intermediate, and very thick bodies have their depths range as 14.7 m, 11.0 m and 8.8 m respectively. On the average, the depths to these mineral rocks from the surface ranged from 6.5 to 19.6 m. However, Peter did not specify from this method the shape of the buried body. He assumed that there could be a deformation of these mineral rocks due to some factors such as tectonic stress, weathering etc. He gave the possible geometry of the mineral rocks through the index values or proportionality factor. The shallower the depth, the thicker the body of the mineral rocks and the anomaly will be more pronounced and vice versa.

IOP Conf. Series: Earth and Environmental Science **173** (2018) 012030 doi:10.1088/1755-1315/173/1/012030

Seven geomagnetic sections of the study area presented on Figs. 5a to 5g were generated based on the weighted average of the geology and the inferences from the geophysical results. Geomagnetic sections 1 to 3 are in N-S azimuth while the geomagnetic sections 4 to 7 are in E-W azimuth. Bedrock elevation and depression noticed across all the geomagnetic sections has revealed the true inhomogeneity nature of the subsurface. Intrusive body was noticed towards the southern part of geomagnetic section 1 (Fig. 5a). The bedrock of geomagnetic sections 5 to 7 was fractured. Two fracture zones that present on Fig. 5e trend in NE-SW and NW-SE orientations, a fracture zone was present on Fig. 5f which trend in NE-SW orientation. Fig. 5g revealed two fracture zones which also trend in NE-SW orientation. The planar features (such as intrusive body and fracture) mapped in the bedrock revealed that the overburden is underlain with some weak zones that would not be suitable for the construction of high-rise buildings. The bedrock is also overlain by thin-to-thick overburden (an overburden thickness ranged from 6.5 to 19.6 m). However, some fracture zones identified along traverses 5 to 7 (Figs. 5e to 5g) would be useful for borehole development in Olupona Housing estate. Competent and weak zones for engineering purposes have been revealed on Fig. 6. These zones were selected based on the fracture orientations and the intrusive body mapped in the study area as shown on Fig. 6. The fracture orientations are paramount in engineering and hydrogeological purposes. It determines the direction of flow of water and buildings placed across fracture zones are safer than those placed parallel to it. Based on Fig. 6, the tip of northeastern, southwestern, southern and southeastern regions of the study area are competent for civil engineering purposes. Other regions are considered unsuitable for the erection of high-rise buildings. Borehole development would be successful if the boreholes are drilled along the fractured zones.

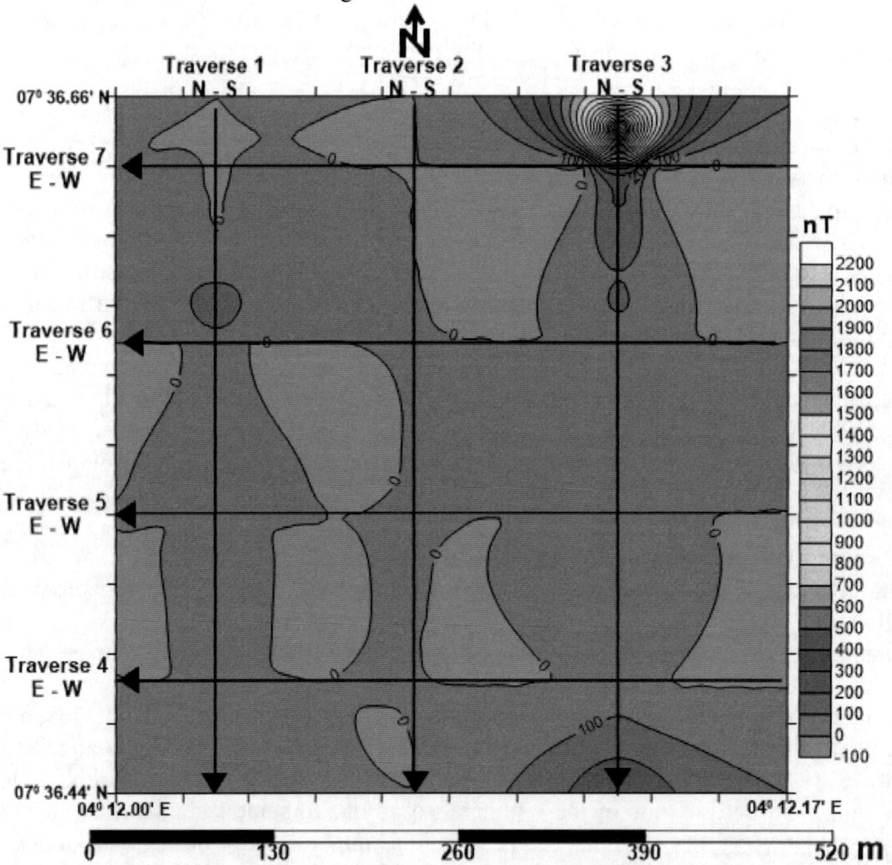

Fig. 3. Total horizontal derivative of Olupona Housing Estate

IOP Conf. Series: Earth and Environmental Science **173** (2018) 012030 doi:10.1088/1755-1315/173/1/012030

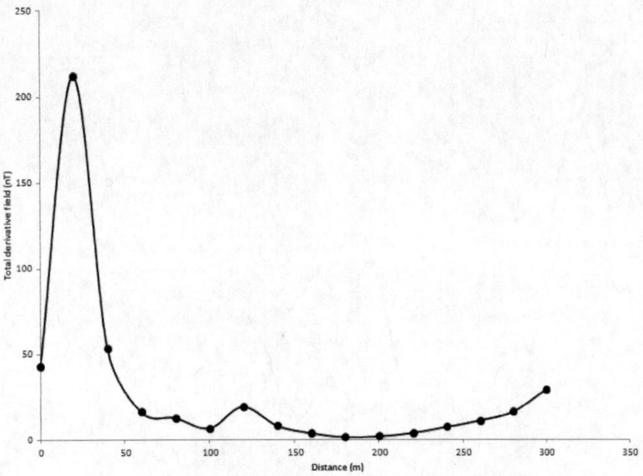

Fig. 4a. Ground magnetic profile along traverse 1 (from north to south)

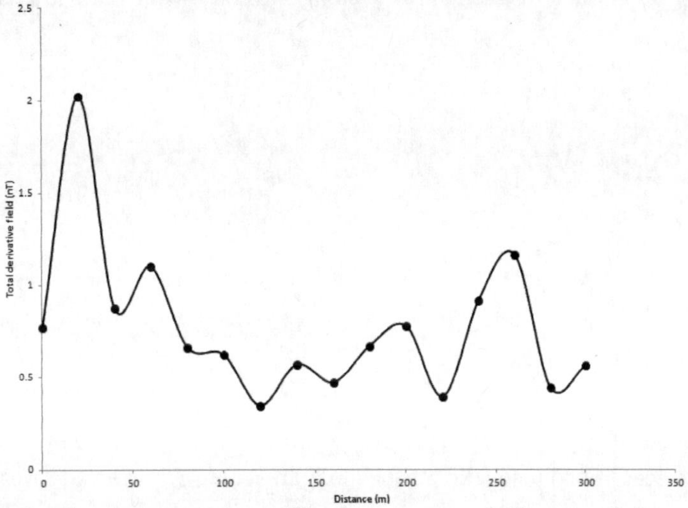

Fig. 4b. Ground magnetic profile along traverse 2 (from north to south)

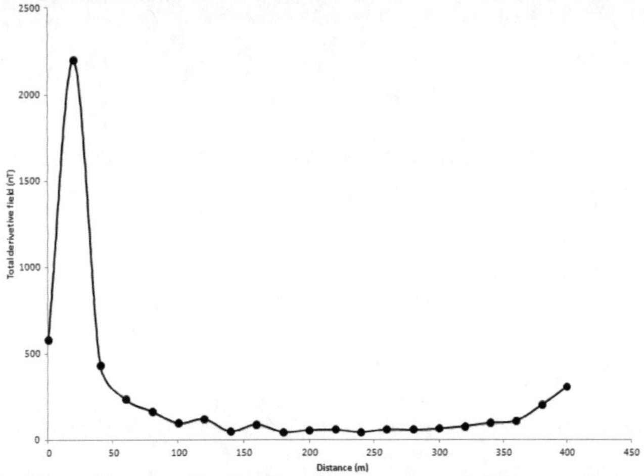

Fig. 4c. Ground magnetic profile along traverse 3 (from north to south)

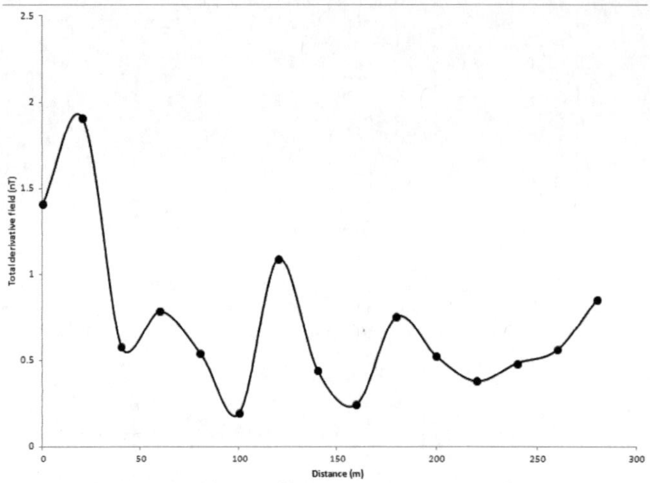

Fig. 4d. Ground magnetic profile along traverse 4 (from east to west)

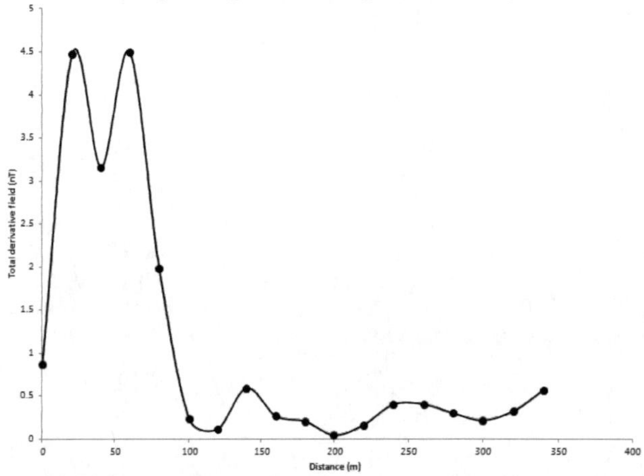

Fig. 4e. Ground magnetic profile along traverse 5 (from east to west)

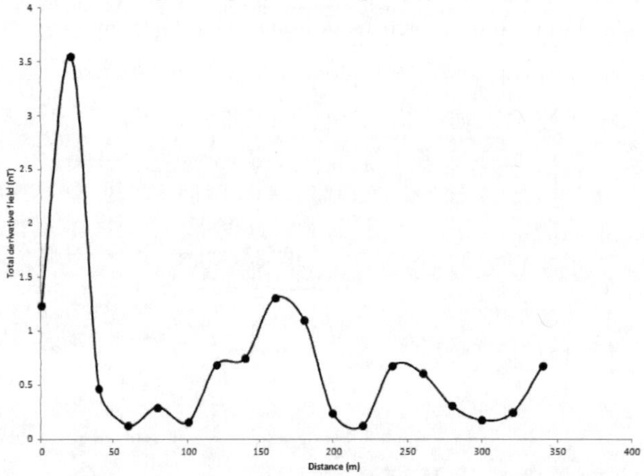

Fig. 4f. Ground magnetic profile along traverse 6 (from east to west)

IOP Conf. Series: Earth and Environmental Science **173** (2018) 012030 doi:10.1088/1755-1315/173/1/012030

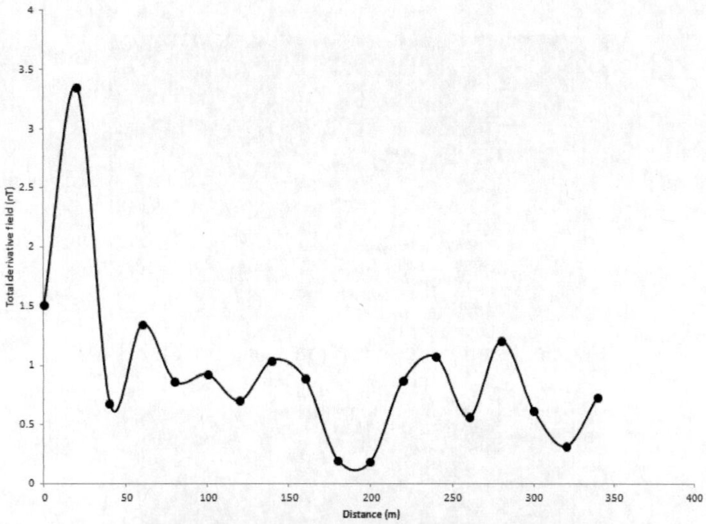

Fig. 4g. Ground magnetic profile along traverse 7 (from east to west)

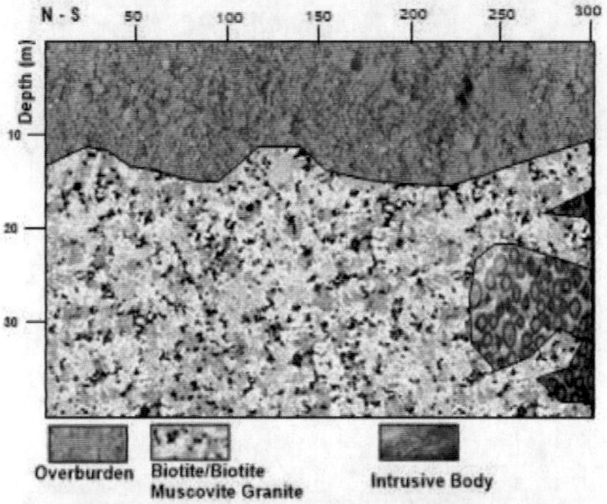

Fig. 5a. Geomagnetic section along traverse 1

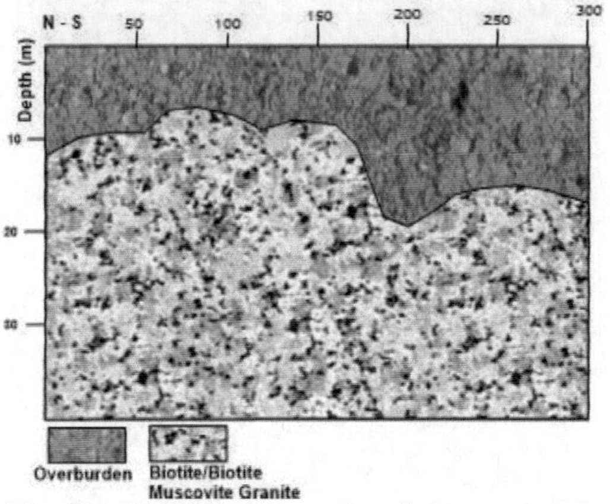

Fig. 5b. Geomagnetic section along traverse 2

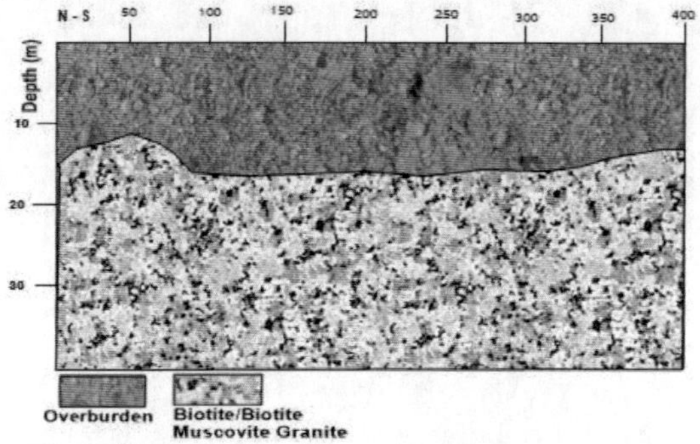

Fig. 5c. Geomagnetic section along traverse 3

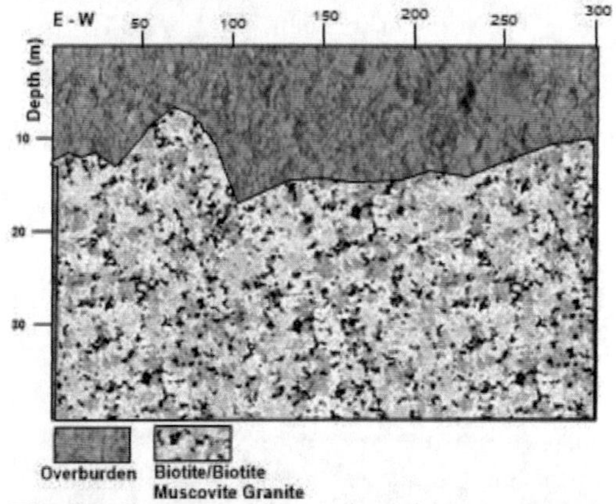

Fig. 5d. Geomagnetic section along traverse 4

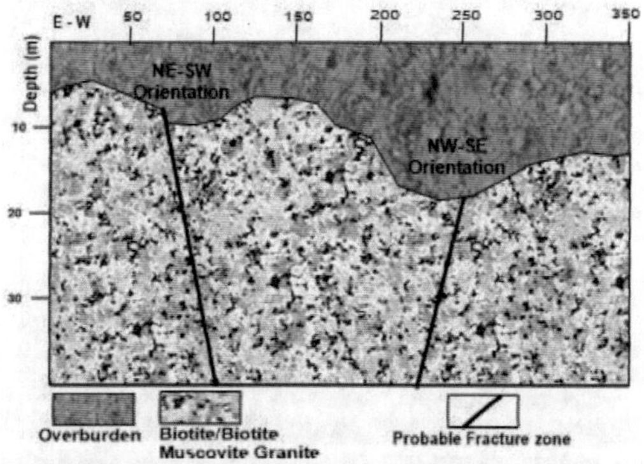

Fig. 5e. Geomagnetic section along traverse 5

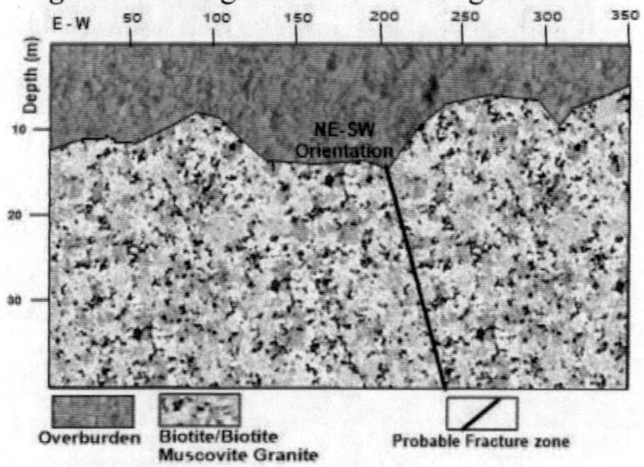

Fig. 5f. Geomagnetic section along traverse 6

IOP Conf. Series: Earth and Environmental Science **173** (2018) 012030 doi:10.1088/1755-1315/173/1/012030

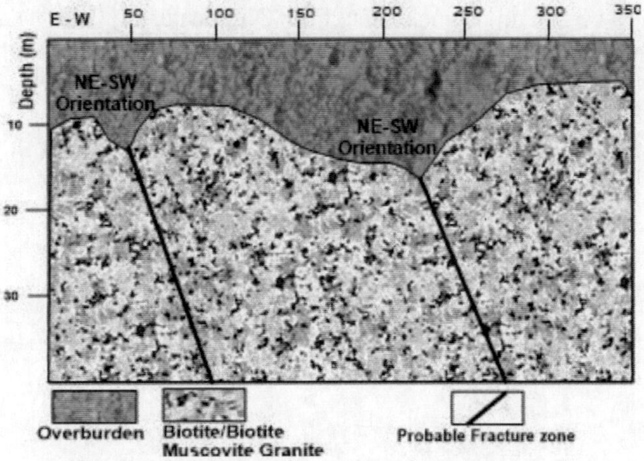

Fig. 5g. Geomagnetic section along traverse 7

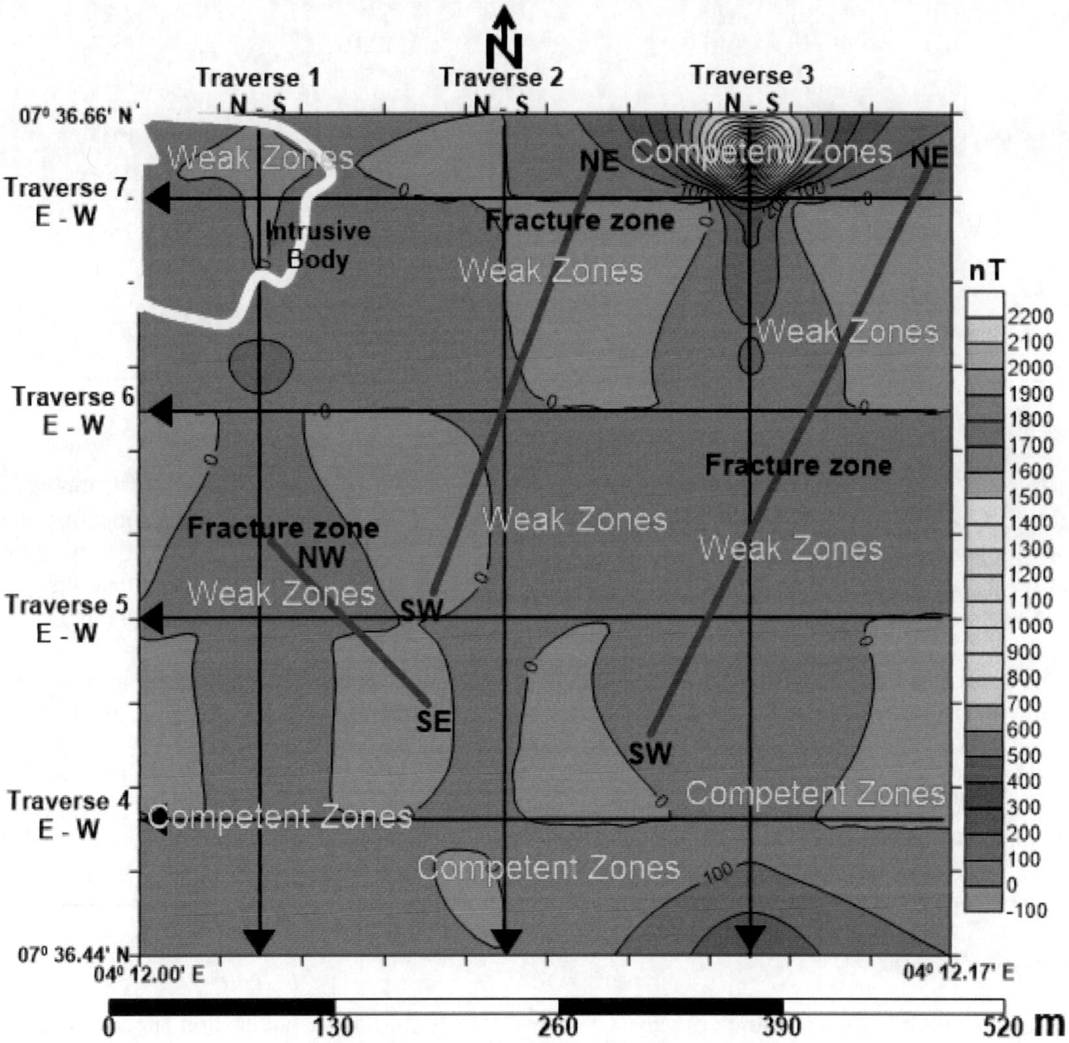

Fig. 6. Composite map of the study area

Table 1. Depth estimates of ground magnetic data from Olupona using Peter's half-slope method

Profiles	Very thin body (m)	Intermediate thickness (m)	Very thick body (m)	Average (m)
1	16.7	12.5	10.0	13.1
	16.7	12.5	10.0	13.1
2	14.2	10.6	8.5	11.1
	10.0	7.5	6.0	7.8
	12.5	9.4	7.5	9.8
	25.0	18.8	15.0	19.6
	21.7	16.3	13.0	17.0
3	17.5	13.1	10.5	13.7
4	15.0	11.3	9.0	11.8
	10.0	7.5	6.0	7.8
	16.7	12.5	10.0	13.1
	16.7	12.5	10.0	13.1
5	8.3	6.3	5.0	6.5
	13.3	10.0	8.0	10.4
	10.0	7.5	6.0	7.8
	20.0	15.0	12.0	15.7
6	14.2	10.6	8.5	11.1
	11.7	8.8	7.0	9.2
	16.7	12.5	10.0	13.1
	12.5	9.4	7.5	9.8
7	12.5	9.4	7.5	9.8
	12.5	9.4	7.5	9.8
	16.7	12.5	10.0	13.1
	16.7	12.5	10.0	13.1
	10.0	7.5	6.0	7.8
Mean	14.7	11.0	8.8	11.5

5. Conclusion

The integrated geophysical technique has proved very useful and cost effective in mapping of fracture type and depth to bedrock estimates which would be useful for civil engineering and hydrogeological purposes in the study area. In order to avoid differential settlement in the buildings that would be constructed there in the future, it is imperative to sound a note of warning that high-rise building constructions should be avoided in the study area. The trends of the identified fractured zones are NE-SW and NW-SE directions.

However, the fracture zones should be maximized for groundwater prospects because if productive boreholes are drilled along fracture zone(s), it would serve other zones that are not underlain by fractured bedrock. In a PreCambrian Basement terrain, the aquifers in most cases are remote and sectionalized, these rocks posses negligible or no primary intergranular porosity and permeability despite being deformed, therefore, groundwater accumulation in the aquifers is basically due to the development of secondary porosity and permeability by weathering and/or fracturing of the parent rocks. The study area could be designed in such a way that water could be channeled to individual house by constructing reservoir(s) that would serve the entire estate for domestic purposes.

Acknowledgement

We appreciate the management of Covenant University, Nigeria for the publication support received from them.

References

[1] George A.M., Abong A.A. and Obi D.A. (2013). Fracture Zone Detection using Very Low Frequency (VLF) Electromagnetic Method in parts of Oban Massif, Southeastern Nigeria. Advances in Applied Science Research, **4**(6): 104-121.

[2] Adagunodo T.A., Adeniji A.A., Erinle A.V., Akinwumi S.A., Adewoyin O.O., Joel E.S., Kayode O.T. (2017). Geophysical Investigation into the Integrity of a Reclaimed Open Dumpsite for Civil Engineering Purpose. Interciencia Journal, **42**(11): 324 – 339.

[3] Ojo A.O., Omotoso T.O. and Adekanle O.J. (2014). Determination of Location and Depth of Mineral Rocks at Olode Village in Ibadan, Oyo State, Nigeria, using Geophysical Methods. International Journal of Geophysics, Article ID 306862: 1-13.

[4] Kayode J.S. and Adelusi A.O. (2010). Ground Magnetic Data Interpretation of Ijebu-Jesa area, Southwestern Nigeria, using Total Component. Research Journal of Applied Sciences, Engineering and Technology, **2**(8): 703-709.

[5] Young C.T. and Droege D.R. (1986). Archaeological Applications of Resistivity and Magnetic Methods at Fort Wilkins State Park, Michigan. Geophysics, **51**(3): 568-575.

[6] Ogungbemi O.S. and Oladapo M.I. (2014). Combined use of Electromagnetic and Magnetic Survey for Hydrogeophysical Characterization of Ijapo Housing Estate, Akure. British Journal of Applied Science and Technology, **4**(7): 1138-1148.

[7] Sunmonu L.A., Adagunodo T.A., Bayowa O.G. and Erinle A.V. (2016). Geophysical Mapping of the Proposed Osun State Housing Estate, Olupona for Subsurface Competence and Groundwater Potential. Journal of Basic and Applied Research, **2**(2): 27-47.

[8] Paal G. (1965). Ore Prospecting Based on VLF Radio Signals. Geoexploration, **3**: 139-147.

[9] Babu V.R., Ram S. and Sundararajan N. (2007). Modeling and Inversion of Magnetic and VLF-EM data with an Application to Basement Fractures: a case study from Raigarh, India. Geophysics, **72**(5): B133-B140.

[10] Gnaneshwar P., Shivaji A., Srinivas Y., Jettaiah P. and Sundararajan N. (2011). Very-Low-Frequuency Electromagnetic (VLF-EM) Measurements in the Schirmacheroasen Area, East Antartica. Polar Science, **5**: 11-19.

[11] Adagunodo T.A., Sunmonu L.A. and Oladejo O.P. (2014). Electromagnetic Investigation into the cause(s) of Road Failure along Takie-Ikoyi Road, Ogbomoso. International Journal of Business and Applied Sciences, **1**(1): 78-85.

[12] Olafisoye E.R., Sunmonu L.A., Ojoawo A., Adagunodo T.A. and Oladejo O.P. (2012). Application of Very Low Frequency Electromagnetic and Hydro-physicochemical Methods in the Investigation of Groundwater Contamination at Aarada Waste Disposal Site, Ogbomoso, Southwestern Nigeria. *Australian Journal of Basic and Applied Sciences*, **6**(8), 401–409.

[13] Adagunodo T.A. and Sunmonu L.A. (2012). Ground Magnetic Survey to Investigate on Fault Pattern of Industrial Estate Ogbomoso, Southwestern Nigeria. Advances in Applied Science Research, **3**(5): 3142-3149.

[14] Adagunodo T.A., Sunmonu L.A. and Adabanija M.A. (2015a). Geomagnetic Signature Pattern of Industrial Layout Orile Igbon. Advances in Architecture, City and Environment, **1**(3): 14-25.

[15] Adagunodo T.A., Sunmonu L.A. and Adeniji A.A. (2015b). An Overview of Magnetic Method in Mineral Exploration. Journal of Global Ecology and Environment, **3**(1): 13-28.

[16] Sunmonu L.A., Olafisoye E.R., Adagunodo T.A., and Alagbe O.A. (2013). Geophysical and Hydro-physicochemical Evaluation of Hand-dug Wells near a Dumpsite in Oyo State, Nigeria. *Archives of Applied Science Research*, **5**(6), 29–40.

[17] Sunmonu L.A., Adagunodo T.A. Olafisoye E.R. and Oladejo O.P. (2012). The Groundwater Potential Evaluation at Industrial Estate Ogbomoso Southwestern Nigeria. *RMZ-Materials and Geoenvironment*, **59**(4), 363–390.

[18] Adagunodo T.A., Sunmonu L.A. and Adeniji A.A. (2015). Effect of Dynamic Pattern of the Saprolitic Zone and its Basement on Building Stability: a Case Study of a High-Rise Building in Ogbomoso. *Journal of Applied Physical Science International*, **3**(3), 106–115.

[19] Oyeyemi K.D., Aizebeokhai A.P., Adagunodo T.A., Olofinnade O.M., Sanuade O.A., Olaojo A.A. (2017). Subsoil Characterization using Geoelectrical and Geotechnical Investigations: Implications for foundation Studies. International Journal of Civil Engineering and Technology, **8**(10): 302 – 314.

[20] Adagunodo T.A., Sunmonu L.A., Adabanija M.A., Suleiman E.A., Odetunmibi O.A. (2017). Geoexploration of Radioelement's Datasets in a Flood Plain of Crystalline Bedrock. Data in Brief, **15C**: 809 – 820. http://dx.doi.org/10.1016/j.dib.2017.10.046

[21] Adagunodo T.A., Hammed O.S., Usikalu M.R., Ayara W.A., Ravisankar R. (2018). Data on the Radiometric Survey over a Kaolinitic Terrain in Dahomey Basin, Nigeria. Data in Brief, **18C**: 814 – 822. https://doi.org/10.1016/j.dib.2018.03.088,

[22] Adagunodo T.A., George A.I., Ojoawo I.A., Ojesanmi K. and Ravisankar R. (2018). Radioactivity and Radiological Hazards from a Kaolin Mining Field in Ifonyintedo, Nigeria. MethodsX, **5C**: 362 – 374. https://doi.org/10.1016/j.mex.2018.04.009.

[23] Adagunodo T.A., Sunmonu L.A., Emetere M.E. (2018). Heavy Metals' Data in Soils for Agricultural Activities. Data in Brief, https://doi.org/10.1016/j.dib.2018.04.115.

[24] Akinwumi I.I., Awoyera P.O., Bello O.O. (2015). Indigenous earth building construction technology in Ota, Nigeria. Indian Journal of Traditional Knowledge, **14**(2): 206–212.

[25] Akinwumi I.I. (2014). Earth building construction processes in Benin city, Nigeria and Engineering classification of earth materials used. Indian Journal of Traditional Knowledge, **13**(4): 686–690.

[26] Arisoy M.O. and Dikmen U. (2013). Edge Detection of Magnetic Sources using Enhanced Total Horizontal Derivative of the Tilt Angle. Yerbilimleri, **34**(1): 73-82.

[27] Cordell L. and Grauch V.J.S. (1985). Mapping Basement Magnetization Zones from Aeromagnetic data in the San Juan Basin, New Mexico, in W.J. Hinze, Ed., The Utility of Regional Gravity and Magnetic Anomaly Maps. SEG, 181-197.

[28] Peters L.J. (1949). The Direct Approach to Magnetic Interpretation and its Practical Application. Geophysics, **14**: 290-320.

IOP Conf. Series: Earth and Environmental Science **173** (2018) 012031 doi:10.1088/1755-1315/173/1/012031

Site characterization of Ayetoro Housing Scheme, Oyo, Nigeria

T.A. Adagunodo[1], L.A. Sunmonu[2], O.P. Oladejo[3], O.S. Hammed[4], K.D. Oyeyemi[1], O.T. Kayode[1]

[1]Department of Physics, Covenant University, Ota, Ogun State, Nigeria
[2]Department of Pure and Applied Physics, Ladoke Akintola University of Technology, Ogbomoso, Oyo State, Nigeria
[3]Department of Physics, Emmanuel Alayande College of Education, Oyo, Oyo State, Nigeria
[4]Department of Physics, Federal University of Oye-Ekiti, Oye-Ekiti, Ekiti State, Nigeria

theophilus.adagunodo@covenantuniversity.edu.ng; taadagunodo@gmail.com

Abstract. Detail knowledge about the subsurface properties and its engineering implications is crucial in pre-foundational studies. Geophysical investigation has been found relevant in probing the subsurface for the purpose of deducing its characteristics before any engineering construction activities would commence because of its non-invasiveness. The deduced soil characteristics are used as preliminary information to determine the suitability of the site under investigation. When the results from such study are properly utilized, it prevents prospective structural failure and loss of valuable asset. Ground magnetic technique and Vertical Electrical Sounding (VES) was used to characterize the subsurface layers of Ayetoro Housing Scheme, Ojongbodu, Oyo West Local Government Area, with a view to determining the competency of the subsurface and hydrogeological prospect of the study area. A total of six traverses were established in the E-W direction and N-S direction in order to map the magnetic signatures in the study area. Thirty VES stations were occupied across the study area using the Schlumberger electrode array configuration with current electrode spacing varying from 130 to 200 m. The data acquired was interpreted qualitatively and quantitatively. The magnetic highs and lows were observed in all traverses. The magnetic highs represent regions of magnetic minerals such as metamorphic or igneous rocks while magnetic lows represent regions with linear geologic features. Power spectrum and analytical signal filtering techniques were used for the ground magnetic interpretation. Depth to magnetic sources from power spectrum analysis (local field) ranged from 5 to 62 m. However, depth to magnetic sources from analytical signal varied from 20 to 160 m with an average variation from 38.3 to 77.6 m. The western part, some part of the northwestern zone, base of the northeastern zone and some part (probably base) of the southern region of the study area were interpreted as regions of magnetic highs while others belong to average and low magnetic distribution. Twelve of the thirty modeled curves were H-type, eight were QH-type, three were HA-type, one was Q-type, two were A-type, two were HK-type and two were KH-type. The overburden thickness ranges from 1.8 m to 44.8 m, the geoelectrical sections obtained from the Sounding curves showed both 3-layer and 4-layer earth model. Nine of the VES stations show fresh bedrock while the remaining twenty one showed fractured basement. The models showed that the subsurface layers categorized into topsoil, weathered/clay layer, and fractured/fresh basement. It was concluded that the subsurface is incompetent for high-rise buildings since the lithology of the area showed that clayey zones, thick overburden and fractured bedrocks constitute two-third of the study area. However, groundwater exploration (hand dug wells and boreholes) for domestic uses would not be problematic in the study area.

Keywords: Site characterization, Structural failure, Competent zone, Incompetent zone, Basement complex.

1. Introduction

Investigation of the subsurface properties for its engineering worthiness by geologists and geophysicists provide vital information that aid the building engineers the choice and design of civil engineering structures to adopt. A civil engineering construction without consideration of geological structures may result to differential settlements which could finally lead to collapse of such building. Linear features such as voids, fracture, and increase in water table are treats to the foundation of buildings. Geophysical methods suitable to provide subsurface information about the nature of the underground strata are quite available. These methods are however not to serve as replacement for geotechnical investigations, but the beauty in adopting these geophysical techniques is that they render cheap and quick preliminary approach to harvesting subsurface information for man's utility [1].
Unceasing rate of structural failure in Nigeria has been so alarming such that both Federal and State Government are perturbed about the ugly incidence. This has led government to instruct the town planning unit to come up with regulatory code for any building above a storey. Oyo state government has not been left out in this instruction too since the quest to make life easy for the masses has been her watchword always. The causes of this collapse can be attributed to many factors among which are poor or inadequate construction materials and incompetent foundation soils. Incautiously, one particular major point that has always not been given serious attention in this part of the world is lack of inadequate information on the nature of subsurface conditions prior to construction exercise. Building foundation may experience failure due to presence of concealed geologic features such as sinkhole/cavity and shear zones which can actually lead to subsurface subsidence. The satisfactory design and construction of any building foundation can be accomplished only when the character of the soil or rock on which it is to be built is ascertained. Therefore, it is often necessary prior to building construction to investigate the physical properties of subsoil/foundation soils and determine its suitability for design and construction of building structures [2].
Geophysical investigation is one of the methods used in probing the soil/subsoil and subsurface for any engineering construction activities. This is because of its non-intrusive approach to civil engineering site investigation, relatively low and time saving advantage [3]. Geophysical methods or techniques are often used in site investigation to determine depths to the basement and map subsurface characterization prior to excavation and construction. Geophysical techniques have also been found very useful in geologic characterization [4 – 6], aquifer characterization [7 – 8], identification of gamma doses emanating from subsurface [9 – 11], contaminant plume identification [12 – 13] and location of buried wastes and other anthropogenic features [14]. Meaningful characterization of the earth's subsurface using geophysics often requires an integrated approach [15].

In this study, ground magnetic method and Electrical resistivity employing Schulumberger electrode array also known as Vertical Electrical Sounding (VES) were used in order to know the geophysical characteristics of Ayetoro Housing Scheme. The purpose of using the two methods is to affirm a convincing result. This study is aimed at determining the characteristics of the subsurface at Ayetoro Housing Scheme in order to assess the stability of the subsurface for residential purpose and its hydrogeological prospects. Ground magnetic method or VES has been successfully used by several authors in order to evaluate the subsurface competency hence the adoption of the integrated techniques. Among the authors are Yusuf et al. [3], Macleod et al. [16], Jeffrey [17], Babu et al. [18], Kasidi and Ndatuwong [19], Adagunodo and Sunmonu [20], Adagunodo and Sunmonu [21], Jatau et al. [22], Fatoba et al. [23], Ibitoye et al. [24], Ibeneme et al. [25] and Ojo et al. [26]. Adewoyin et al. [27] posited that geophysical technique safes cost and time for private residential building developers when the area of consideration is large.

2. The study area and its Geology

Ayetoro is an area in Oyo town, Oyo West LGA of Oyo state, Southwest Nigeria (Figure 1). It is situated about 10 km west of Emmanuel Alayande College of Education and lies within Latitude N07° 50′ 00″ to N07° 53′ 00″ and Longitude E003° 52′ 00″ to E003° 55′ 00″ approximately with an average elevation range of about 292 m. The study area is one of the major towns in Southwestern Nigeria. It has gained influx of citizens since the establishment of St. Andrew's College, Oyo (now Emmanuel Alayande College of Education, Oyo) in 1896. Presently, Oyo has up to five tertiary institutions which has made the town to be part of urban renewal project for the government of Oyo state. Hence, the study area (Ayetoro Housing Scheme) is one of the new areas in the state undergoing positive turn around.

The geology of study area is crystalline rocks, which is of Precambrian in age. These rocks are composed of metamorphic and igneous rocks, which are found even beyond the study area. Some metasediments that are found in the surrounding environments of the study area such as Kajola, Ifedapo and Ifeloju local government area are: quartzofelds pathic biotite schists, quartzites and marble. Larger quantities of quartzites are found in Oyo town (Figure 2). Laterites are also found visible in some parts of Oyo, Oyo state, Nigeria

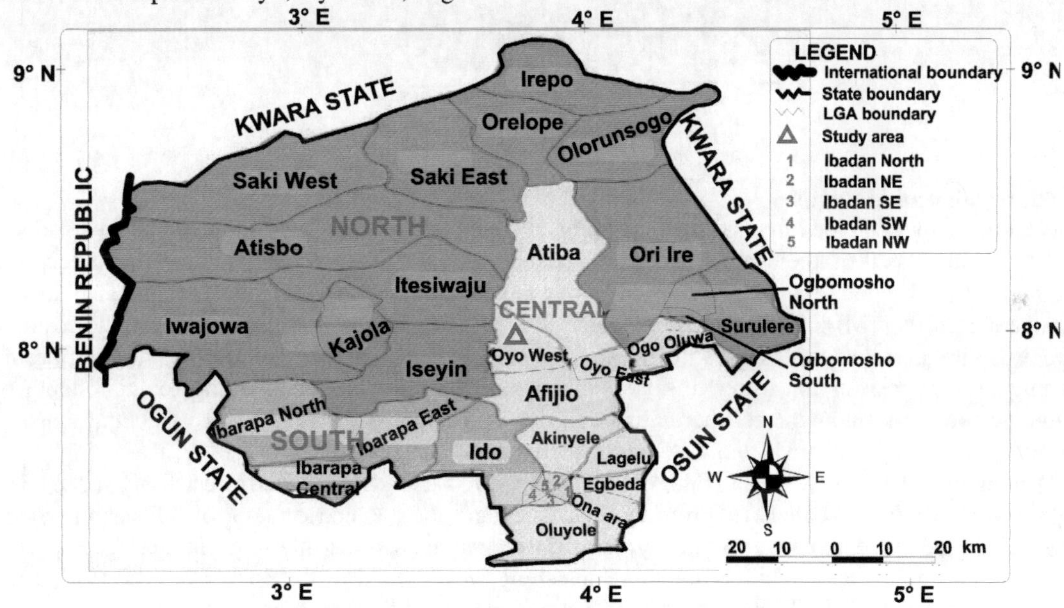

Figure 1: Map of Oyo State Indicating the Study Area

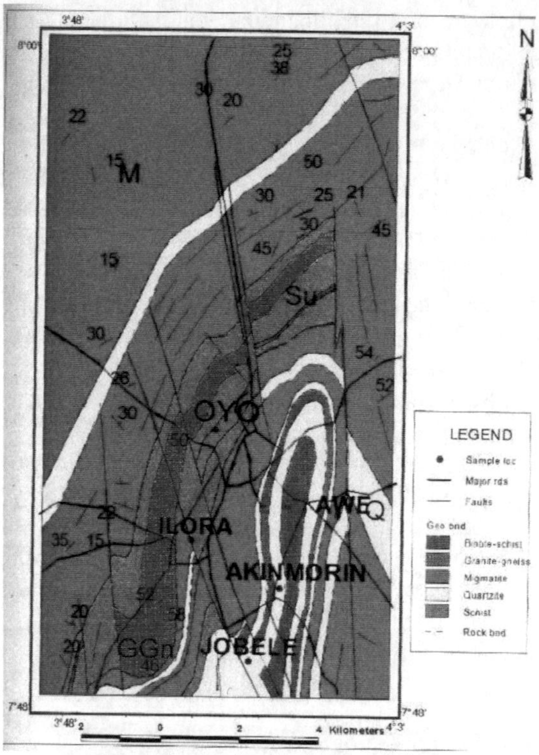

Figure 2: Geological map of Oyo

3. Materials and Methods

The study area was divided into six traverses (Figure 3), on which the inter station spacing was chosen to be ten meters, then the magnetic field of the area was measured using a proton precession magnetometer. The total magnetic intensity (TMI) was recorded from the field, the regional field was removed from the TMI during data processing. After the removal of the regional field, the residual data was further enhanced using analytic signal [28 – 30] and power spectrum [31, 32] filtering techniques. The reason for using this two enhancement techniques is to compare their depth to magnetic sources results whether there are variations in the results or not. The geostructural maps of the study area were generated from the analytic signal data.

A total of thirty VES data (Figure 3) were acquired using PZ-02 Earth resistivity meter. The apparent resistivity was calculated from the resistance and the geometric factor of VES as presented by Koefoed [33]. The apparent resisitivity was then plotted against the inter-current electrode spacing on a bi-logarithmic paper. The partial curve matching was applied to the curves in order to get the layers parameters (i.e. the layers' resistivity and thickness). The curve types gotten from the partial curve matching were then classified according to Koefoed [33] method of VES curves classifications.

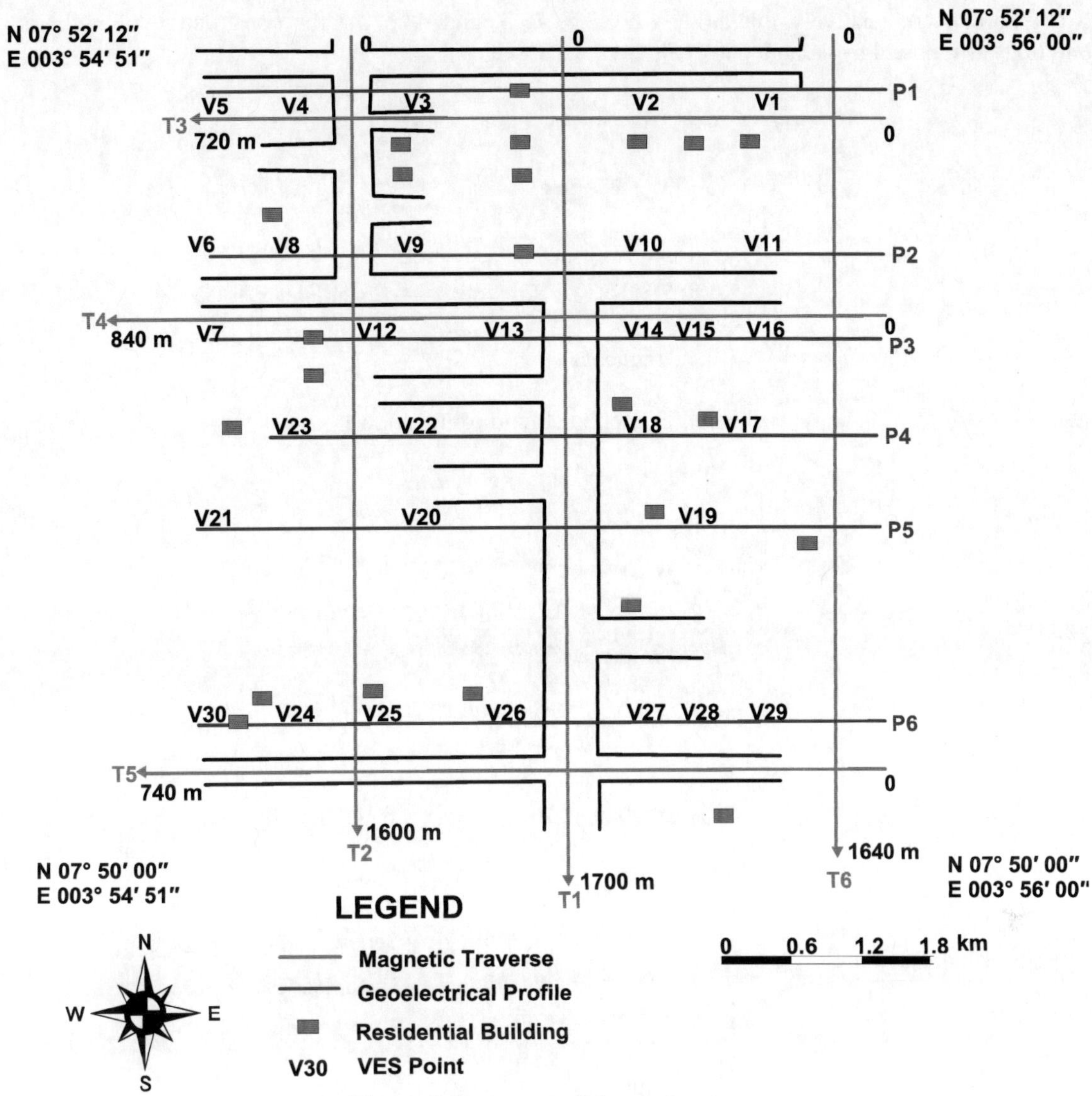

Figure 3: Base map of the study area.

4. Results and Discussion

The results of the ground magnetic data were presented in log power spectrum and analytic signal profiles. The depths to magnetic sources were however calculated from the features on these profiles. VES results were presented as curves and geoelectrical sections.

The observed power spectrum curves were obtained from plotting the log power against the frequency for the six magnetic traverses (figures 4.1 to 4.12). It is divided into two parts the regional field (figures 4.1, 4.3, 4.5, 4.7, 4.9, 4.11) and the local field (figures 4.2, 4.4, 4.6, 4.8, 4.10, 4.12).

4.1 Log Power Spectrum

Based on the models of Spector and Grant [31], depths to magnetic basement or sources were calculated and presented in Table 1. The depth from regional field ranged from 0.0620 to 0.1500 km, the depth from local field ranged from 0.0050 to 0.0620 km while the total depth ranged from 0.0695 to 0.1935 km. Traverses 2 and 4 showed thin overburden (i.e. 5 and 12 m) while, traverses 1, 3, 5 and

6 overburden are relatively thick (i.e. 62, 43.5, 43.5, and 43.5 m) for construction of high rise buildings as reported by Adagunodo [15].

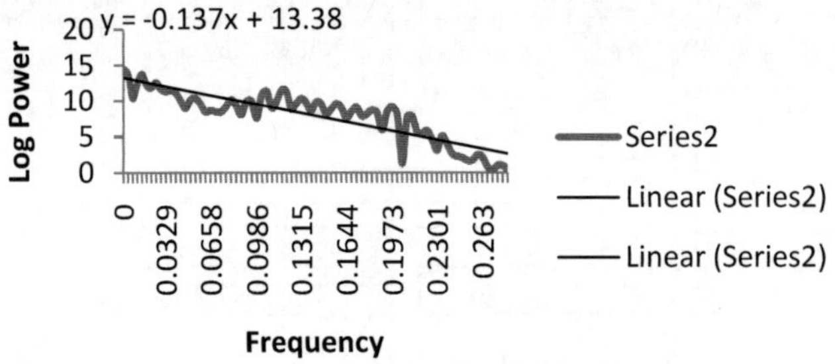

Figure 4.1: Regional field for traverse 1

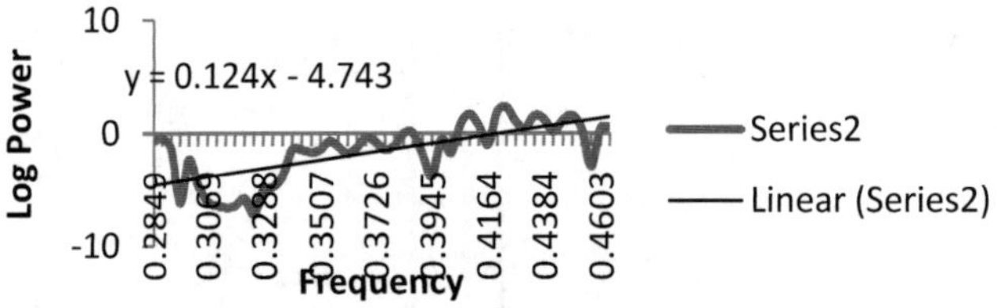

Figure 4.2: Local field for traverse 1

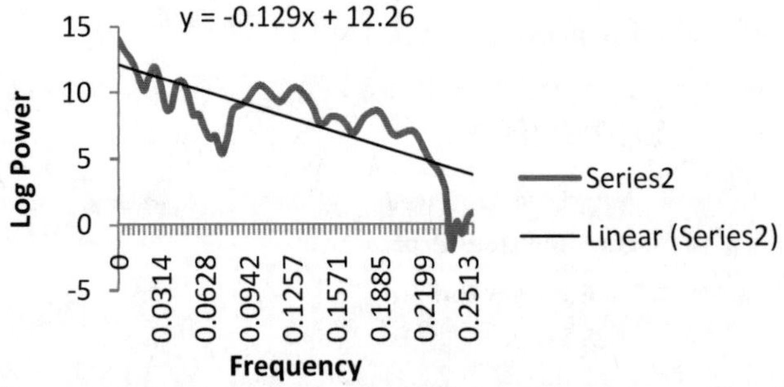

Figure 4.3: Regional field for traverse 2

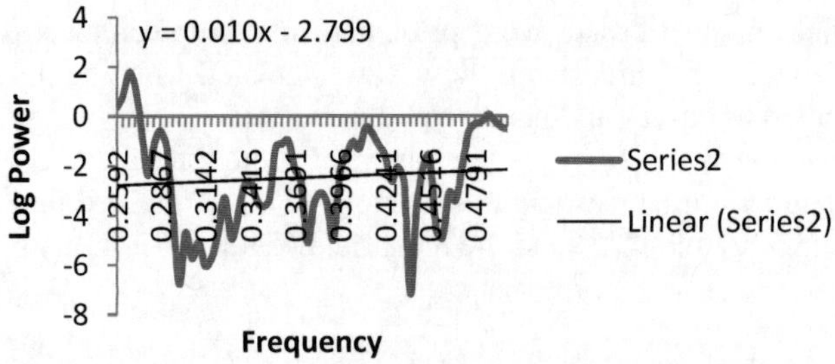

Figure 4.4: Local field for traverse 2

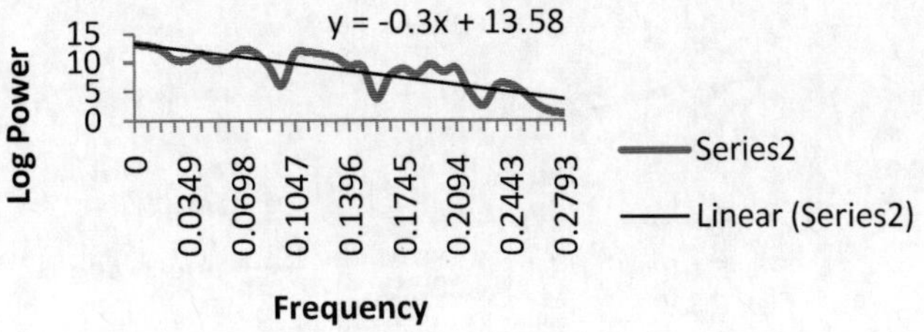

Figure 4.5: Regional field for traverse 3

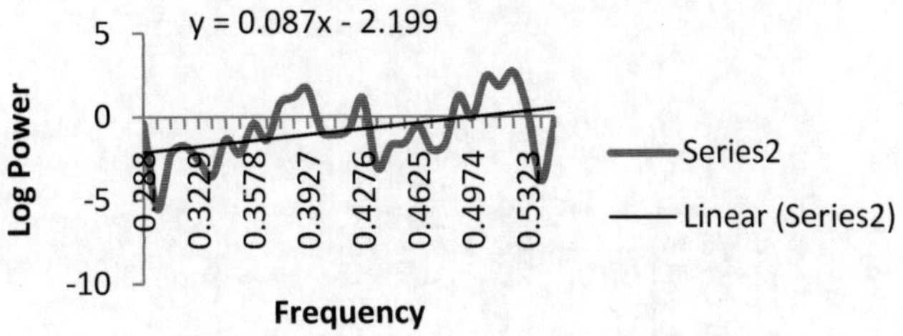

Figure 4.6: Local field for traverse 3

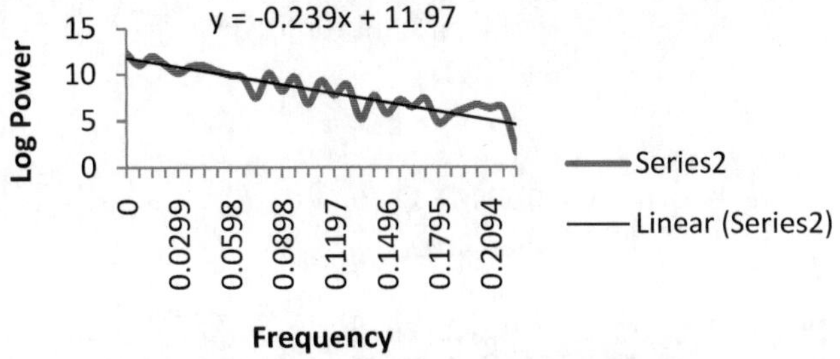

Figure 4.7: Regional field for traverse 4

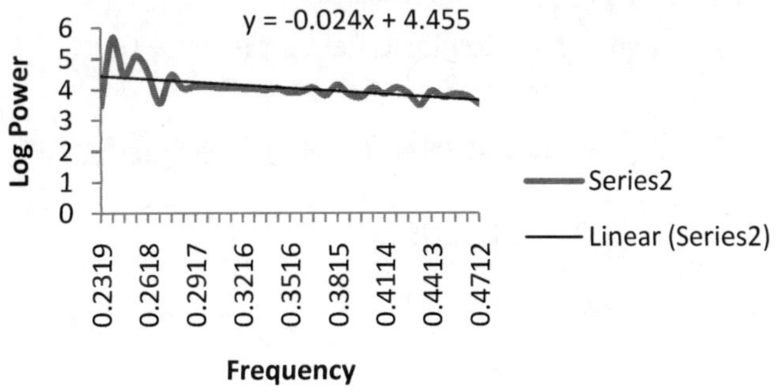

Figure 4.8: Local field for traverse 4

IOP Conf. Series: Earth and Environmental Science **173** (2018) 012031 doi:10.1088/1755-1315/173/1/012031

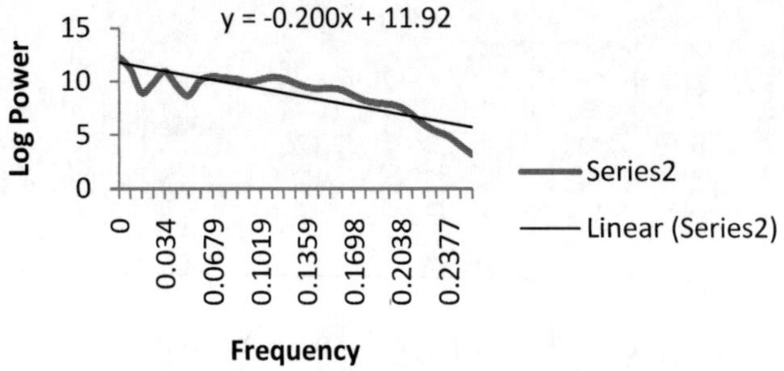

Figure 4.9: Regional field for traverse 5

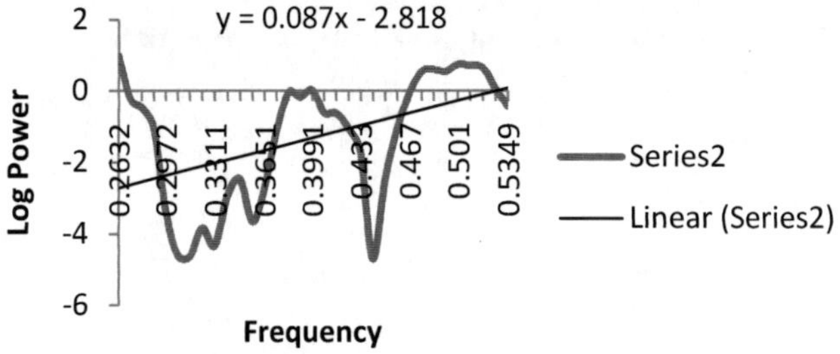

Figure 4.10: Local field for traverse 5

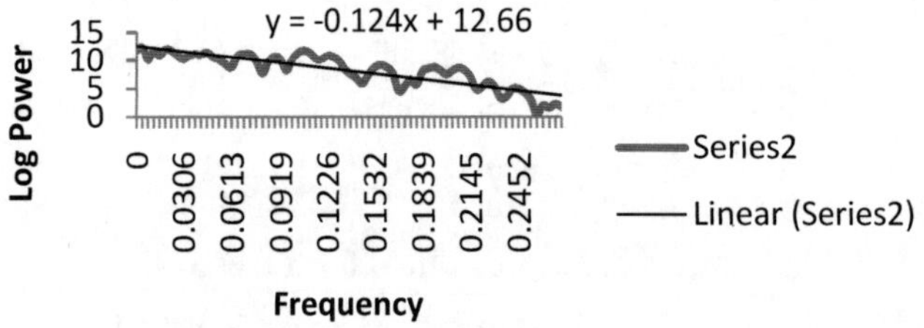

Figure 4.11: Regional field for traverse 6

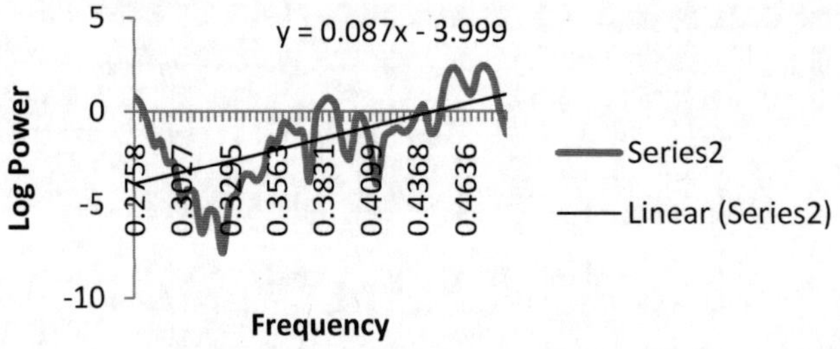

Figure 4.12: Local field for traverse 6

Table 1: Depth to magnetic sources at different traverses from the power spectrum curves.

	Regional (km)	Local (km)	Total (km)	Thickness of the basement (km)
Traverse 1	0.0685	0.0620	0.1305	0.0065
Traverse 2	0.0645	0.0050	0.0695	0.0595
Traverse 3	0.1500	0.0435	0.1935	0.1065
Traverse 4	0.1195	0.0120	0.1315	0.1075
Traverse 5	0.1000	0.0435	0.1435	0.0565
Traverse 6	0.0620	0.0435	0.1055	0.0185
Average	0.0941	0.0349	0.1290	0.0592

4.2 Analytical Signal Curves

Analytic signal is the second enhancement technique used for this study and its principles have been reported by Nabighian [28] and [29]. The depth to the anomaly was calculated using half-width method and the results were presented in Table 2. The analytic signal curves are presented in Figures 5.1 to 5.6. The magnetic highs and lows were observed in all the traverses. Generally, the magnetic highs represent regions of magnetic minerals such as metamorphic or igneous rocks while magnetic lows represent regions with linear geologic features. The total distance covered by each traverse is 1.7, 1.6, 0.72, 0.84, 0.72, and 1.6 km respectively.

Table 2: Depth to Magnetic Sources for Analytic Signal.

Traverse	Depth to magnetic sources for the different peaks in each traverse (m)								Mean (m)
1	40	30	20	100	40	30	60	160	60
2	60	40	40	40	120	-	-	-	60
3	35	35	20	45	55	40	-	-	38.33333
4	50	55	40	-	-	-	-	-	48.33333
5	90	68	75	-	-	-	-	-	77.66667
6	40	25	30	30	90	80	-	-	49.16667
	Mean of all depth to magnetic sources (m)								55.58333

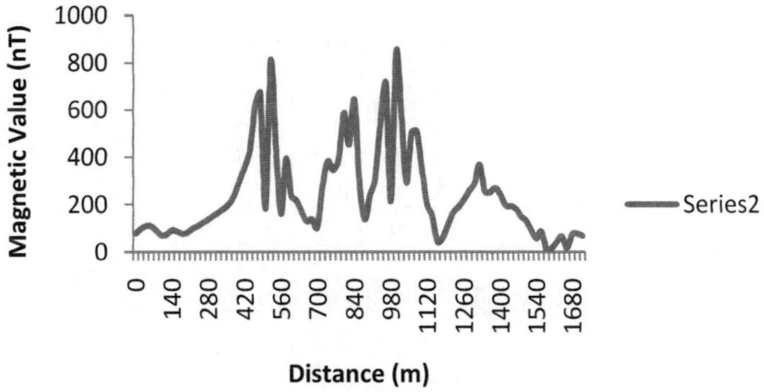

Figure 5.1: Analytic Signal curve for traverse 1.

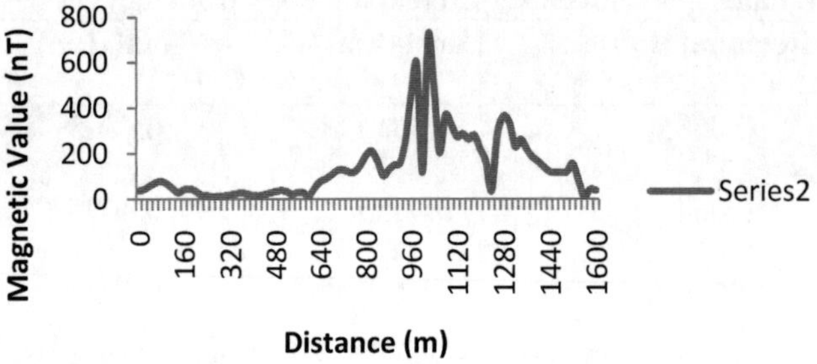

Figure 5.2: Analytic Signal curve for traverse 2

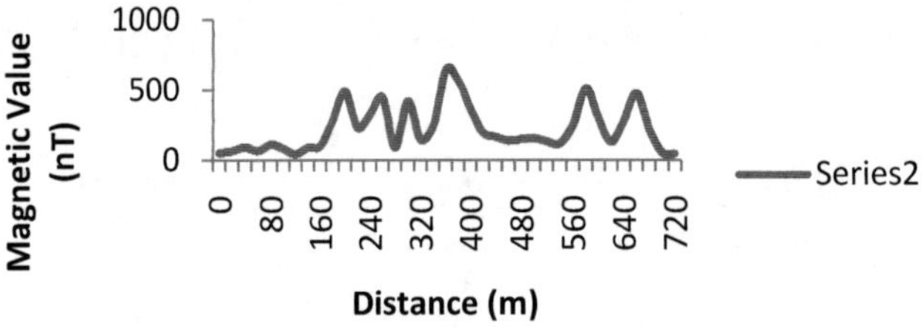

Figure 5.3: Analytic Signal curve for traverse 3

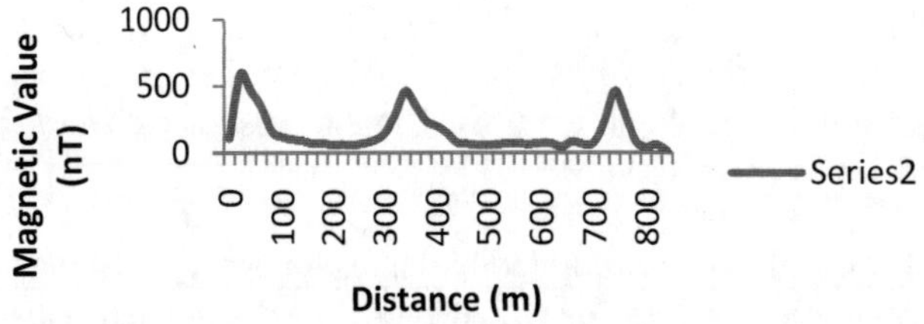

Figure 5.4: Analytic Signal curve for traverse 4

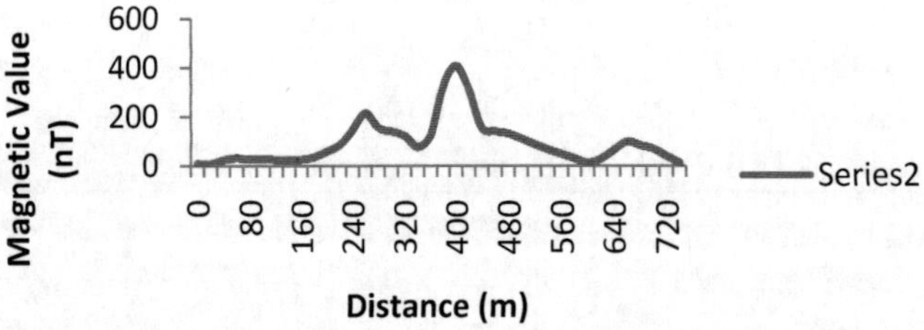

Figure 5.5: Analytic Signal curve for traverse 5

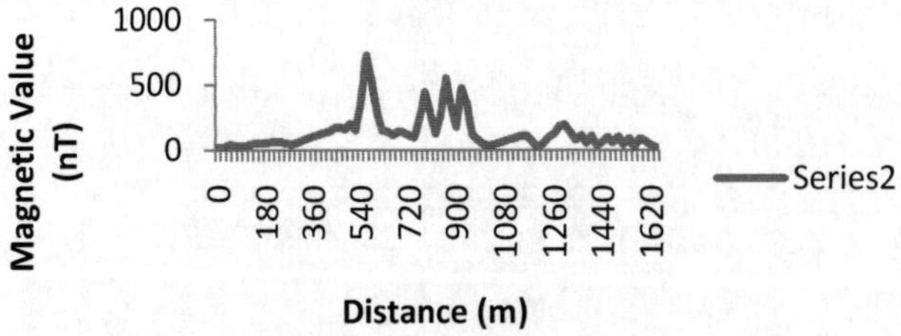

Figure 5.6: Analytic Signal curve for traverse 6

IOP Conf. Series: Earth and Environmental Science **173** (2018) 012031 doi:10.1088/1755-1315/173/1/012031

4.3 Geomagnetic Imaging of Ayetoro Housing Scheme

The contour map generated by surfer software showed areas with magnetic highs as well as areas with magnetic lows. Areas with magnetic highs are said to be competent for high-rise buildings while areas with magnetic lows are said to be incompetent. Magnetic values greater than 150 nT were experienced at the Western part, some part of the Northwest zone, base of the Northeast zone and some part (probably base) of the Southern region of the study area (Figure 6). These zones are considered to be competent for civil engineering activities irrespective of the shape and size. Since the magnetic signatures are not evenly distributed, it might not necessarily be confirmed that the study area is suitable for the construction of high-rise buildings.

The surface map further enhances the points made by the contour map because the thickness of the magnetic signatures are revealed (Figure 7) unlike the 2-D map that was only represented by contour lines. If high-rise building foundations were laid on this terrain, variations in the zones of the basement depressions (subsurface heterogeneity) might result into failed foundations which shall give rise to cracks in the columns of the building. This might further result to structural failure in the future.

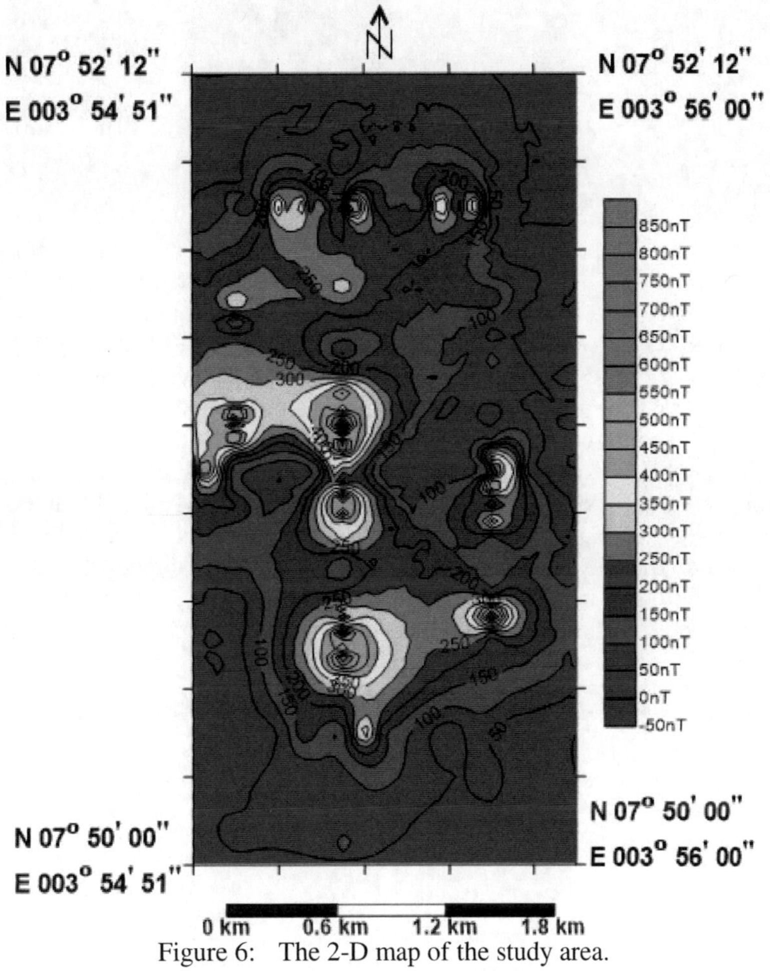

Figure 6: The 2-D map of the study area.

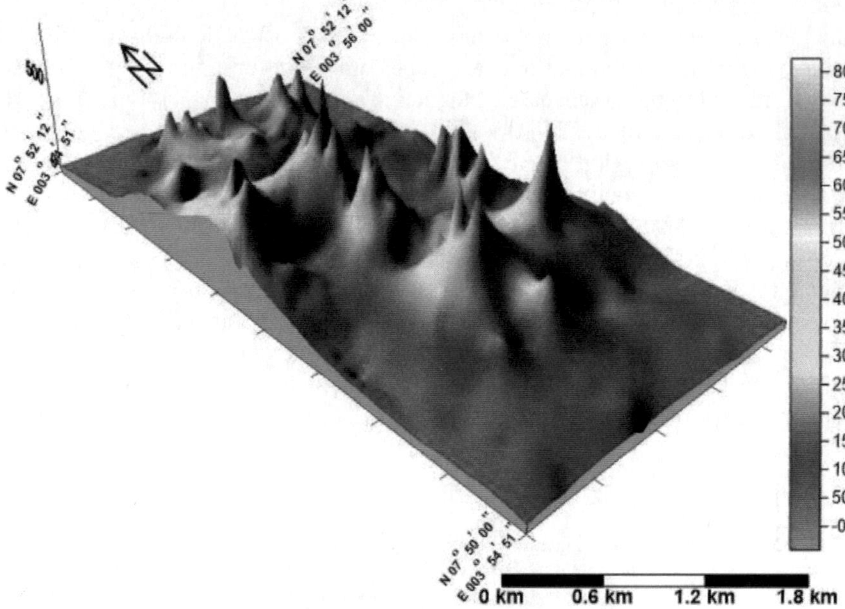

Figure 7: 3-D map for surface area.

4.4 Vertical Electrical Sounding

Thirty VES stations were occupied in the study area. The overburden thickness ranged from 1.8 to 44.8 m. Seven curve types were identified in the study area which are 12 H-type (VES 2, 3, 10, 12, 16, 17, 19, 21, 23, 28, 29, and 30), 8 QH-type (VES 1, 4, 5, 6, 9, 11, 18, and 24), 3 HA-type (VES 14, 15, and 22), 1 Q-type (VES 7), 2 A-type (VES 13 and 20), 2 HK-type (VES 25 and 26) and 2 KH-type (VES 8 and 27). Nine of the VES stations showed fresh bedrock while the remaining twenty one showed fractured bedrock. The VES interpretation reveals that the study area is not competent for high-rise building due to its fracture distribution and thick overburden present in some parts of the study area. The VES curves are grouped into six profiles as shown in Figure 3. Profile 1 has VES 1, 2, 3, 4 and 5 which runs in the E-W direction, profile 2 has VES 11, 10, 9, 8 and 6 in the E-W direction, profile 3 has VES 16, 15, 14, 13, 12 and 7 in the E-W direction, profile 4 has VES 17, 18, 22 and 23 also in the E-W direction, profile 5 has VES 19, 20 and 21in the E-W direction, profile 6 has VES 29, 28, 27, 26, 25, 24 and 30 also in the E-W direction. Seven different curve types experienced in the study area were presented in Figures 8.1 to 8.7. The VES curves' summary is presented in Table 3.

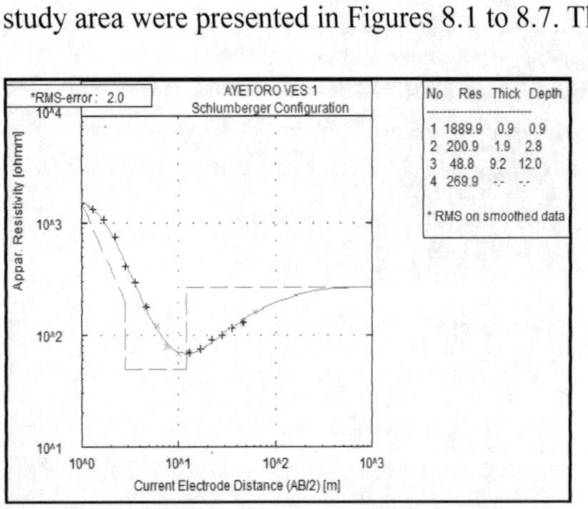

Figure 8.1: Modeled curve for VES 1 (QH).

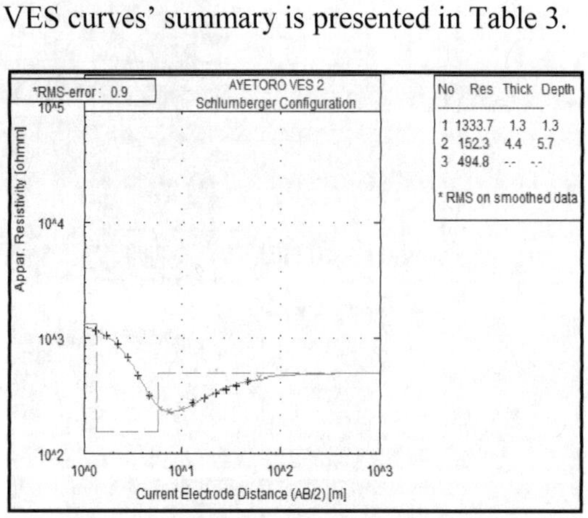

Figure 8.2: Modeled curve for VES 2 (H).

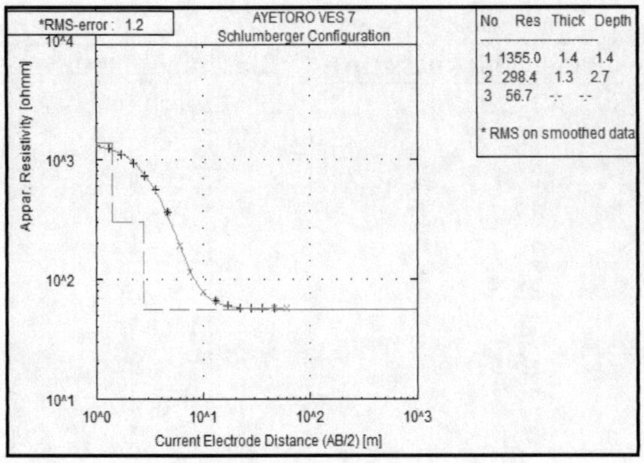

Figure 8.3: Modeled curve for VES 7 (Q).

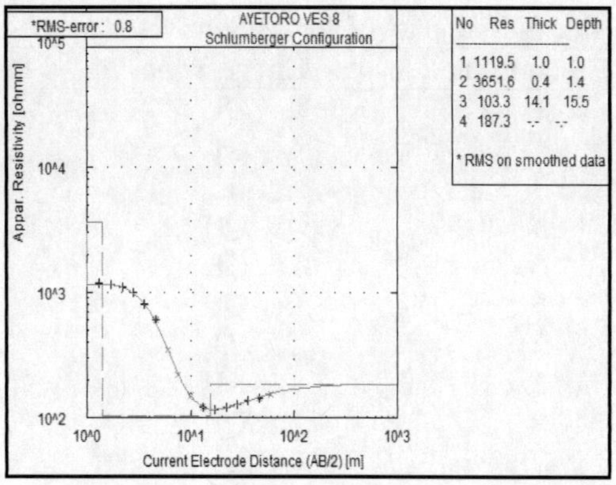

Figure 8.4: Modeled curve for VES 8 (KH).

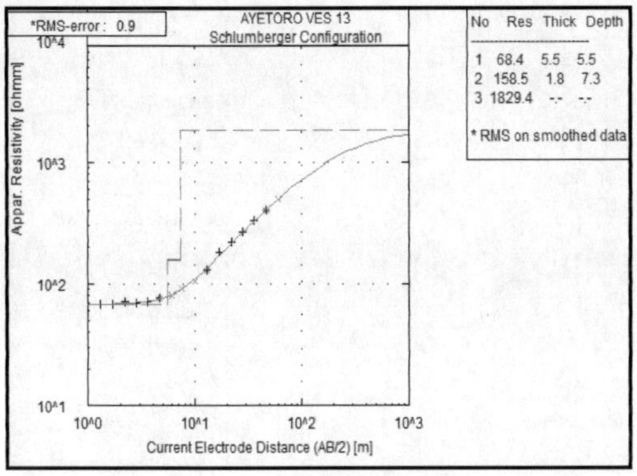

Figure 8.5: Modeled curve for VES 13 (A).

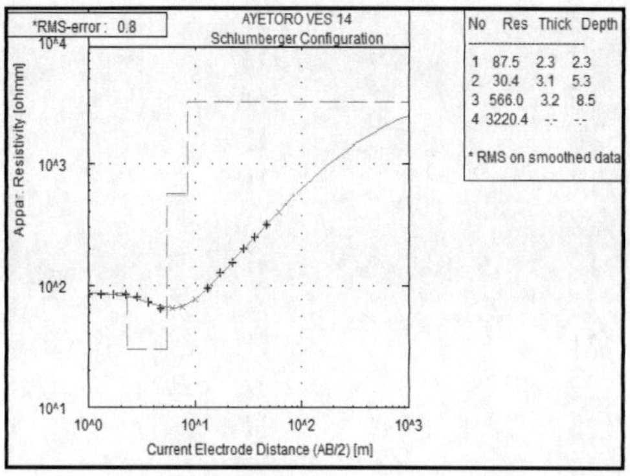

Figure 8.6: Modeled curve for VES 14 (HA).

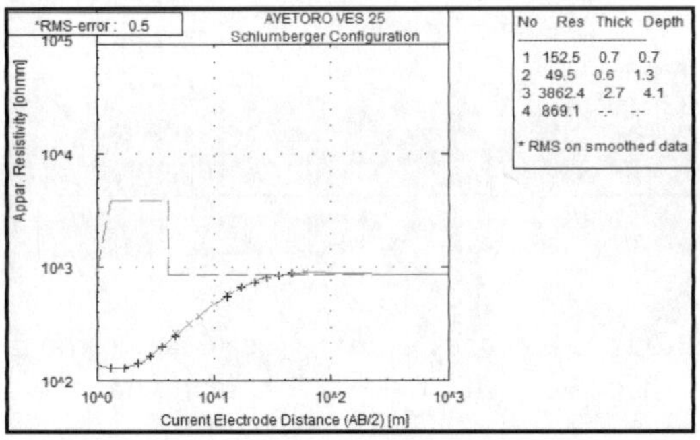

Figure 8.7: Modeled curve for VES 25 (HK).

IOP Conf. Series: Earth and Environmental Science **173** (2018) 012031 doi:10.1088/1755-1315/173/1/012031

Table 3: Summary of the VES curves.

VES Number	Curve Type	Overburden Thickness (m)	Basement Nature
1	QH	12.0	Fractured
2	H	5.7	Fractured
3	H	8.8	Fractured
4	QH	27.1	Fresh
5	QH	29.1	Fractured
6	QH	25.3	Fractured
7	Q	2.7	Fractured
8	KH	15.5	Fractured
9	QH	22.3	Fresh
10	H	7.1	Fresh
11	QH	12.1	Fresh
12	H	18.8	Fractured
13	A	7.3	Fractured
14	HA	8.5	Fresh
15	HA	5.1	Fractured
16	H	11.5	Fresh
17	H	6.4	Fractured
18	QH	17.8	Fractured
19	H	9.2	Fractured
20	A	15.7	Fractured
21	H	1.8	Fractured
22	HA	30.0	Fractured
23	H	8.1	Fractured
24	QH	44.8	Fresh
25	HK	4.1	Fractured
26	HK	4.9	Fractured
27	KH	7.9	Fresh
28	H	5.9	Fresh
29	H	7.3	Fractured
30	H	5.8	Fractured

4.5 Geoelectrical Section.

The 30 VES stations were grouped into 6 profiles according to how they could be located on a straight line, the result of the VES station in each profile were then interpreted to see a 2-D representation of the subsurface.

4.5.1 Profile 1. From the geoelectric section of profile one it was revealed that the topsoil which is less than 2 m contains lateritic formation (Figure 9.1). The second layer which was interpreted as weathered layer i.e. mixture of clay and sand which extends up to the depth of about 25 m towards the western region of the profile, also noticed on the profile was clayey zone which was present at the eastern and central part of the profile. Directly under this formation lies fractured bedrocks with the shoot-out of fresh bedrock or basement towards the western region of the profile. This profile is incompetent for high-rise buildings.

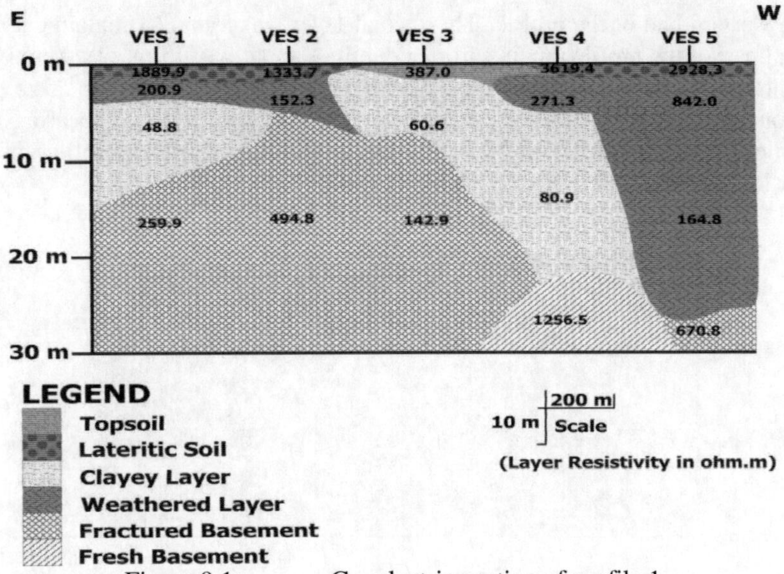

Figure 9.1: Geoelectric section of profile 1

4.5.2 *Profile 2***.** It was revealed from the geoelectric section that it has topsoil that extends to a depth of about 2 m in the eastern part, 5 m in the central part, and 2 m in the western region (Figure 9.2). The western part of the topsoil was also found to contain lateritic formation. The second layer was found to be made up of clayey layer on the eastern and west central region up to depth of about 15 m. Weathered layer was found in the central and part of the western region of the profile. The profile was found to be underlain by fractured basement on the eastern and western part while the central part is underlain with fresh basement. Only the central part of this profile is competent for high-rise building constructions.

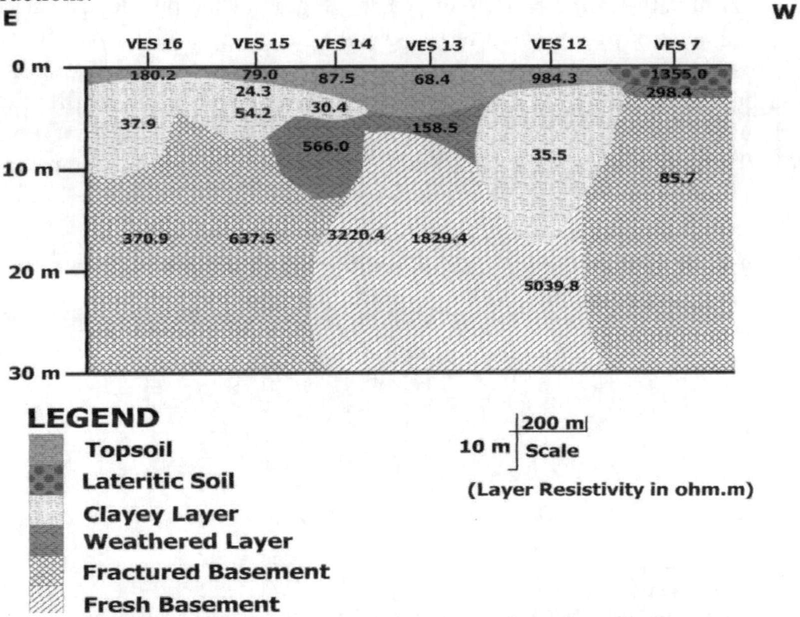

Figure 9.2: Geoelectric section of profile 2

4.5.3 *Profile 3*. The topsoil extends to a depth of about 2 m on the eastern part, and 5 m on the western part (Figure 9.3). The topsoil is mostly made up of lateritic formations from about a distance

IOP Conf. Series: Earth and Environmental Science **173** (2018) 012031 doi:10.1088/1755-1315/173/1/012031

of 100 m to the western end of the profile. The second layer was found to contain clayey zone in the eastern to central part of the profile ranging from a depth of about 7 to 15 m, weathered layer was also found in the central to western part of the profile with a depth ranging from 2 to 25 m. The basement was found to contain majorly fractured basement with fresh basement only appearing in the western part of the profile. Due to the thick overburden of the fresh bedrock in this profile, the entire profile will be good for low-rise building construction.

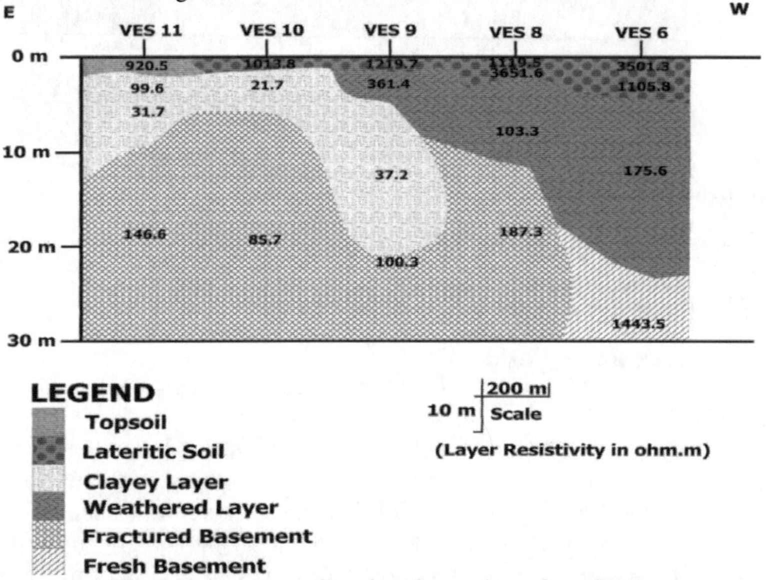

Figure 9.3: Geoelectric section of profile 3

4.5.4 *Profile 4*. The topsoil has a depth of less than 2 m throughout the profile and contains laterite for about 200 m between the eastern and central region of the profile. The second layer was found to contain weathered layer in the eastern and west central part of the profile with a depth of about 6 m in the eastern region and 28 m in the west central part, the second layer was also found to contain clayey layer to a depth of 15 m in the central and western part of the profile (Figure 9.4). The profile is majorly underlain with fractured basement with fresh basement shooting up in the eastern part of the profile. The profile would be prosperous for borehole drilling and incompetent for high-rise building.

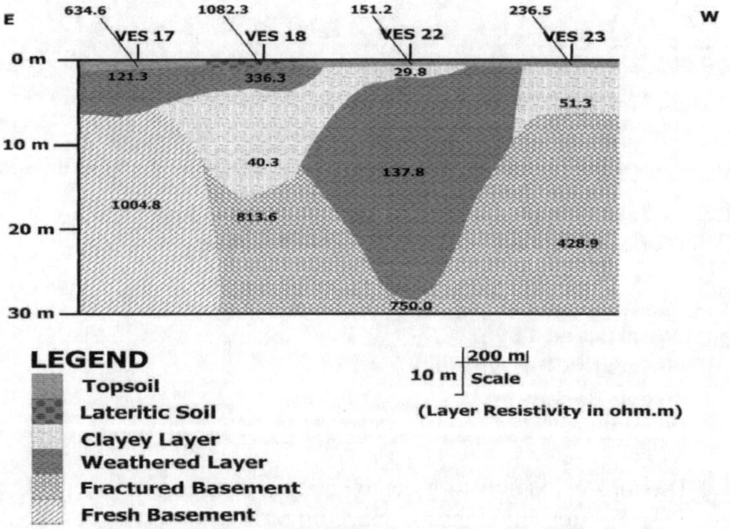

Figure 9.4: Geoelctric section of profile 4

4.5.5 *Profile 5.* This profile has a topsoil of about 2 m, the second layer contains majorly clayey zone which ranges in depth from 5 m in the western part to 10 m in the eastern part. It also contains weathered layer in the central part of the profile with a depth of about 15 m, the profile showed fractured bedrock throughout (Figure 9.5). Hence, it is incompetent for civil engineering activities.

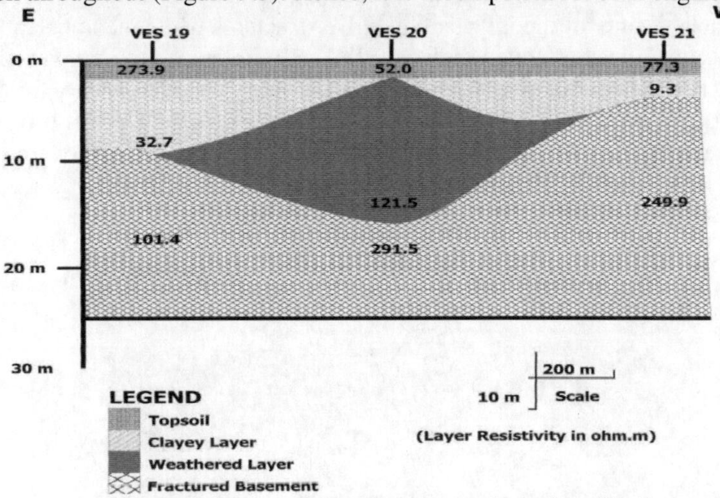

Figure 9.5: Geoelectric section of profile 5.

4.5.6 *Profile 6.* It has a topsoil of about 2 m deep and shows lateritic formation on the eastern part. The second layer has a depth of about 6 m and is made up of weathered layer on the eastern and western end of the profile. It also consists lateritic formation in the central part of the profile and a clayey layer of about 400 m wide above the lateritic layer (Figure 9.6). The basement is majorly intercalations of fractured and fresh bedrocks. Inhomogeneity in this profile makes it incompetent for civil engineering purposes but fractured bedrocks would be productive for groundwater if explored properly.

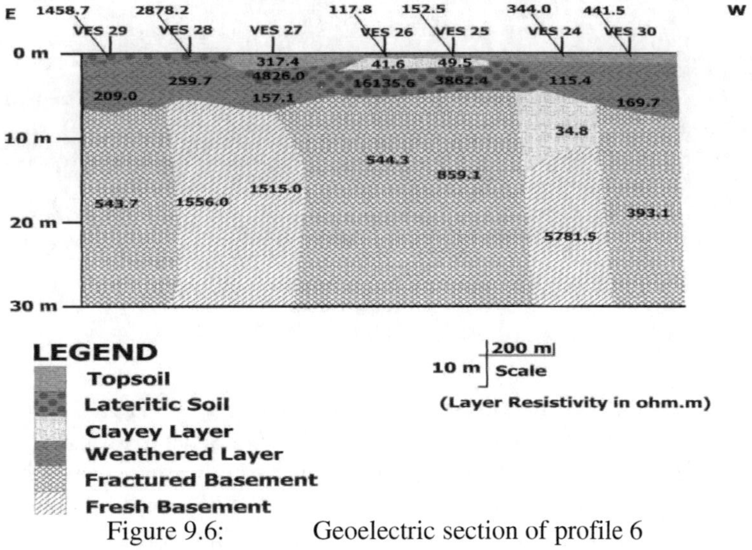

Figure 9.6: Geoelectric section of profile 6

4.6 Site Characterization

In a complex terrain such as crystalline bedrock, it is unadvisable to delineate the subsurface features via single geophysical technique. An integrated approach is recommended as the information lost in one technique would be revealed through the other method. Figure 10 shows the subsurface characterization of Ayetoro Housing Scheme. The weighted averages of all the parameters considered in this study (geoelectrical and geomagnetic features) were used to classify the study area into competent, moderate, and incompetent zones. Competent and moderately competent zones are recommended for high-rise 'residential buildings', moderate zone needs an experts' advice in other to choose the most suitable design for the intended high rise structures while incompetent zone should be avoided for such task. Incompetent zone trends in NW – SE direction with a digression towards the south. If the building developers that would be employed in this study area keenly follow the outcome of this study and make use of appropriate building materials, it would greatly avert any form of structural failure that might occur in this allocate expanse of land by Oyo State government, Nigeria.

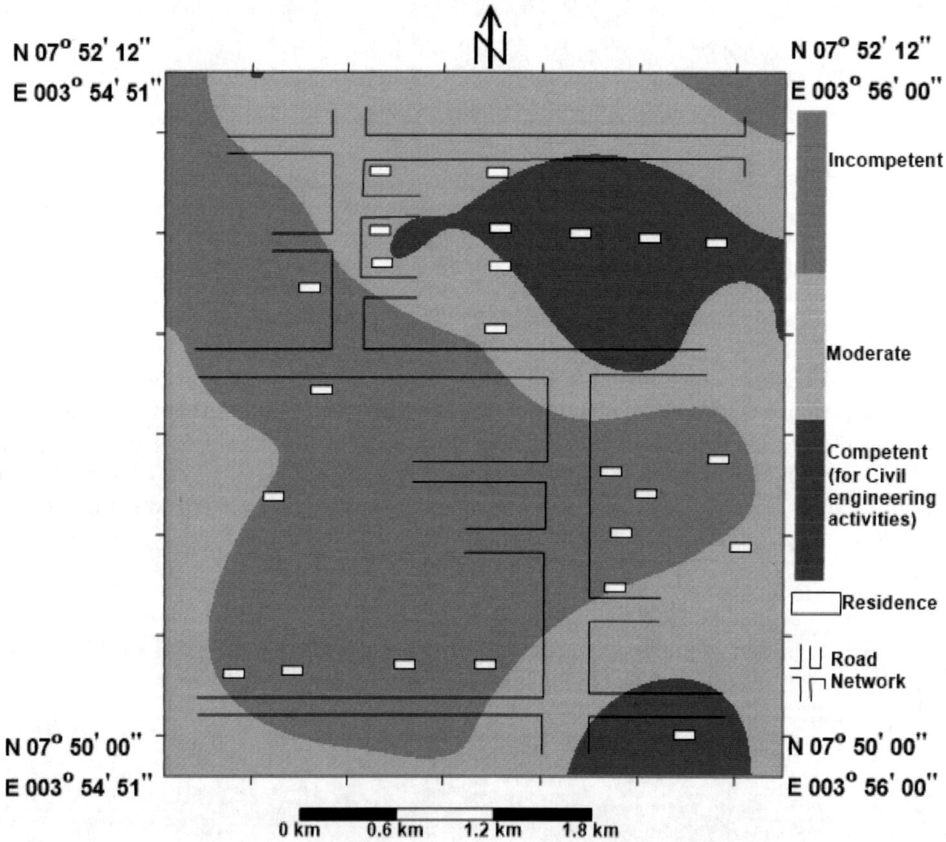

Figure 10: Site characterization map of the Ayetoro.

5. Conclusion

The study has been able to affirm that justifiable conclusion needs an integrated approach. From the qualitative and quantitative characterization of Ayetoro Housing Scheme using ground magnetic and vertical electrical sounding techniques, it is concluded that the study area is generally not suitable for the construction of high-rise buildings, and such structures should be avoided. However, the study area is suitable for low-rise buildings' constructions. Drilling of boreholes instead of wells is also recommended in this terrain due to the nature of the bedrock. This would also reduce water scarcity in the study area during dry season. It is further recommended that building constructions in the study area should strictly be handled by professionals.

Acknowledgement
We appreciate the management of Covenant University, Nigeria for the publication support received from them.

References

[1] Obasi I.A., Onwe I.M. and Igwe E.O. (2015). Geoelectrical Subsurface Characterization for Foundation purposes in the College of Agricultural Sciences (CAS) Campus, Ebonyi State University, Abakaliki, Southwestern Nigeria. Journal of Environment and Earth Science, **5**(18): 42 – 52.

[2] Fajana A.O., Olaseeni O.G., Bamidele O.E. and Olabode O.P. (2016). Geophysical and Geotechnical Investigation for post foundation studies, Faculty of Social Sciences and Humanities, Federal University Oye Ekiti. FUOYE Journal of Engineering and Technology, **1**(1): 62 – 66.

[3] Yusuf G.A., Akinrinade O.J. and Ojo J.S. (2015). An Engineering Site Characterization using geophysical methods: a case study from Akure, Southwestern Nigeria. Journal of Earth Sciences and Geotechnical Enginerring, **5**(4): 57 – 77.

[4] Adagunodo T.A., Adeniji A.A., Erinle A.V., Akinwumi S.A., Adewoyin O.O., Joel E.S., Kayode O.T. (2017). Geophysical Investigation into the Integrity of a Reclaimed Open Dumpsite for Civil Engineering Purpose. Interciencia Journal, **42**(11): 324 – 339.

[5] Oyeyemi K.D., Aizebeokhai A.P., Adagunodo T.A., Olofinnade O.M., Sanuade O.A., Olaojo A.A. (2017). Subsoil Characterization using Geoelectrical and Geotechnical Investigations: Implications for foundation Studies. International Journal of Civil Engineering and Technology, **8**(10): 302 – 314.

[6] Adewoyin O.O., Joshua E.O., Akinyemi M.L., Omeje M. and Joel E.S. (2017). Investigation to determine the vulnerability of reclaimed land to building collapse using near surface geophysical method. Journal of Physics: Conference Series, **852**(1): 012030.

[7] Adagunodo T.A., Sunmonu L.A., Ojoawo A., Oladejo O.P. and Olafisoye E.R. (2013). The Hydro Geophysical Investigation of Oyo State Industrial Estate Ogbomosho, Southwestern Nigeria Using Vertical Electrical Soundings. *Research Journal of Applied Sciences, Engineering and Technology*, **5**(5), 1816–1829.

[8] Oladejo O.P., Sunmonu L.A., Ojoawo A., Adagunodo T.A. and Olafisoye E.R. (2013). Geophysical Investigation for Groundwater Development at Oyo State Housing Estate Ogbomosho, Southwestern Nigeria. *Research Journal of Applied Sciences, Engineering and Technology*, **5**(5), 1811–1815.

[9] Adagunodo T.A., Sunmonu L.A., Adabanija M.A., Suleiman E.A., Odetunmibi O.A. (2017). Geoexploration of Radioelement's Datasets in a Flood Plain of Crystalline Bedrock. Data in Brief, **15**C: 809 – 820. http://dx.doi.org/10.1016/j.dib.2017.10.046.

[10] Adagunodo T.A., Hammed O.S., Usikalu M.R., Ayara W.A., Ravisankar R. (2018). Data on the Radiometric Survey over a Kaolinitic Terrain in Dahomey Basin, Nigeria. Data in Brief, **18**C: 814 – 822. https://doi.org/10.1016/j.dib.2018.03.088.

[11] Adagunodo T.A., George A.I., Ojoawo I.A., Ojesanmi K. and Ravisankar R. (2018). Radioactivity and Radiological Hazards from a Kaolin Mining Field in Ifonyintedo, Nigeria. MethodsX, https://doi.org/10.1016/j.mex.2018.04.009.

[12] Olafisoye E.R., Sunmonu L.A., Ojoawo A., Adagunodo T.A. and Oladejo O.P. (2012). Application of Very Low Frequency Electromagnetic and Hydro-physicochemical Methods in the Investigation of Groundwater Contamination at Aarada Waste Disposal Site, Ogbomoso, Southwestern Nigeria. *Australian Journal of Basic and Applied Sciences*, **6**(8), 401–409.

[13] Adagunodo T.A. (2017). Groundwater Contamination: Performance, Effects, Limitations and Control. Chapter 3 in Book: Groundwater Contamination: Performance, Limitations and Impacts, 1 – 135. ISBN: 978-1-153611-017-3; 978-1-53611-003-6. Editor: Anna L. Powell © 2017 Nova Science Publishers, Inc. Pp. 33 – 64.

[14] Adagunodo T.A., Sunmonu L.A., Oladejo O.P. and Ojoawo I.A. (2013). Vertical Electrical Sounding to Determine Fracture Distribution at Adumasun Area, Oniye, Southwestern Nigeria. IOSR Journal of Applied Geology and Geophysics, **1**(3): 10–22.

[15] Adagunodo A. and Sunmonu A. (2013). The Study of Basement Pattern of an Industrial Estate. ISBN: 978-3-659-35930-9. Published by LAP Lambert Academic Publishing GmbH & Co. KG, Heinrich-Bocking. Str. 6-8, 66121 Saarbrucken, Deutschland, Germany.

[16] Macleod I.N., Jones K., and Dai T.F. (1993). 3-D analytic signal in the Interpretation of total magnetic field data at low magnetic latitudes. Exploration Geophysics, **24**: 679 – 687.

[17] Jeffrey D.P. (2000). Locating magnetic contacts: A comparison of the Horizontal gradient, Analytic signal, and Local wavenumber methods. Society of Exploration Geophysicist, Abstract with programs, Calagry: 50-70.

[18] Babu V.R. Ram S. and Sundararajan N. (2007). Modeling and Inversion of Magnetic and VLF-EM data with an Application to Basement Fractures: a case study from Raigarh, India. Geophysics, **72**(5): B133-B140.

[19] Kasidi, S. and Ndatuwong L.G. (2008). Spectral Analysis of Aeromagnetic data over Longuda plateau and Environs, North-Eastern Nigeria. Continental Journal of Earth sciences, **3**: 28 – 32.

[20] Adagunodo T.A. and Sunmonu L.A. (2012). Estimation of overburden thickness of industrial estate Ogbomoso, Southwestern Nigeria. Advances in Applied Science Research, **3** (5): 3129-3141.

[21] Adagunodo T.A. and Sunmonu L.A. (2012). Groundmagnetic survey to investigate on the fault pattern of industrial estate Ogbomoso, Southwestern Nigeria. Advances in Applied Science Research, **3**(5): 3142-3149.

[22] Jatau B.S., Patrick N.O., Baba A., and Fadele S.I. (2013). The use of vertical electrical sounding (VES) for subsurface geophysical investigation around Bomo area, Kaduna state, Nigeria, IOSR Journal of Engineering (IOSRJEN), **3**(1): 10-15.

[23] Fatoba J. O., Salami B. M. and Adesida A. (2013). Structural failure investigation using electrical resistivity method: A case study of Amafor Ihuokpala, Enugu, Southeastern Nigeria. Journal of Geology and Mining Research, **5**(8): 208-215.

[24] Ibitoye Folahan and Ipinmoroti Felix (2013). Salami Mudasiru, Akinluwade Kunle, Adeyinka Taiwo, Adelana Adetunji, Application of Geophysical Methods to Building Foundation Studies. International Journal of Geosciences, **4**(1): 1256-1266.

[25] Ibeneme S.I., Okereke C.N., Iroegbu C., Etiefe E.O. (2014). Vertical Electrical Sounding for Aquifer Characterization around the Lower Orashi River Sub-Basin Southeastern Nigeria. Communications in Applied Sciences, **1**(2): 36-51.

[26] Ojo A.O., Omotoso T.O. and Adekanle O.J. (2014). Determination of Location and Depth of Mineral Rocks at Olode Village in Ibadan, Oyo State, Nigeria, using Geophysical Methods. International Journal of Geophysics, Article ID **306862**: 1-13.

[27] Adewoyin O.O., Joshua E.O., Akinwumi I.I., Omeje M. and Joel E.S. (2017). Evaluation of Geotechnical parameters using geophysical data. J. Eng. Technol. Sci., **49**(1): 95 – 113.

[28] Nabighian M.N. (1972). The analytic signal of two-dimensional magnetic bodies with polygonal cross-sections: Its properties and use for automated anomaly interpretation. Geophysics, **37**: 507-517.

[29] Nabighian M.N. (1984). Towards a three dimensional automatic interpretation of potential field data via generalized Hilbert transform Fundamental relations. Geophysics, **47**: 780-786.

[30] Roest W.R., Verhoef J., and Pilkington (1992). Magnetic interpretation using the 3D Analytic signal. Geophysics, **57**: 116-125.

[31] Spector A. and Grant F.S. (1970). Statistical Model for Interpreting aeromagnetic data. Geophysics, **2**(25): 293 – 303.

[32] Olusola O.O. (2010). Depth estimation from the Aeromagnetic data of Wuyo, using Matched filtering and Log power spectrum. Unpublished undergraduate Thesis, Federal University of Agriculture Abeokuta, Nigeria.
[33] Koefoed O. (1979). Geosounding principles, I. Resistivity sounding measurements. Elsevier Scientific Publishing, Comp. Amsterdam.

Assessment of background radiation in some parts of Canaanland Ogun state, southwestern Nigeria using integrated geophysical techniques

Olusegun Adewoyin*[1], Maxwell Omeje[1], Emmanuel Joel[1]

[1]Department of Physics, College of Science and Technology Covenant University PMB 1023, Ota, Ogun state, Nigeria

*[1] Corresponding author's email address: segadot@yahoo.com

Abstract. Radioactivity in soil occurs naturally due to the presence of ^{238}U, ^{232}Th and ^{40}K. Undue exposure of humans to these radiations can result in some serious health problems. There is need to have adequate information on the level of radiation that is available around us in order to take adequate measures for protection. Electrical resistivity and radiometric methods were used in this study. The results by the radiometric method revealed the activity concentrations of U, Th and K to vary from 12.658 Bq kg^{-1} - 42.299 Bq kg^{-1}, 44.9645 Bq kg^{-1} - 128.702 Bq kg^{-1} and 31.3Bq kg^{-1} - 453.85 Bq kg^{-1}, respectively. The result of the electrical resistivity method varied between 75 and 2025 Ωm. Finally, it was observed that region of high resistivity was prone to higher level of radiation, which could be as a result of some radioactive geological features in the subsurface.

Keywords: Radioactivity, exposure, resistivity, geological features

1. Introduction

Our environment is constantly dominated with naturally occurring radionuclides to which man is regularly exposed. This is because these radioactive substances are present in soil and air around us [1]. Most prominent among these radionuclides are 238U (226Ra), 232Th and 40K which can pose serious hazard to man if their presence around us is more than the world recommended standard [2-4]. Although, naturally occurring radionuclides have natural origin, but their concentrations can be influenced by the activities of man. Such activities may include industrial processes such as cement production, coal mining, oil and gas exploration, fertilizer production (phosphate) and fertilizer application and so on [5-6].

Radiation has some useful applications in different areas including agriculture, medicine, mining, geology, archaeology, biology, etc. But there is a minimum permissible level of this radiation that man should be exposed to. The intensity of radiation depends on the amount of naturally occurring radioactive minerals (NORM) present in the soil and the time of exposure [7-8]. Food crops grown on the soil with high level of radionuclide content can pose serious health hazard such as the destruction of the gene pool [9-10]. Long term exposure can cause radiation sicknesses such as diarrhea, fever, nausea, vomiting, dizziness and disorientation, hair loss, blood stools, fatigue, infections, poor wound healing, low blood pressure.

Since these radionuclides originate in nature, it implies that there is no place in the world that is totally free from the influence of radioactive materials. Therefore, it is important to assess the level of radiation that people are exposed to from time to time so as to ascertain that the level of these radionuclide concentration is not beyond the world recommended safe limit.

2. Geology of the Study Area

The area of study is in the Eastern part of Dahomey Basin (Fig. 1), extending from the Volta Delta (Southeastern Ghana) to the Western flank of the Niger Delta in Nigeria [11]. The basin is separated

IOP Conf. Series: Earth and Environmental Science **173** (2018) 012032 doi:10.1088/1755-1315/173/1/012032

from the Niger Delta in the Eastern section by the Benin Hinge Line and Okitipupa Ridge and marks the continental extension of the chain fracture zone [12-13]. The stratigraphy of the basin consists of six formations which comprises of Abeokuta, Ewekoro, Akinbo, Oshosun, Ilaro and Benin Formations. The Abeokuta Formation is Cretaceous. The Abeokuta formation is mainly made up of large grained sand that spreads over the whole basin. It also contains mudstones, siltstones and shale-clay with slim limestone beds formed due to marine transgression [14-15]. It is poorly sorted with a sequence of continental grits and pebbly sands. Overlying the Abeokuta formation is the Ewekoro formation. It is predominantly comprised of shallow marine limestone as a result of the contamination of the marine transgression. The Ewekoro limestone is Palacocene in age. Overlying the Ewekoro Formation is the Akinbo formation. It is dominated by shale and it is of Late Palaeocene to Early Eocene. The Akinbo Formation is overlain by the Oshosun Formation. It is composed of Eocene shale. The Oshosun formation is overlain by the Ilaro formation. It is predominantly a sequence of coarse sandy estuarine, deltaic, and continental beds. The Ilaro Formation displays rapid lateral facies changes. Overlying the Ilaro Formation is the Benin Formation. It is predominantly composed of coastal plain sands and tertiary alluvium deposits [16]. Canaanland, the site of the study is in Ogun state, southwestern Nigeria. This site is the seat of Faith Tabernacle, the world acclaimed largest church auditorium in the world, where over 400,000 people worship every Sunday.

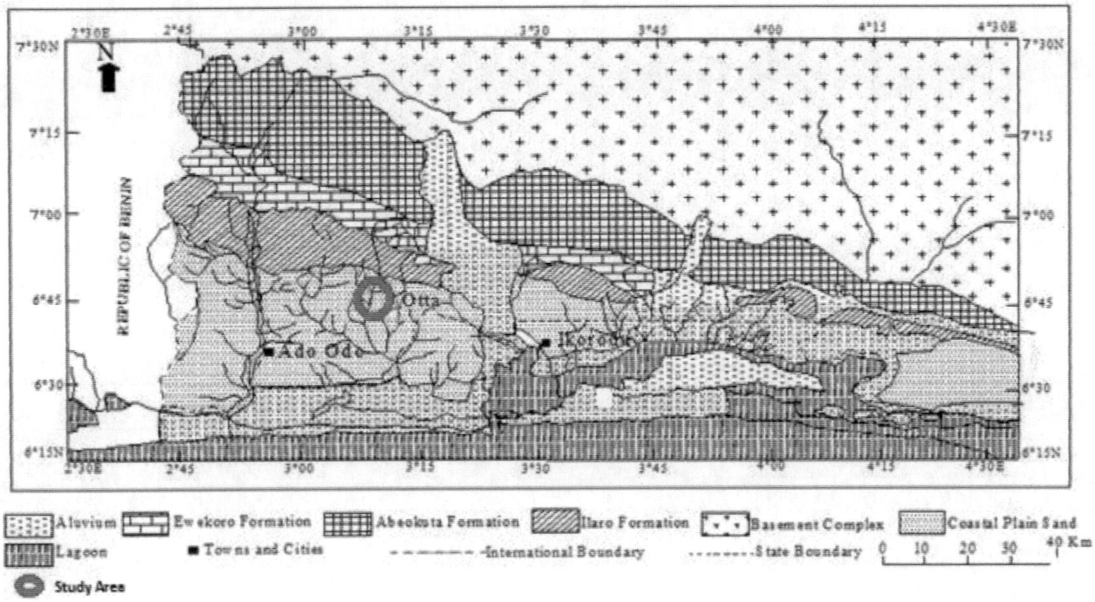

Figure 1. Geological Map of the Study Area

3. Materials and Methods

In this study, electrical resistivity was integrated with radiometric method. The resistivity method was used to determine both vertical and lateral variation in the resistivity of the study area. For this study, an ABEM Terrameter 1000/4000 (SAS) was engaged. Wenner electrode configuration was used. The length of each profile was 100 m and an electrode spacing of 5 m was used. The measurement was conducted by arranging the four electrodes such that the two outer electrodes are for injecting current into the subsurface while the two internal electrodes are required to estimate the potential difference. The value of the resistance is measured from the application of ohms law [17-18]. In order to determine the resistivity, the value of the resistance is multiplied with the geometric factor of this electrode arrangement. The four electrodes move laterally along the profile, maintaining constant distance, from one end to the other. Similarly, radiometric method was used in this study to determine the background radiation of the study area. A hand held gamma spectrometry was used in this study. This equipment detect gamma rays emitted from the earth. The gamma rays are estimated by counting the number of times each gamma ray of particular energy intersects the spectrometer. In this survey the radionuclides detected were 238U, 232Th and 40K present in the soil. The background radiation measurement was carried out along the same profile along which the resistivity measurement was done (Figure 2).

The electrical resistivity data collected were formatted in a manner that could be fed into DIPRO software in order to produce a 2-D image of the area surveyed. Information on the local geology and hand dug well available at the time of survey was of great assistance in interpreting the 2-D images [17] [19]. In a similar manner, the data obtained from gamma spectrometry were used to calculate some important radiological parameters required to achieve the aim of this study.

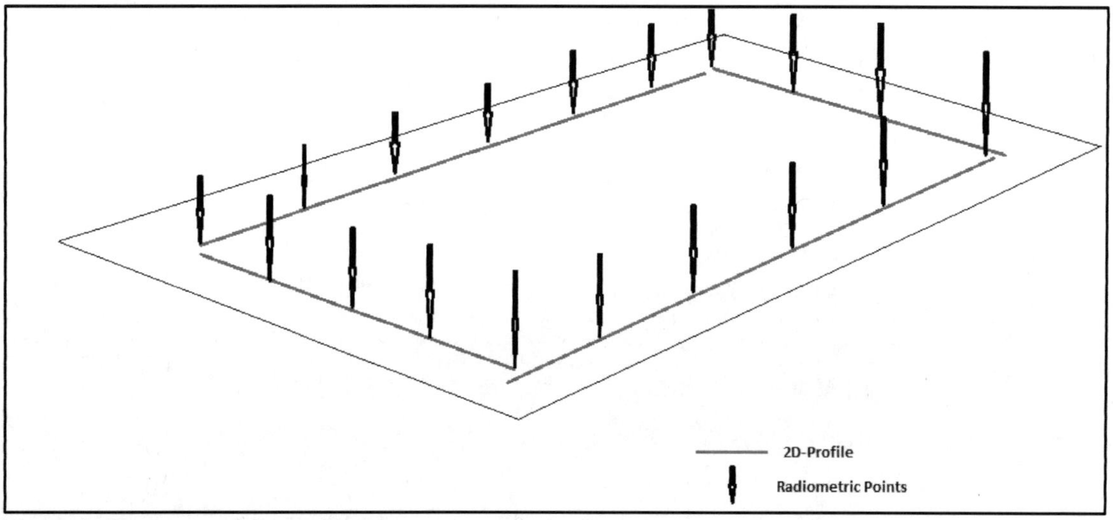

Figure 2. Base map of the study area

4. Results and Discussion

4.1. Radioactivity Concentrations

The detected naturally occurring radionuclides (226Ra, 232Th and 40K) of 14 locations within the study site are presented in Table 1. The activity concentrations ranged between 14.52 and 35.82, 44.96 and 128.70 and 31.30 and 453.85 Bq kg^{-1} in Radium, Thorium and Potassium respectively. It was also observed that the variations between the minimum and maximum values were reported between locations 1 and 6 for ^{226}Ra, locations 1 and 13 for ^{232}Th and between locations 2, 5, 8 and 13 for ^{40}K. The rampant re-occurrence of locations 1 and 13 could be as a result of the underlying geological features in which Uranium and Potassium are more dominant.

4.2. Radium Equivalent Activity (Raeq)

The radium equivalent activity represents a weighted sum of activities of ^{238}U, ^{232}Th and ^{40}K. It is based on the fact that 370 Bq kg^{-1} of ^{238}U, 259 Bq kg^{-1} of ^{232}Th, and 4810 Bqkg^{-1} of ^{40}K produce the same gamma radiation dose rate. This is calculated by using equation (1) [20][3].

$$Raeq\ (Bgkg^{-1}) = C_U + 1.43C_{Th} + 0.077C_K \qquad (1)$$

4.3. External Absorbed Dose Rate (D)

This is the estimation of the amount of radiation energy absorbed or deposited per unit mass of substance [10]. The absorbed gamma dose rates due to gamma radiations in air at 1 m above the ground surface for the uniform distribution of the naturally occurring radionuclides (^{238}U, ^{232}Th and ^{40}K) were calculated using equation (2).

$$D(nGyh^{-1}) = 0.462C_U + 0.604C_{Th} + 0.041C_K \quad (2)$$

The results obtained as shown in Table 1 revealed that the variation in the estimated values varied between locations 1 and 13, with location 1 having the lowest value. The results ranged between 40.52 and 115.35 nGy h^{-1}. The very high value noticed in location 13 could be as a result of some embedded geological features which are more pronounced in location 13.

4.4. Annual Effective Dose (AED)

The annual effective dose rate (AEDR) in mSvy^{-1} resulting from the absorbed dose values (D) was calculated using equation (3) [21-22].

$$EDE\ (mSvy^{-1}) = D(nGyh^{-1}) \times O_c \times F_c \times 8760 \times 10^{-6} \quad (3)$$

From the table of results, it could be seen that the minimum value of 0.049 mSv y^{-1} was noticed in location 1, while the highest value of 0.142 mSv y^{-1} was observed in location 13.

4.5. External Hazard Index (Hex)

The external hazard index (Hex) can then be estimated using equation (4)

$$H_{ex} = \frac{C_U}{370} + \frac{C_{Th}}{259} + \frac{C_K}{4810} \le 1 \quad (4)$$

This index must be less than unity (1) for the radiation hazard to be considered insignificant. In all the readings, the external hazard index was far less than 1, therefore, external hazard index may be regarded as non-significant in the area of study.

Table 1: Shows the estimated radiological parameters from the study area

S/N	^{226}Ra Bg kg^{-1}	^{232}Th Bq kg^{-1}	^{40}K Bq kg^{-1}	Ra$_{eq}$ Bq kg^{-1}	D(out) nGy h^{-1}	AED (mSv y^{-1})	H$_{ex}$
1	14.51	44.96	93.90	86.04	40.02	0.049	0.232
2	21.00	46.89	31.30	90.46	41.36	0.051	0.244
3	27.79	55.83	86.08	114.25	52.54	0.064	0.309
4	28.41	55.32	70.43	112.93	51.79	0.064	0.305
5	23.47	58.87	31.30	110.06	50.34	0.062	0.297
6	35.82	58.36	54.78	123.49	56.30	0.069	0.334
7	19.74	86.58	242.58	162.23	76.22	0.094	0.438
8	21.00	45.37	31.30	88.29	40.35	0.050	0.238
9	22.54	50.85	39.13	98.27	44.98	0.055	0.265
10	17.60	53.19	101.68	101.48	47.12	0.058	0.274

11	12.66	73.49	211.23	134.01	63.19	0.078	0.362
12	21.30	104.65	406.75	202.27	95.94	0.118	0.546
13	24.70	128.70	453.85	243.69	115.35	0.142	0.658
14	23.77	50.34	273.88	116.85	55.31	0.068	0.316

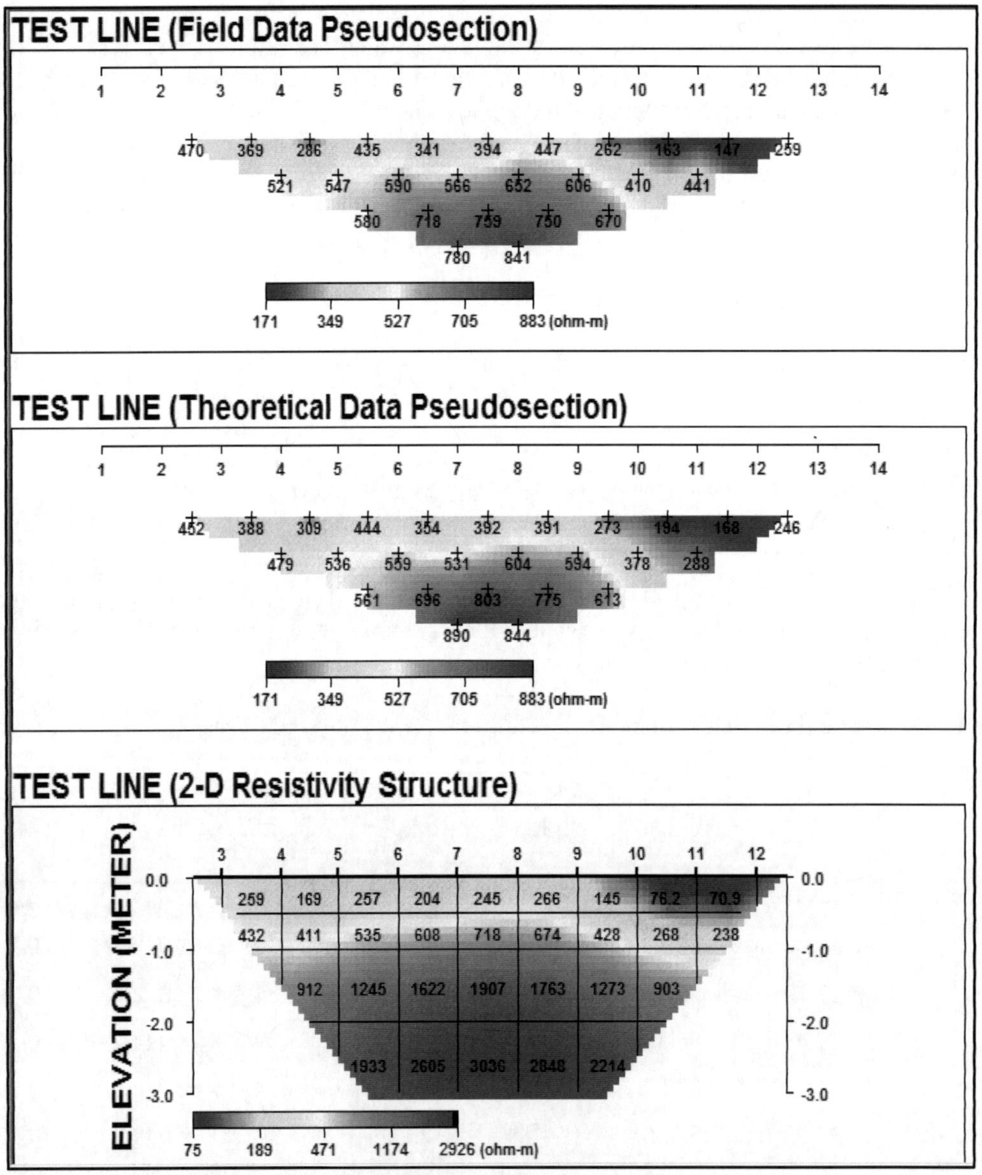

Figure 3: 2-D resistivity image of the surveyed profile in the area of study

Within the depth of investigation, two distinct layers were delineated by the electrical resistivity method. The first layer or the top soil is composed of sandy clay, this is characterized by an unconsolidated material of low resistivity ranging between 75 and 432 Ωm. The thickness of the top soil is about 1.5 m in the area of study. The second layer is characterized by a geologic formation of relatively high resistivity ranging between 903 and 2926 Ωm. This layer is described as a sandy mud stone, as observed in Figure 3. The variation in the resistivity within this lithological unit could be as a result of the age of deposition or the level of compaction of the unit [23]. The thickness of the second layer could be above 1.5 m captured by the 2-D image due to the depth of investigation.

It could be seen from the above that the variation in the activity concentration measured in this study, especially at location 13, could be the influence of highly resistive geologic formation that dominated the second geologic layer in the study area. There are no obvious activities carried out in the study area that could trigger undue increase in the background radiation in the area of study.

5. Conclusion

Integration of electrical resistivity and radiometric methods were employed to assess the background radiation of some parts of Canaanland and also to delineate the geological features that are contributing to this radiation. The results by the radiometric method revealed the activity concentrations of U, Th and K to vary from 12.658 Bq kg^{-1} - 42.299 Bq kg^{-1}, 44.9645 Bq kg^{-1} - 128.702 Bq kg^{-1} and 31.3Bq kg^{-1} - 453.85 Bq kg^{-1}, respectively. The results of the Radium equivalent activity varied between 86.04 and 243.69 Bq kg^{-1}. Furthermore, the result of the external absorbed dose rate varied between 40.02 and 115.35 nGy h^{-1}. However, the result of the electrical resistivity method varied between 75 and 1025 Ωm. Finally, it was observed that region of high resistivity was prone to higher level of radiation, which could be as a result of some radioactive geological features in the subsurface.

Acknowledgement

The authors wish to appreciate the management of Covenant University for providing sponsorship for this project.

References

[1] Arabi AM, Ahmed NK, Salahel DK (2008) Assessment of terrestrial gamma radiation doses for some Egyptian granite samples. RadiatProtDosim. 128(3): 382–385.

[2] International Atomic Energy Agency, IAEA (1989) Construction and use of calibration Facilities for Radiometric Field Equipment. *Technical Reports Series no.309, IAEA,Vienna.*

[3] EC (European Commission) (1999) Radiological protection principles concerning the natural radioactivity of building materials. Radiation protection 112. Directorate General Environment,Nuclear Safety and Civil Protection (Geneva: EC).

[4] United Nations Scientific Committee on the effects of Atomic Radiation, UNSCEAR (2000)

[5] Erees F S, Dayanıklı SA &Çam S (2006) Natural radionuclides in the building materials used in Manisa city, Turkey. *Indoor and Built Environment*, *15*(5), 495-498.

[6] Ghose S, AsaduzzamanKh, Zaman N (2012) Radiological significance of marble used for construction of dwellings in Bangladesh. Radioprot. 47(1): 105–118.

[7] Internal Commission on Radiological Protection. (1990). Annual Limits on Intake of Radionuclides by Workers Based on the Recommendations, Annals on the ICRP, ICRP publication 67, *Oxford Press.*(ICRP, 1999)

[8] ICRP (1994). *Protection against Rn-222 at home and at work.* ICRP publication 65; Ann. ICRP;23 (2): 1-48.

[9] Avwiri GO, Nte FU, Olanrewaju AI (2011) Determination of radionuclide concentration of landfill at Eliozu, Port Harcourt, Rivers State. Scientia Africana. 10 (1).

[10] Omeje M, Wagiran H, Ibrahim N, Lee SK, Soheil, S (2013) Comparison of ^{238}U, *Protection Dosimetry* , pp 1-7. ^{232}Th, and ^{40}K in different layers of subsurface structuresion Dei-Dei and Kubwa, Abuja.

[11] Ogbe FAG (1970) Stratigraphy of strata exposed in the Ewekoroquary, western Nigeria. In: Dessauvagie TFJ, Whiteman AJ (eds) African geology. University of Ibadan Press, Nigeria, pp. 305–324
[12] Jones, H. A. and Hockey R. D, (1964): The geology of part of southwestern Nigerian. *Bulletin of the Geological Survey of Nigeria*, 31 (4): 101pp
[13] Onuoha KO (1999) Structural features of Nigeria's coastal margin: an assessment based on age data from wells. J Afr Earth Sci 29(3): 485–499
[14] Omatsola ME, Adegoke OS (1981) Tectonic evolution and cretaceous stratigraphy of the Dahomey Basin. Nigerian J Min Geol 18(1): 130–137
[15] Olabode SO (2006) Siliciclastic slope deposits from the cretaceous Abeokuta group, Dahomey (Benin) basin, southwestern Nigeria. J Afr Earth Sci 46:187–200
[16] Okosun EA (1990) A review of the cretaceous stratigraphy of the Dahomey embayment, West Africa. Cretac Res 11:17–27
[17] Loke MH, Barker RD (1996) Practical techniques for 3D resistivity surveys and data inversion. Geophys Prospect 44:499–524
[18] Rubin Y, Hubbard S (2005) Hydrogeophysics, water science and technology library 50. Springer, Berlin, p. 523
[19] Loke MH, Chambers JE, Rucker DF, Kuras O, Wilkinson PB (2013) Recent developments in the direct-current geoelectrical imaging method. J ApplGeophys 95:135–156
[20] Beck HL (1980) Exposure rate conversion factors for radionuclides deposited on the Ground. Department of Energy, New York (USA). Environmental Measurements Lab.
[21] Rucker D, LokeMH, LevithMT, Noonan GE (2010) Electrical resistivity characterization of an industrial site using long electrodes. Geophysics 75(4):WA95–WA104
[22] Juvanbot DI., MM Izam, G.G. Nyam, M. Yusuf (2014.)Indoor and outdoor Gamma Dose rate Exposure Levels in Major Commercial Building materials Distribution Outlets in Jos, Plateau State Nigeria. Asian Review of Environmental and Earth Sciences. Vo1, No 1, 5-7 June 2014. Asian Outline Journal Publishing Group.
[23] Adewoyin O.O., Joshua E.O., Akinwumi I.I., Omeje M., Joel E.S. (2017). Evaluation of Geotechnical Parameters using Geophysics Data. Journal of Engineering and Technological Sciences 49(1): 95 – 113.

Interpretation of high resolution aeromagnetic data for mineral prospect in Igbeti-Moro area, Southwestern Nigeria

[1*]N. K. Olasunkanmi, [2]L. A. Sunmonu, [2]M. A. Adabanija, and [3]P. O. Oladejo

[1]Physics and Materials Science Department, Kwara State University, Malete Nigeria.
[2]Pure and Applied Department, Ladoke Akintola University of Technology, Ogbomoso Nigeria.
[2]Earth Science Department, Ladoke Akintola University of Technology, Ogbomoso Nigeria.
[3]Physics Department, Emmanuel Alayande College of Education, Oyo Nigeria

Correspondence: nurudeen.olasunkanmi@kwasu.edu.ng

Abstract. High resolution aeromagnetic data that covers Igbeti-Moro area, within southwestern Nigeria, has been subjected to data enhancement processes and interpreted to establish significant geologic features associated with occurrence of marble, gabbro and muscovite mineralization in the area. These were established by applying 2D Fast Fourier transform filters for reduction to equator (RTE), lineament enhancement of vertical and total horizontal gradient, edge detection of analytic signal amplitude (ASA), and magnetic source depth parameter imaging (SPI) and average power spectrum (APS); using Oasis Montaj software. The RTE residual magnetic intensity of the area ranged between - 338.3 nT to - 2.4 nT, 9.9 to 27.7 nT and 32.1 to 208.0 nT for low, intermediate and high magnetic susceptibility respectively which indicate contrasting basement rocks (namely; granite-gneiss) and intrusion of quartz and amphibolite schists. The lineament development and its disjoint revealed fractures and faults associated with emplacement of the quartz-schist and muscovite-schist, which would have formed synchronously with isoclinal fold and hosts marble, gabbro, amphibolites and muscovite mineralogy at shallow depth that ranges from 0 to 200 m in the northeastern, western, north central sections of the area. The geologic features observed in the study area have been established to be associated with the solid minerals that can be of social-economic benefit.
Keywords: geologic features, lineament, Fast Fourier transform, magnetic anomalies, Mineralization

1. Introduction

Minerals are key element to nations building and require several methods to recognise its areas of deposit. The deposition of these minerals can be in tabular or shell-like form, filling a fracture in a host rock controlled by hydraulic of magmatic water that carries large amount of dissolved minerals which it deposits in veins as it travels to the earth surface [19];[9]. However, minerals are not easily determined except with the expertise of geologists and application of geophysical tools. Geophysical method such as aeromagnetic data has been chiefly used for geologic structural mapping and mineral exploration [10];[15]. The elementary geophysical understanding behind this is that; different rock types host different magnetic responses and as a result several analytical data processing tools is needed to enhance the output images of these magnetic responses for better geologic interpretation. Among the filtering and analytical tools employed are upward and downward continuations, vertical derivative (VDR), total horizontal derivative (THDR), analytic signal, tilt (TDR) and theta derivatives etc. These tools reveal the variation in magnetic susceptibility of the basement and that of the overlying sediments [8];[1]. In this present study, airborne magnetic technique has been used to delineate geologic features that are associated with solid minerals in part of Igbeti-Moro area, Southwestern Nigeria.

1.1. Description and geological settings

The study area is located between Oyo and Kwara, Southwestern Nigeria and within about 3064 square kilometer between longitudes 4°00′ E to 4°30′ E and latitudes 8°30′ N to 9°00′ N. It covers Igbeti (northern part of Oyo State) and Moro (southeastern Kwara State) in Southwestern Nigeria and it is traversed by a major road linking the south-western part to northern regions of Nigeria. The climate of the study area is characterised by both the wet and dry seasons. The rainy season begin in late March and end in October, while the dry season begin in November and end in early March. The natural vegetation is guinea savanna characterized with scanty forest or averagely thick forest mainly along the stream's path. Its major lithological units are of the precambrian basement complex of Nigeria [14] comprising the older granite series, metasediments and migmatitic gnesis complex (Fig. 1).

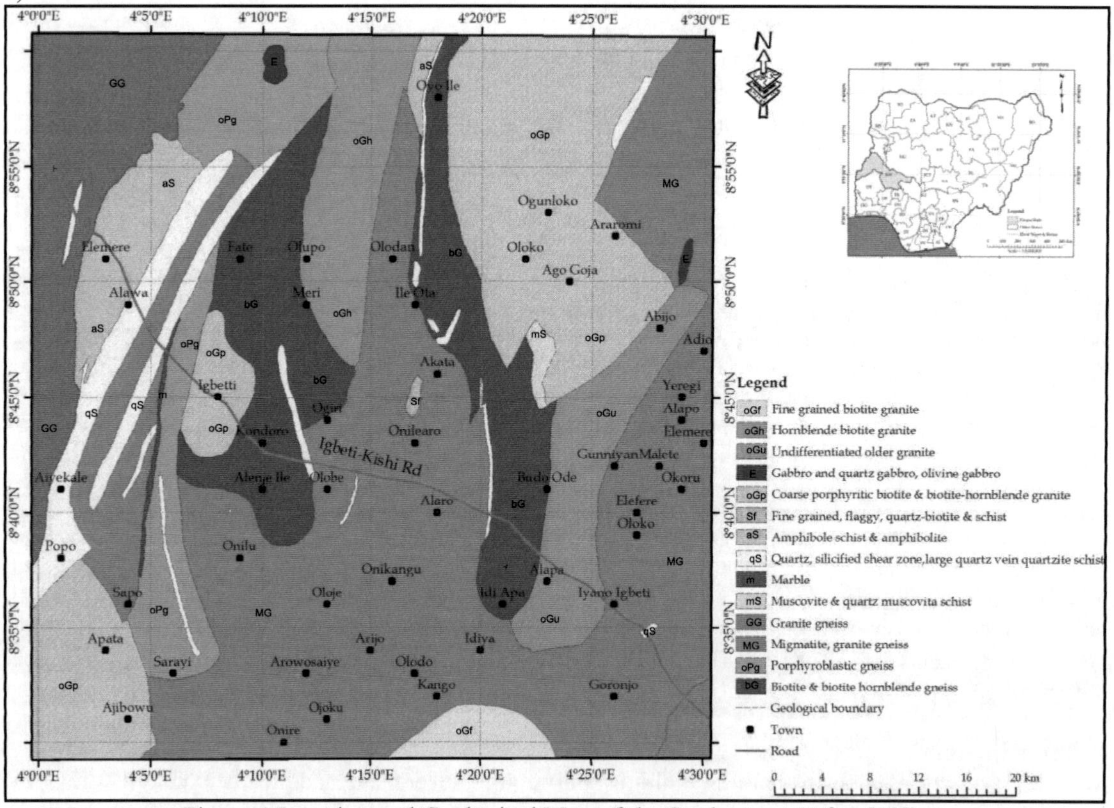

Figure 1 Location and Geological Map of the Study Area (After [13])

2. Data acquisition and processing

The aeromagnetic data was acquired from Nigeria Geological Survey Agency (NGSA), as part of the data set of total field magnetic intensity obtained during airborne geophysical survey of Nigeria within 2003 and 2009 by Fugro Survey Limited. The survey was aimed at promoting mineral exploration in Nigeria [12]. Igbeti-Moro has been designated as sheet 201of the gridded aeromagnetic data of Nigeria. It was de-cultured, leveled and corrected for international Geomagnetic Reference Field (IGRF). The grid enhanced anomaly details and reduced possible noise and latitude effects [18].

IOP Conf. Series: Earth and Environmental Science **173** (2018) 012033 doi:10.1088/1755-1315/173/1/012033

2.1. Data reduction and processing

In mineral exploration, to infer meaningful interpretation from aeromagnetic data, processing and reduction of potential field are essential tools [24]. In this study, the data was filtered using the fast Fourier transform, conversion of data from one form to another by reducing the data to magnetic equator (RTE) and producing residual magnetic intensity map. Edges of magnetic source body were enhanced by vertical (VDR) and total horizontal derivatives (THDR) while depth to magnetic sources estimate using source parameter imaging (SPI) and average power spectrum. Data processing and reduction methods are most effective when dealing with shallow source depths and rock types with relatively large contrasts in magnetization [3]. The asymmetric and lateral shift of measured magnetic total field (Fig. 2a) was corrected by applying RTE and the resultant residual anomaly map (Fig. 2b) showed magnetic responses of rock masses or mineral deposits, their shapes and magnitude, the direction and strength of its magnetization with respect to their global location. The rate of change of magnetic field spatially in vertical and horizontal directions is quantified with their derivatives [21] by enhancing local anomalies obscured by broader regional trends. THDR is related to vertical derivative (VDR) by mathematical expression in Equation 1[17] and its map (Fig. 3) delineated limits of intrusive bodies, faults and other lateral changes [11].

$$Tilt\ derivative\ (\theta) = \tan^{-1}\left(\frac{VDR}{THDR}\right) \qquad \text{Eqn. 1}$$

The SPI submenu of Oasis montaj's geo-software was applied for an automatic location and depth determination of causative bodies from the gridded magnetic data using the local wavenumber of analytic signal [20];[23].

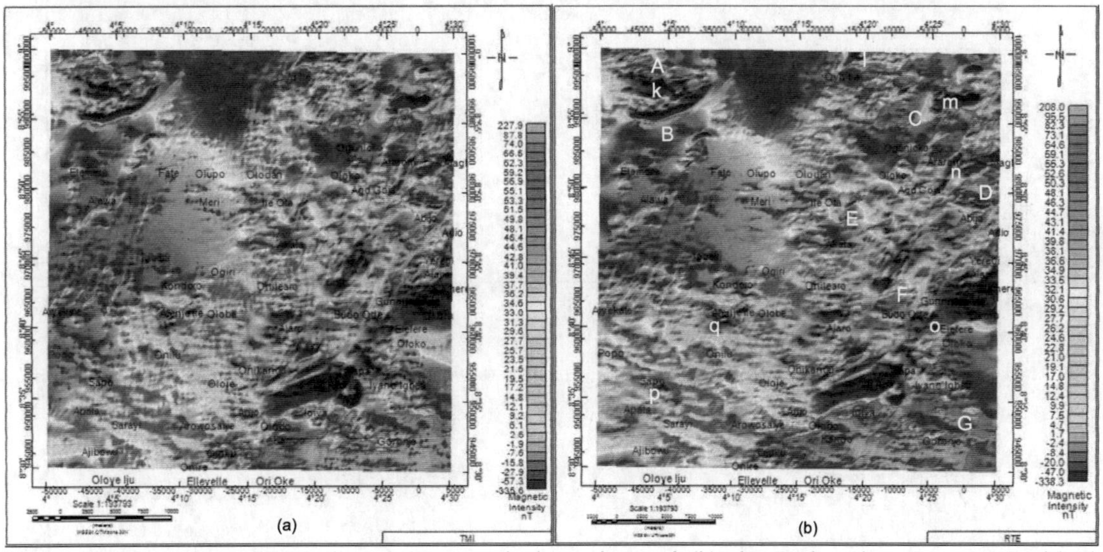

Figure 2 Comparison of (a) the total magnetic intensity and (b) the Reduced to Equator Residual Magnetic Intensity maps.

3. Results and Discussion

The RTE map show no significant modification of the original magnetic anomalies on total magnetic intensity (TMI) map (Figure 2a), marked as positive and negative anomalies A, B, C, D, E, F, G, and k, l, m, n, o, p, q, r respectively. The amplitude of anomalies range from – 338.3 nT to 208.0 nT (Figure 2b) which were consistent in pattern, trend and amplitude of the total magnetic anomalies within – 335.6 nT to 227.7 nT shows that the data has been well filtered. The southeast and northwestern regions of the RTE map is characterized by positive (high) magnetic intensity value range between 32.1 nT to 208.0 nT, the west and north – north central regions host intermediate

IOP Conf. Series: Earth and Environmental Science **173** (2018) 012033 doi:10.1088/1755-1315/173/1/012033

magnetic value range within 9.9 to 27.7 nT while the northeast to east, central and southwestern regions have negative anomalies range between - 338.3 nT to - 2.4 nT. The THDR map defined the edge of these lithologic bodies and linear features with low anomaly gradient within -1.47 and 0.04 nT/m.

3.1. Lithologic mapping

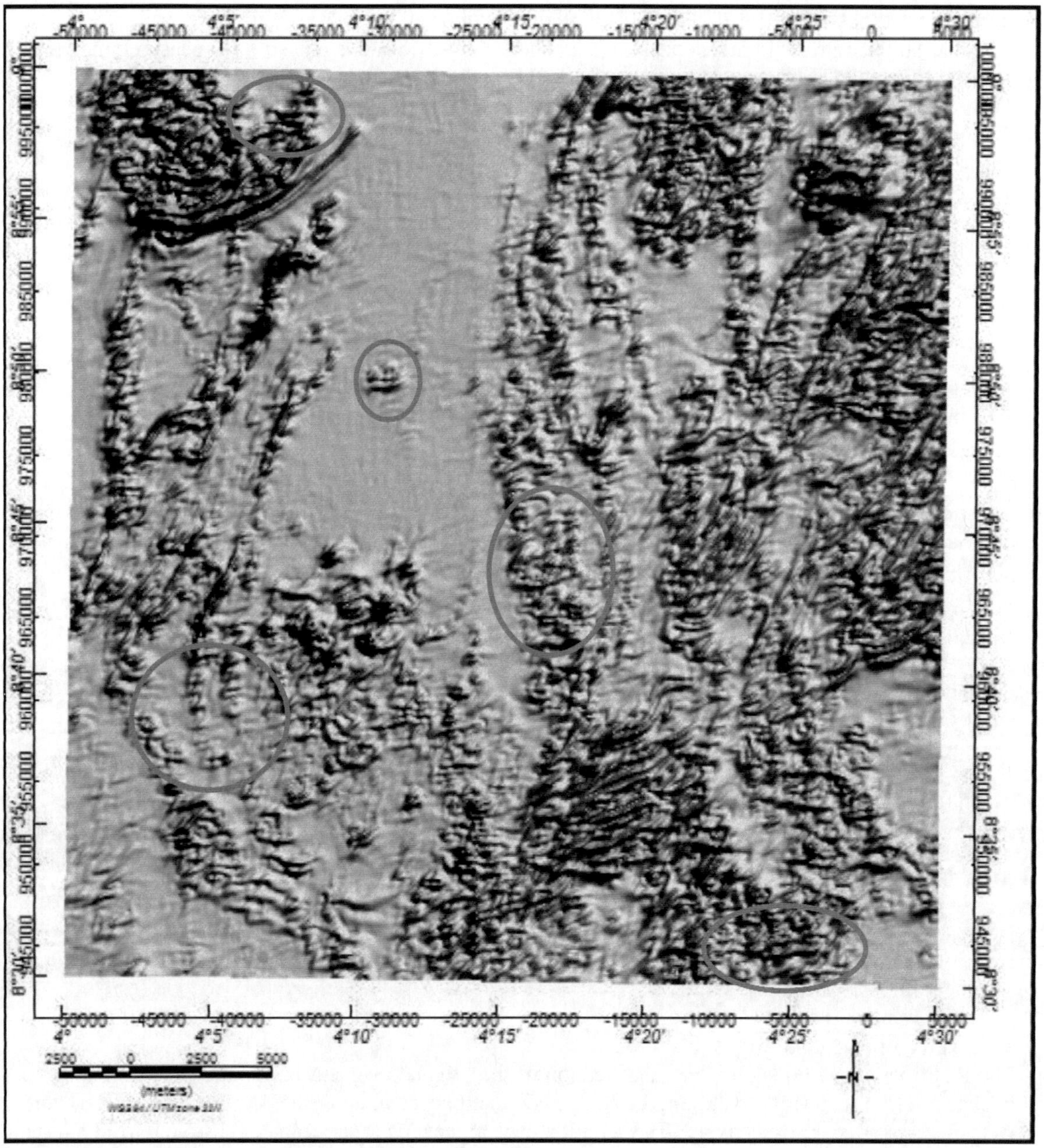

4.

Figure 3 Total horizontal Derivative map

Careful examination of the RTE map (Fig. 2b) with the geological map (Fig. 1) showed that the alternating positive and negative magnetic values suggest rock contrast in the basement of the study area. The intermediate magnetic anomalies are connected with intrusion of quartz and amphibolite schists bounded by relatively low magnetic intensity; which is associated with older granitoids [2]. The negative anomalies are associated with mining and prospecting areas for minerals [4];[16] like the marble, gabbro, amphibolites and muscovite mineralogy within the granite-gneisses in the west and north – north central regions, which commonly carry small amounts of magnetite and dark minerals like biotite and hornblende (paramagnetic). The minerals have possibly been deposited in the quartz vein [5];[6] or have resulted from the intrusion of hydrothermal fliuds which can deposit less magnetic materials in the fractured rock [7]. The positive magnetic anomalies amplitude is allied with gneiss complex rocks namely; granite gneiss, migmatite gneiss and biotite gneiss which makes the various lithologic units in the area [22].

The SPI map (Fig. 4) showed shallow to deeper magnetic source depth range of - 335.6 to 227.9 m. The region of extreme topography coincide notable hills that stands around 330 m above the country rocks in the north-northwest, sparingly in southwest and southeastern parts. These showed that the host rock is more magnetic than the intrusive rocks and can be due to reversed remanent magnetization [5] or associated with weakly magnetic rocks. The area of revealed mineralogy banding falls within shallow mineral exploration depth range classification of [6].

4.1. Structural mapping

The THDR map (Figure 3) revealed prominent NE – SW structural trend which is in concordance with the generally known deformation trend (Kibarian orogeny) in the basement complex rock of Nigeria. The scanty east – west lineaments in concordance with the east – west magnetic anomalies (Fig. 2b) enclosed in red oval (Fig. 3) indicate ductile deformation which could exhibit structural features such as faults and fractures. The faults indicate the emplacement of quartz-schist and muscovite-schist which host the marble, gabbro, amphibolites and muscovite mineralogy in the eastern, western and north central parts of the study area.

IOP Conf. Series: Earth and Environmental Science **173** (2018) 012033 doi:10.1088/1755-1315/173/1/012033

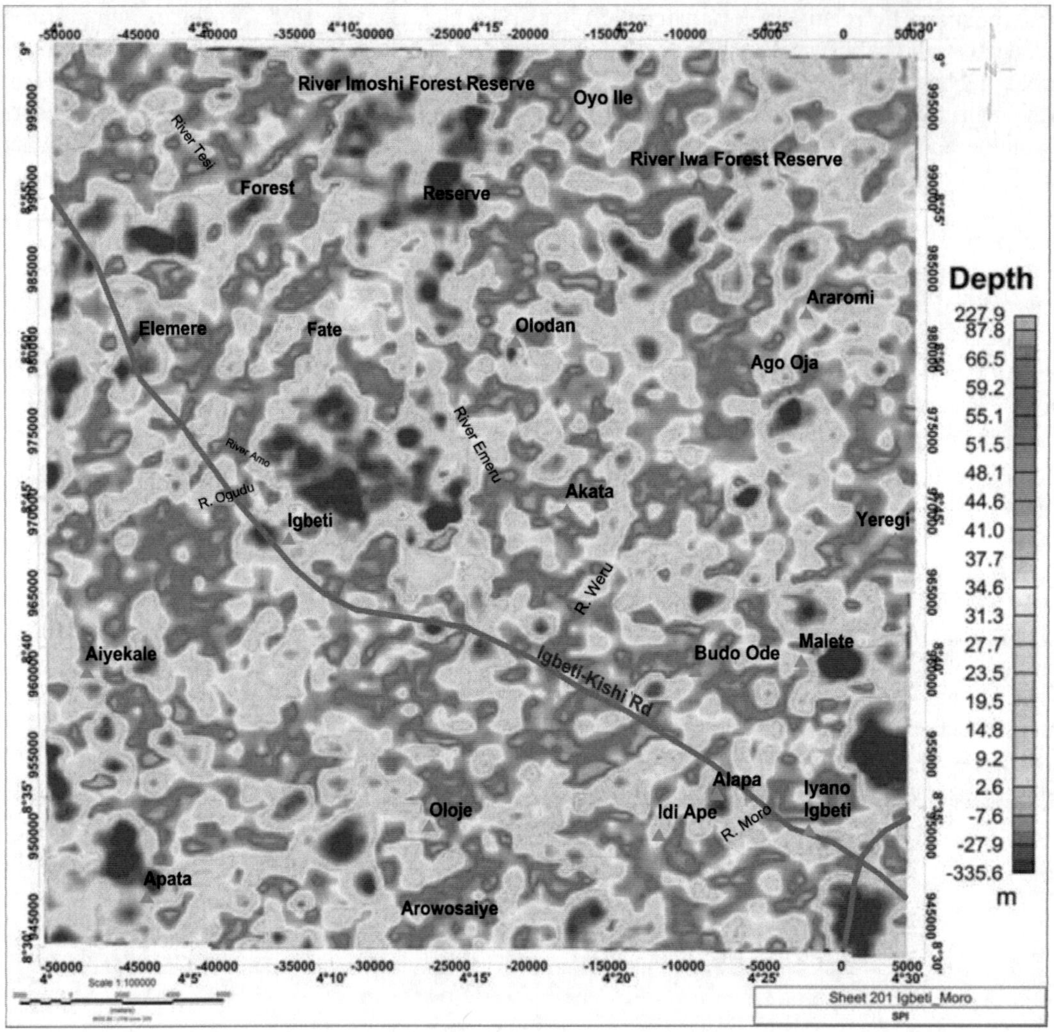

Figure 4 Source Parameter Imaging depth estimate maps

5. Conclusion

High resolution aeromagnetic data of Igbeti-Moro has been processed, enhanced and analysed using Oasis Montaj geo-software in order to map geologic structures related to mineral exploration in the area. From the results of interpretation, lithological boundaries and contacts, intersection of geological structures, faults and magnetic sources depths were established. The negative anomalies of the reduced-to-equator residual magnetic intensity indicated the marble, gabbro, amphibolites and muscovite mineralogy which are distributed within the western, northern and northeastern parts of the study area and bounded by positive anomalies associated with gneiss and older granitoids; which formed the major lithologic unit of the area. The minerals exist at shallow depth range of 0 – 200 m and have possibly been deposited by intrusion of hydrothermal fluids along quartz schist veins during the deformation that led to the formation of resultant faults. The pattern of occurrence and associated geologic structures showed that the minerals deposits in the study area are structurally controlled; which is critical in mining operation.

References

[1] Amigun, J. O., Afolabi, O., & Ako, B. D. (2012). Application of airborne magnetic data to mineral exploration in the Okene Iron Ore Province of Nigeria. *International Research Journal of Geology and Mining (IRJGM)*, 132-140.

[2] Biyiha-Kelaba, W., Ndougsa-Mbarga, T., Yene-Atangana, J., Ngoumou, P. C., & Tabod, T. C. (2013) 2.5D Models Derived from the Magnetic Anomalies Obtained by Upwards Continuation in the Mimbi Area, Southern Cameroon. *Journal of Earth Sciences and Geothecnical Engineering, 3*(4), 175-199.

[3] Cooper, G. R., & Cowan, D. R. (2004). Filtering using variable order vertical derivatives. *Computers & Geosciences* , 455-459.

[4] Hanna, W. F. (1969). Negative aeromagnetic anomalies over mineralized areas of the boulder batholith, Montana. *Gelo. Survey Prof. Paper, 650*, D155 - D167.

[5] Hans, T. (2012). *Magnetic survey at Nesodden - survey and interpretation report.* Lulea: GeoVista AB.

[6] Intrepid, 2012. *Estimating Source Depth.* Intrepid Geophysics.

[7] Jayeoba, A., & Odumade, D. (2015). Geological and Structural Interpretation of Ado-Ekiti Southwest and its Adjoining Areas Using Aeromagnetic Data. *Pacific Section AAPG, SEG and SEPM Joint Technical Conference.* Oxnard, California: AAPG.

[8] John, M. J., Marianne, E. P., & Antony, D. P. (2003). Improving geologic understanding with gravity and magnetic data: Examples from Gabon, Nigeria and the Gulf of Mexico. *EAGE*, pp. 57-62.

[9] McGraw, H. (2003). *Dictionary of geology and mineralogy* (2nd ed.). United State of America: McGraw-Hill Companies.

[10] Moghaddam, M., Mohammad, S., Saeid, M., & Nasim, H. (2015). Interpretation of Aeromagnetic Data to Locate Buried Faults in North of Zanjan Province, Iran. *J Remote Sensing & GIS, 4*(2), 1 - 5.

[11] Nabighian, M. N., Grauch, J. S., Hansen, R. O., LaFehr, T. R., Li, Y., Pierce, J. W., . . . Ruder, M. E. (2005). The historical development of the magnetic method in exploration. *Geophysics*, 33 - 61.

[12] Nigerian Geological Survey Agency, (2008). *Airborne Geophysical Survey: Total Magnetic Intensity of Igbeti (Sheet 201).*

[13] Nigerian Geological Survey Agency, (2006). Geological map of Igbeti-Moro *(Sheet 201) Area.*

[14] Obaje, N. G. (2009). Geology and Mineral Resources of Nigeria. In *Lecture Notes in Earth Sciences,* Berlin Heidelberg: Springer.

[15] Oladunjoye, M. A., Olayinka, A. I., Alaba, M., & Adabanija, M. A. (2016). Interpretation of high resolution aeromagnetic data for lineaments study and occurence of banded iron formation in Ogbomoso area, Southwestern Nigeria. *Journal of African Earth Sciences*, 43-53.

[16] Olasunkanmi, N. K., Olufemi, S. B., & Aina, A. (2017). Exploration for iron ore in Agbado-Okudu, Kogi State, Nigeria. *Arab J Geosci , 10* (541). doi.org/10.1007/s12517-017-3250-3

[17] Paananen, M. (2013). Complete lineament interpretation of the Oikiluoto region. *Posiva, 2013-02 ISBN 978-951-652-234-3.*

[18] Reeves, C. (2005). *Aeromagnetic Surveys: Principles, Practice and Interpretation.* Geosoft.

[19] Richard, A. M. (2001). *dictionary of Geophysics, Astrophysics and astronomy.* Boca Raton London New York Washington, D.C.: CRC Press.

[20] Robinson, E. A., & Treitel, S. (2000). Geophysical Signal Analysis. Tulsa, USA: Society of Exploration Geophysics.

[21] Ross, C. B. (2002). *Geophysical and Remote Sensing Methods for Regolith Exploration.* Australia: Geoscience.

[22] Telford, W. M., Geldart, L. P., Sheriff, R. E., & Keys, D. A. (1990). *Applied Geophysics.* Cambridge: Cambridge University Press.

[23] Thurston, J. B., Guillon, J. C., & Smith, R. S. (1999). Model independent depth estimation with the SPITM method. *69th Annual international meeting* (pp. 403 - 406). Society of Exploration Geophysics.

[24] Wynn, J. (2002). Evaluating ground water in arid lands using airborne magnetic and airborne electromagnetic methods- and example in the southwestern U.S. and northern Mexico. *The Leading Edge*, 62 - 65.

IOP Conf. Series: Earth and Environmental Science **173** (2018) 012034 doi:10.1088/1755-1315/173/1/012034

Depth Estimate and Mapping of source geometries from High Resolution Aeromagnetic (HRAM) data of Benisheikh, Nigeria

***[1]Lawal, T.O., [2]Sunday, J. A., [3]Nwankwo, L.I., [1]Orosun, M. M., [1]Yusuf, K. A and [1]Ige, S. O.**

[1]Department of Physics, University of Ilorin, Ilorin Nigeria
[2]Department of SLT, Kwara State Polytechnique, Ilorin.
[3]Department of Applied Geophysics, University of Ilorin, Ilorin Nigeria

***taofeeqlawal4u@gmail.com**

Abstract. Depth to magnetic basement and mapping of source geometries were estimated from High Resolution Aeromagnetic (HRAM) data of Benisheikh using wavelet transformation technique. This technique was chosen because it has been proven to be a powerful and efficient tool in interpreting potential field data. In this work, a non-orthogonal wavelet function with a good symmetry and higher vanishing moment, Morlet was chosen as the analyzing wavelet. Wavelet power spectrums of the aeromagnetic anomalies were obtained using the scaled normalized analyzing wavelet in order to estimate the depth to the magnetic basement. Also, the square of the absolute value of the wavelet coefficient were plotted against the period in other to identify and map variations of pronounced and least energy values which can be regarded as magnetic source geometric features found from the HRAM data profiles. The results obtained from the analysis have shown the ability of wavelet transform as a tool in depth estimation and mapping of geological features of HRAM data of this part of Nigeria.
KEYWORDS: Wavelets transform, High Resolution Aeromagnetic data (HRAM), Morlet Wavelet and geometrics features.

1.0 Introduction

Magnetic technique is one of the geophysical survey techniques that exploit the considerable differences in the characteristics of magnetic minerals on the foundation of anomalies in the Earth's magnetic field which is as a result of the properties of magnetic rocks beneath earth surface [1]. These anomalies can be analyzed to estimate depth to magnetic basement and mapping of geological features. The importance of basement depth geological features for minerals signatures in basins cannot be overemphasized. This is because of the impact of the basement on the geology of the sedimentary rocks found above it and also the subsequent control on the development of hydrocarbon pools [2]. When the magnetic depth to the basement is contoured, it will provide an illustration of the spatial variation about the sedimentary thickness of the area under consideration. These illustrations correlate meaningfully to the known sedimentary activities in the region of study. The usefulness of these studies is to establish a reconnaissance survey using HRAM data for hydrocarbon or mineral signatures of an area so that Seismic techniques can be applied to these areas for hydrocarbon accumulation [3], [4].

The HRAM method was rediscovered 1990s to provide valuable data to solve petroleum exploration problems in the oil and gas industries [4]. This is evident from many reputable publications[2]; [3]; [4]; [5]; [6]; [7]. Some papers on the analysis of HRAM anomaly for hydrocarbon or mineral exploration are those of Western Canada [3]; [4], oil traps in China [5]; [8], over the Chad lineament (North-Central Africa) [9]. etc.

Geologically, Chad basin has been described as a broad sediment-filed depression stranding Northeastern Nigeria and adjourning parts of Chad Republic (Fig.1).The sedimentary rocks have a cumulative thickness of over 3.6 km and consist of thick basal continental sequence and transitional calcareous deposit. The stratigraphic sequence (Fig.2) shows that Chad, Kerri-kerri and Gombe formations have an average thickness of 130 to 400 m. Below these formations are the Fika shale with a dark grey to black in color, with an average thickness of 430 m. Others are Gongila and Bima formations with an average thickness of 320 m and 3.5 km respectively [10].

In this work, we have applied wavelet analysis using Morlet as the analyzing wavelet to the High Resolution Aeromagnetic (HRAM) data of part of Chad basin, Nigeria in other to estimate the depth to the magnetic basement and mapping of source geometries from High Resolution Aeromagnetic (HRAM) data of Benisheikh, Nigeria.

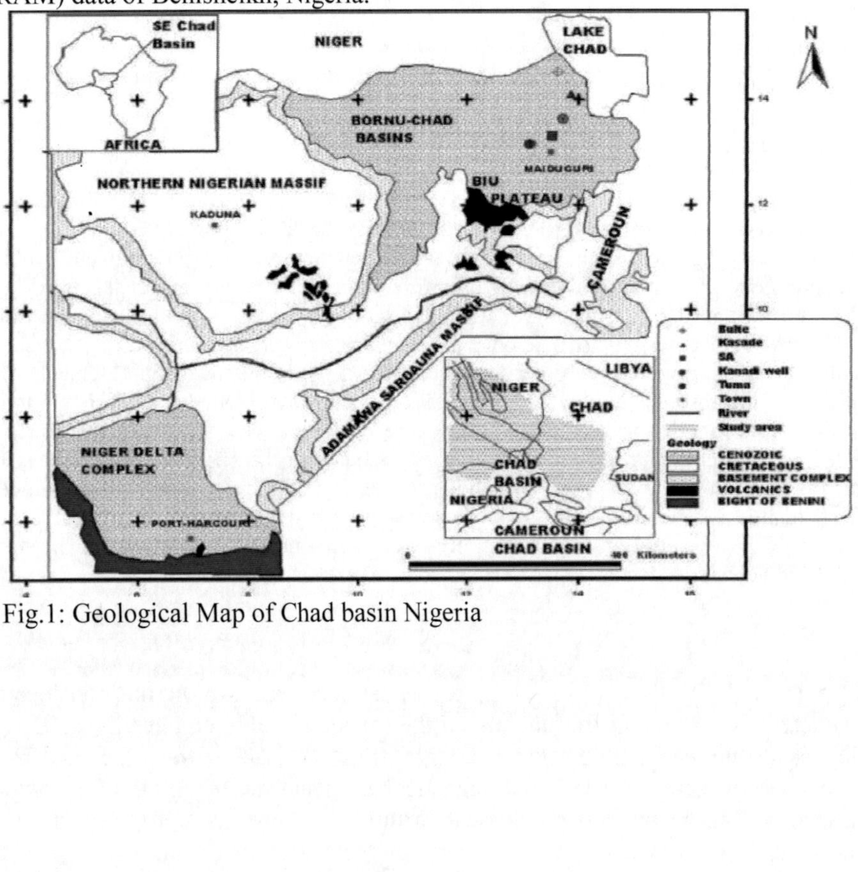

Fig.1: Geological Map of Chad basin Nigeria

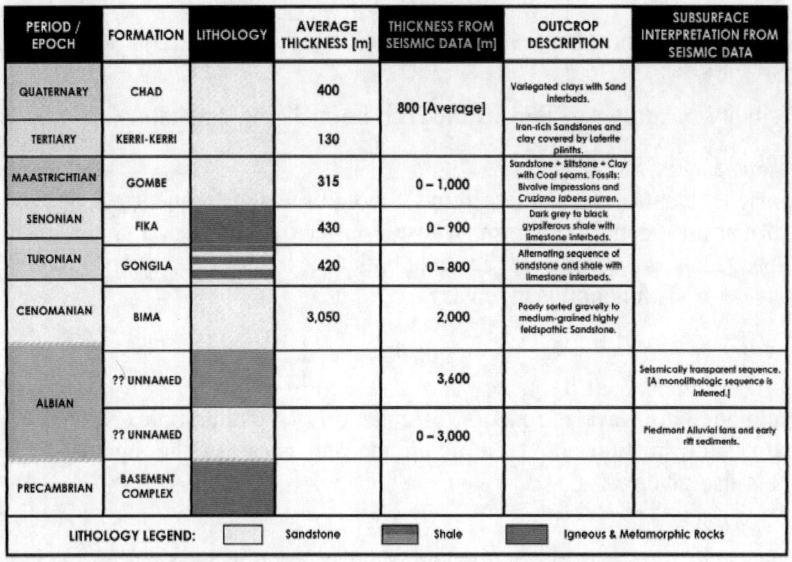

Fig.2: Generalized Stratigraphy sequence of Chad basin Nigeria [10].

2.0. Materials and Methods

2.1 Theory of wavelet transform

The Continuous Wavelet Transform (CWT) of a discrete sequence X_u is defined as the convolution of X_u with a scale s and translated version of the wavelet basis or mother wavelet $\psi_{s,t}(u)$. The CWT is given as [11] and [12]

$$W_u(s,t) = \sum_{u=0}^{N-1} X_u \overline{\psi}_{s,t}(u) \qquad s > 0 \qquad (1)$$

where

$$\psi_{s,t}(u) = \frac{1}{\sqrt{s}} \psi\left(\frac{u-t}{s}\right) \qquad (2)$$

is the dilated and translated wavelet basis or mother wavelet function, s > 0 is the scale factor, t is the translation parameter, and $\overline{\psi}_{s,t}(u)$ is the complex conjugate $\psi_{s,t}(u)$. [11] also interpreted the wavelet transform as a mathematical microscope, where the magnification is given by $1/s$ and the optics are given by the choice of wavelet.

In frequency domain, equation (1) can be equivalently expressed as

$$W_u(s) = \frac{\sqrt{s}}{2\pi} \sum_{k=0}^{N-1} X_k \psi(sw_k) e^{jw_k udt} \qquad (3)$$

From equation (3), the wavelet transform of a discrete profile data X_k can be viewed as the coefficient of the CWT at different scales. Generally, it has been observed that a large-scale wavelet transform can be used to obtain the spectral depth estimation [13]. But the choice of wavelet function depends on the nature of the signal to be considered. It is important to note that the choice of the wavelet function $\psi_{s,t}(u)$ is neither unique nor arbitrary. This function $\psi_{s,t}(u)$ is normalized to have unit energy (i.e., $\int_{-\infty}^{+\infty} |\psi(u)|^2 \, du = 1$) so that it has a compact support, or sufficiently fast decay, to obtain localization in space and zero mean (i.e., $\int \psi(u)du = 0$). The requirement of zero mean is called the admissibility condition of the wavelet.The normalizing constant $1/\sqrt{s}$ is chosen so that $\psi_{s,t}(u)$ has the same energy for all scales s.

In aeromagnetic data analysis, resolution in both time and frequency is necessary. Therefore Morlet wavelet, a non orthogonal transform, whose Fourier transform is a Gaussian function and is fairly ideal band-pass filter, will give a resolution in both time and frequency are adopted in this work.

The Morlet wavelet basis function is given as

$$\psi(t) = \pi^{-1/4} e^{iw_0 t} e^{-t^2/2} \quad (4)$$

Where w_0 is a non dimensional frequency and is taken to be 6 in this work in other to satisfy the admissibility condition. This wavelet function of equation (4) is a complex valued (Fig. 3a) which enable one to extract information about the amplitude and phase of the signal being analyzed. Its Fourier transform is also given as

$$\overline{\psi}(k) = e^{-(k-k_0)/2} \quad (5)$$

Its spectrum character is shown in Fig.3b. The wavelet function in equation (5) is approximately zero for k<0. This is particularly attractive for certain analyses where one needs to eliminate the interference of negative frequencies for the interpretation of results.

a

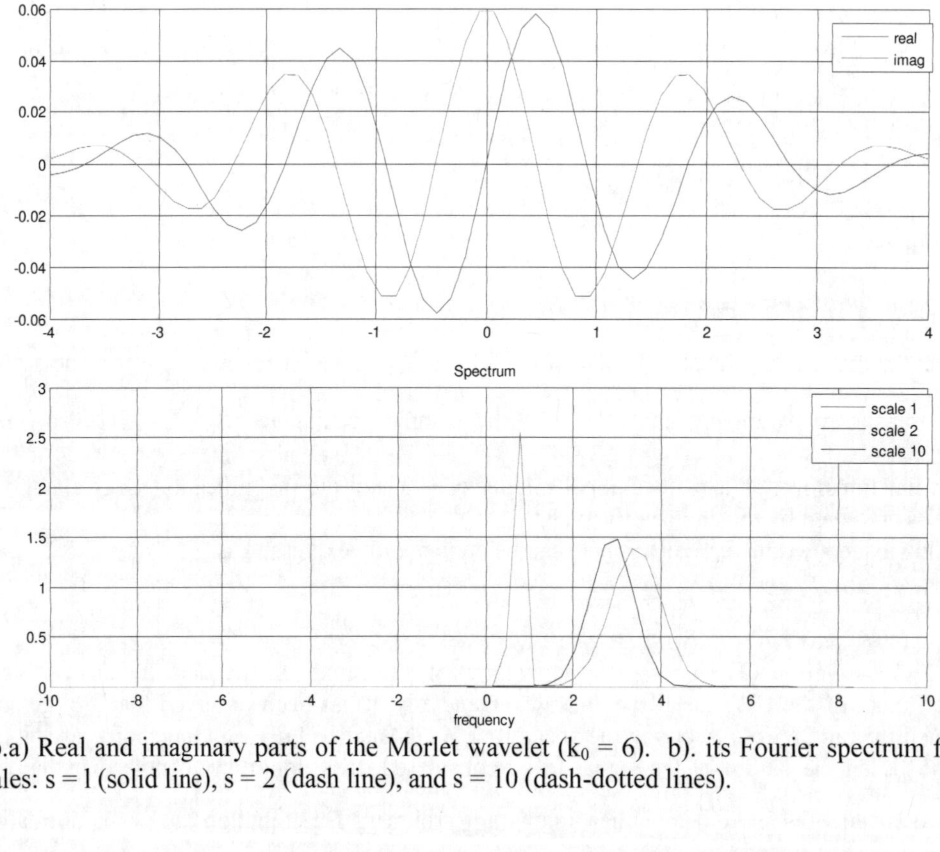

Fig. 3a-b.a) Real and imaginary parts of the Morlet wavelet ($k_0 = 6$). b). its Fourier spectrum for different scales: s = 1 (solid line), s = 2 (dash line), and s = 10 (dash-dotted lines).

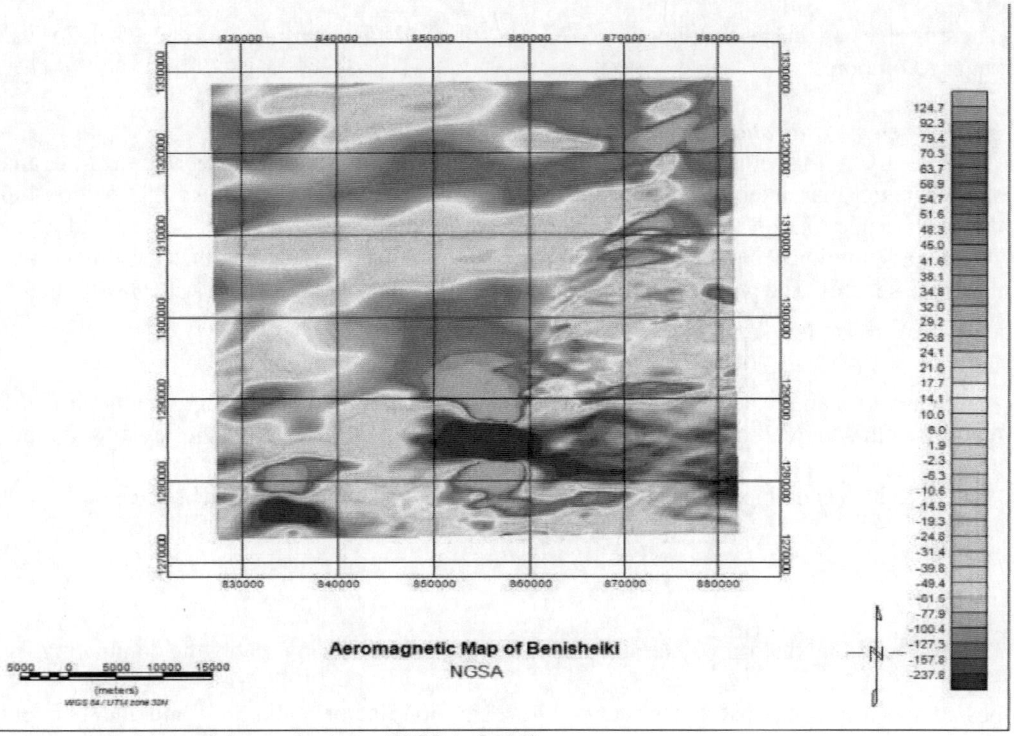

Fig. 4: Aeromagnetic Map of the study Area (NGSA, 2010)

From the definition of the Continuous wavelet transform, the coefficient can be computed by convolution in space domain or in Fourier space using discrete Fourier transform.

We computed the coefficient of the CWT in the frequency domain because it is considerably faster to do calculations in Fourier space and the implementation steps are listed below

1. We computed the Fourier transform *H(k)* of the original signal *h(x)*.
2. We multiply the Fourier transform H (k) with the Morlet wavelet $\overline{\psi(k)}$ in the frequency domain and obtain the wavelet transform at scale s = 1.75 that is, $\overline{W(k)} = \overrightarrow{H(k)} \times \overrightarrow{\psi(k)}$
3. We computed the inverse Fourier transform of $\overline{W(k)}$ and obtain the coefficient of wavelet transform result $W_u(t)$ in space domain.
4. We calculated the wavelet transform at different scales with the dilated wavelet basis and obtain the result of the wavelet transform at different scales following steps 2 and 3.

To obtain the wavelet power spectrum from the implementation steps mention above, the modulus square of the coefficient of the CWT is computed for every profile data that is

$$P(k) = |W_u(t)|^2 \quad (6)$$

The maximum decomposition scale relates the dimension of the original data and the scale can take continuous values with a maximum of half of the data dimension [13]; [14]. Here we also adopted [13] scaling relation which is defined as:-

$$s_j = s_0 2^{j\partial j}, \qquad j = 0,1.........,J \quad (7)$$

$$J = \partial j^{-1} \log_2(N\partial t / s_0) \quad (8)$$

Where s_0 is the smallest resolvable scale and J determines the largest scale. ∂j depends on the width in spectral- space of the wavelet function. For Morlet wavelet, a ∂j of 0.5 is the largest value that still gives adequate sampling in scale. a ∂j of 0.125 is appropriate because smaller values gives better resolution.

2.2 Depth estimation

Generally a mathematical basis for the application of power spectrum analysis to aeromagnetic map interpretation has been developed by [15]. They studied the relationship between the power spectrum of aeromagnetic anomalies and the average depth of source bodies using some statistic assumption. This provides a foundation for anomaly source parameter estimation of depth to magnetic sources. The power spectrum of aeromagnetic anomalies can be written as-:

$$\langle P(k)\rangle = 4\pi^2 M^2 \langle e^{-2hk}\rangle\langle 1-e^{-tk}\rangle\langle S^2(k)\rangle \quad (9)$$

Where $\langle\ \rangle$ stands for ensemble average, M is the magnetic moment/unit volume, h is the depth of the top of the source body, t is the thickness of the source body, k is the radial wave number and $\langle S^2(k)\rangle$ which is defined in equation (9) represent the horizontal size of the source body.

$$\langle S^2(k)\rangle = \frac{1}{\pi}\int_0^{\pi}\langle S^2(k,\theta)\rangle d\theta \quad (10)$$

[15] found that the $\langle e^{-2hk}\rangle$ term which is the depth factor is invariably the dominating factor in the power spectrum and that the effect of the extension factor $\langle 1-e^{-tk}\rangle$ and the horizontal factor $\langle S^2(k)\rangle$ is comparatively small, especially in low-frequency bands. From equation (8), the third and the last are relatively small, therefore it becomes

$$P(r) \approx Ce^{-2hk} \quad (11)$$

Taking the log of equation (11), we have

$$\ln(P(r)) \approx -2hk + C' \quad (12)$$

$$h = -\frac{s}{2\pi}, \quad (13)$$

Where C and C' are constant coefficients and h is the average depth of the source body. From equation (12), the plot of log power spectrum against the radial wave number gives results of different spectrum segments. The slopes obtained from the above plot indicate the average depth to the source bodes.

2.3 Mapping of source geometries from HRAM data of the study Area

High Resolution Aeromagnetic map with sheet number 88 was procured from the Nigeria Geological Survey Agency (NGSA). The study area is bounded by the UTM coordinate Northings, 1270000 mN and 1330000 mN and Easting, 830000 mE and 880000 mE. The map was windowed from the National sheet index data base delivered in Oasis Montaj format. The contour interval is variable at 5, 10, 25, 50 and 100nT.The average magnetic inclination across the survey area was from 9° in the north to 0^oin the south. The map was carefully gridded and digitized using Oasis Montaj at an equal spacing of 0.875km yielding an approximate eleven (11) profiles which cut across the entire study area.

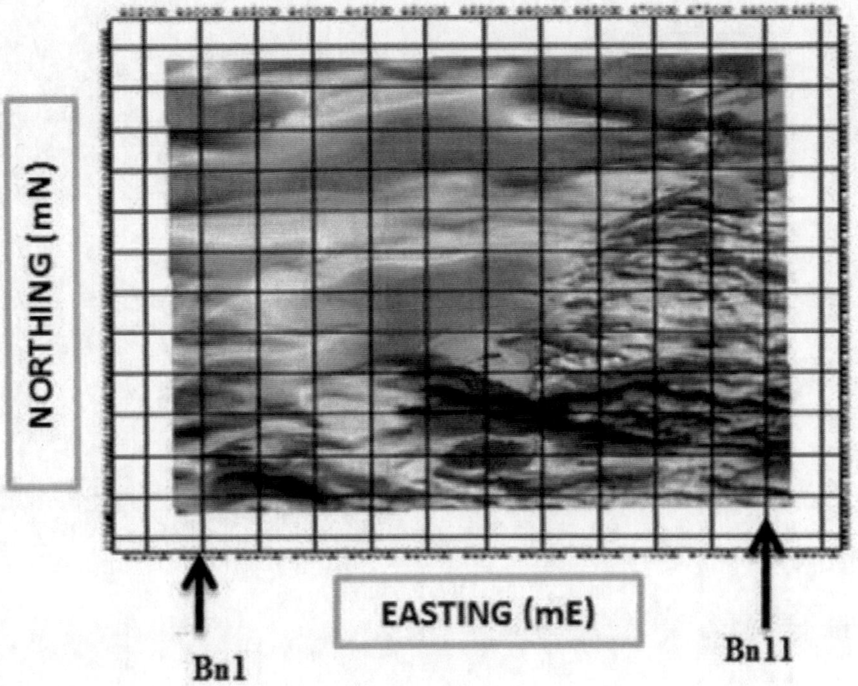

Fig. 5: Gridded Aeromagnetic Map showing Profile (P_1) to Profile (P_{11})

3.0 Results and Discussion

For the purpose of estimating depth to the magnetic sources which is a quantitative interpretation technique, profile lines labelled (1-11) were drawn on the aeromagnetic map shown in the Fig. 5 and it would be observed that these profile lines cut across various anomaly of interest. Therefore, the implementation steps of CWT in the frequency domain described above was applied to each of the profiles and the wavelet spectral analysis was computed using equation (12). From the wavelet spectral analysis, logarithms of the spectral energies against the frequencies for four out of eleven profiles on a map are shown in Fig. 6. Two linear segments could be drawn from each graph, the gradients of each linear segment were evaluated and equation (13) was used to estimate the depth to the shallow and deeper basements. The wavelet power spectral of the anomaly is mainly concentrated in the low frequency (0.0-0.3 km^{-1}) region and linear segment from this region represent contributions from the deep-seated causative bodies obtained from the profile graph and the high frequency region (0.35-0.5 km^{-1}) represents contributions from the shallow bodies. The gradient of the linear segment was evaluated and depth to magnetic sources was determined from each of the profile plots. The result obtained from the deeper sources varies from a thickness of 0.41 to 6.10 km while the shallow source depth ranges from 0.10 to 1.51 km respectively.

In other to map the geometric source, the results of the implementation steps described above as applied to some of the profiles and scalogram plots are shown in (Fig.7). The results shows two types source geometries, namely spheres and horizontal dykes. Profile 5 has a dyke like source geometry while other profiles are predominantly spheres. It can also be observed that majority of these sources are found at the low frequency portion of the scalogram plots and are located a varying distances in the profiles. This is an evidence to show that magnetic sources identified from this study are deep-seated structures and they can be regarded as sources responsible for hydrocarbon prospects.

IOP Conf. Series: Earth and Environmental Science **173** (2018) 012034 doi:10.1088/1755-1315/173/1/012034

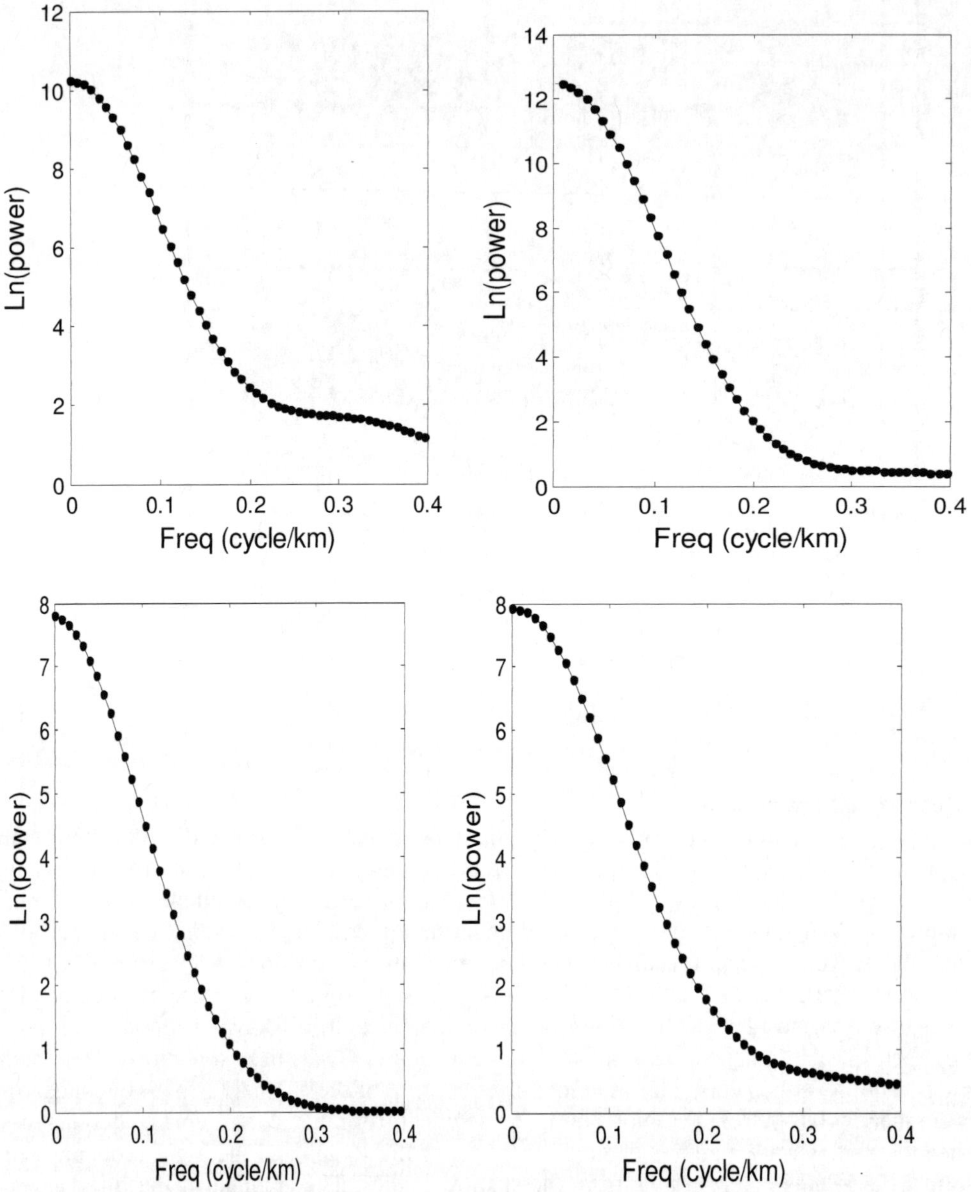

Fig. 6: Depth estimate from the wavelet spectral analysis of some selected profiles

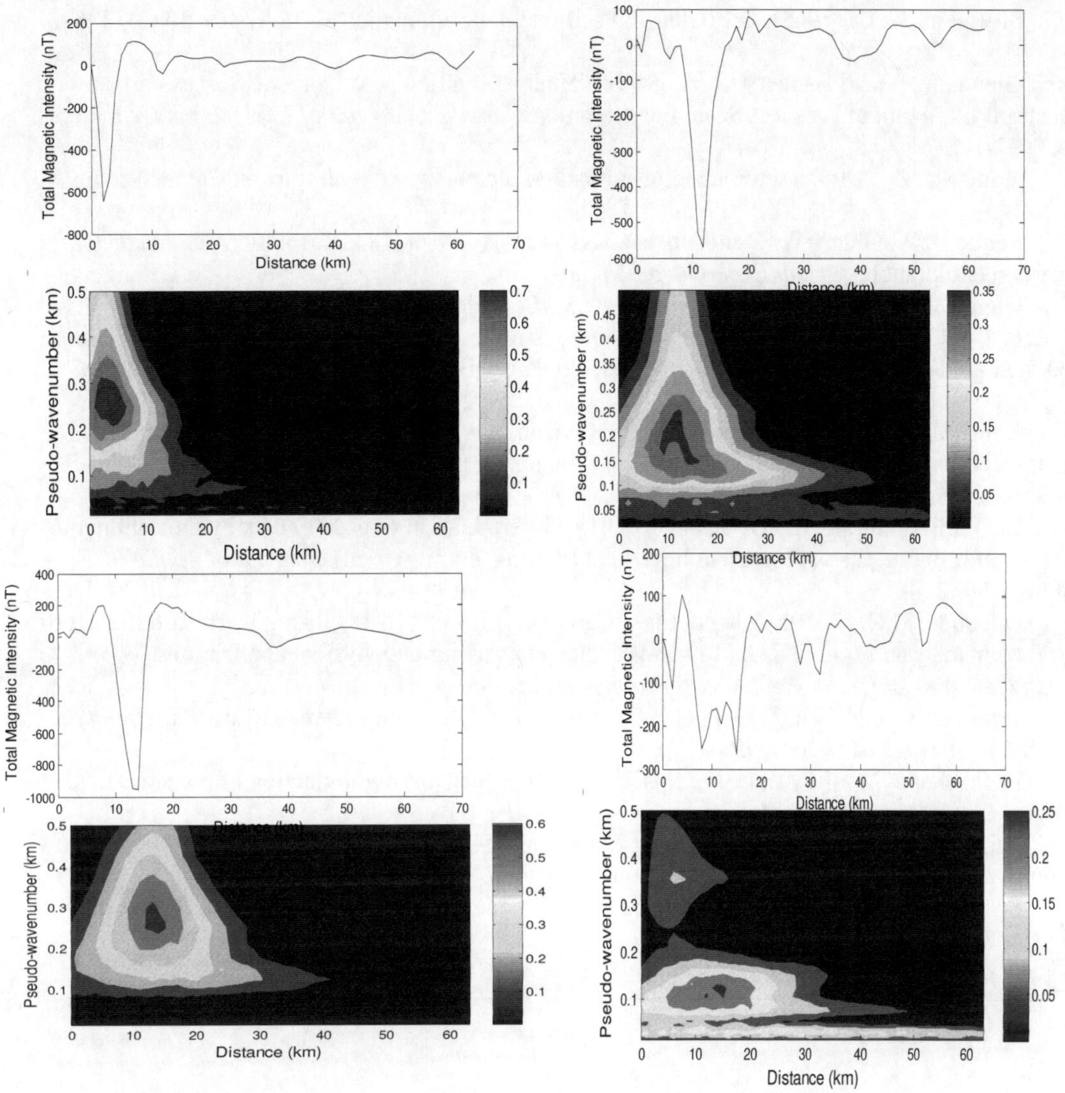

Fig. 7: Some Selected Profiles (P_1, P_3, P_5, and P_{10}) and their Scalogram plots.

4.0 Conclusions

Wavelet transform has been applied to the HRAM data of Benisheikh, Nigeria. It has efficiently analyzed and interprets both geometries and locations of the magnetic sources. It can therefore be suggested that these sources are responsible for hydrocarbon prospect. The technique can equally be applied to other part of the study area in other map the distribution of magnetic sources and this will serve as a guide before other methods such as Seismic can be applied for hydrocarbon exploration.

5.0 References

[1] Ross, C. B. (2002): Airborne and Ground magnetics. (Editor) ,Geophysical and remote sensing methods for regolith Exploration. **144**33-45.

[2] Andersen, H. T and Bodger, T. R. (2003): High Resolution Airborne Magnetic Surveying for hydrocarbon Exploration. Airborne geophysical Services. Canada.

[3] Steenland, N. C. (1965): Oil fields and aeromagnetic anomalies. Geophysics, **30** (5), 206-239.

[4] Glenn, W. E. and Badgery, R. A. (1998): High Resolution aeromagnetic Surveys for Hydrocarbon Exploration: Prospect Scale Interpretation.*Canadian Journal of Exploration Geophysics*. **34** (1&2) 97-102

[5] Yong-xia, Z. (1994): Aeromagnetic anomalies and perspective oil traps in China. Geophy. **59**. 1492-1499

[6] Peirce, J. W. Glenn, W. E and Brown, K. C. (eds) (1998): High Resolution aeromagnetic for hydrocarbon exploration. *J. Can. Expl. Geoph.***34**106.

[7] Glenn, W. E., Rhodes, J. A., Goussev, S. A., Benedict, M. L.,Charters R. A.,Peirce J. W. andGenereux, J. (2002): Aeromagnetic Depth Solutions Indicate Numerous MagneticSources within the Sedimentary Section of the Western CanadaSedimentary Basin. *CSEG Geophysics Conference Proceedings.*

[8] Yang, Y.S., Li, Y.Y. and Liu, T.Y. (2010): Continuous wavelet transforms, theoretical aspectsand application to aeromagnetic data at the Huanghua Depression, Dagang Oilfield, China. *Geophysical Prospecting*.**58**, 669– 684

[9] Li, Y., Braitenberg, C. and Yang, Y (2013): Interpretation of gravity data by the continuous wavelet transform: The case of the Chad lineament (North-Central Africa). *Journal of Applied Geophysics.* **90**62-70.

[10] Odebode, M.O. (2010): A handout on Geology of Boron (Chad) Basin, Northeastern Nigeria.

[11] Kumar, J and Foufoula, G. (1997): Wavelet analysis for geophysical applications. *The American Geophysical Union. Reviews of Geophysics*. **35.**385– 412.

[12] Torrence, C. and Compo G. P. (1998): A practical guide to wavelet analysis. *Bulletin of the American Meteorological Society*. **79**:61–78

[13] Fedi, M and Rapolla A. (1997): Space-Frequency analysis and reduction of potential field ambiguity. *Annali Di Geofisica.***XL** (5).

[14] Xu,Y; Tianyo H, Zhiwei L, Qiuliang D and Zhang L. (2009): Regional gravity anomaly separation using wavelet transform and spectrum analysis. *J. Geophys. Eng.***6** 279-287.

[15] Spector, A. and Grant. F.S. (1970): Statistical models for interpreting aeromagnetic data Geophysics.**35** 293–302.

Effects of Physico-chemical attributes of forest-based herbal soap on human skin in southwestern Nigeria

Elizabeth O.A. OLUWALANA

Department of Agricultural Economics and Farm Management, Federal University of Agriculture, Abeokuta, Ogun State, Nigeria

oluwalanaeoa@funaab.edu.ng; +2348037158122

Abstract. There is a growing awareness of medicinal benefits of renewable forest resources that can be used as natural cosmetics for the benefits of human skin health. Today, more rural dwellers in forest reservation have resorted into utilizing more Non-Timber Forest Products (NTFPs) and agro-processing wastes products (APWPs) into African Black Soap (ABS) otherwise called Herbal Soap (HS). Some of the NTFPs that fall under natural cosmetics are herbal soap (black soap) produced from Cocoa Pod Husks (CPH); Shea-butter; Coconut Oil; Aloe-Vera Gel, and Natural Fragrance Oil. This is an excellent natural body-care soap in the original fashion and suitable for all skin types. The major raw materials that are used to produce herbal soap are from NTFPs and APWPs such as CPH; Palm Kernel Oil (PKO); cassava peels (CP) and, Plantain Peels (PP). 5g weight of each sample was dissolved in 50ml of solution (10% Concentration) to get the results. The results of the physico-chemical analysis (PCA) on Herbal soap and photographic impressions of ten (10) samples collected during field survey from different locations in South western, Nigeria revealed the following: The pH range of the herbal soaps (10% solution) is between 9.89 to 10.18 which is a good indication that the herbal soaps fall in the mid alkaline pH range - (that is, ideal soap pH range that have little or no irritation on mildest skin texture. Some synthetic soap that bleaches gives a pH range above 10.50 and other acidic soap below 6.98. pH range of the soap (10%) lies between (981-3110) x 10 mg/L which is a good indication. The Total Dissolved Solid (TDS) mg/L of the herbal soaps falls between 981-3110 mg/L which is a good indication of inertness of the soap to atmosphere when expose to moisture. Industrial PKO (981-1775) melted but traditional processed PKO (1800 -3110) stayed dried. The Conductivity (US/cm) of the herbal soaps ranges from 18,200-47,150 (US/cm) ^{x10}. This measured parameter indicates the extents of dissolved ions present in the herbal soap materials that when releases nourishes the skin on use. The higher the dissolved ions, the better for skin absorption of the desirable ions which result into healthy skin in human being

Keywords: Forest Based Herbal soap, Non-Timber Forest Products (NTFPs), Physico-chemical attributes, pH, Healthy Skin

1. Introduction

Researchers have identified some wastes generated from agro-processing enterprises being utilized by small-scale industry to produce consumer products such as black/herbal soap [1, 2]. Some of these raw materials from forest and agriculture are classified as non-timber forest products (NTFPs) and agro-processing wastes products (APWPs). The NTFPs being utilized in FB-SSEs are usually classified into firewood and sawdust (off-cuts and dust from sawn timbers); medicinal plants (bark; leaves; roots; seeds); ropes, foods and fruits. The agricultural-processing wastes generated by the rural entrepreneurs include cocoa-pod husks (CPHs); Palm-kernel Oil (PKO); plantain peels; cassava peels and palm kernel bunches (PKBs).

African black soap (ABS) is all-natural soap handcrafted that have more than 100 varieties of black soap [3]. Each region and tribe has his or her own recipe that has been handed down through generations. Ghanaians and Nigerians have used black soap to help relieve acne, oily skin, clear blemishes and various other skin problems, including eczema and psoriasis. African women use black soap during pregnancy and afterwards to keep them from getting stretch marks and to protect them from dry skin that often accompanies pregnancy [3]. Black or herbal soap is used in many traditional medicine formulations, for the cure of uterine fibroid, vaginal infections, and charms. Herbal soap is unique in that it contains no preservatives and colour enhancers. In recent year, there has been a grassroots return to making "natural" soap in the home [4]. In modern times, the use of herbal soap has become universal due to a better understanding of the role of personal hygiene and medicinal importance on skin in particular. Tables 1 and 2 below shows the medicinal components as reported by Yusuf [1] and mineral composition of Cocoa Pud husk (CPH) by Ozung [5].

1.1 Research Questions

(i) What are the Non-Timber Forest Products (NTFPs), Agro-processing waste products (APWPs), Value-Added Products (VAPs) being utilized by FB-SSEs Herbal Soap processing?

(ii) What are the health potentials in terms of medicinal uses and Physico-chemical attributes of herbal soap?

1.2 Objective

The main objective is to analyze the physico-chemical attributes of herbal soap that influence the healthy human skin.

Table 1: Medicinal Potential Components of Cocoa Pod Husk (Moisture free)

Constituents	Percentage
Crude Protein	5.69-9.69
Fatty Substances	0.03-0.15
Glucose	1.16-3.92
Sucrose	0.02-0.18
Pectin	5.3-7.08
Nitrogen free extract	44.21-51.27
Crude fibre	33.19-39.45
Theobromine	0.20-0.29
Ash	8.83-10.18

Source: Yusuf and Okoruwa (1995)

IOP Conf. Series: Earth and Environmental Science **173** (2018) 012035 doi:10.1088/1755-1315/173/1/012035

Table 2: Mineral Composition of Ash and of Cocoa Pods

Mineral (g/kg) RCPHM FCPHM HCPHM S.E.M	Mineral (g/kg) RCPHM FCPHM HCPHM S.E.M	Mineral (g/kg) RCPHM FCPHM HCPHM S.E.M	Mineral (g/kg) RCPHM FCPHM HCPHM S.E.M	Mineral (g/kg) RCPHM FCPHM HCPHM S.E.M
Calcium (Ca) 11.16a 5.65b 7.23b 1.64	Calcium (Ca) 11.16a 5.65b 7.23b 1.64	Calcium (Ca) 11.16a 5.65b 7.23b 1.64	Calcium (Ca) 11.16a 5.65b 7.23b 1.64	Calcium (Ca) 11.16a 5.65b 7.23b 1.64
Magnesium (Mg) 2.79a 2.44a 1.53b 0.38	Magnesium (Mg) 2.79a 2.44a 1.53b 0.38	Magnesium (Mg) 2.79a 2.44a 1.53b 0.38	Magnesium (Mg) 2.79a 2.44a 1.53b 0.38	Magnesium (Mg) 2.79a 2.44a 1.53b 0.38
Phosphorus (P) 0.51a 0.30b 0.47a 0.06	Phosphorus (P) 0.51a 0.30b 0.47a 0.06	Phosphorus (P) 0.51a 0.30b 0.47a 0.06	Phosphorus (P) 0.51a 0.30b 0.47a 0.06	Phosphorus (P) 0.51a 0.30b 0.47a 0.06
Potassium (K) 1.98a 1.07b 1.15b 0.29	Potassium (K) 1.98a 1.07b 1.15b 0.29	Potassium (K) 1.98a 1.07b 1.15b 0.29	Potassium (K) 1.98a 1.07b 1.15b 0.29	Potassium (K) 1.98a 1.07b 1.15b 0.29
Sodium (Na) 0.35b 0.48b 0.74a 0.11	Sodium (Na) 0.35b 0.48b 0.74a 0.11	Sodium (Na) 0.35b 0.48b 0.74a 0.11	Sodium (Na) 0.35b 0.48b 0.74a 0.11	Sodium (Na) 0.35b 0.48b 0.74a 0.11

Source: Ozung etal, 2016

2. Materials and Method

2.1 Analytical Techniques

Identification of the Non-Timber Forest Products (NTFPs), Agro-Processing Waste Products (APWPs) and Value-Addition Products (VAPs) used in ashes and/or lye production for making herbal soap. There were many forest plant species used as fuel-wood, ashes and lye production by Forest-Based Small-Scale Enterprises (FB-SSEs) in herbal soap production. The parts of the plants being used are the off-cuts, sawn wood, branch, stem, log waste. They are classified as Non-Timber Forest Products (NTFPs) used as fuel-wood for cooking, baking in the bakery, lye production and soap production. Saw dusts ashes from sawmill are also utilized by the herbal soap processors.

The ashes collected were used to produce lye for production of herbal soap. The part of the tree used are usually classified as wastes, hence, the utilization of these wastes is desirable to generate more income, thereby improve the processors income generating activities leading to poverty reduction and improvement in the livelihood status of the entrepreneurs involved in herbal soap production. Furthermore, some of the plants are in abundant, scarce or not common in the study areas.

Furthermore, agricultural wastes generated during commercial activities such as cocoa processing (Cocoa pod husks), cassava processing (cassava peels), plantains processing (plantains peels), palm oil processing (palm oil bunches) etc. Identification of the wastes that can be used to make herbal soap was carried out during this study as wastes-to-wealth strategy to generate more income to improve the livelihood status of the entrepreneurs.

Some Value-Addition Products (VAPs) that were identified to add value to herbal soap produced by producers or final consumers of value-added herbal soap such Dudu Osun and UNAAB Herbal Soap. Some of these products are pure honey, sheabutter, cocoa butter, coconut oil, camwood *(osun),* lime juice, lime rind, sulphur *(imi-ojo),* Aloe-vera and Ethiopian pepper.

2.2 Physico-Chemical Analysis of Herbal Soap Samples

Chemical analysis was carried out in the laboratory on ten (10) herbal soap samples to confirm the chemical attribute of herbal soap that influence its effect on human skin. Five grams (5gm) weight of each sample was dissolved in 50ml of solution. (10% Concentration) to get pH range of the herbal soaps. An ideal soap should have little or no irritation on mildest skin texture.

The Total Dissolved Solid (TDS) mg/L of the herbal soaps was analysis to get indication of inertness of the soap to atmosphere when expose to moisture. Herbal soap processed with industrially processed PKO and traditionally processed PKO were tested.

The Conductivity (μS/cm) was the parameter used to measured the extents of dissolved ions present in the herbal soap materials that when released nourishes the skin on use. The higher the dissolved ions, the better for skin absorption of the desirable ions that result into healthy and smooth skin. The chemical analysis results confirmed the age long medicinal attributes associated with herbal soap on the skin.

3. Results and Discussion of Findings

3.1 Utilization of Agricultural-Wastes Products in Herbal Soap Processing Enterprise

There are many agro-processing wastes being utilized in herbal soap production. Some of the products and processing are stated in Figure 1. The first stage involved the collection of dry cocoa pod husks with or without palm kernel bunches and cassava wastes. The second stage entailed the burning into ashes the agro-wastes. The ashes are put into perforated clay pots, water, usually rain water or stream water added and lye is produced. The third stage which is called saponification of lye *(aaro)* and palm kernel oil *(adin-agbon)* or palm oil *(epo pupa)* to produce herbal soap.

The last stage is the value-addition stage which involves packaging herbal soap with or without wastes from generated cassava or yam commercial processing. The wastes used as *'fillers'* are classified as cassava granules (*Garri)* waste, cassava flour *(lafun)* waste and yam flour *(elubo)* wastes. Fillers are meant to make the herbal soap look bulky and cheap to the customers. The health implication of the inclusion of the filler has not been ascertained but the fillers likely reduce the potency and lathering effect of the soap.

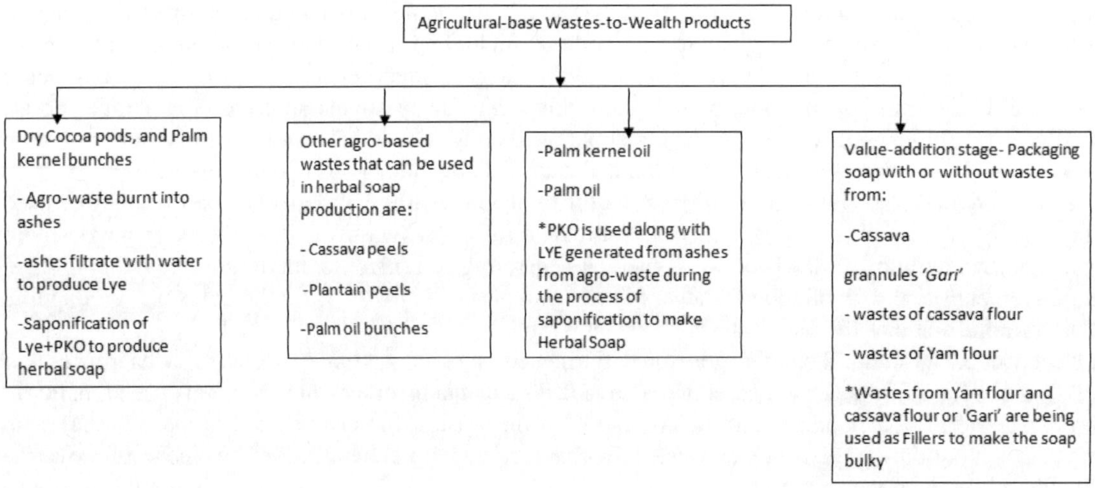

Figure 1: Value chain in herbal soap production and processing (Source: Field Survey, 2015)

3.2 Physico-Chemical Attributes of Herbal soaps from Southwest Nigeria

The result of the Physico-chemical analysis on herbal soaps is presented in Table 3. The photographic impressions of ten (10) samples analyzed are in Figures 2 to 15. Five grams (5gm) weight of each sample was dissolved in 50ml of solution. (i.e. 10% Concentration) to get the result. The pH range of the herbal soaps (10% solution) ranged between 9.8 and 10.18. This is a good indication that the herbal soaps are alkaline with little or no irritation on any skin texture. The Total Dissolved Solid

IOP Conf. Series: Earth and Environmental Science **173** (2018) 012035 doi:10.1088/1755-1315/173/1/012035

(TDS) mg/L of the herbal soaps falls between 981 and 3110 mg/L which is a good indication of inertness of herbal soap to atmosphere when expose to moisture. When herbal soap is manufactured with industrially processed palm kernel oil (PKO) with TDS mg/L ranged from 981-1775, it easily melted when expose to moisture. But, when herbal soap is manufactured with traditional processed palm kernel oil (PKO) with TDS mg/L ranged from 1800 -3110) stayed dried when expose to moisture.

Table 3: Physico-Chemical Attribute of Herbal soaps Produced by Entrepreneurs from Southwest Nigeria.

ID	LGA/ State	Vegetation type	Description	pH	TDS (mg/c) ^{x10}	Conductivity (US/cm) ^{x10}
A	Odeda / Ogun	Forest	Cocoa pod Ash + Palm Kernel oil (Traditional/Industrial) (No filler)	10.0	1339/ PKO industrial (melted)	475X5=23,750
B	Ibarapa / Oyo	Derived Savannah	Cocoa pod Ash + Palm Kernel oil (Traditional/Industrial) + (filler-garri)	9.95	1135/ PKO industrial	422X5=21,100
C	Afijo/Oy o	Derived Savannah	Wood Ash + Palm Kernel oil (Traditional/Industrial) + (filler-elubo)	10.10	1723/ PKO industrial	643X5=32,150
D	Iseyin / Oyo	Derived Savannah	Cocoa pod Ash + Palm Kernel oil (Traditional/Industrial) (No filler)	9.99	981/ PKO industrial	382X5=19,100
E[1]	Aawe/ Oyo	Derived Savannah	Cocoa pod Ash + Palm Kernel oil (Traditional/Industrial) (No filler)	9.90	1589/ PKO industrial	546X5= 27,300
E[2]	Aawe/ Oyo	Derived Savannah	Cocoa pod Ash + Palm Kernel oil (Traditional/Industrial) (No filler)	10.04	1833/ PKO traditional	639X5= 31,950
F[1]	Akinyele Oyo	Forest	Cocoa pod Ash + Palm Kernel oil (Traditional/Industrial) (No filler)	10.08	1775/ PKOindustrial	617X5=30,850
F[2]	Akinyele Oyo	Forest	Cocoa pod Ash + Palm Kernel oil (Traditional/Industrial) (No filler)	10.13	1919/ PK traditional	636X5=31,800
G	Osogbo / Osun	Forest	Wood Ash + Palm Kernel oil (Traditional/Industrial) (No filler)	9.89	3110/ PKO traditional	943X5=47,150
H	Agege/ Lagos	Control /Imported from Ghana	Cocoa pod Ash + Palm Kernel oil (Traditional/Industrial) (No filler)	10.18	1092/ PKO industrial	364X5=18,200

Source: Physico-Chemical field survey Analysis, 2015 Note: 5g sample dissolved in 50 ml of solution. (10% Concentration).

Figure 2: Waste-to-wealth initiative (Utilization of Agricultural-wastes such as Cocoa pod husks in herbal soap making)

Figure 3: Waste-to-wealth initiative (Utilization of Agricultural -wastes such as cassava peels in herbal soap making)

Figure 4: Waste-to-wealth initiative (Utilization of Agricultural - wastes such as Palm-oil bunches in herbal soap making)

Figure 5: Waste-to-wealth initiative (Utilization of Agricultural - wastes such as Plantain peels in herbal soap making)

Figure 6: Herbal soap (Sample A) Cocoapod ash/lye + fried palm kernel oil (no filler added) found in Osiele, Ogun State, Nigeria, Forest Area.

Figure 7: Herbal soap (Sample B) Cocoa pod ash/lye + fried palm kernel oil (filler added) found in Igboora, Oyo State, Derived Savannah Area.

IOP Conf. Series: Earth and Environmental Science **173** (2018) 012035 doi:10.1088/1755-1315/173/1/012035

Figure 8: Herbal soap (Sample C) Wood ash/lye + fried palm kernel oil (filler added) found in Ipako, Oyo State, Savannah Area.

Figure 9: Herbal soap (Sample D) Cocoa pod ash/lye + fried palm kernel oil (filler added) found in Iseyin, Oyo State, Derived Savannah Area

Figure 10: Herbal soap (Sample E) Cocoapod ash/lye + white palm kernel oil (no filler added) found in Aawe, Oyo State, Derived Savannah Area.

Figure 11: Herbal soap (Sample F) Cocoapod ash/lye + fried palm kernel oil (no filler added) found in Aawe, Oyo State, Derived Savannah Area

Figure 12: Herbal soap (Sample G) Cocoa pod ash/lye + fried palm kernel oil (filler added) found in Igboora, Oyo State, Derived Savannah Area

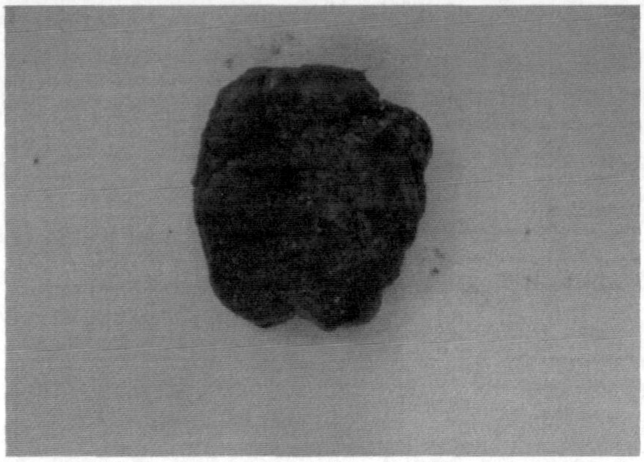

Figure 13: Herbal soap (Sample H) Cocoa pod ash/lye + white palm kernel oil (filler added) found in Akinyele, Oyo State, Forest Area

Figure 14: Herbal soap (Sample I) Wood ash/lye + roasted palm kernel oil (no filler added)

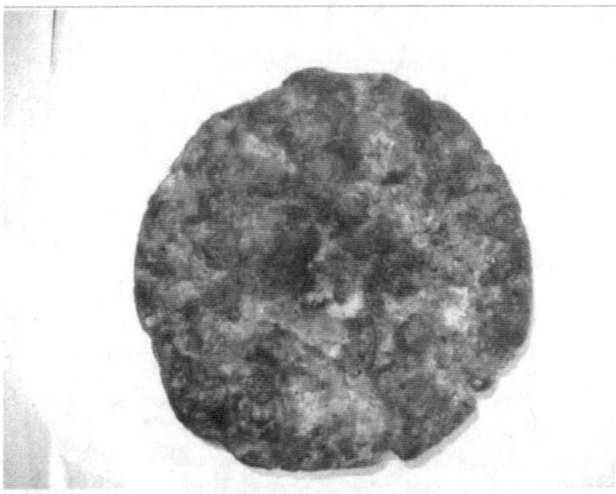

Figure 15: Herbal soap (Sample J) Cocoa pod and plantain ashes/lye + fried palm kernel oil (filler added) found in Agege market, Lagos State, (Control Specimen imported from Ghana)

The conductivity (µS/cm) of the herbal soaps range from 18,200- 47,150 (µS/cm). This measured parameter indicated the extents of dissolved ions present in the herbal soap materials that when released nourishes the skin on use. The higher the dissolved ions, the better for skin absorption of the desirable ions from herbal soap and this result into healthy and beautiful skin. The results of chemical attributes confirm the age long attributes associated with herbal/black soap which is in line with findings of Yusuf and Adeogun [1, 2].

Also, table 4 shows some of the uses to which herbal soap is put into the study area. Majority of the respondents (87.64%) indicated the use of soap as medicine. The medical uses include using the soap for stomach pain, wherein the soap is liquidified in water and drank. It is also used in curing skin diseases such as eczema, prosiasis, ringworm, sores, and wounds. For eczema, prosiasis, and ringworm, the affected part is washed, sometimes scrubbed with a rough object or leaf such as the leaf of *Ficus exasperata* ("eepin" in Yoruba) – sand paper leaf, and the soap is applied. For sores and wounds, affected parts are washed using the soap. Herbal soap is also used as an ingredient in the formulation of anti-fibroid drug. The soap is burnt with other materials, pulverized and taken in daily. The use of herbal soap as cosmetics percentage was 84.46% and daily needs had 81.67%.

The high percentages recorded among the respondents may be as a result of resent awareness of the medicinal benefit derived from uses of natural products as cosmetics or daily needs. Also, the emerging market for natural/herbal products because of healthy living style people are trying to maintain by purchasing and consuming natural products. Furthermore, the use of herbal soap as cosmetics involved additional processes which involved adding other natural/herbal products such as natural honey, shea-butter, cocoa butter, coconut-oil, camwood *(Osun),* sulphur *(Imi-ojo)* and others herbs such as Ethiopian pepper, Aloe-vera, lime juice and rind into the soap before use. Individual

IOP Conf. Series: Earth and Environmental Science **173** (2018) 012035 doi:10.1088/1755-1315/173/1/012035

and packaging industries usually use these products to produced value-added herbal soaps such Dudu Osun and UNAAB herbal soaps. Cultural purposes (28.08%) recorded the lowest rate, meaning that few people are still involved in using herbal soap for cultural purposes, maybe based on religious believe.

Table 4: Distribution of Respondents with the Uses of Herbal Soap *

Uses	Ekiti n = 30	Lagos n = 33	Ogun n = 56	Ondo n = 32	Osun n = 16	Oyo N = 335	Total n = 502	%	Mode	Mean
Medical	-	33	56	-	16	335	440	(87.64)	Medical	
Cosmetic	-	33	56	-	-	335	424	(84.46)		
Cultural purposes	-	10	15	-	16	100	141	(28.08)		
Daily Needs	-	30	40	-	2	300	410	(81.67)		

Source: Field Survey, 2015
* Multiple responses.
Values in parentheses are in percentages

4. Conclusion

In conclusion, there are many outstanding policy findings from the study on Forest-based Small-scale Enterprises (FB-SSEs) processing and marketing herbal soap which can guide all stakeholders in policy making for FB-SSEs in South West Nigeria, Nigeria and the world at large. There are opportunities to utilize NTFPs and ABWPs identified in this study to generate wealth and good health. The wealth so generated from wastes can lead to reduction of poverty among the rural entrepreneur generally especially the women processing the herbal soap in particular. This concept is classified as ''*waste-to-wealth*'' initiative to improve the economic and health status of the beneficiaries. Furthermore, the emerging business opportunity in ash production in Ondo and Ekiti States will provide subsidiary occupation for cocoa farmers who usually dispose the cocoa pod husks as ''wastes'' on the farm. There is the need for Ondo and Ekiti States Government to utilize the comparative advantage of the abundance of cocoa in these states.

5. Policy Recommendation

1. There is the need for all stakeholders – university researchers, governments, NGOs and individuals to promote appropriate technology that will enhance the utilization of indigenous knowledge especially in herbal soap production. It is in this area that the Asian "Tigers" have excelled in using technology to upgrade indigenous knowledge. It is most unlikely that herbal soap production will eventually be able to withstand technological constraints from industrially produced soaps. The herbal soap production subsector should be vigorously promoted and advertised by all stakeholders.
2. The problem of purchasing modern equipment to process and package products (value-addition process) should be addressed by concerned stakeholders to solve the problems of inefficient resource-use and marketing of products. Wastes-to-wealth technology should be encouraged to enhance the livelihood status of rural entrepreneurs, women processors in particular.
3. The problem of purchasing modern equipment to process and package (value-addition process) should also be solved by providing incentives to the entrepreneur in form of subsidized processing and packaging equipment such as cocoa pod husk drier, PKO pressing machine, soap cutting machine, storage facility, so as to reduce the high cost of labour being encountered by entrepreneurs.
4. There is the need for the communities where Agro-business waste and NTFPs abound to form themselves into cooperative bodies in order to benefit from various credit assistance of State and Federal governments. The assistance can enable them to purchase capital assets such as modern

equipment to remove the drudgery involved in Herbal soap production. Furthermore, the quality and packaging of Herbal soap will be enhanced from the usage of modern machines. More jobs will be created, thereby reducing poverty and increasing standard of living of beneficiaries.

References

[1] Yusuf S. A. and Okoruwa V. O. (1995). Transforming Wastes into Money: The case of Black soap making using Cocoa Pod Husks. In: Cocoa Revolution in Nigeria. Proceedings of a National Seminar on Revolutionizing Nigeria's Cocoa Industry. University of Ibadan. Nigeria. 101-112.

[2] Oluyole, K. A. and Adeogun, S. O. (2005). Determining the Profitability level of Black Soap production from Cocoa Pod Husk in Lagelu Local Government Area of Oyo state. Journal of Agriculture, Forestry and Social Sciences (JOAFSS),Vol 3No 1.

[3] Kiltz, W. (2011). An ADUNAGOW Magazine, 2011. There is nothing basic about African's black soap. February 18, 2011. http://www.adunagow.net/2009/?p=1067

[4] Good Scents Candies, 2006. History and use of Soap: Soap in Ancient Times. http://www.goodscentsscandies.us/soaphistory.php.

[5] Ozung, P.O, O. O. Kennedy Oko, and E.A Agiang. (2016). Chemical Composition of Different Treated Forms of Cocoa Pod Husk. Asian Journal of Agricultural Sciences 8(2): 5-9.

Radiological Safety of Water from Hadejia River

[1]Orosun. M. M, [1]Alabi A. B,[1]Olawepo .A. O, [2]Orosun. R. O., [1]Lawal T. O. and [1]Ige. S. O

[1]Department of Physics, University of Ilorin, Ilorin, Nigeria.

[2]Factory Department, Savannah Sugar Company Limited, Numan, Nigeria.

Email: muyiwaorosun@yahoo.com

Abstract. Water samples from Hadejia River in the north eastern part of Nigeria were studied in order to ascertain the suitability and radiological safety of water from the river for human consumption. Radioactive measurements, using a lead shielded sodium iodide detector coupled to a multichannel analyzer were used to estimate eight radiological parameters of water samples from Hadejia River. The results show that the values of all the parameters fall within the minimum universal standard, indicating that consuming the water pose no serious radiological hazard, especially for adults. The values of the Absorbed Dose Rate (50.10511 nGyh-^{1}) and AEDE (12.458 mSvy^{-1}) for the infant however portend that infant consumer could be susceptible to radiation hazard on consuming water from the River. It is however recommended that activities that are capable of enhancing the radiological content of the River be avoided within the area.

Keywords: Radioactivity, Gamma Spectrometry, Specific Activity, Radiological Impact Parameters.

1. Introduction

Water is of prime importance to man [16]. Man uses water for washing, drinking, cooking and for several other purposes which together make water an indispensable commodity to man. There are several sources of water; these include rain, river, stream, water-wells among others. In civilized societies, pipe-borne drinking water is made available hygienically, after being subjected to treatment. However, rural dwellers in developing countries such as Nigeria still depend on rivers and streams as sources of drinking water. Drinking water directly from these sources is, however, not without hazards. Possible health hazards posed by drinking water directly from sources such as streams and rivers include cholera, ring worm, cancer, typhoid etc. Health hazards are due mainly to pollutants in such water bodies. A good example of water pollutant is radionuclide. Radionuclides are elements with unstable nuclides which, as a result of their instability can disintegrate by emitting ionizing radiations such as alpha, gamma and beta radiations. Radionuclides such as isotopes of potassium, K-40; uranium, U-238 and Thorium, Th-232 are present in the human environment (e.g. soil, atmosphere) as a result of both natural and anthropogenic activities. These isotopes can find their ways into water sources through rainfall and weathering activities thereby constituting pollutants thus contaminating the water [15]. When water from such rivers is used for drinking, these elements are ingested into the human systems, leading to health risk since such element will eventually disintegrate, emitting ionizing radiations into the human system depending on their half life periods. The biological effect of ionizing radiations on human systems can be immediate or delayed, stochastic or non-stochastic. The degree and magnitude of hazard usually depends on the amount absorbed and the rate of absorption by body cells and tissues [25].

Effects of exposure of human cells to radiations from these nuclides can lead to cancer of the lung, pancreas, hepatic, bone, skin and kidney and other diseases such as cataracts, sterility, atrophy of the kidney and leukaemia [1, 2, 4 and 22]. Several authors have studied the level of radiological contamination in rivers within and outside Nigeria with the objective of quantifying the radiological health hazard posed by drinking water from such rivers, using different methods [3, 5, 8, 13, 17 and 18]. Results from such studies showed that radiological values in some rivers are lower while some are higher when compared with recommended standards by the International Commission on Radiological Protection [9]. For instance, the results of [13] and [15] showed higher values of radiological content in the rivers studied. [5] surveyed the radionuclide concentration of soil, sediments and water in Aba River, Abia state and reported higher concentrations compared to previous studies already reported by [12] and [23] for the same River. Thus the level of radiological hazards in a River may change with time. These changes over a period of time could be attributed to the growth of human activities around the river. According to [4], the growth of human population and rapid industrialization can lead to increased pollution of water sources. Since rivers still serve as the major source of water for people in the rural areas especially in Africa, it becomes imperative to know the level of radiation hazards to which humans are exposed by consuming such water hence the objective of this study is to investigate the level of radiological contamination in the Hadejia river in order to ascertain the level of the radiological health hazard posed to human life by consuming water from the river. This is necessary because over 80 percent of the locals living around the river depend on it for drinking, fishing, irrigation and general domestic use. The wetlands support extensive wet-season rice farming, flood-recession agriculture and dry-season irrigation [7, 21]. The river thus supports economy and domestic activities among hundreds of thousands of Nigeria.

2. Materials and methods

2.1. The study area

Fig.1 is the map of the study area. Hadejia River is located in Jigawa state (Latitude 12.64° – 12.65°N and Longitude 10.63° – 10.64°E) bounded by Kano in the south east and Yobe state in the north eastern part of Nigeria. The geology of Jigawa state principally comprises of crystalline and sedimentary rocks, underlain by basement complex rocks. The crystalline rocks are represented by older granites found in pockets part of the study area. The older granite is Precambrian in origin consisting of metamorphic structures of gneiss and amphibolites. The sedimentary rocks that are found in most parts of the area were uncomfortably deposited on the basement crystalline rock [7, 14 and 21].

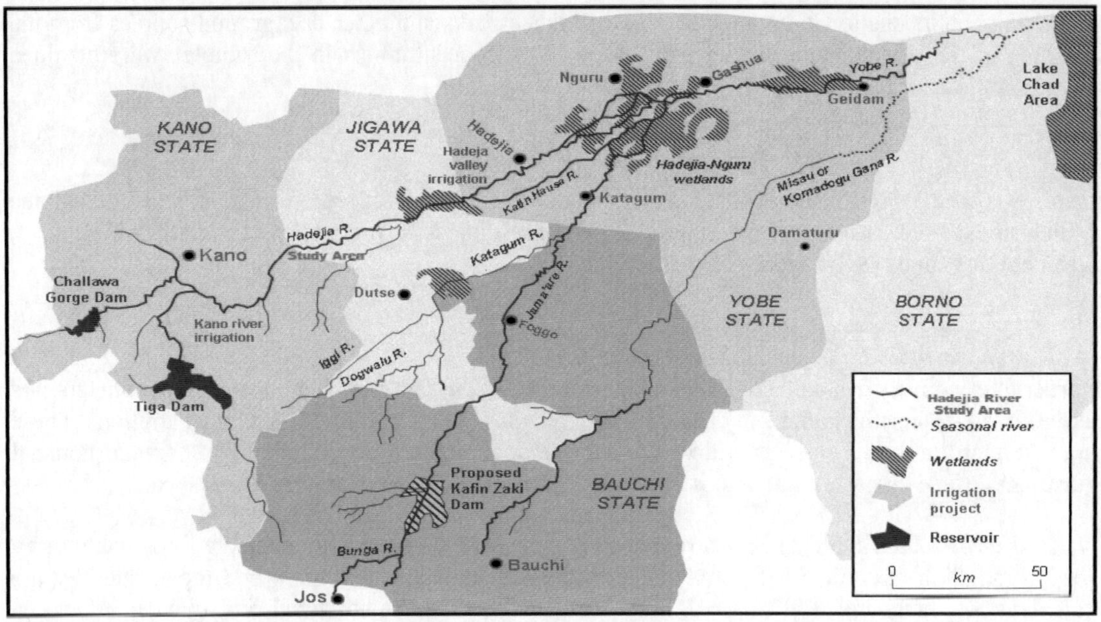

Figure 1: Map of the study area.

2.2. *Method of water sampling and preparation*

A total of nine water samples were collected using 1 litre polythene bottles and labeled KW1-KW9. Samples were collected randomly in 2 liters polythene bottles with tight covers which were carefully washed in the laboratory and rinsed three times with the sample water to be sure that the samples collected are representative of the bulk. These samples were collected at the bank of the river; an area where there is less dilution of the washout from the surrounding environment. The samples were acidified with11 M HCl, at a rate of 10 ml per liter to prevent the water from biological growth and chemical action with the surface of the container and transferred to laboratory. The samples were then sealed and stored in the laboratory for 28 days before being analyzed. This was done in order to allow radon and its short-lived progenies to reach secular equilibrium prior to gamma spectroscopy [5]. The gamma spectrometric measurement was carried out using a well calibrated Sodium Iodide NaI (Tl) detector enclosed in 5 cm thick lead shield to reduce background radiation. The spectra were analyzed and activity concentration of the detected radionuclide was computed directly and compared with the recommended standards.

2.3. *Measurement of Radionuclide Concentration*

Radioactivity measurements were done using a lead-shielded 76 mm × 76 mm NaI (TI) detector crystal coupled to a Canberra Series 10 plus Multichannel Analyzer (MCA) through a preamplifier. It has a resolution (FWHM) of about 8% at energy of 0.662 MeV (Cesium, ^{137}Cs) which is considered adequate to distinguish the gamma ray energies of interest. Each sealed sample was placed on the shielded NaI (TI) detector and counted for 18,000 s. This is to ensure that the photons emitted by the samples would only be sufficiently discriminated if their emission probability and their energy were high enough, and the surrounding background continuum low enough. The activity concentration of Bismuth, ^{214}Bi (determined from its 1.120 MeV and 0.609 MeV γ-ray peaks) were chosen to provide an estimate of Radium isotope, ^{226}Ra (Uranium, ^{238}U) in the samples, while that of the daughter radionuclide Actinium, ^{228}Ac (determined from its 0.911 MeV γ-ray peak) was chosen as an indicator of Thorium, ^{232}Th. Potassium, ^{40}K was determined by measuring the 1.460 MeV γ-rays emitted during its decay. The samples were placed symmetrically on top of the detector and measured for a period of 18, 000 s. The net area under the corresponding peaks in the energy spectrum was computed by subtracting counts due to Compton scattering of higher peaks and other background sources from the total area of the peaks. From the net area, the activity concentrations in the samples were obtained using the following equation:

$$C = \frac{A}{\varepsilon V s P \gamma t c} \quad (1)$$

where, A is the net area of the peak, ε is efficiency of the detector for radionuclide n, V_s is the volume of the water sample for measurement in liter, P_γ is gamma emission probability (or branch ratio), and t_c is the counting time [18, 24 and 25].

2.4. Evaluation of Radiological Hazard Parameters

In order to effectively evaluate the radiological impacts on the samples, there are eight parameters that were estimated using the gamma ray spectrometry results for ^{40}K, ^{238}U and ^{232}Th concentrations. These parameters are absorbed dose rate, annual effective dose, radium equivalent index, annual gonadal equivalent dose, external and internal hazard indices, gamma index and excess lifetime cancer risk.

2.4.1. Absorbed Dose Rate. The absorbed dose is a measure of the energy deposited in a medium by ionizing radiation per unit mass. It may be measured as joules per kilogram and represented by the equivalent S.I. unit, gray (Gy) or rad. The health risk due to radiological and clinical effects of radiation is directly related to the absorbed dose rate hence, estimation of this parameter is the first step to evaluating health risk [18, 19]. The unit of absorbed dose rate is nano Gray per hour (nGy/h). The absorbed dose rate D (nGy/h), due to activity concentration of ^{238}U, ^{232}Th and ^{40}K was calculated using;

$$D = C_U A_U + C_{Th} A_{Th} + C_K A_K \quad (2)$$

where A_U, A_{Th}, A_K are the radioactivity concentration in Bq/L and C_U, C_{Th}, and C_k are dose conversion factors which are 0.462, 0.604 and 0.0417 for ^{238}U, ^{232}Th and ^{40}K respectively. Average value is given as 57nGy/h [24, 25].

2.4.2. Annual Effective Dose (AED). Effective Dose is a dose quantity defined by the International Commission on Radiation Protection to monitor and control human exposure to ionizing radiation. It is the tissue-weighted sum of the equivalent doses in all specified tissues and organs of the body and represents the stochastic health risks to the whole body. It takes into account the type of radiation and the nature of each organ or tissue being irradiated, and enables summation of organ doses due to varying levels and types of radiation, both internal and external, to produce an overall calculated effective dose. This combines both internal and external exposures. The Annual Effective Dose (AED) is the sum of the effective dose over a year.

(a) Annual Effective Dose for Ingested Radionuclide ($AED_{\text{Internal Exposure}}$): The annual effective dose rate for all the ingested radionuclides from water was calculated using equation (3) below.

$$\mathbf{AED}_{\textbf{Internal Exposure}} = \sum_i I_i \times 365 \times D_i \quad (3)$$

where I_i is the daily intakes of radionuclide. Intake (Bq/d) = (concentration of radionuclide in food or water in Ci/lb or Bq/kg) × (consumption rate of water or food in lb/day or kg/day) and the ingestion dose coefficient (dose conversion factor) Di for adults for ^{40}K, ^{232}Th and ^{238}U is 6.2×10^{-9}, 2.3×10^{-7} and 4.5×10^{-8} Sv/Bq, respectively [10, 24]. The annual effective dose resulting from the ingestion of water was estimated based on the assumption that a daily intake of water per person is 2 l/d for adults and 1 l/d for lower ages and 0.5 l/d for infants [4, 27]. The Dose Conversion Factors for ingestion of radionuclides for members of the public for all ages provided by [10] is giving in Table 1.

Table 1: Dose Conversion Factors for ingestion of radionuclides for members of the public to 70 years of age [10].

Radionuclides	$T_{1/2}$ (years)	DCF (Sv/Bq) Other ages					
		Infants	1year	5years	10years	15years	Adults
^{40}K	$1.2x10^9$	$5.2x10^{-8}$	$4.2x10^{-8}$	$2.2x10^{-8}$	$1.3x10^{-8}$	$7.6x10^{-9}$	$6.2x10^{-9}$
^{232}Th	$1.405x10^1$	$1.6x10^{-6}$	$4.5x10^{-7}$	$3.5x10^{-7}$	$2.9x10^{-7}$	$2.5x10^{-7}$	$2.3x10^{-7}$
^{238}U	$4.468x10^9$	$1.4x10^{-7}$	$1.2x10^{-7}$	$8.0x10^{-8}$	$6.8x10^{-8}$	$6.7x10^{-8}$	$4.5x10^{-8}$

(b) Annual Effective Dose for External Exposures ($AED_{external\ Exposure}$): The annual effective dose received outdoor and indoor by a member of the public is calculated from the absorbed dose rate by applying dose conversion factor of 0.7Sv/Gy and occupancy factor for outdoor and indoor was 0.2 and 0.8 respectively [4]. AED is determined using the following equations [1, 4 and 11].

$$AED_{outdoor}\,(\mu Sv/y) = D(nGy/h)\times 8760h\times 0.7\ (Sv/Gy)\times 0.2\times 10^{-3} \quad (4)$$

$$AED_{indoor}\,(\mu Sv/y) = D(nGy/h)\times 8760h\times 0.7\ (Sv/Gy)\times 0.8\times 10^{-3} \quad (5)$$

The AED_{indoor} occurs within a house whereby the radiation risks due to use of the soil as building material is taken into consideration. $AED_{outdoor}$ involves a consideration of the absorbed dose emitted from radionuclide in the environment such as ^{238}U, ^{232}Th and ^{40}K.

2.4.3. Radium Equivalent Activity Index (Ra_{eq}). The radium equivalent (Ra_{eq}) activity represents a weighted sum of activities of ^{238}U, ^{232}Th and ^{40}K. It is based on the estimation that 1 Bq/kgof ^{238}U, 0.7 Bq/kg of ^{232}Th and 13 Bq/kg of ^{40}K produce the same radiation dose rates. This allows a single index or number to describe the gamma output from different mixtures of ^{238}U, ^{232}Th and ^{40}K in a material. It was calculated using the formula [11, 18]:

$$Ra_{eq} = C_U + 1.43C_{Th} + 0.077C_K \quad (6)$$

Where C_U, C_{Th}, C_K are the radioactivity concentration in Bq/Kg of ^{238}U, ^{232}Th and ^{40}K.

2.4.4. Radiation Hazard Indices. The external radiation hazard (H_{ext}) and the internal radiation hazard (H_{int}) was calculated as follows:

$$H_{ext} = \left(\frac{C_U}{370}\right) + \left(\frac{C_{Th}}{259}\right) + \left(\frac{C_K}{4810}\right) \quad (7)$$

$$H_{int} = \left(\frac{C_U}{185}\right) + \left(\frac{C_{Th}}{259}\right) + \left(\frac{C_K}{4810}\right) \quad (8)$$

H_{int} should be less than unity for the radiation hazard to be negligible. Internal exposure to radon is very hazardous which can lead to respiratory diseases like asthma [4]. Natural radionuclide in soil, sediment and rocks produce an external radiation field to which all humans are exposed. H_{ext} must be less than unity for this external radiation hazard to be negligible. H_{ext} equal to unity corresponds to the upper limit of radium equivalent dose (370 Bq/kg) [4, 6, 11, 17, 24, 25 and 27].

2.4.5. Excess Lifetime Cancer Risk (ELCR). The Excess Lifetime cancer risk (ELCR) was calculated using the following equation [4]:

$$ELCR = AED \times DL \times RF \quad (9)$$

Where, AED is the Annual Equivalent Dose Equivalent, DL is the average duration of life (estimated to 70 years), and RF is the Risk Factor (Sv^{-1}), i.e. fatal cancer risk per Sievert. For stochastic effects, ICRP uses RF as 0.05 for public [4]. Average value of ELCR is given as 0.2×10^{-3} [24, 25].

2.4.6. Annual Gonadal Equivalent Dose (AGED). The gonads, the bone marrow and the bone surface cells are considered as organs of interest by [24] because of their sensitivity to radiation. An increase in AGED has been known to affect the bone marrow, causing destruction of the red blood cells that are then replaced by white blood cells. This situation results in a blood cancer called leukemia which is fatal. The AGED for the resident using such material for building was evaluated by the following equation;

$$AGED\ (\mu Sv/y)\ C = 3.09C_U + 4.18C_{Th} + 0.314C_K \quad (10)$$

Where, C_U, C_{Th}, and C_K are the radioactivity concentration of ^{238}U, ^{232}Th and ^{40}K in soil and water samples.

2.4.7. Representative Gamma Index (I_γ). This is used to estimate the gamma radiation hazard associated with the natural radionuclide in specific investigated samples. The representative gamma index was estimated as follow [4]:

$$I_\gamma = C_U/150 + C_{Th}/100 + C_K/1500) \leq 1 \quad (11)$$

Where, C_U, C_{Th}, and C_K are the radioactivity concentration of ^{238}U, ^{232}Th and ^{40}K in soil and water samples. The I_γ is correlated with the annual dose rate due to the excess external gamma radiation caused by superficial material. An increase in the representative gamma index greater than the universal standard of unity may result in radiation risk leading to the deformation of human cells thereby causing cancer. Values of $I_\gamma = 1$ corresponds to an annual effective dose of less than or equal to 1 mSv, while $I_\gamma = 0.5$ corresponds to annual effective dose less or equal to 0.3 mSv [4, 11]. Thus, I_γ should be used only as a screening tool for identifying materials that might be of concern to be used as construction materials, though materials with $I_\gamma > 1$ should be avoided [18, 20] since these values correspond to dose rates higher than 1 mSv/y, which is the highest value of the dose rates recommended for humans [24].

3. Results and Discussions

The result of the gamma ray spectrometry of the water samples is presented in Table 2. The radionuclide observed with reliable regularity belonged to the decay series chain headed by Uranium, ^{238}U and Thorium, ^{232}Th as well as the non-series Potassium, ^{40}K. The ^{40}K activity concentration dominated over the ^{238}U and ^{232}Th elemental activities as expected. The activity concentration of ^{40}K ranges between 25.42±9.65 and 93.95±32.57 Bql^{-1} with a mean of 70.13667 Bql^{-1} in the water samples. The activity concentration for Uranium, ^{238}U in the water samples ranged between 41.78±12.47 and 66.94±18.72 Bql^{-1} with a mean of 55.41444 Bql^{-1}. For Thorium, ^{232}Th the activity concentration ranged between 6.64 ± 2.94 and 56.87 ± 61.94 Bql^{-1} with a mean of 35.68778 Bql^{-1}(see Table 2 and Fig. 2). The activity concentration of ^{40}K, ^{238}U and ^{232}Th in all the samples exceed the world safe limit of 10.0, 10.0 and 1.0 Bql^{-1} respectively for water samples by [24] and [26] standards and hence poses a serious health effect.

Table 2: Activity concentration of ^{40}K, ^{238}U and ^{232}Th in the water samples.

Code	K-40 (Bql^{-1})	Th-232 (Bql^{-1})	U-238 (Bql^{-1})
kw1	25.42	12.18	63.66
kw2	77.86	51.01	44.92
kw3	37.21	31.63	43.97
kw4	91.86	33.91	64.82
kw5	85.60	50.25	41.78
kw6	47.28	34.54	66.94
kw7	84.49	44.16	46.36
kw8	93.95	6.64	65.61
kw9	87.56	56.87	60.67
Min	25.42	6.64	41.78
Max	93.95	56.87	66.94
Mean	70.14	35.69	55.41
Median	84.49	34.54	60.67
Stdv	26.12	17.26	10.78

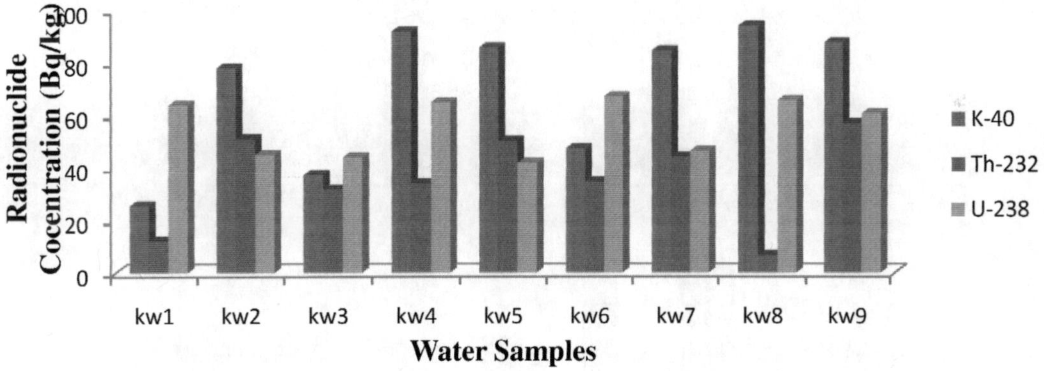

Figure 2: Activity Concentration of ^{238}U, ^{232}Th and ^{40}K in the selected water samples from Hadejia River.

The mean Absorbed Dose Rate for the water samples is 50.105nGyh^{-1} (see Table 3). The mean Absorbed Dose Rate for the water is slightly below the general average value of 57nGyh^{-1} [24]. The mean AEDE for the ingested radionuclide in drinking water from the area for infants, 1year, 5years, 10years, 15years and adults is 0.012458, 0.009332, 0.006734, 0.005481, 0.0048 and 0.00807mSvy^{-1} respectively (see Table 3 and Fig. 3). These average values of AEDE for the different age groups are within the acceptable safe limits of 0.1 mSv/year, for the general public. Although it should be noted that daily intake of water per person of 2 ld^{-1} for adults, 1 ld^{-1} for lower ages and 0.5 ld^{-1} for infants was used as the intake of water varies among these age groups. From Fig 3, it is clear that the infants as expected are more susceptible to radiation hazards in the area followed by 1year then adults, 5years, 10years and then 15years respectively. They are all within the safe limits of 0.1 $mSvy^{-1}$.

The mean Ra_{eq} for the water samples is 109.290Bql^{-1}. The estimated average value is lower than the recommended maximum value of 370 Bql^{-1} for the safe use of materials in the construction of buildings. This means that the water from this area can safely be used for building and other purposes without much fear of radiological hazards.

Table 3: AEDE for different age categories for the people in the study area.

SAMPLE	AEDE ($mSvy^{-1}$)					
	ADULTS	INFANTS	1YEAR	5YEARS	10YEARS	15YEARS
KW1	0.00425	0.00542	0.00517	0.00361	0.00298	0.00273
KW2	0.01000	0.01670	0.01150	0.00845	0.00688	0.00596
KW3	0.00692	0.01070	0.00769	0.00562	0.00461	0.00406
KW4	0.00824	0.01240	0.00981	0.00696	0.00563	0.00493
KW5	0.01010	0.01650	0.01130	0.00832	0.00676	0.00584
KW6	0.00821	0.01220	0.00932	0.00674	0.00554	0.00491
KW7	0.00932	0.01480	0.01050	0.00767	0.00662	0.00539
KW8	0.00369	0.00450	0.00540	0.00351	0.00277	0.00247
KW9	0.01190	0.01890	0.01330	0.00973	0.00794	0.00691
MEAN	0.00807	0.012458	0.009332	0.006734	0.005481	0.00480

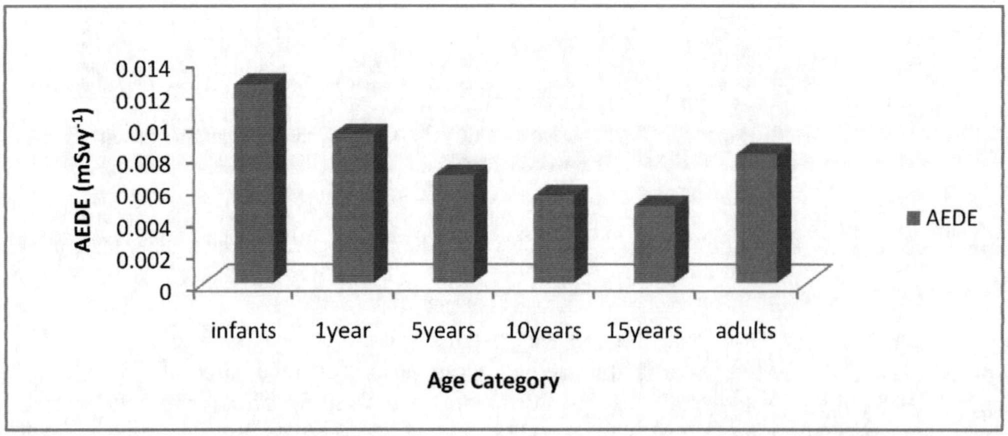

Figure 3: AEDE ($mSvy^{-1}$) for different age category/group of people in the study area

Table 4: Radiation hazard parameters for the water samples

SAMPLE CODE	D ($nGyh^{-1}$)	AED (ADULTS) (μSvy^{-1})	RA_{eq} (Bql^{-1})	H_{ext}	H_{int}	ELCR	AGED (μvy^{-1})	I γ
KW1	37.835	0.00425	83.034	0.224	0.396	0.0665	255.604	0.563
KW2	54.833	0.01000	123.859	0.334	0.456	0.1225	376.473	0.861
KW3	40.982	0.00692	92.066	0.249	0.368	0.0875	279.764	0.634
KW4	54.287	0.00824	120.384	0.325	0.500	0.1050	370.882	0.832
KW5	53.248	0.01010	120.229	0.325	0.438	0.1155	366.023	0.838
KW6	53.774	0.00821	119.973	0.323	0.505	0.1050	366.072	0.823
KW7	51.661	0.00932	116.015	0.314	0.440	0.1085	354.371	0.807
KW8	38.269	0.00369	82.339	0.223	0.401	0.0560	259.990	0.566
KW9	66.057	0.01190	125.712	0.402	0.566	0.1440	452.681	1.031
MIN	37.835	0.00369	82.339	0.223	0.368	0.0560	255.604	0.563
MAX	66.057	0.01190	125.712	0.402	0.566	0.1440	452.681	1.031
MEAN	50.105	0.00807	109.290	0.302	0.452	0.1012	342.429	0.773

The estimated hazard indices H_{int} and H_{ext} for the water samples are 0.452 and 0.302 respectively. (See Table 4). These values of H_{int} and H_{ext} in water are less than unity which follows that hazardous effects of these radionuclides and their short-lived progenies are negligible. The estimated ELCR for the water samples is 0.1012 (1.012×10^{-3}) (See Table 4). The estimated value of ELCR is higher than the average value of 0.2×10^{-3}. The high value of the ELCR index for the water samples is due to high AEDE caused by high specific activity of ^{40}K radionuclide in the samples. This high value implies that the probability of developing cancer over a lifetime considering seventy years as the average life span of humans is high.The estimated mean value for the AGED in the water samples is 342.429 μSvy^{-1}. The estimated mean Iγ for the water samples is $0.773 Svy^{-1}$. These values are within the safe limits.

4. Conclusion

A radiological study of the water from Hadejia river have been carried out by the estimation of the eight radiological parameters using data obtained from the gamma ray spectrometry of the samples drawn from the river. The results showed trends that are generally low for most radiation hazard indices calculated except for few indices whose values are above the UNSCEAR recommended thresholds. Therefore, there may be no serious immediate radiological effects to the general populace as a result of consuming water from the river. However, considering the fact that artificial radionuclides are synthetic radionuclide produced by human activities like high use of fertilizers and agrochemicals, it is hereby recommended that all activities that enhance or introduce radio-nuclides to the environment be discouraged or prohibited within the vicinity of the river in order to avoid any possibilities of radiological hazards.

References

[1] Adagunodo T.A., George A.I., Ojoawo I.A., Ojesanmi K. and Ravisankar R. (2018). Radioactivity and Radiological Hazards from a Kaolin Mining Field in Ifonyintedo, Nigeria. MethodsX, 5C: 362 – 374. https://doi.org/10.1016/j.mex.2018.04.009.

[2] Adagunodo T.A., Sunmonu L.A., Adabanija M.A., Suleiman E.A., Odetunmibi O.A. (2017). Geoexploration of Radioelement's Datasets in a Flood Plain of Crystalline Bedrock. Data Brief, 15C: 809 – 820. http://dx.doi.org/10.1016/j.dib.2017.10.046.

[3] Akinyose F. C., Tchokossa P., Orosun M. M., Oluyide S. O., Olatunji O., and Martins G. (2018). A Study of Natural Radioactivity and Gamma Radiation Hazard in Tobacco Leaves and Cigarettes in Oyo State, Nigeria. Manila Journal of Science **11** (2018), pp. 104-114

[4] Avwiri G.O, Ononugbo C.P, Nwokeoji I.E. (2014): radiation hazard indices and excess lifetime cancer risk in soil, sediment and water around mini-okoro/oginigba creek, port harcourt, rivers state, nigeria. "Comprehensive Journal of Environment and Earth Sciences" Vol. 3(1), pp. 38-50 Dec. 2014 ISSN-2315-7488

[5] Avwiri, G. O., Owate, I. O. and Enyinna, P. I. (2005): "Radionuclide Concentrations Survey of Soil, Sediments and Water in Aba River, Abia State, Nigeria". Scientia Africana, 4 (1&2): 67-72.

[6] Farai I. P. and Isinkaye M. O. (2009): Radiological Safety Assessment of Surface Water Dam Sediments Used as Building Materials in Southwestern Nigeria, Journal of Radiological Protection, Vol. 29, No. 1, 2009, pp. 85-93. doi:10.1088/0952-4746/29/1/006

[7] Goes B. J. M. (2001): Effects of damming the Hadejia River in semiarid northern Nigeria—lessons learnt for future management Regional Management of Water Resources (Proceedings of a symposium held during the Sixth IAHS Scientific Assembly at Maastricht, The Netherlands, July 2001). IAHS Publ. no. 268, 2001.

[8] Hasan, M. M., Ali, M. I., Paul, D., Haydar, M. A. and Islam, S. M. A. (2014): "Natural Radioactivity and Assessment of Associated Radiation Hazards in Soil and Water Samples Collected from in and around of the Barapukuria 2×125 MW Coal Fired Thermal Power Plant, Dinajpur, Bangladesh." Journal of Nuclear and Particle Physics, 4(1): 17-24.

[9] ICRP (1991): "International Commission on Radiation Units and Measurements". International Commission on Radiological Protection(ICRP)Report No.60. New York, Pergamon Press, 1991.

[10] ICRP (2012): ICRP Compendium of dose coefficients based on ICRP publication 60. Annals of the ICRP, Vol. 41 supplement 1, 2012.

[11] Issa S.A.M., Uosif, M.A.M. and Elsaman R. (2013). "Gamma radioactivity measurements in Nile River sediment samples".Turkish J Eng Env Sci (2013) 37: 109 – 122. doi:10.3906/muh-1207-21.

[12] Jibiri, N. A. O., Mbawonku, A. O., Oridale, A. A. and Ujiagbadion, C. (1999): "Natural Radionuclide Concentration Levels in Soil and Water Around Cement Factory, Ewekoro, Ogun State". Nig. Journal of Physics, II: 12 – 16.

[13] Najat, K. M. and Mohamed S. M. (2013): "Natural Radioactivity in Soil and Water from Likuyu Village in the Neighborhood of Mkuju Uranium Deposit".International Journal of Analytical Chemistry.Article ID 501856, 4 pages.

[14] NGSA (2004): Geological and Minerals Maps of Kano and Jigawa States. Nigerian Geological Survey Agency 1: 100,000, 2004 Edition.Nigeria.

[15] Nwankwo. L. I. (2013): Determination of Natural Radioactivity in Groundwaterin Tanke-Ilorin, Nigeria. *West African Journal of Applied Ecology,* vol. 21(1), 2013

[16] Oladejo O.P., Sunmonu L.A., Ojoawo A., Adagunodo T.A. and Olafisoye E.R. (2013). Geophysical Investigation for Groundwater Development at Oyo State Housing Estate Ogbomosho, Southwestern Nigeria. *Research Journal of Applied Sciences, Engineering and Technology*, 5(5), 1811–1815.

[17] Ononugbo CP, Avwiri GO, Egieya JM (2013): Evaluation Of Natural Radionuclide Content in Surface and Ground Water and Excess Lifetime Cancer Risk Due To Gamma Radioactivity. Academic Research International. 4(6): 636-647.

[18] Orosun M. M., Lawal T. O., and Fisayo A. C. (2016): Natural Radionuclide Concentrations and Radiological Impact Assessment of Soil and Water in Tanke-Ilorin, Nigeria. *Zimbabwe Journal of Science & Technology* pp 158- 172 Vol. 11 [2016] e-ISSN 2409-0360

[19] Ramasamy, V., Suresh, G., Meenakshisundaram, V. and Ponnusamy, V. (2011): Horizontal and Vertical Characterization of Radionuclides and Minerals in River Sediments", Applied Radiation and Isotopes, 69, 184-195, 2011.

[20] Ravisankar, R.,Vanasundari, K., Chandrasekaran, A., Rajalakshmi, A., Suganya, M., Vijayagopal, P. and Meenakshisundaram, V. (2012) "Measurement of Natural Radioactivity in Building Materials of Namakkal, Tamil Nadu, India Using Gamma-Ray Spectrometry", Applied Radiation and Isotopes, 70, 699–704, 2012.

[21] Sobowale A.1, Adewumi J. K., Otun J. A. and Adie D.B. (2010): "Water Resources Potentials of Hadejia River Sub-Catchment of Komadugu Yobe River Basin, Nigeria" Agricultural Engineering International: the CIGR Ejournal. Manuscript 1065. Vol. XII, April, 2010.

[22] The Taskin H., Karavus, M., Ay P., Topuzoglu A., Hindiroglu S. and Karahan, G. (2009). "Radionuclide Concentrations inSoil and Lifetime Cancer Risk Due to the Gamma Radioactivity in Kirklareli, Turkey", Journal of Environmental Radioactivity, 100; 49-53.

[23] Tchokossa, P., Olomo J.B. and Balogun F.A. (2011): "Assessment of Radionuclide Concentrations and Absorbed Dose from Consumption of Community Water Supplies in Oil and Gas producing Areas in Delta State, Nigeria". World Journal of Nuclear Science and Technology, 77-86.

[24] UNSCEAR (2000): United Nations Scientific Committee on the Effects of Atomic Radiation (UNSCEAR). Sources and effects of ionizing radiation. UNSCEAR 2000, Report to the General Assembly, with Scientific Annexes. Volume I: Sources. New York: United Nations (2000).

[25] UNSCEAR (2008): United Nations Scientific Committee on Effect of Atomic Radiation. Sources and Effect of Ionising Radiation . Report vol.1 to the General Assembly with scientific Annexes: United Nations Sales Publications,United Nations, New York.

[26] WHO (2008): World Health Organisation. Guidelines for Drinking Water Quality in Cooperating first Addendum vol.1, Recommendations, third Edition. Radiological Aspect Geneva: World Health Organiztion. WHO Geneva.

[27] WHO (2011): World Health Organisation Guidelines for Drinking Water Quality. Fourth edition incorporating the 1st and 2nd Agenda vol.1 Recommendations; WHO Geneva. WHO, World Health Organisation (2008). Guidelines for Drinking Water Quality in Cooperating first Addendum vol.1, Recommendations, third Edition. Radiological Aspect Geneva: World Health Organiztion.WHO Geneva.

Natural Radioactivity Concentration and Its Health Implication on Dwellers in Selected Locations of Ota

M. R. Usikalu, C. A. Onumejor, A. Akinpelu, J. A. Achuka, M. Omeje and O. F. Oladapo
Department of Physics, Covenant University, P.M.B 1023, Ota, Ogun State

Correspondence email: moji.usikalu@covenantuniversity.edu.ng

Abstract. Elevated background ionizing radiation has its health effects on people who reside in such areas, this necessitate the need for constant monitoring. The activity concentrations of K-40, Th-232 and U-238 were measured in three different selected study areas in Ota using RS-230 gamma spectrometer. The highest activity concentrations for the three radionuclides were recorded in the Industrial Estate. The mean dose rate recorded was 45.37 $nGyh^{-1}$, 37.12 $nGyh^{-1}$ and 33.33 $nGyh^{-1}$ for Industrial Estate, Obasanjo Estate and Atan respectively. The mean outdoor annual equivalent dose estimated was 0.056 $mSvy^{-1}$, 0.045 $mSvy^{-1}$ and 0.041 $mSvy^{-1}$ respectively for the three locations. The estimated excess lifetime cancer risk ranged from 0.14 x 10^{-3} – 0.277 x 10^{-3} for the study areas. The radiological variables estimated in this study were all within world average recommended limit. The study concluded that the locations considered in this work are safe for dwellers and that industrial activity has influence on the background radiation.
Keywords: Radioactivity, Radiation dose, Health implication, Ota

1. Introduction

Naturally occurring radionuclides in the environment has its health implications on humans that reside in areas with high concentrations. Natural radionuclides are part of the earth's original crust (primordial radionuclides), they also occur through the interaction of cosmic rays (cosmogenic radionuclides) and human activities such as industrial, research and domestic activity (artificial or anthropogenic radionuclide) [1, 2]. Cosmogenic and Primordial radionuclides naturally found in the environment are over sixty in number, including terrestrial background radiation from nucleo-synthesis of stars. A good example of some common primordial radionuclides that forms the major background radiation and activity concentrations in our environment are the non-serial potassium-40, the series uranium-238, Thorium-232 and their progenies. Some industrial activities make use of radionuclides as part of technological processes for productions or service deliveries, quality control and lots more, this increase radiation burden above the background level in the environment. Radioactivity in the environment is one of the major sources of exposure to human. Exposure to radiation found in the environment is inevitable [1, 2, 3]. The soil and parent rocks could give out radiation because they possess the ability to retain and accumulate radionuclides of primordial origin. As the retained radionuclides in the soil/parent rocks decays, it releases radiation to the environment and serves as a continuous exposure source to human [4]. According to [5], most of the radioactivity concentrations in the environment come from igneous rock like granite, followed by the sedimentary rocks, while the least contribution comes from phosphate rocks and some shale. The natural and artificial radiation sources add up to form background radiation that gets to human on daily basis. If the background radiation in an environment is higher than the set limit it can lead to adverse radiological effects [3]. The knowledge of radiation level in an area is very important for effective radiological impact assessment from both cosmogenic and terrestrial exposure sources. The surest way to know if an environment is free of radiological hazard is to monitor and measure the radiation levels that can be subjected to radiological evaluations using the required parameters. Health risks related to natural radioactivity are of great concern and require assessment in order to estimate the risks. Assessment of radionuclides concentrations in water, rocks and soils in different parts of the world and in Nigeria have been on for several decades and it is on the increase due to increased human activities, town and city expansion. Developmental activities of human have the tendency to also increase the exposure rate, including the health risk associated to radiation exposure. Some of the indigenous studies on radiological assessment include the work of [1, 6-12] to mention a few. The need to monitor our environment at interval for possible radiation risk studies cannot be over emphasized, especially in growing regions and industrial areas where the detailed production activities of the industries are not known. It is paramount to monitor the radiation dose in order to evaluate the radiological implications of the radionuclides on the dwellers in these areas. Hence, the present study measured the activity levels three promodial radionuclides and and evaluated their health implications to those in the areas.

2. Material and Methods

Ota is one of the most popular towns in Ogun State, western Nigeria. It has population estimate of 163,783 dwellers living around and inside the town [13]. Ota is the capital of the Awori Yoruba ethnic group and also the local government head-quarter of Ado-Odo/Ota local Government Area. As of 1999, it has the third largest concentration of industries in Nigeria. Ota is also well known as the home of former Nigerian president Olusegun Obasanjo's farm, the Canaanland compound of the mega church Winners' Chapel, and the Africa Leadership Forum. The main occupation of Ota dwellers is trading and farming. Three locations in Ota town were selected for the study to represent Industrial, Mixed (mixture of residential and industrial) and residential areas. Figure 1 shows the map of Ota town with the study areas highlighted. Selected area name and code are Industrial Estate-IE, Obasanjo Estate-OE, and Atan-AT, representing industrial, Mixed and residential areas respectively.

RS 230 BGO Super Spec gamma ray spectrometer was used for the measurement. The performance level of the BGO (Bismuth Germanate Oxide) detector in the spectrometry used is 3 times higher than the performance of the common sodium iodide (NAI) crystal used in larger portable units. The spectrometer is auto-stabilizing on naturally occurring (40-K, 232-Th and 238-U) radioactivity and does not require any test sample or sources. This portable hand held radiation survey spectrometer was held at about 1m above the ground at each time of measurement. In order to measure the activity concentration, the features of the device was taken to assay mode of operation. At each study area, five (5) locations were selected five (5) measurements were taken at each sampling point. Once the setup is completed, a five (5) minutes waiting was observed to allow for auto-stabilizing of the device on naturally radio element of the study area background. The activity concentration of K-40, Th-232 and U-238 were recorded as displayed on the radiation detector along with the dose rate DR_A. This allows a readout of K in percentage (%), Th and U in part per million (ppm). This was converted to $Bqkg^{-1}$ using the standard conversion factor.

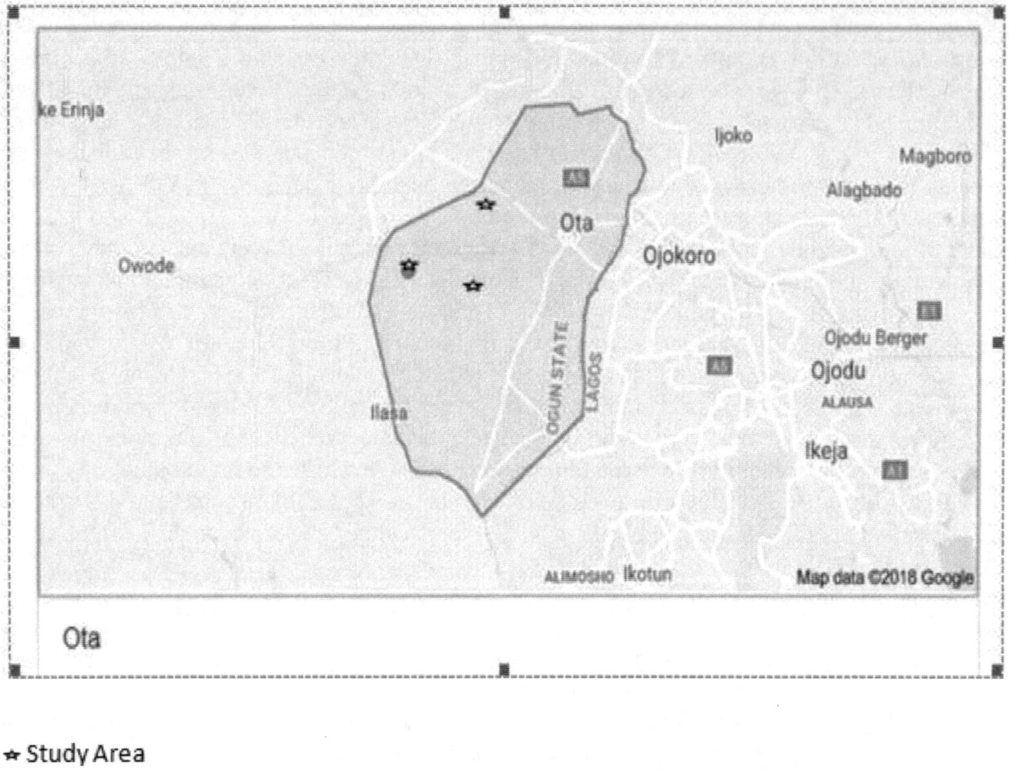

Figure 1 Map of study area

2.1. Dose calculations

2.1.1 Annual outdoor effective dose rate

The radiation dose rate that gets to the members of the public within an exposed area is referred to as outdoor effective dose rate (AEDR). It can be calculated using two main factors; one of the factors is used to convert DR_A ($nGyh^{-1}$) in air to biological outdoor effective dose rates (Svy^{-1}), popularly referred to as outdoor occupancy factor given as 0.2, while the other factor consider the total exposure time spent by an individual within the outdoor radiation area [14-15]. The AEDR in this current study was calculated using equation 1

$$AEDR = T * f * Q * D_A * \varepsilon \qquad 1$$

where T is the time given as 8766h/y, f is the outdoor occupancy factor given as 0.2, Q is the quotient of the effective dose rate given as 0.7(SvG/y), D_A is the absorbed dose rate in air (nGyh-1) and ε is a factor converting nano (10^{-9}) into the micro (10^{-6}).

2.1.2 Excess lifetime cancer risk ELCR

ELCR is defined as the probability that an individual will develop cancer over his lifetime of exposure to radiation and it is given as equation 2

$$ELCR = AEDR \times DL \times RF \qquad 2$$

where DL is the average lifespan (70 years) and RF is risk factor (Sv^{-1}) which is 0.057, for stochastic effects from low-dose background radiation [17].

3. Results and discussions

The range, mean activity concentration measured from the study areas is presented in Table 1. At the Industrial Estate the mean activity concentration ranged from 15.56 $Bqkg^{-1}$ to 23.71 $Bqkg^{-1}$, 40.92 $Bqkg^{-1}$ to 52.86 $Bqkg^{-1}$ and 62.60 $Bqkg^{-1}$ to 350.56 $Bqkg^{-1}$ for U-238, 232-Th and K-40 respectively. At the Obasanjo Estate (OE) the mean activity concentration in $Bqkg^{-1}$ ranged between 15.56 to 19.51, 30.86 to 48.88, and 31.31 to 93.90 for U-238, 232-Th and K-40 respectively. In Atan area the range for the mean activity concentration of U-238, Th-232 and K-40 in $Bqkg^{-1}$ were 12.60-21.24, 30.61-43.06 and 25.04-131.46 respectively. The highest mean activity concentration and radiation dose rate was recorded in Industrial Estate, while Atan had the least concentration and radiation dose. The elevation in the activity measured Industrial Estate may be due to contribution from anthropogenic sources due to several companies sited in the area. K-40 has the highest activity concentration but 232-Th has the highest contribution to the radiological implication on health of the dwellers in the three study locations. The mean activity concentration measured in all the locations are within the recommended limit except for mean concemtration of 232-Th in the Industrial Estate that exceeded the world average value of 45 $Bqkg^{-1}$ in soil and its environs [15, 16]. Figure 2 is the spatial variation of the three radionuclides in the study areas. The dose rate (DR) world average limit is given as 59 $nGyh^{-1}$ [15], the dose rate measured was found to be within the recommended limit for all the study locations. The radiological parameters estimated, the outdoor annual equivalent dose rate (AEDR) and excess lifetime cancer risk (ELCR) for the study areas are presented on Table 2 and Figures 3 and 4 are pictorial representation of the variation of the parameters within the study area. The mean AEDR estimated in $mSvy^{-1}$ is 0.056, 0.045 and 0.041 for Industrial Estate, Obasanjo Estate and Atan respectively. The mean AEDR obatined for all the locations are lower than the recommended limit of 0.08 $mSvy^{-1}$ [15]. The mean ELCR obtained for the study areas was 0.22 x 10^{-3}, 0.18 x 10^{-3} and 0.16 x 10^{-3} for the Industrial Estate, Obasanjo Estate and Atan respectively which is also lower than the world average limit of 0.29 x 10^{-3}. This implies that dwellers in these areas are safe for now but these values can increase with time as the industrial activities increase in the area.

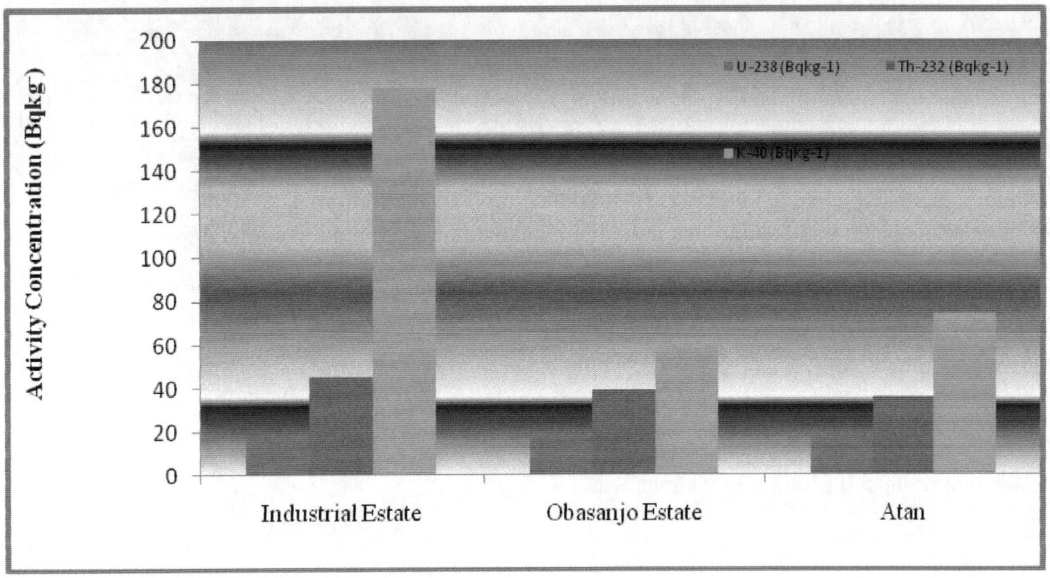

Figure 2 Spatial Variation of the mean activity concentration for the three locations

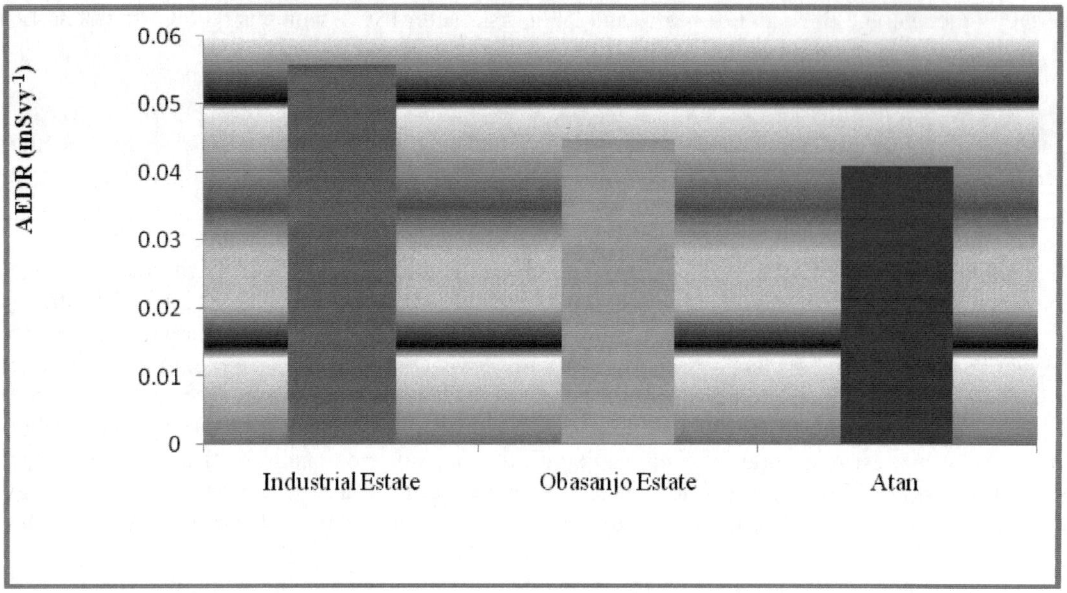

Figure 3 Mean absorbed dose rate

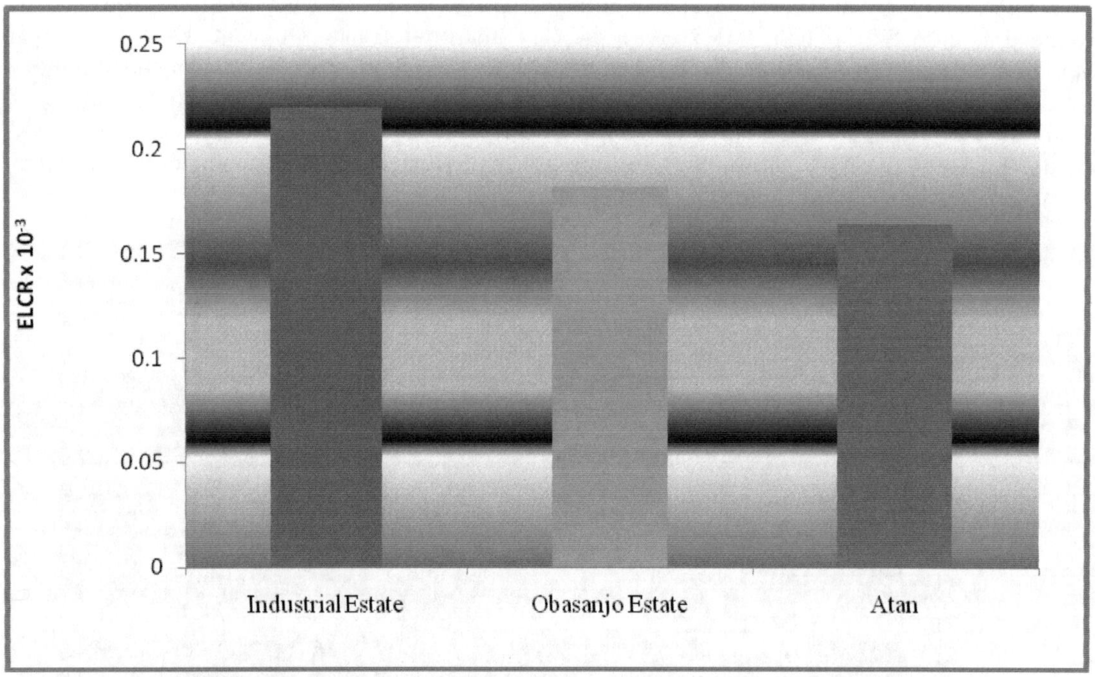

Figure 4 Mean estimated lifetimedose cancer risk

Table 1 Measured activity concentration

IOP Conf. Series: Earth and Environmental Science **173** (2018) 012037 doi:10.1088/1755-1315/173/1/012037

Sampling Areas	Range Mean	U-238 ($Bqkg^{-1}$)	Th-232 ($Bqkg^{-1}$)	K-40 ($Bqkg^{-1}$)
IE	Range	17.29-35.82	30.86-53.59	62.60-93.90
	Mean	23.71	40.92	93.90
	Range	8.65-29.64	32.07-51.16	0.00-93.90
	Mean	20.50	41.41	62.60
	Range	0.00-30.88	46.28-58.87	187.80-532.1
	Mean	15.56	52.86	350.56
	Range	7.41-24.70	32.48-58.87	62.60-156.50
	Mean	15.81	47.97	100.16
	Range	2.47-33.35	33.70-48.72	219.10-375.60
	Mean	17.29	42.31	281.7
OE	Range	3.71-33.35	41.41-54.00	31.30-125.20
	Mean	19.51	48.88	93.90
	Range	3.71-45.70	19.89-39.38	0.00-93.90
	Mean	18.03	31.42	50.08
	Range	4.94-37.05	24.36-36.95	0.00-187.8
	Mean	19.51	30.86	93.90
	Range	12.35-41.99	29.23-47.10	0.00-31.30
	Mean	19.51	39.22	31.31
	Range	6.18-23.47	31.26-57.65	0.00-93.90
	Mean	15.56	45.55	43.82
Atan	Range	0.00-34.00	22.74-92.30	0.00-125.20
	Mean	15.31	43.06	56.34
	Range	7.41-34.58	23.14-41.82	0.00-93.90
	Mean	20.50	30.61	25.04
	Range	8.65-43.23	25.58-47.50	93.90-187.80
	Mean	19.27	38.49	131.46
	Range	0.00-23.47	28.01-39.79	31.30-62.60
	Mean	12.60	33.05	43.82
	Range	2.47-33.36	28.014-41.82	0.00-187.80
	Mean	21.24	35.00	112.68

Table 2 Estimated radiological parameters

IOP Conf. Series: Earth and Environmental Science **173** (2018) 012037 doi:10.1088/1755-1315/173/1/012037

Sampling Areas	Range Mean	DR ($nGyh^{-1}$)	AEDR (Outdoor) $mSvy^{-1}$	ELCR x 10^{-3}
IE	Range	39.10-41.10	0.048-0.050	
	Mean	41.24	0.051	0.2019
	Range	33.70-43.80	0.041-0.054	
	Mean	39.12	0.048	0.1916
	Range	51.90-56.00	0.064-0.069	
	Mean	56.50	0.069	0.2767
	Range	33.70-48.50	0.050-0.056	
	Mean	42.64	0.052	0.2088
	Range	40.70-56.10	0.041-0.670	
	Mean	47.36	0.058	0.2319
OE	Range	41.80-46.70	0.051-0.057	
	Mean	44.98	0.055	0.2203
	Range	24.40-40.10	0.030-0.049	
	Mean	30.76	0.038	0.1506
	Range	32.60-35.90	0.040-0.044	
	Mean	32.84	0.040	0.1608
	Range	31.00-38.10	0.038-0.047	
	Mean	36.16	0.044	0.1771
	Range	31.00-39.90	0.038-0.049	
	Mean	40.88	0.050	0.2002
Atan	Range	28.40-38.00	0.029-0.046	
	Mean	30.28	0.037	0.1483
	Range	22.50-37.60	0.028-0.046	
	Mean	30.46	0.037	0.1492
	Range	34.00-37.00	0.042-0.045	
	Mean	39.16	0.048	0.1918
	Range	26.40-29.80	0.032-0.037	
	Mean	29.42	0.036	0.1441
	Range	32.80-39.20	0.040-0.048	
	Mean	37.34	0.046	0.1828

4. Conclusion

The in-situ activity concentrations of U-238, Th-232 and K-40 and dose rate were measured in Industral Estate, Obasanjo Estate and Atan using RS-230 Super Spec gamma ray spectrometer. The measured mean values of U-238, Th-232 and K-40 are all within the world average recommended limits except for concentration of Th-232 that exceeded the recommneded in Industrial Estate. Industrial area recorded the highest mean concentration of K-40, U-238, Th-232 and dose rate values. Obansajo area, which is a mixture of industrial and domestic areas have the next, while Atan area being residential zone recorded the least activity concentration. The same trend was observed in the estimated radiological health parameters. Industrial Estate recorded the highest values of annual equivalent dose rate and excess lifetime cancer risk. It is radiological safe for the dwellers in these study areas for now but this can increase as the industrial activities increase in the area which can elevate the radiation burden of the dwellers. Thus, it can be concluded that industrial activities have influence on the background radiation, activity concentration, concentration of NORM (naturally occurring radioactive materials) and radiation dose rate in an environment.

Acknowledgment
The authors thank Covenant University for conference support

References

[1] S.O Felix, C.A Onumejor , A. Akinlua , O. K. Owoade "Geochemistry and health burden of radionuclides and trace metals in shale samples from the North-Western Niger Delta,". J Radioanal Nucl Chem (2013) 295:871–881
[2] USEPA (2007) EPA "Occupational Health Effects of Uranium Mining, United States Environmental Protection Agency," http://www. epa.for/radiation/ternorm
[3] M. R Eyebiokin, A. M Arogunjo, A. B Rabiu, G. Oboh and F. A Balogun "Activity concentration of commonly consumed vegetables in Ondo State," (2005) *Nigerian Journal of Physics*, **17S**, 187-191
[4] J.B Olomo "The invisible tool," An inaugural lecture delivered at Oduduwa Hall, OAU Ile-Ife, Nigeria on Tuesday 27th June 2006. ISSN0189-7848
[5] A. S Alaamer "Assessmenet of human exposures to natural resources of radiation in soil of Riyadh, Saudi Arabia," (2008) *Turkish J. Eng. Env Sci.* 32: 229-234
[6] S. K Alausa "Radiological Assessment of Soils on the Waysides of the Road Underconstruction in Ijebu-Ode, Ogun State, Southwestern Nigeria" (2014). *Journal of Natural Sciences Research,* **l4 (15):** 80-84.
[7] M.R Usikalu, P. P. Maleka, M. Malik, K.D. Oyeyemi , O.O. Adewoyin "Assessment of geogenic natural radionuclide contents of soil samples collected from Ogun State, South western, Nigeria," (2015) *International Journal of Radiation Research*, 13(4): 355-361
[8] M.R. Usikalu, I. A. Fuwape, S. S. Jatto, O. F. Awe, A. B. Rabiu and J. A. Achuka "Assessment of radiological parameters of soil in Kogi State, Nigeria," (2017). *Environmental Forensics*, **18(1):** 1–14
[9] M. R. Usikalu, A. B Rabiu, K. D Oyeyemi, J. A Achuka and M. Maaza "Radiation hazard in soil from Ajaokuta North-central Nigeria", (2017) *International Journal of Radiation Research*, **15(2):** 119-224
[10] Adagunodo T.A., Sunmonu L.A., Adabanija M.A., Suleiman E.A., Odetunmibi O.A. (2017). Geoexploration of Radioelement's Datasets in a Flood Plain of Crystalline Bedrock. Data in Brief, 15C: 809 – 820. http://dx.doi.org/10.1016/j.dib.2017.10.046.
[11] Adagunodo T.A., Hammed O.S., Usikalu M.R., Ayara W.A., Ravisankar R. (2018). Data on the Radiometric Survey over a Kaolinitic Terrain in Dahomey Basin, Nigeria. Data in Brief, 18C: 814 – 822. https://doi.org/10.1016/j.dib.2018.03.088.
[12] Adagunodo T.A., George A.I., Ojoawo I.A., Ojesanmi K. and Ravisankar R. (2018). Radioactivity and Radiological Hazards from a Kaolin Mining Field in Ifonyintedo, Nigeria. MethodsX, 5C: 362 – 374. https://doi.org/10.1016/j.mex.2018.04.009.
[13] A.J Ruhollah "Ota: The Biography of the Foremost Awori Town," (1999). Penink & Co. 16
[14] N.N Jibiri and O.S Bankole "Soil radioactivity and radiation absorbed dose rates at roadsides in high-traffic density area in Ibadan Metropolis, southwestern, Nigeria," (2006) *Radiation Protection Dosimetry*, 118: 453 – 458

[15] UNSCEAR "Source, effect and risk of ionizing radiation," (2000) New York, United Nations
[16] UNSCEAR "United Nations Scientific Committee on the Effects of Atomic Radiation Report to the General Assembly on the Sources and Effects of Ionizing Radiation. Report to the General Assembly with Scientific Annexes." (2008) United Nations, New York Vol. 1. New York: United Nations;
[17] ICRP. "The 2007 Recommendations of the International Commission on Radiological Protection" (2007). Annals of the ICRP Publication 103 .2-4. Elsevier.

Cancer risks from head radiography procedures

Achuka J. A.[a*], Aweda M. A.[b], Usikalu M. R.[a]

[a]Department of Physics, Covenant University Ota, Ogun State, Nigeria
[b]Department of Radiation Biology, Radiotherapy, Radiodiagnosis and Radiography, College of Medicine, Lagos University Teaching Hospital, Idi-Araba, Lagos, Nigeria

justina.achuka@covenantuniversity.edu.ng

Abstract. The goal of this study is to evaluate the risk of cancer induction in head radiography procedures with a view to promote dose optimization and enhance patient safety. Thermoluminescent dosimeter (TLD 100) was used to determine the entrance surface dose (ESD) of 20 patients presented for head radiography in two tertiary healthcare institutions in Southwest Nigeria. The corresponding effective dose and doses to the brain, oral mucosa and salivary gland were evaluated using PCXMC software. Incidence cancer risks were evaluated using BEIR VII model. The total entrance surface dose (ESDT) for mandible, paranasal sinuses and skull radiography ranged between 3.01-19.12 mGy with a mean of 7.52 mGy. The resulting effective dose, brain dose, oral mucosa dose and salivary gland dose has a mean of 0.25 mGy, 2.84 mGy, 3.06 mGy, and 4.97 mGy respectively. The least incidence of cancer risk obtained in this study is 1: 7000. Failure in the adoption of complete optimization technique was responsible for the increased risk. Periodic dose audit and enforcement of radiation protection policy will help to checkmate the lapses and alleviate patient risk.

Keywords: Head radiography, Entrance surface dose, Incidence cancer risk

1. Introduction

Head radiography is used to visualize the cranium, facial bones and jaw bones for fractures. Included under head radiography are skull examinations, mandible examinations and postnasal or paranasal sinuses examinations. The examination is performed to investigate cases such as head injury, head pain, sinus infection, hypertrophy of the adenoids, inflammatory diseases of the sinus cavities, tumors, facial fractures among others [1-3].

In the developed society, higher imaging technique such as computed tomography is employed in the diagnosis of head injuries and diseases. However, in the study environment, plain radiography is the most easily accessible radiodiagnostic tool for head conditions due to its availability and cost effectiveness. The use of plain radiography for diagnosis in place of higher imaging techniques is a common phenomenon in developing countries as reported in literature [1-2, 4-5].

During head radiography procedures 2-3 views are projected thus, making the radiation dose delivered to patients to increase. Imaging of the head exposes several organs located in the head region to the risks of ionizing radiation. Cancer induction in low dose radiation cannot be overruled. Even, microwave radiation is known to induce DNA damage, produce chromosomal aberration, histological changes, genotoxic effects and others in tissues of exposed rats [6-9]. It is based on this fact that this study was embarked upon in order to estimate the risks of cancer induction in head radiography with a view to promote dose optimization and enhance patient safety.

2. Materials and Methods

This study was conducted in the radiology departments of two tertiary healthcare institutions in Southwest Nigeria, designated as centre E and centre F. A total of 10 adult human subjects who visited the x-ray unit for head radiography were selected in each centre. The study was conducted for a period of one month. Consent was obtained from each patient before the commencement of the examination. Institutional consent was also obtained from each of the hospital used and also from the Nigerian Institute of Medical Research (NIMR). Thermoluminiscent dosimeter (TLD-100: LiF: Mg, Ti) chips were used to obtain the entrance surface dose (ESD) during the procedure. The TLD chips were obtained from RadPro International GmbH, Poland. The chips were oven-annealed according to specification using Carbolite oven made in England. Irradiation was conducted at the Secondary Standard Dosimetry Laboratory (SSDL) of the National Institute of Radiation Protection and Research (NIRPR), Ibadan. Calibration of TLD chips and reader were conducted and TLD signal was read using Harshaw Reader (Model 3500) at the Department of Physics, Obafemi Awolowo University Ile-Ife. Each of the TLD was enclosed in labelled black polythene pack. A total of three coded chips were used to measure the entrance surface dose (ESD) of each projection view during each head procedure in order to obtain the mean and enhance precision. The chips were attached to an elastic tape and placed in the primary beam of x-rays where the beam intercepted with the irradiated part of the patient.

The quality control of the x-ray machines were conducted using MagicMax quality control kits (IBA Dosimetry, Germany). Patient's clinical information and exposure parameters were noted and recorded using self-structured form. The effective dose, brain dose, oral mucosa dose and salivary gland dose was evaluated from the measured entrance surface dose (ESD) using PCXMC software (version 20Rotation). Thereafter, BEIR VII model was used to estimate the incidence cancer risk. Statistical analysis was done using SPSS (Version 23).

3. Results and Discussion

Twenty (20) adult human subjects underwent head radiography under three (3) different examinations; mandible (6), paranasal sinuses (6) and skull (8). The total entrance surface dose (ESDT) delivered to each patient; the corresponding effective doses; brain dose; oral mucosa dose; salivary gland dose; incidence cancer risks and the patients' indices are presented in Table 1. Descriptive analysis of parameters in Table 1 is presented in Table 2. Analysis of variance for entrance surface dose for each projection is presented in Table 3. Table 4 displayed the coefficients of predictors and Table 5 revealed the Pearson correlation. Figures 1, 2, and 3 compares the exposure parameters and entrance surface dose between the two hospitals for mandible, paranasal sinuses and skull examinations respectively. Figure 5 compares the entrance surface dose from this study with international standards. Sample of radiographs of three views is depicted in Figure 4.

Patient dose assessment is encouraged worldwide due to increased knowledge of the health effects of ionizing radiation. High dose implies high risks. The total entrance surface dose (ESDT) for this study ranges from 3.01-19.12 mGy with a mean of 7.52 mGy. The resulting effective dose varies from 0.09-0.70 mSv with a mean of 0.25 mSv. Total dose to the brain; oral mucosa; and salivary gland ranges from 1.01-7.77 mGy; 1.02-9.03 mGy; and 1.82-13.46 mGy respectively in ascending order of dose. The order of organ dose increase is similar to that reported for Head LAT by [10]. The least incidence cancer risk is about 14 per 100,000 (1:7000); this is considered a low risk and the highest incidence cancer risk is about 156 per 100,000; this is categorized as moderate risk [10]. Large variation in radiation dose values as recorded in this study is well documented in literature [11-14].

Table 1: Distribution of radiation doses and the associated incidence cancer risk from head radiography

Centre	Examination	No of views	Age (y)	Sex	BMI	ESD_T (mGy)	ED (mSv)	BD (mGy)	OMD (mGy)	SGD (mGy)	ICR
E1	Mandible	3	20	F	24.22	9.48	0.35	3.88	4.24	6.60	156.04
E2	Mandible	3	33	F	25.39	10.25	0.36	4.11	4.44	6.94	99.24
E3	Mandible	3	41	F	28.13	10.30	0.35	3.88	3.98	6.61	89.03
E4	PNS	3	54	M	24.38	18.7	0.65	7.28	7.78	12.52	102.33
E5	PNS	3	60	M	26.18	19.12	0.64	7.03	6.96	11.80	93.50
E6	PNS	3	47	M	21.51	18.84	0.70	7.77	9.03	13.46	118.45
E7	Skull	2	65	M	27.97	3.69	0.11	1.25	1.25	2.23	13.63
E8	Skull	2	49	M	25.46	3.08	0.10	1.11	1.12	1.95	18.57
E9	Skull	2	55	F	29.69	3.22	0.10	1.12	1.16	1.97	21.66
E10	Skull	2	19	F	21.48	3.34	0.12	1.38	1.68	2.70	57.87
F1	Mandible	3	24	M	22.59	4.81	0.16	1.80	2.04	3.18	39.16
F2	Mandible	3	56	F	28.91	4.99	0.15	1.70	1.76	3.01	31.33
F3	Mandible	3	36	M	20.08	3.86	0.14	1.55	1.78	3.14	27.79
F4	PNS	3	65	M	25.10	8.01	0.25	2.74	2.76	5.02	29.59
F5	PNS	3	50	M	23.67	7.88	0.26	2.93	3.42	4.97	46.57
F6	PNS	2	52	F	29.69	8.03	0.24	2.59	2.68	4.69	57.14
F7	Skull	2	38	F	27.34	3.08	0.09	1.05	1.06	1.93	28.73
F8	Skull	2	41	M	27.97	3.06	0.09	1.01	1.02	1.82	19.35
F9	Skull	2	21	F	20.70	3.01	0.10	1.16	1.40	2.35	47.19
F10	Skull	2	43	M	23.83	3.73	0.13	1.44	1.66	2.54	22.48

BMI = body mass index; ESDT = Total entrance surface dose; ED = effective dose; BD = brain dose; OMD = oral mucosa dose; SGD = Salivary gland dose; ICR = incidence cancer risk

Radiation risk models are a function of patient dose and age at exposure. Also, exposure parameters are key factors that determine the patient dose. It is therefore important that these factors are considered during irradiation of patients. The impact of exposure parameters to entrance (ESD) in this study is significant as shown in Table 3 and 4. Literature has shown that the use of appropriate exposure parameters is a means of dose reduction strategy [11, 15]. It is evidenced in this study that the choice of appropriate exposure parameters was responsible for the low ESD recorded in centre F. This implies that the implementation of appropriate dose reduction methods are more pronounced in centre F compared to centre E as shown in Figures 1, 2 and 3. This gap might be attributable to the fact that centre F has more trained professionals compared to centre E. Ultimately, there will be a higher incidence cancer risks accrue to patients who underwent head radiography in centre E. Training on optimization of dose and procedures is essential in centre E. This will help to improve the dose outcome and minimize risk to patient.

Though low ESD enhances minimal risks, the influence of age factor to incidence cancer risks should not be compromised. According to the risk models, age at exposure is proportional to risks [10, 16]. But age grouping by the world standards in radiology considers age 16 and above as adults [12, 17]. This implies using the same exposure parameters for all adults; the outcome will be higher risk to young adults compared to older adults. It is important that the adult age bracket should be revisited such that exposure parameters will be age dependent within the adult age bracket just as we have for paediatrics. This will enhance low risk to young adults. At centre F, though the ESD is low but the age at exposure increased the incidence cancer risk for younger patients. Same was also observed in centre E for patients with low ESD and young in age.

Table 2: Statistical analysis of radiation doses and incidence cancer risks

		Age	ESD_T	ED	BD	OMD	SGD	ICR
N	Valid	20	20	20	20	20	20	20
	Missing	0	0	0	0	0	0	0
Mean		43.45	7.52	.25	2.84	3.06	4.97	55.98
Median		45.00	4.90	.16	1.75	1.91	3.16	42.87
Std. Deviation		14.51	5.52	.20	2.19	2.38	3.69	40.12
Variance		210.58	30.48	.04	4.81	5.66	13.59	1608.64
Skewness		-.336	1.302	1.350	1.352	1.435	1.376	1.067
Std. Error of Skewness		.512	.512	.512	.512	.512	.512	.512
Kurtosis		-.863	.593	.711	.707	1.196	.841	.327
Std. Error of Kurtosis		.992	.992	.992	.992	.992	.992	.992
Range		46.00	16.11	.61	6.76	8.01	11.64	142.41
Minimum		19.00	3.01	.09	1.01	1.02	1.82	13.63
Maximum		65.00	19.12	.70	7.77	9.03	13.46	156.04
Percentiles	25	33.75	3.25	.10	1.18	1.29	2.26	23.81
	50	45.00	4.90	.16	1.75	1.91	3.16	42.87
	75	54.75	10.06	.35	3.88	4.18	6.61	92.38

ESDT = Total entrance surface dose; ED = effective dose; BD = brain dose; OMD = oral mucosa dose; SGD = Salivary gland dose; ICR = incidence cancer risk

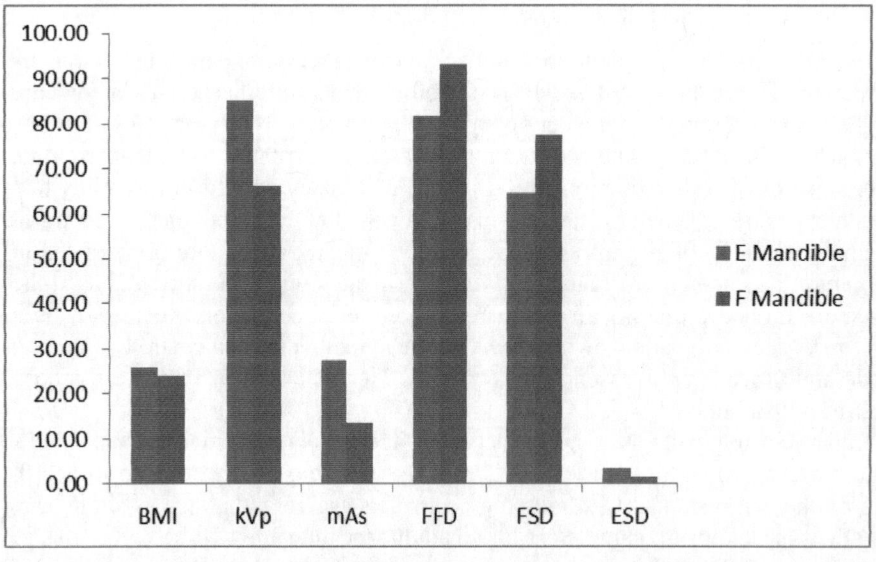

Figure 1: Comparison of exposure parameters for mandible examination in the two centres

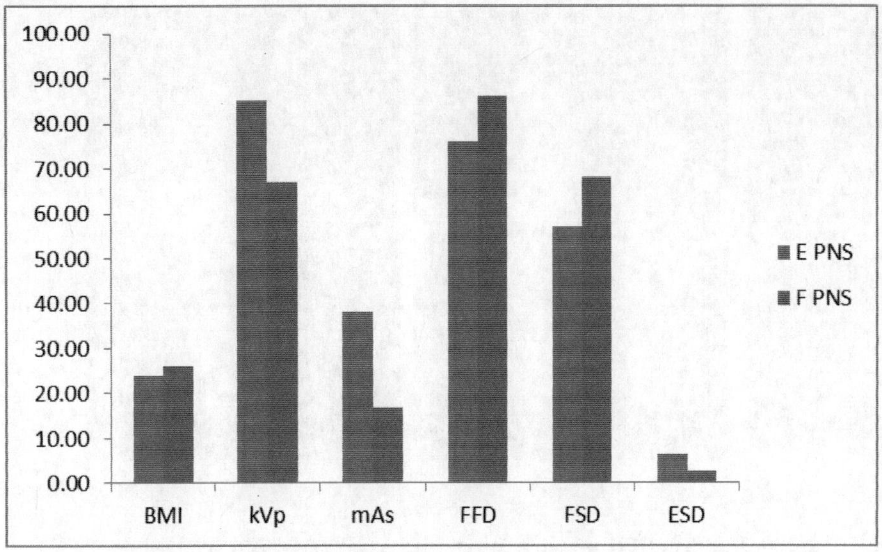

Figure 2: Comparison of exposure parameters for PNS examination in the two centres

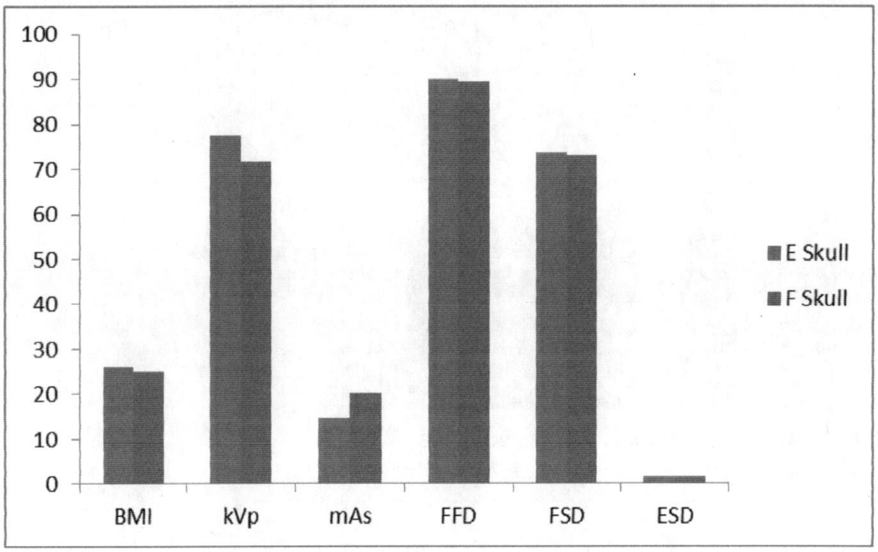

Figure 3: Comparison of exposure parameters for Skull examination in the two centres

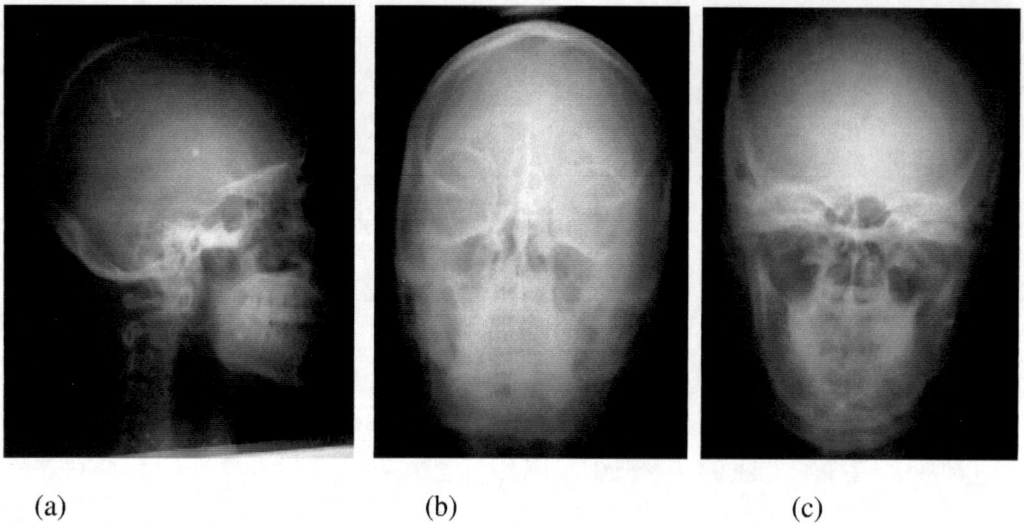

Figure 4: Radiographs of head examination: (a) lateral (LAT) view (b) occipitomental view (OMV) (c) posterior-anterior (PA) view

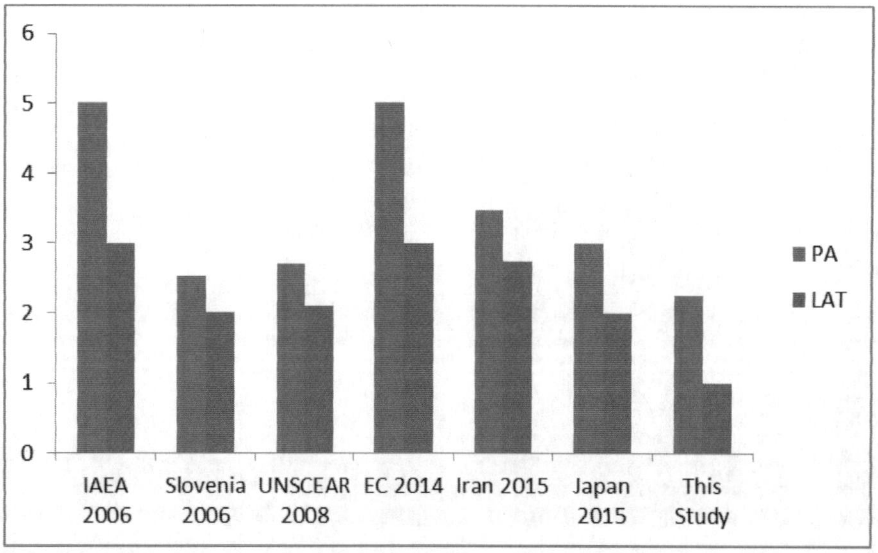

Figure 5: Comparison of entrance surface dose for skull with other studies

Table 3: Analysis of variance for entrance surface dose

Model		Sum of Squares	df	Mean Square	F	Sig.
1	Regression	169.203	5	33.841	41.639	$.000^b$
	Residual	37.385	46	.813		
	Total	206.588	51			

Table 4: Model summary showing coefficients of predictors

	Model	B	Unstandardized Coefficients Std. Error	Standardized Coefficients Beta	t	Sig.
1	(Constant)	7.257	2.322		3.125	.003
	BMI	-.051	.045	-.074	-1.138	.261
	kVp	-.023	.021	-.111	-1.072	.289
	mAs	.114	.020	.641	5.678	.000
	FFD	.125	.049	.446	2.575	.013
	FSD	-.212	.049	-.823	-4.368	.000

Table 5: Pearson Correlation showing correlation between variables

		ESD	BMI	kVp	mAs	FFD	FSD
Pearson Correlation	ESD	1.000	-.063	.613	.828	-.575	-.739
	BMI	-.063	1.000	.120	.053	.163	.100
	kVp	.613	.120	1.000	.790	-.365	-.472
	mAs	.828	.053	.790	1.000	-.450	-.582
	FFD	-.575	.163	-.365	-.450	1.000	.924
	FSD	-.739	.100	-.472	-.582	.924	1.000

ESD = entrance surface dose; BMI = body mass index; kVp = kilo-voltage peak; current time product; FFD = focus film distance; FSD = focus skin distance

Though skull radiography is considered out dated [5], comparison with other studies as depicted in Figure 5 showed that it is still relevant in similar society such as ours. The entrance surface (ESD) for skull examination for this study was lower compared to others because the mean ESD was used while other studies used the third quartile. Also, large variation in ESD is bound to occur due to different methods employed in its evaluation. The importance of head radiography in diagnosis of injuries and diseases cannot be over emphasized especially in an environment where higher imaging equipment for such examination is scarce. The relevance of head radiography in a developing society cannot be neglected in view of its availability and accessibility for skull imaging, it is important that the procedure is evaluated in order to minimize risk to patient.

4. Conclusion

Incidence cancer risk was evaluated in head radiography procedures. Entrance surface dose (ESD), age at exposure, professional expertise and choice of exposure parameters were factors that influenced the incidence cancer risks. Failure in the adoption of complete optimization technique was responsible for the increased risk. It is important that age of adults and dose reduction strategies are considered during irradiation of patients. This will enhance optimization of dose and minimize radiation risk to patients.

5. Acknowledgment

The authors are grateful to the Center for Research, Innovation and Discovery (CUCRID), Covenant University Ota, Nigeria for sponsoring this research. Special thanks to all the Radiographers, Radiologists and the healthcare institution used for this study.

References

[1] Cantalupo B.L.V.C., Xavier A.C.S., Silva C.M.L., Andrade M.E.A., Barros V.S.M., Khoury H.J. (2016). Dosimetric evaluation of x-ray examinations of paranasal sinuses in pediatric patients. Radiologia Brazileira, 49 (2): 79-85.

[2] Kolo E.S., Salisu A.D., Tabari A.M., Dahilo E.A., Aluko A.A. (2010). Plain radiographic

evaluation of the nasopharynx: do raters agree? International Journal of Pediatric Otorhinolaryngology, 74 (5): 532-534.

[3] American Society of Radiologic Technologists (ASRT, 2009). Skull radiography. American Society of Radiologic Technologists, Albuquerque. www.asrt.org.

[4] Adedeji T.O., Amusa Y.B., Aremu A.A. (2016). Correlation between adenoidal nasopharyngeal ratio and symptoms of enlarged adenoids in children with adenoidal hypertrophy. African Journal of Paediatric Surgery, 13 (1): 14-19.

[5] Hitesh C., Malhotra R., Yadav R.K., Griwan M., Paliwal P.K., Aggarwal A.D. (2015). Diagnostic utility of conventional radiography in head injury. Journal of Clinical and Diagnostic Research, 9 (6): TC 13-TC 15.

[6] Usikalu M.R., Aweda M.A., Alimba C.G., Achuka J.A. (2016). Chromosomal aberration after exposure to 2.45 Ghz microwave radiation. Research Journal of Applied Sciences, 11(5): 232-234.

[7] Usikalu M.R., Obembe O.O., Akinyemi M.L., Zhu J. (2013). Short duration exposure to 2.45 Ghz microwave radiation induces DNA damage in Sprague Dawley rat's reproduction systems. African Journal of Biotechnology, 12 (2): 115-122.

[8] Usikalu M.R., Rotimi S.O., Oguegbu A.E. (2012). Effect of exposure of 900 MHz radiofrequency radiation on rat brain. European Journal of Experimental Biology, 2 (6): 2499-2504.

[9] Tice R.R., Hook G.G., Donner M., McRee D.I., Guy A.W. (2002). Genotoxicity of radiofrequency signals: Investigation of DNA damage and micronuclei induction in cultured human blood cells. Bioelectromagnetics, 23: 113-126.

[10] Wall B.F., Haylock R., Jansen J.T.M., Hillier M.C., Hart D., Shrimpton P.C. (2011). Radiation risks from medical x-ray examinations as a function of the age and sex of the patient. Health Protection Agency, HPA-CRCE-028: 1-70.

[11] Alzimami K., Sulieman A., Babikir E., Alsafid K., Alkhorayef M. and Omer H. (2015). Estimation of effective dose during hysterosalpingography procedures in certain hospital in Sudan. Applied Radiation and Isotopes, 100: 2-6.

[12] European Commission (EC, 2014). Diagnostic reference levels in thirty six European Countries. Radiation Protection No 180. Luxembourg, European Union.

[13] Abdelhalim M.A.K. (2011). Diagnostic radiographic examinations in Saudi Arabia based on thermoluminescent dosimetery. African Journal of Biotechnology, 10 (48): 9817-9823.

[14] International Atomic Energy Agency, Applying radiation safety standards in diagnostic radiology and interventional procedures using x-rays, Safety standards series No. 39, IAEA, Vienna, (2006).

[15] Gholami M., Maziar A., Khosravi H.R., Ebrahimzadeh F., Mayahi S. (2015). Diagnostic reference levels (DRLs) for routine x-ray examinations in Lorestan province, Iran. International Journal of Radiation Research, 13 (1): 85-90.

[16] Biological Effects of Ionizing Radiation (BEIR VII, 2006). Health risks from exposure to low level of ionizing radiation. National Research Council BEIR VII Phase, National Academies, Washington D.C.

[17] United Nations Scientific Committee on the Effects of Atomic Radiation (UNSCEAR, 2008). Sources and effects of ionizing radiation, Report to the General Assembly with scientific annexes.

Radiation from Visual Display Unit

M. R Usikalu, I. O Babarimisa, S. A. Akinwumi, M. L. Akinyemi, T. A. Adagunodo and W. A. Ayara

Department of Physics, Covenant University, P.M.B 1023, Ota, Ogun State

Correspondence email: moji.usikalu@covenantuniversity.edu.ng

Abstract. The radiation from different types of Visual Display Units (VDU) (that is, Liquid Crystal Display (LCD) and Cathode Ray Tube (CRT)), projector and interactive screen were examined using a Cell Sensor Extremely Low Frequency (ELF) detection meter. The study revealed that ELF radiation in CRT VDUs is eight (8) times greater than in LCD VDUs. The mean ELF obtained from interactive screen at 30 cm was 1.6 mG and 1.9 mG at start-up and after 3 hours of use respectively with a mean value of 1.03 mG for a distance of 100 cm at start up. The mean ELF radiation from the projector screen at a distance of 10 cm is 0 mG, while that of interactive screen is 2.4 mG at the same distance. Thus, the study suggested that, projector and screen is better than interactive screen when making presentation considering the closeness of user to the interactive screen throughout the course of presentation.

Keywords: Radiation, Projector, Interactive screen, LCD, CRT

1. Introduction

Radiation is the the transmission or emission of energy that is in form of rays, waves or particles through vacuum or through a medium. Electromagnetic (EM) radiation is a radiation in the form of waves. The EM waves can be distinguished from one another through their wavelength or frequency of oscillations in the formation of the EM spectrum, which is arranged in magnitude of increasing frequency and decreasing wavelength: radio waves, microwaves, infrared, visible light, ultraviolet radiation, X-rays and gamma rays [1]. Radiation can either be ionizing or non-ionizing. Ionizing radiation possesses energy that is high enough to knock off electrons from molecules or atoms. Non-ionizing radiation on the other hand does not possess energy high enough to ionize molecules or atom. Non-ionizing do not produce charged ions when in moving through medium the energy just enough to excite the molecules [2, 3]. EM radiation can be from a natural or man-made source. Natural sources are sunlight and radioactive materials that arises from the earth crust. Examples of artificial sources of ionising radiation are X-rays equipment, diagnostic and therapeutic radio chemicals and nuclear energy generation. Television, lasers, radio broadcasting, household appliances, projectors, and interactive boards etc, are artificial sources of non-ionizing radiation [4, 5]. The radiation from screens and projectors belongs to extremely low frequency range of EM spectrum. Extremely low frequency (ELF) is the electromagnetic radiation having accompanying wavelengths from 100,000 to 10,000 mega metres, and frequencies between 3 and 30 Hz [6].

Computer Visual Display Units (VDU) is of two major types: Cathode Ray Tube (CRT) and Liquid Crystal Display (LCD). The CRT is a vacuum tube that contains one or more electron guns, and used phosphorescent screen to view images. It has a way of accelerating and deflecting the electron beam(s) onto the screen to form the images. The images can be in form of pictures (television, computer monitor), electrical waveforms (oscilloscope), radar targets etc. LCD is the technology used for displays in smaller computers and notebooks. Radiation from VDUs can cause eye irritation and strain. Moreover, the electromagnetic radiation also causes dryness and irritation in our eyes. ELF radiation can produce, various health effects ranging from allergic reactions and sleep disorder, to heart disease, Alzheimer's disease and cancer [6]. Higher rates of miscarriages and possibly birth defects were associated with using CRT screens for long periods by pregnant women. The magnetic component is considered as the dangerous part because of its deep penetration into the human body. It was noted that ELF radiation decay naturally and quickly with increasing distance from source. According to [7], CRT VDUs emit radiation of 3 mG at 30cm, measured from the front and 4 mG at the same distance from sides. The box-shaped cathode ray tube (CRT) computer VDUs usually emit high levels of radiation even at 30 cm compared to the modern low-radiation flat screens [8]. An Uninterruptible Power Supply, (UPS) emits radiation of 20 mG at 30 cm and over 1 mG at 1 m, even when still connected to power during charging of the battery but switched off. Also, small desktop computer printers produce about 0.5 mG at 60 cm in standby mode and double the amount when printing is in progress. Wi-Fi information networks, modems, wireless routers and other wireless devices equally emit radio frequency and microwave electromagnetic radiation. They are not totally safe and as such can result in cell/ DNA damages, cause infertility and affect the biological process [7, 9]. Large sub-woofers used as part of computer's sound system emit 20 mG of ELF radiation at 0 cm, 3 mG at 60 cm, 0.5 mG at 90 cm when just powered ON even when not producing sound. Through the use of all these electronic gadgets, human beings are constantly being exposed to this ELF radiation. Therefore, it is expedient to investigate exposure from all these devices.

2. Materials and Methods

The device used in measuring the ELF radiation from VDUs, projectors and interactive monitors is the Cell Sensor manufactured by Action Electronic, USA. The cell sensor is a radio frequency detection meter which is also capable of being used as an extremely low frequency, ELF detection meter. Cell sensor possesses a remote probe that when connected to the unit makes it an extremely low frequency detection meter. It is used primarily for radio frequency detection. When the probe is connected to the main unit, it allows one to measure ELF fields in all places, while making it easy to watch the readings on the display unit of the meter. The device measures ELF fields in milligauss (mG) and is shown on the bottom scale of the display marked 'Power (ELF)' in green print. The meter measures in two different ELF scales. The device has a switch on the side that makes it possible to switch between high and normal sensitivity. The high sensitivity scale is of 1 to 5 mG while the normal sensitivity scale is of 1 to 50 mG. Taking the readings involved first isolating the source of ELF. The probe is then positioned next to the point at which reading is to be taken from. Once close to the source, the needle on the display deflects. To get accurate readings, the probe is rotated in different directions; horizontally, vertically and sideways and the mean reading is recorded. The ELF radiation from different VDUs: LCD and CRT; interactive displays and projectors used in College of Science and Technology classrooms were taken from 0 - 100 cm at intervals of 5 cm for 0 - 20 cm and intervals of 10 cm for 20 - 100 cm.

3. Results and Discussion

The ELF radiation from LCD and CRT visual display units at start up and after 3 hrs of usage, in order to compare the radiation from the two types of the display unit, the column charts for LCD and CRT VDUs at both instances of start-up and after 3 hours are shown as Figures 1 and 2 respectively. It was observed that the ELF radiation in CRT VDUs is eight (8) times greater than that measured in LCD VDUs. It was also observed that the ELF radiation increases over time in the case of both types of VDUs that are in use. In the case of a typical VDU, the ELF radiation detected at 0 cm at start-up was 3.7 mG and it increased to 4.0 mG after it had been switched on for 3 hours. The same increase in ELF radiation occurred in all other VDUs as usage time increases. It can be said that the ELF radiation is cumulative and increases with time or with prolonged use but there was no significance difference

IOP Conf. Series: Earth and Environmental Science **173** (2018) 012039 doi:10.1088/1755-1315/173/1/012039

between the radiation measure between startup and 3 hr of operation. The reason for low radiation from LCDs than CRTs may be because CLDs use much less power than LED and gas-plasma display displays, they also work on the principle of blocking light rather than emitting it. For LCD screens, display is more composed of colour or monochrome pixels that are filled with liquid crystals. The liquid crystals are lit by a backlight or reflector. LCD screens also utilise very small amounts of electric power. This study is in consonance with [10, 11] that CRT screens emit a lot more radiation than LCD screens. This is a may also be the reason why CRT VDUs are gradually been faced out and LCD VDUs have taken over the market.

The ELF measured from the projectors and screens are displayed in Figures 3 and 4, it was observed that the ELF radiation from the projector screen is very low. However, the ELF radiation from the projector is very high compared to that of its screen with ELF readings as high as 1.0 mG at 20 cm from the projector. Interactive screens and projectors are often used for making presentations. Seeing that they are used for a similar purpose, comparing the ELF radiation from both of them to determine which of the two is better with low radiation incurred during presentation desirable. When using a projector, users often stay closer to the projector screen than to the projector itself, especially when it is hung overhead. For the interactive screen, this does not have a separate device that projects the desired image, presenters stay close to the screen while making presentations. The mean ELF obtained from interactive screen at 30 cm was 1.6 mG and 1.9 mG at start-up and after 3 hours of use respectively. It also has a mean value of 1.03 mG for a distance of 100 cm at start up. It was also observed that the ELF radiation from the interactive screen increases with time. After 1 hour, it has a mean value of 2.36 mG at 0 cm which was 2.0 mG at start-up; after 3 hours, the mean ELF radiation from the interactive screen at 0 cm was 2.56 mG. This proves that the ELF radiation increases with time of use. It is observed that the mean ELF radiation from the projector screen at a distance of 10 cm is 0 mG, while the ELF of the same instance and distance from interactive screen is 2.4 mG as shown in Figure 4. This revealed that the ELF radiation from interactive screen is much higher than that from the projector screen. This suggests that projector screens are better than interactive screens when considering the radiation incurred by the user during a given period of time, although the display quality of the latter is far better than that of the former. The high ELF observed in interactive screen may be attributed to the fact that the CPU, the visual unit and other electronic parts are coupled together which gives rise to cumulative ELF obtained from the device compare to the projector and the screen that are position separately.

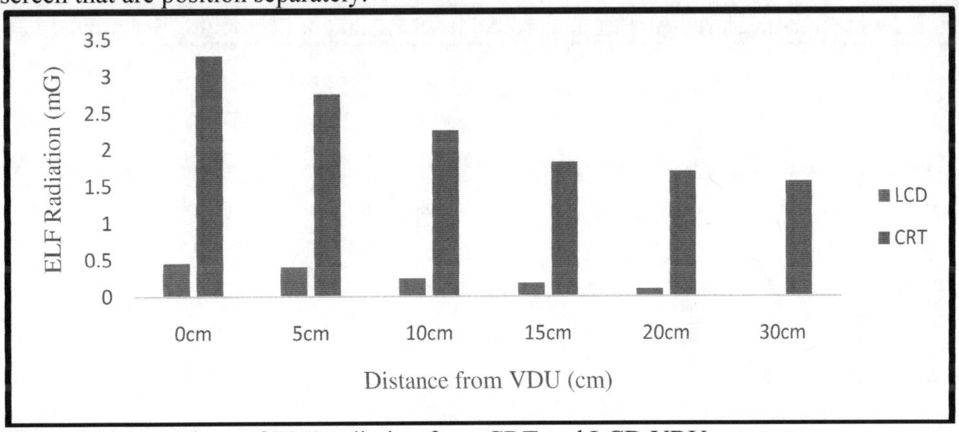

Figure 1: Comparison of ELF radiation from CRT and LCD VDUs at start up

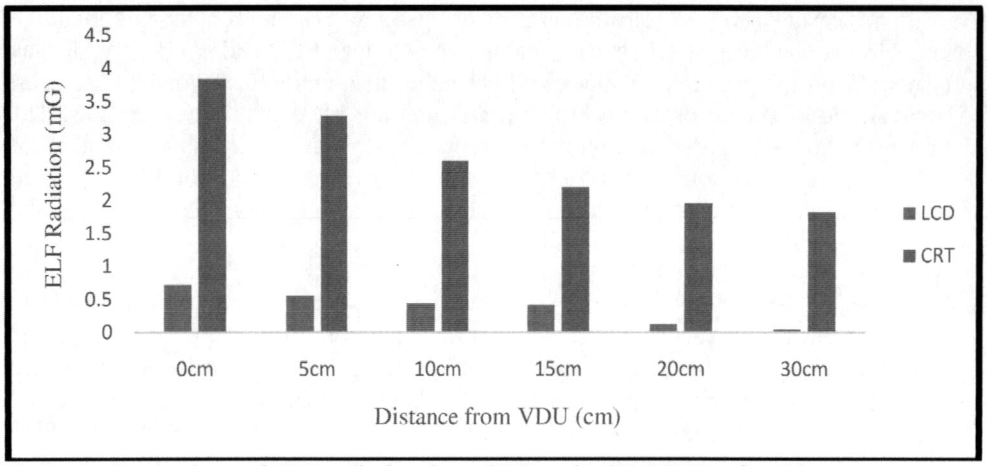

Figure 2: Comparison of ELF radiation from CRT and LCD VDUs after 3 hours

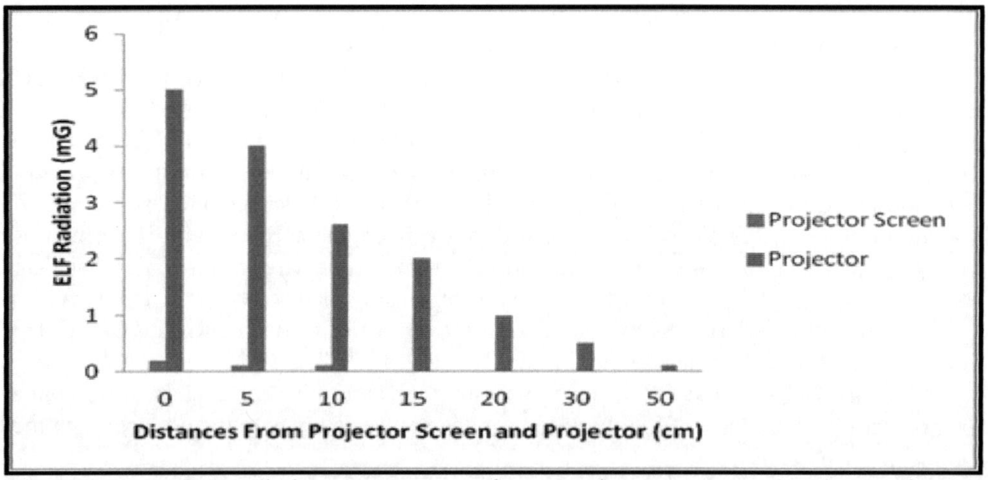

Figure 3: Mean ELF radiation measurements from a projector screen and projector after 1 hour

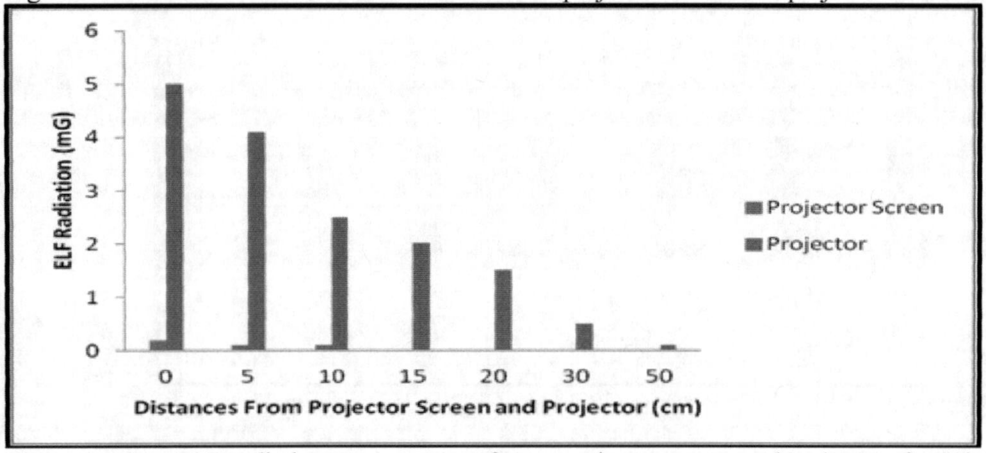

Figure 4: Mean ELF radiation measurements from a projector screen and projector after 3 hours

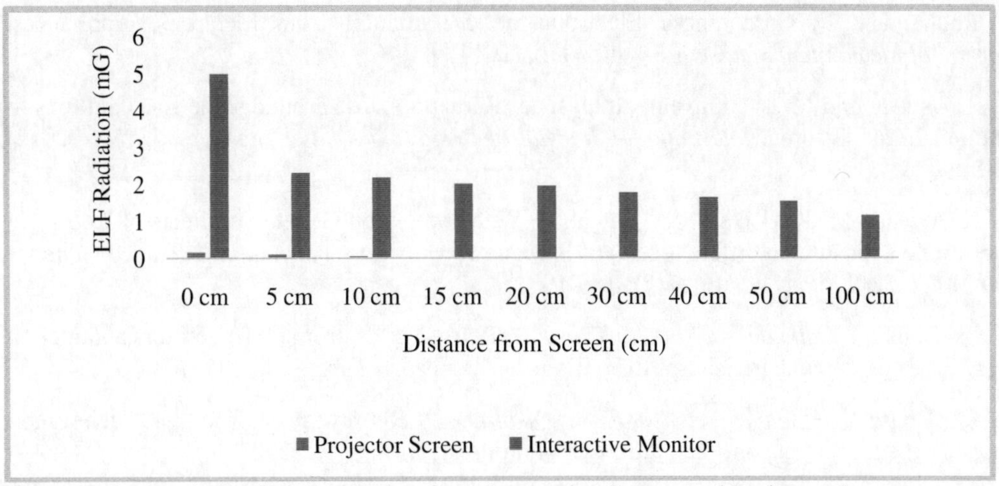

Figure 5: Comparison of ELF radiation from Interactive Screen and Projector Screen

4. Conclusion

The study has confirmed the presence of ELF radiation from VDUs (that is, LCD and CRT), projector and interactive screen. It was observed that the ELF radiation in CRT VDUs is eight (8) times greater than that measured in LCD VDUs. It was also observed that the ELF radiation increases over time in the case of both types of VDUs that are in use. It was observed the ELF radiation from interactive screen is much higher than that from the projector screen. It is therefore radiation prudent to use projector and screen for presentation than interactive screen. This is necessary in order to create awareness so that users can reduce the time spent with the interactive screen.

Acknowledgment

We appreciate the management of Covenant University for sponsoring this work.

References

[1]. J. Lucas *Electromagnetism*. Retrieved from Livescience.com: http://www.livescience.com/38169-electromagnetism.html, (2015)

[2]. World Health Organization, WHO's Electromagnetic fields (EMF) http://www.who.int/peh-emf/about/WhatisEMF/en/index3.html, accessed 23rd, April, 2018.

[3]. M. R. Usikalu, T. T. Ikeh and C. O. Olawole (2014). Safe Distance to Extremely Low Frequency Radiation Associated with Power Transmission Lines Located in Ota, Southwest, Nigeria, *International Journal of Engineering Technology, IJENS* 14(2):118-121

[4]. Radiation Protection Program *Radiation from Computer Monitors*. Retrieved from Environment, Health & Safety: https://ehs.mit.edu/site/content/radiation-computer-monitors, (2010).

[5]. M. L. Akinyemi, J. S. Kayode and M. R. Usikalu Investigation of Extremely Low Frequency (ELF) Hot Spots in the College of Science and Technology, Covenant University, Ota, *Turkish Journal of Physics,* (2011), **35(3):** 359-361.

[6]. International Agency for Research on Cancer Non- Ionizing Radiation, Part 1: Static and extremely low-frequency (ELF) electric and magnetic fields. IARC Monographs on the Evaluation of carcinogenic Risks to Humans, (IARC, 2002), Volume 80, Lyon: IARC Press

[7]. R. Radha, and P. Gurupranesh Electromagnetic Radiation from Electronic Appliances. *Journal of Mechanical and Civil Engineering*, (2014), 41-46.

[8]. M. R. Usikalu and M. L. Akinyemi Analysis of Radiation Dose around some Base Stations in Ota and Lagos Environ, *International Journal of Basic and Applied Sciences, IJENS,* (2012) **12(5):** 7-12.

[9]. M. A. Aweda, M. R. Usikalu, N. Ding, J. H. Wan and J. Zhu Genetoxic effects of 2.45 GHz microwave exposure on different cells of Sprague Dawley rats *International Journal Genetics and Molecular Biology,* (2010), **2(9):** 189-197.

[10]. M. Belis *Cathode Ray Tube*. Retrieved from Inverntors.about.com: http://inventors.about.com/od/cstartinventions/a/CathodeRayTube.htm, (2015)

[11]. M. Rouse *Liquid Crystal Display*. Retrieved from TechTarget: http://whatis.techtarget.com/definition/LCD-liquid-crystal-display, (2005)

Natural dye extracts from fruit peels as sensitizer in ZnO-based dye-sensitized solar cells

Oluwaseun Adedokun*, Mojoyinola Kofoworola Awodele, Yekinni Kolawole Sanusi and Ayodeji Oladiran Awodugba

Department of Pure and Applied Physics, Ladoke Akintola University of Technology, Ogbomoso, Nigeria.

*Corresponding Author's Email Addresses: oadedokun@lautech.edu.ng adedokunoluwaseun2@gmail.com

Abstract
As a result of twin problems of environmental pollution and continuous depletion of fossil fuels, a search for renewable and clean energy sources is necessary. This paper discusses the simple extraction techniques without any further purification of natural dyes from some fruit peels and their performance in DSSCs. Natural dyes extracted from fruit peels of *Carica papaya*, *Citrus lanatus*, *Persea americana* and *Solanum melongena* were used as sensitizers to fabricate ZnO-based dye-sensitized solar cells (DSSCs). The extracts of the natural dyes were characterized and studied using UV-Vis absorption spectroscopy, Fourier transforms infra-red (FTIR) and Photoluminescence (PL) spectroscopy. The photo-anode semiconductor material was prepared and characterized. The photo-voltaic characteristics of the fabricated DSSCs were measured under simulated solar light (power of incident light 100 $mWcm^{-2}$ from Air Mass 1.5G). These studies indicated the presence of pheophytin 'a' in most of the extracts studied as the main pigments with other accessory pigments. The solar to electric conversion efficiencies for the pheophytin 'a' dye based solar cells are estimated as 0.017%, 0.013%, 0.010% and 0.011% for *Carica papaya*, *Citrus lanatus*, *Persea americana* and *Solanum melongena*, respectively. The dye extracts from *Carica papaya* exhibits higher photo-sensitized performance compared to the other extracts and this is due to the better charge transfer between the dyes of *Carica papaya* and ZnO photo-anode surface. Conversion of visible light into electricity was accomplished with natural dyes and ZnO-based DSSCs, resulting in excellent photo-electric properties compared to those in literature.
Keywords: Fruit peels, natural dye, pheophytin a, photoanode, DSSCs, conversion efficiency

1. Introduction

The considerable high attention and extensive research efforts have been devoted over the last decades to dye-sensitized solar cells (DSSCs). Since a new type of solar cells is developed in 1991, these have attracted considerable much attention due to their environmental friendliness and low cost of production [1]. DSSCs are regarded as promising low-cost solar cells with high light-to-energy conversion efficiency. Although, many photovoltaic cells that assure the conversion of solar energy from sunlight to electricity, have already been unfolded in the past years. However, their practicable implementation is still limited by two significant problems namely; conversion efficiency and cost [2,3].

DSSC's are a photo-electrochemical system which consists of four main parts, such as, the porous nano-crystaline wide band gap semiconductor electrode, sensitizer, counter electrode and electrolyte [3,4]. The conversion of electricity from sunlight occurs in DSSC due to sensitization of semiconductor photo-anode. This is based on absorption of visible light characteristics of dye extract, the nature of the locality of the semiconductor conduction band relative to the dye excited state and the attachment of the molecules of dye unto the surface of the semiconductor porous layer [5,6].

The photo-sensitizer used in DSSCs can be divided into two (2) broadly types according to dye structure: metal complex such as polypyridyl complexes of ruthenium (inorganic dyes) and natural or synthetic organic dyes (organic dyes). Ruthenium polypyridyl complex synthetic inorganic dyes have been widely employed as photo-sensitizers in DSSCs with high efficiency up to 11-13% [7,8,9,10]. But ruthenium dyes are expensive, extremely rare, and undesirable considering its impact on the environment.

In recent times, there is significant interest toward the usage of extracted natural dyes from various plants, leaves, flowers and fruits as photo-sensitizer in DSSCs which sometimes are biodegradable and easily available [3,6]. Several natural occurring pigments like chlorophyll [11,12,13,14], carotenoid [15,16], anthocyanin [17,18] and betalains [19] have been successfully utilized and reported as photo-sensitizers for DSSCs. The photo-sensitizer from naturally occurring pigments proffer many advantages such as large absorption coefficient, highly light-harvesting efficiency, non-toxic, environmentally friendliness, full biodegradable, availability and cheap cost [3,19]. Usage of natural occurring materials in DSSC was recently investigated and reported with extracts from *Citrus paradisi, Citrus sinensis, Citrus limon* and *Citrus tangelo* peels utilized as a photo-sensitizer resulting to 0.63%, 0.36%, 0.10% and 0.51% efficiency, respectively [20].

In this study, natural pigments from fruit peels waste of *Carica papaya*, *Citrus lanatus*, *Persea americana* and *Solanum melongena,* were extracted. To the best of our knowledge, natural pigments from many of these fruits peels have not been previously reported as photo-sensitizers for DSSCs. The optical spectra characteristics of the extracted pigments of *Carica papaya*, *Citrus lanatus*, *Persea americana* and *Solanum melongena* in ethanol solution were primarily studied by using UV-Vis absorption, FT-IR and PL spectroscopy techniques. The ZnO nanoparticle as photo-anode material was structurally and compositionally characterized. The photo-electrochemical properties of DSSCs fabricated with the natural pigments from fruit peels of *Carica papaya*, *Citrus lanatus*, *Persea americana* and *Solanum melongena* were studied.

2. Materials and Methods

2.1 Preparation of natural photo-sensitizers

Carica papaya (pawpaw), *Citrus lanatus* (watermelon), *Persea americana* (avocado) and *Solanum melongena* (eggplant), were purchased from nearby market. The peels from the fruits were taken and washed severally with distilled water, followed by vacuum-drying at 60°C until the weight become invariant. The already dried peels were broken-down into fine powder. In a particular experiment, the natural photo-sensitizers in this research were extracted by dissolving fine powder of 1 g into ethanol of 50 ml and left undisturbed in the darkness for 48 hrs at room temperature. The residue was filtered and the resulting dye solutions were concentrated and then protected from exposure to direct light. The natural pigments were further analyzed and utilized as natural photo-sensitizers without any other purification.

2.2 Fabrication of DSSCs

A highly conducting fluorine-doped tin oxide (FTO) films deposited on glass plate were purchased (Solaronix, SA) with 8 Ω/sq resistance was employed as received. FTO substrates were washed in a beaker of solution of detergent inside ultrasonic bath for 30 min and then washed with mixture of acetone and ethanol for the next 30 min.

The smooth and homogenous paste was acquired using ZnO powder (May & Baker Ltd Dagenham, England) and was deposited unto surface of FTO glass by employing doctor blade method with an effective area of 1 cm × 1 cm. The sample was dried at 125 °C for 15 min in an oven. After repeating the process, the pre-selected thickness was achieved and finally, the sample was sintered for 1 hr at 450 °C in a furnace then cool down to 80 °C, before being immersed into natural dye solution kept at room temperature at a specific interval to guarantee complete sensitizer uptake. Platinum

nanoparticles-coated FTO and Iodolyte AN-50 which were purchased from Solaronix, SA were employed as counter electrode and electrolyte respectively, as supplied.
The DSSCs were fabricated by sandwiched the dyed sensitized ZnO electrode and counter electrode formed with platinum together. The details of the cell fabrication and paste making are followed as reported elsewhere [21]. The properties of the fabricated cells were then studied.

2.3 Characterization and measurements

The ZnO powder was characterized for structural characteristics using X'pert pro MPD PANanalytical X-ray diffraction (XRD) with CuαK radiation (λ = 1.5406 Å). The morphology and compositional of ZnO powder sample has been analyzed with scanning electron microscope, SEM (LEO 430i, Carl Zeiss). The microstructure of the ZnO powder was analyzed by transmission electron microscopy (TEM; Tecnai G2 30ST operating at an acceleration voltage of 300 kV).

The UV-Vis absorption characteristics of the natural pigments were measured and the wavelength of all the measurements were scanned from 400-800 nm using UV-Vis-NIR spectrometer (Shimadzu UV-3600). The typical peaks of the natural pigments were measured between 4000 and 400 cm^{-1} with a scan of 200 at interval on a NICOLET 380 FT-IR spectrometer. The room-temperature PL spectra for the natural pigments were taken on a spectrofluorometer (QM-40, Photon Technology International, PTI) using a Xenon lamp (150 W) as an excitation source, at 420 nm and 2 nm wavelength and band-pass, respectively.

The photo-current (I) and photo-voltage (V) characteristics were obtained with solar simulator (Newport model No: 96000) at 100 mW/cm^2 (1 sun AM1.5) and 25°C, coupled with source measure unit (Kethley 2400). A mask was used for the measurement with the cell active area of 0.5 cm^2. The photo-electrochemical results presented were average of measurements taken on three repeated experiments for the sample under test.

3. Results and Discussion

3.1 X-ray Diffraction (XRD) studies of ZnO sample

The room temperature XRD characterization was performed to establish the phase evolution of the ZnO powder sample as shown in Fig. 1. It is observed that XRD patterns matched well with reported reference patterns of wurtzite ZnO (hexagonal phase, JCPDS file 75-0576). The mean crystallite size, D calculated from the highest diffraction peak along the <101> plane for ZnO powder sample is 34.2 nm.

3.2 Microstructural and compositional studies of ZnO sample

The SEM microstructure and EDX spectra for ZnO powder samples are revealed in Fig. 2. The particle morphology is spherical and homogenous in nature with small aggregation of the particles. In addition to spherical shape in ZnO nanoparticles, hexagonal structure could also be found. The EDX results also showed a corresponding peak Zinc (Zn) and Oxygen (O).

In order to strengthen structural characterization, TEM characterization was carried out on the sample of ZnO nanopowder. The TEM images of the sample in ethanol are displayed as shown in Fig. 3. A spherical shape nanoparticle was formed as shown in Fig. 3(a). The diameter of the nanoparticles varied in the range of 20 nm to 50 nm. The high-resolution transmission electron microscopy (HRTEM) image confirms the single crystalline nature of individual nanoparticles, where the spacing corresponds to (101) reflection as shown in Fig. 3(b). The selected area diffraction (SAED) pattern shown in Fig. 3(c) indicates (101), (102) and (200) planes of ZnO.

3.3 Optical Absorption Studies of Extracted Natural Photo-sensitizers

The UV-Vis optical absorption spectra of *Carica papaya*, *Citrus lanatus*, *Persea americana* and *Solanum melongena* in ethanol as a solvent are revealed in Fig. 4. Also the corresponding colour of the extracted natural dye is shown in the inset of Fig. 4. It is observed that, all the selected natural sensitizers have similar absorbance peaks. The two major intense absorption peaks of natural extract as shown in Fig. 4 (a-d) are centered mainly on wavelengths of about 411 nm, 666 nm; 410 nm, 666 nm;

411 nm, 665 nm and 411 nm, 666 nm for *Carica papaya*, *Citrus lanatus*, *Persea americana* and *Solanum melongena* nature dyes, respectively. The Soret bands for the entire extract (Fig. 4) exhibits bathochromic shift except *Citrus lanatus* extract, and in addition, three very weak bands were observed around 506 nm, 536 nm and 608 nm. The observed intense Soret and Q band, and weak bands located in the spectral range between the Soret and Qy bands are in accordance with the corresponding literature data reported for Pheophytin 'a' [14,22].

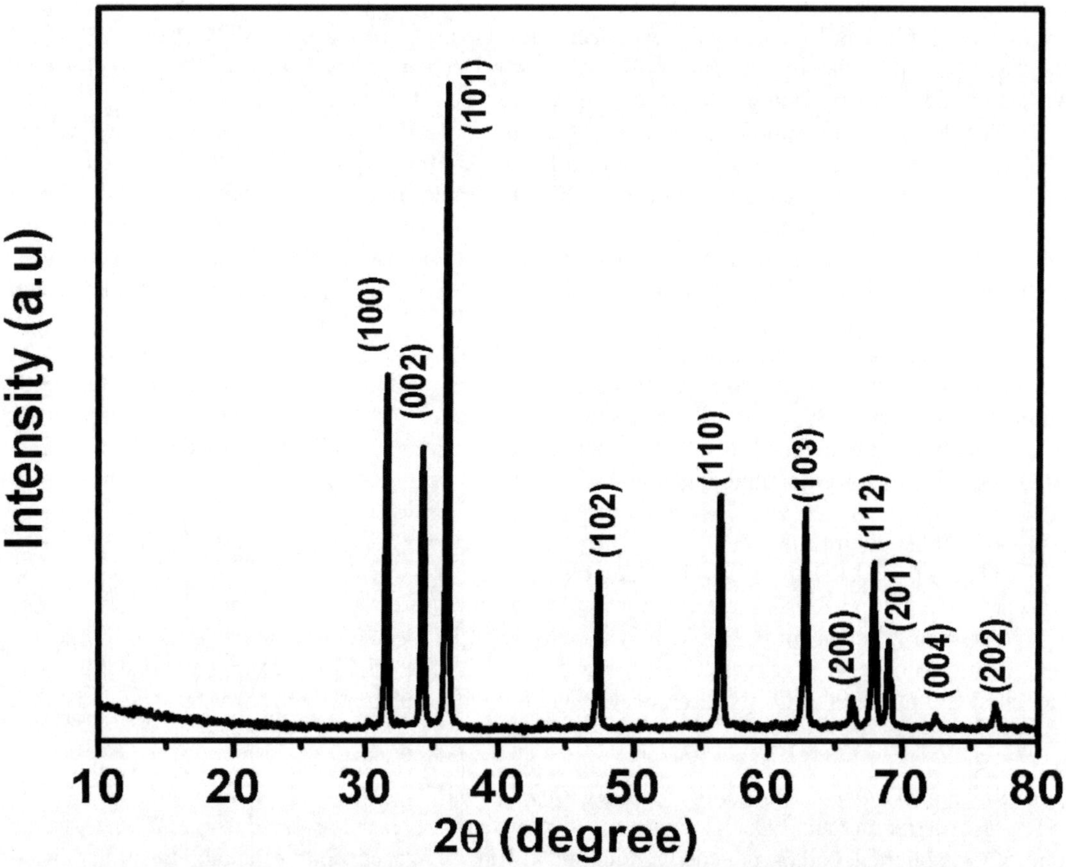

Figure 1. XRD structure of ZnO powder sample with their specific reflection planes

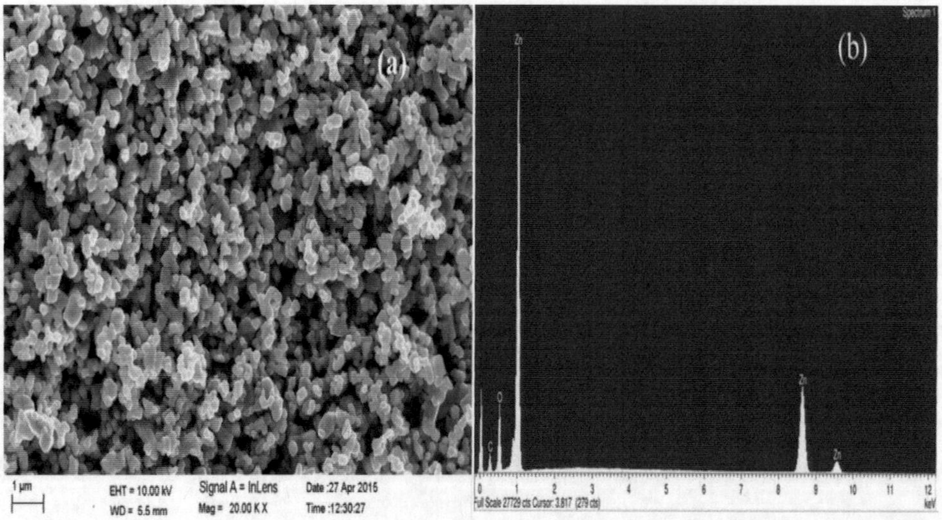

Figure 2. SEM and EDS images of ZnO powder sample

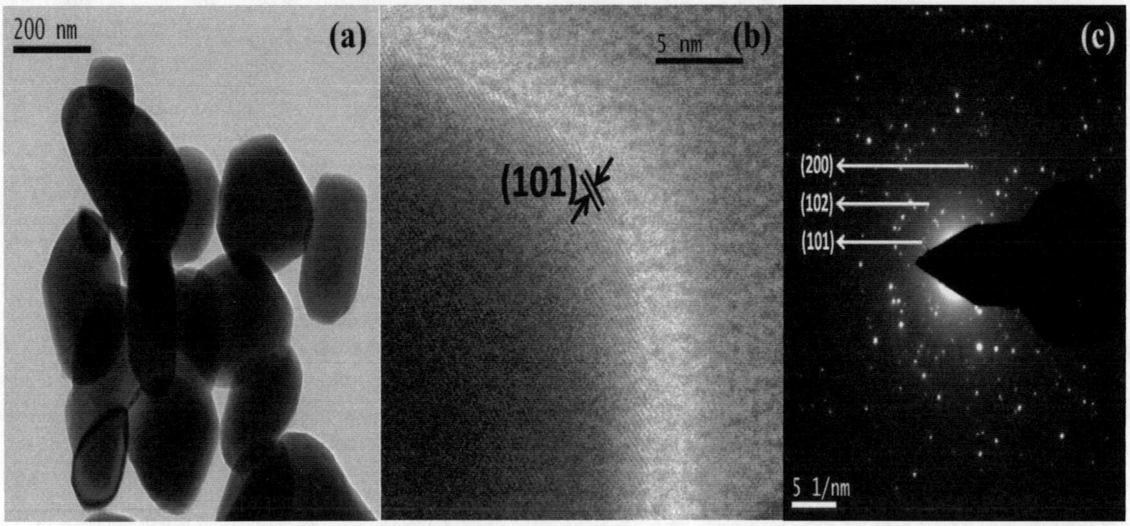

Figure 3: TEM images of ZnO nanoparticles (a) separated nanoparticles (b) HRTEM image of a single nanoparticle, and (c) SAED pattern.

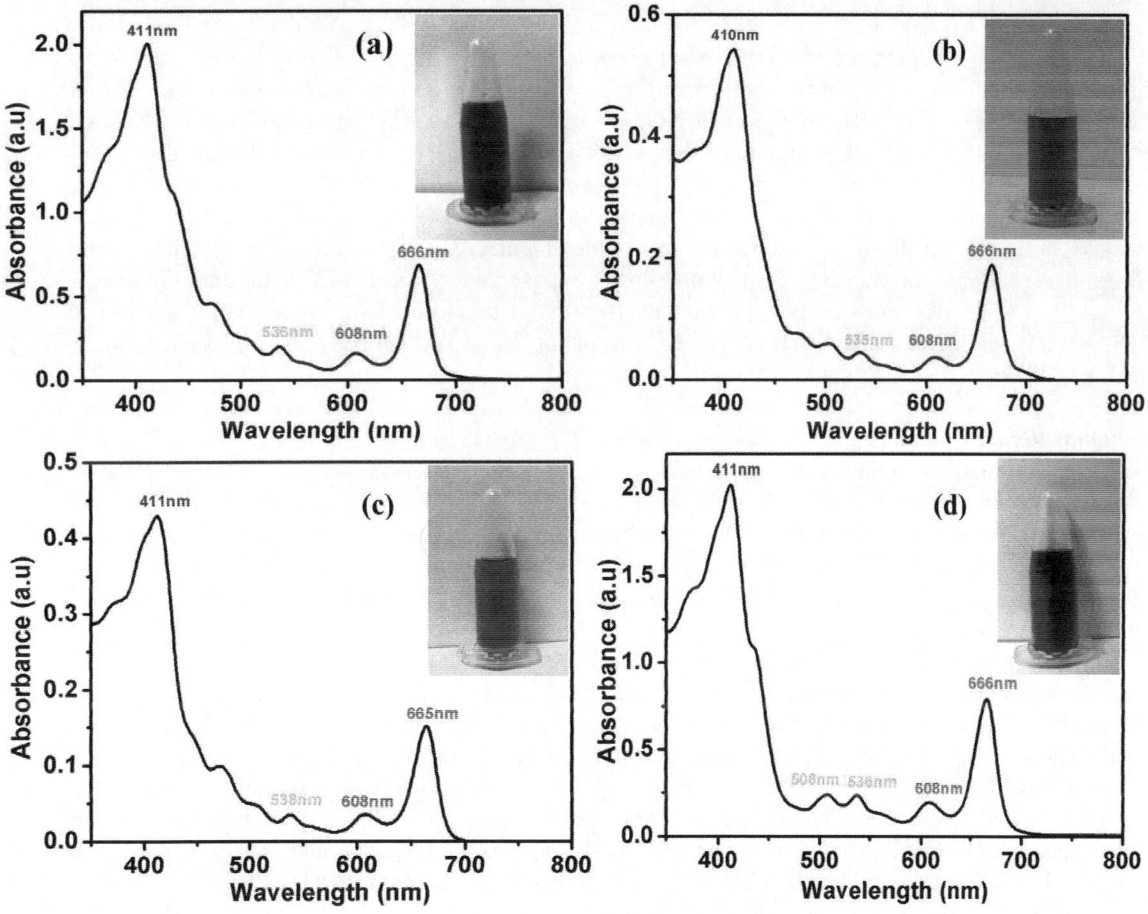

Figure 4: UV-vis absorption spectra for light harvesters obtained from fruit peels (a) *Carica papaya (b) Citrus lanatus* (c) *Persea americana* and (d) *Solanum melongena* alcoholic extracts with the inset showing digital image of natural dye solution.

3.4 Photo-luminescence studies of Extracted natural pigments

The study of photoemission of all the extracted dyes is necessary because there may be possible correspondence between the occurrence of photoelectron emission characteristics and the photoelectric conversion of natural photo-sensitizers based DSSC. The steady-state PL measurements were conducted on all the alcoholic extracts of natural pigment. Fig. 5 shows the emission spectra of *Carica papaya, Citrus lanatus, Persea Americana* and *Solanum melongena* dye extracts in alcohol. They all have identical trend of maximum emission intensity located at 674 nm with an arm at 726 nm and very low emission at around 480 nm with no emission in green region of the spectrum.

Typically, chlorophylls and pheophytins have bright maximum fluorescence intensity between 670-700 nm and with a shoulder of approximately around 710-740 nm indicating their characteristics response [22,23]. The emission intensity of all the dyes exhibited a little red shift compared with that of the pheophytin 'a'. The chlorophyll and its derivative molecules may tend to aggregate in polar solvents and upon aggregation, both absorption and PL spectra exhibit bathochromic shift [24,25]. The photoemission spectra agree well with the result of optical absorption spectra, indicating the main characteristic response of the pheophytin 'a' that masked all other non-chlorophyll that might be present.

3.5 Fourier Transform-Infrared (FT-IR) studies of natural photo-sensitizers

The plotted FTIR data of the natural dyes from *Carica papaya*, *Citrus lanatus*, *Persea Americana and Solanum melongena* fruit peels shown in Fig. 6(a-d) exhibit most of the feature peaks of chlorophyll derivatives [22,24]. Chlorophylls and their derivatives have characteristic infrared band corresponding to N-H, C-H, C=O and C=C groups [12,13,14]. As observed from Fig. 6(a-d), N-H stretching vibration, saturated $C\text{-}H_3$ vibration and $C\text{-}H_2$ vibration are observed respectively at 3228 cm^{-1}, 2923 cm^{-1} and 2845 cm^{-1} for all the four extracted dyes. The characteristic C=O band corresponding to the ketone in the cyclopentanone ring of pheophytin 'a' is observed between 1731 cm^{-1} and 1745 cm^{-1}. Similarly the C=C vibrations of pheophytin 'a' is observed between 1631 $cm^{-1}$and 1646 cm^{-1}. The FTIR spectral analysis also confirms the presence of pheophytin 'a' in all the four fruit peels extracts, which is well correlated with the UV-Vis and PL spectral results.

3.6 Photovoltaic Performance

The cell performances of the natural pigments in the photo-electrochemical solar cells were tested under 100 mW/cm^2 illuminations (AM 1.5). The photo-current density-voltage (J-V) characteristic plots of the DSSCs sensitized with the four natural extracts are shown in Fig. 7 and corresponding photo-electrochemical parameters for the fabricated cells are summarized in Table 1. The open-circuit voltage (V_{OC}) varies from 0.303 to 0.373 V, and the short-circuit photocurrent density (J_{SC}) changes from 0.111 to 0.149 $mAcm^{-2}$.

The efficiency of the DSSCs fabricated with natural pigments from *Carica papaya*, *Citrus lanatus*, *Persea Americana and Solanum melongena* peels are 0.017%, 0.013%, 0.010% and 0.011%, respectively. Precisely, a higher open-circuit voltage (V_{OC} = 0.373 V) and short-circuit photo current density (J_{SC} = 0.149 $mAcm^{-2}$) were obtained from fabricated DSSC sensitized by *Carica papaya* extract, and its efficiency reached 0.017%. The energy conversion efficiency parameters of 0.017% for ZnO based DSSCs sensitized by extracts of Carica papaya is significantly higher than those of the DSSCs sensitized with other natural dyes in this work and showed a comparable performance to the ZnO based DSSCs prepared from other natural dyes used by other groups [21,26,27,28]. This is possibly due to a broader range of light absorption of natural dye adsorbed onto ZnO film, and higher interaction between ZnO and chlorophyll derivative (pheophytins) in these extracts which leads to a better charge transfer. Furthermore, pheophytins structure in the extracts of *Carica papaya* may have the shorter distance between the dye skeleton and a point connected to ZnO surface compared to that of other extracts in this work. And this could facilitate an electron transfer from pheophytins molecules in the extracts to ZnO surface and may be accounted for a better performance of *Carica papaya* extract. From this result, it is obvious that the interaction between the sensitizer and ZnO film

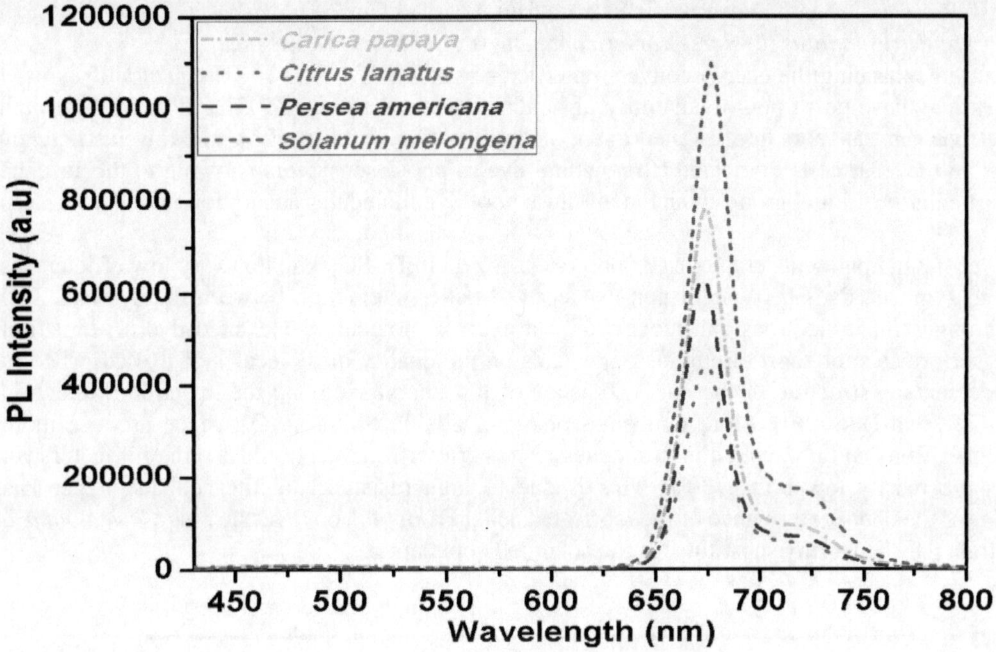

Figure 5: Fluorescence emission spectra of natural dye obtained from fruit peels

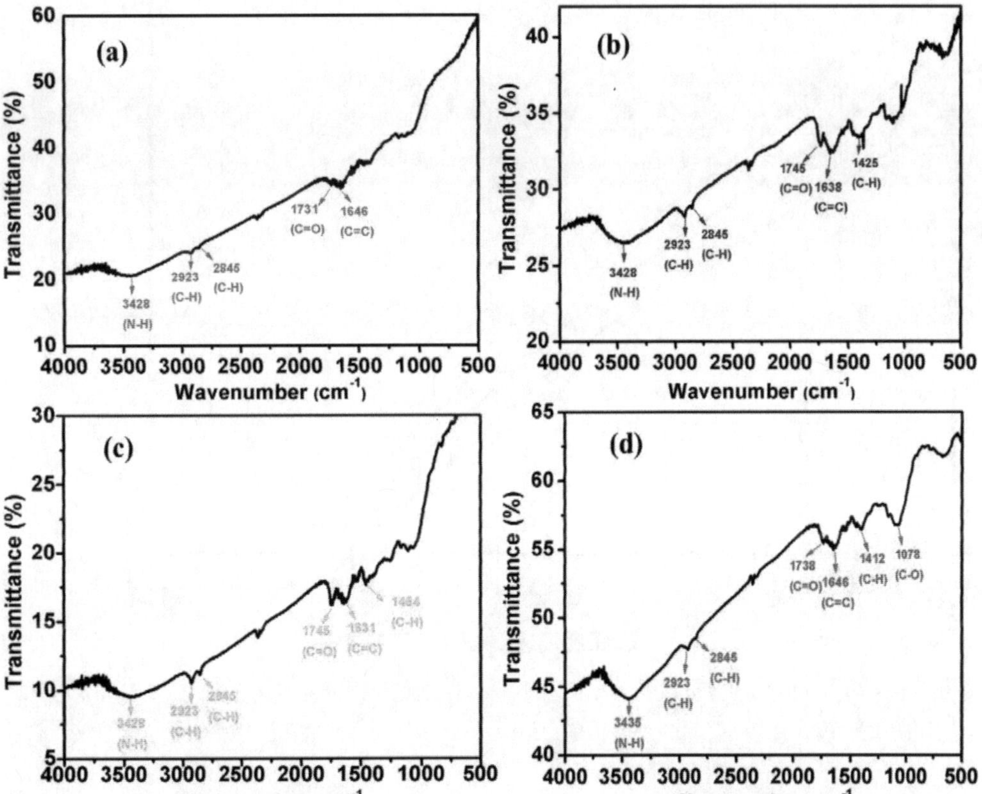

Figure 6: FTIR spectra of natural dye obtained from fruit peels (a) *Carica papaya* *(b) Citrus lanatus* (c) *Persea americana* and (d) *Solanum melongena*

is significant in enhancing the energy conversion efficiency of DSSCs. Some of the natural dyes used in this work may have no or poor availability of bonds between the dye and ZnO molecules through which electrons can transport from excited dye molecules to the ZnO film [26]. The lower efficiency recorded from the solar cells fabricated from natural dye as photo-sensitizers is owing to the absence of some particular functional groups which may cause poor dye molecules adsorption onto the surfaces of ZnO.

The natural pigments employed as photo-sensitizers in DSSC exhibited very low efficiencies compared to synthetic dyes because of non-existence of better interaction between the dye and ZnO surfaces through which electrons can move from the part of molecules of excited dye to ZnO film [3,4]. The performance of short-circuit (Jsc), depends on the quality of molecules of dye adsorbed to photo-anode surface, structure of the dye, efficiency of light harvesting and the injection of electron capacity of dyes in DSSC [6]. More molecules of dye available on the ZnO surface produce more photon number from sunlight, which in turn generates faster electron injection. Also the fill factors for these cells are mostly lower than 30%. This is due to high resistance in the cell and it perhaps responsible for the poor performance of the fabricated cells. From the above results, the *Carica papaya* natural extract may be worthy substitute for the solar cell application.

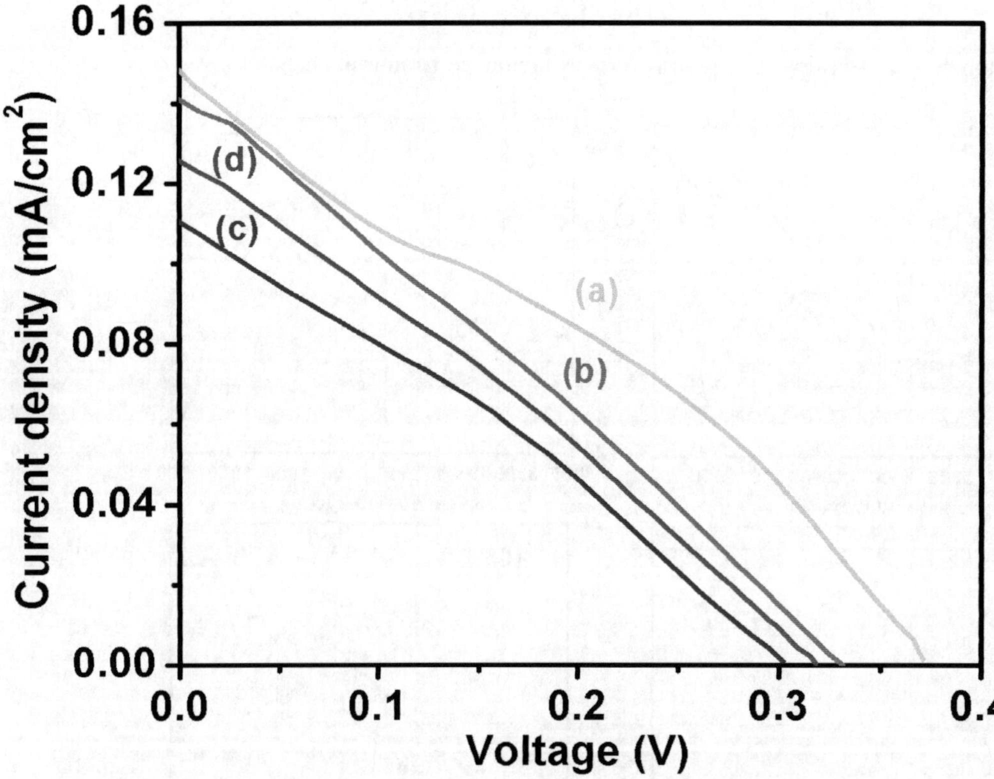

Figure 7: Current density versus voltage curves for ZnO based DSSCs sensitized by (a) *Carica papaya* (b) *Citrus lanatus* (c) *Persea Americana* (d) *Solanum melongena* alcoholic extracts

Table 1: Photo-electrochemical parameters for natural dye DSSCs sensitized by *Carica papaya*, *Citrus lanatus*, *Persea americana* and *Solanum melongena* alcoholic extracts with ZnO as photo-anode

Natural Dye	V_{OC} (V)	J_{SC} (mA/cm^2)	FF (%)	Efficiency, η (%) ± 0.05
Carica papaya	0.373	0.149	30.37	0.017
Citrus lanatus	0.333	0.140	26.78	0.013
Persea americana	0.303	0.111	29.51	0.010
Solanum melongena	0.321	0.126	28.24	0.011

4. Conclusion

The natural pigments from *Carica papaya*, *Citrus lanatus*, *Persea Americana and Solanum melongena* fruit peels were successfully employed as natural photo-sensitizer in DSSCs. The UV-Vis absorption and PL properties of extracted pigments show maximum absorption and photoemission identical to the pheophytin 'a' along with small identifiable parts. The FT-IR data for the natural dye resemble that of structural properties of pheophytin 'a'. The photovoltaic performance reveals that the maximum efficiency acquired for the fabricated cell with *Carica papaya* peels extract is 0.017%, while the fabricated cells efficiency with *Citrus lanatus*, *Persea Americana and Solanum melongena* extracts are 0.013%, 0.010% and 0.011%, respectively. The higher efficiency observed for DSSC with *Carica papaya* peels is owing to better adsorption of dye molecules available to ZnO film and enormous opposition to charge transfer at ZnO-dye-electrolyte interface. Hence the results acquired are encouraging and promising as a consequence of their natural abundance, simplicity of preparation and environmental conviviality. Therefore, further work is recommended to justify the constituents of *Carica papaya* dye and more optimization of the cells.

Acknowledgement
OA gratefully acknowledges CSIR-TWAS for TWAS-CSIR Postgraduate Fellowship and Director CSIR-CGCRI for hosting OA as a TWAS-CSIR fellow at CGCRI, India. OA also acknowledges the help rendered by Dr. P. Sujatha Devi, for providing all the facilities for the synthesis and characterization of materials and for technical discussion while carrying out the work during the stay of OA in CGCRI. OA also acknowledges the central instrumentation facilities of CSIR-CGCRI and Mr. Anurag Roy, for their assistance.

References

[1] B.Q. Liu, X.P. Zhao and W. Luo," The synergistic effect of two photosynthetic pigments in dye-sensitized mesoporous TiO_2 solar cells", *Dyes pigments*, vol. 76, pp. 327-331, 2008.

[2] B. O'Regan and M. Gratzel, "A low cost, high efficiency solar cell based on dye-sensitized colloidal TiO_2 films", *Nature,* vol. 353, pp. 737-740, 1991.

[3] O. Adedokun, K. Titilayo and A.O. Awodugba, "Review on natural dye-sensitized solar cells (DSSCs)", *International Journal of Engineering Technologies*, vol. 2, pp. 6-13, 2016.

[4] M. Gratzel, "Review Dye-sensitized solar cells" *Journal of Photochemistry and photobiology-photochemistry Rev.*, vol. 4, pp. 145-153, 2003.

[5] M. Alhamed, A.S. Issa and A.W. Doubal "Studying of Natural Dyes properties as photosensitizer for Dye-sensitized solar cells (DSSC)", *Journal of Electron devices*, vol. 16, pp. 1370-1383, 2012.

[6] R. Jose, V. Thavasi, S. Ramakrishna, "Metal oxides for dye-sensitized solar cells", *J. Am. Ceram. Soc.*, vol. 92, pp. 289-301, 2009.

[7] S. Mathew, A. Yella, P. Gao, R. Humphry-Baker, F.E. Curchod Basila, N. Ashari-Astani, I. Tavernelli, U. Rothlisberger, Md. K. Nazeeruddin and M. Gratzel, "Dye-sensitized solar cells with 13% efficiency achieved through the molecular engineering of porphyrin sensitizers", *Nature Chemistry*, vol. 6, pp. 242-247, 2014.

[8] T. Bessho, S.M. Zakeeruddin, C.-Y. Yeh, E.W.-G. Diau, M. Gratzel, "Highly efficient mesoscopic dye-sensitized solar cells based on donor-acceptor-substituted porphyrins", *Angew.Chem*, vol. 122, pp. 6796-6799, 2010.

[9] W.M. Campbell, K.W. Jolley, P. Wagner, K. Wagner, P.J. Walsh, K.C. Gordon, L. Schmidt-Mende, Md.K. Nazeerudin, Q. Wang, M. Gratzel and D.L. Officer, "Highly efficient porphyrin sensitizers for dye-sensitized solar cells", *J. Phys. Chem. C*, vol. 111, pp. 11760-11762, 2007.

[10] H. Choi, S.O. Baik, S.O. Kang, J. Ko, M.-S. Kang, Md.K. Nazeerudin and M. Gratzel, "Highly efficient and thermally stable organic sensitizers for solvent-free dye-sensitized solar cells", *AngewChem.*, Vol. 120, pp. 333-336, 2008.

[11] G.R.A. Kumara, S. Kaneko, M. Okuya, B. Onwona-Agyeman, A. Konno and K. Tennakone, "Shiso leaf pigments for dye-sensitized solid-state solar cells", *Sol. Energy Mater. Sol. Cells*, vol. 90, pp. 1220-1226, 2006.

[12] X.-F. Wang, Y. Koyama, O. Kitao, Y. Wada, S.-I. Sasaki, H. Tamiaki and H. Zhou, "Significant enhancement in the power-conversion efficiency of chlorophyll co-sensitized solar cells by mimicking the principles of natural photo-synthetic light-harvesting complexes", *Biosens. Bioelectron*, vol. 25, pp. 1970-1976, 2010.

[13] G. Calogero, I. Citro, G. Di Marco, S. Armeli Minicante, M. Morabito and G. Genovese, "Brown seaweed pigment as a dye source for photoelectrochemical solar cells", *Spectrochim. Acta Part A*, vol. 117, pp. 702-706. 2014.

[14] V. Shanmugam, S. Manoharan, A. Sharafali, S. Anandan, R. Murugan, "Green grasses as light harvesters in dye sensitized solar cells", *Spectrochim. Acta Part A: Molecular and Biomolecular spectroscopy*, vol. 135, pp. 947-952, 2015.

[15] E. Yamazaki, M. Murayama, N. Nishikawa, N. Hashimoto, M. Shoyama and O. Kurita, "Utilization of natural carotenoids as photosensitizers for dye-sensitized solar cells", *Sol. Energy*, vol. 81, pp. 512-516, 2007.

[16] N.M. Gomez-Ortiz, I.A. Vazguez-Maldonado, A.R. Perez-Espadas, G.J. Mena-Rejon, J.A. Azamar-Barrios and G. Oskam, "Dye-sensitized solar cells with natural dyes extracted from achiote seeds", *Sol. Energy mater. Sol. Cells*, vol. 94, pp. 40-44, 2010.

[17] N.J. Cherepy, G.P. Smestad, M. Gratzel and J.Z. Zhang, "Ultrafast electron injection: implications for a photoelectrochemical cell utilizing an anthocyanin dye-sensitized TiO_2 nanocrystalline electrode", *J. Phys. Chem. B*, vol. 101, pp. 9342-9351, 1997.

[18] V. Shanmugam, S. Manoharan, S. Anandan, R. Murugan, "Performance of dye-sensitized solar cells fabricated with extracts from fruits of ivy gourd and flowers of red frangipani as sensitizers", *Spectrochimica Acta Part A*, vol. 104, pp. 35-40, 2013.

IOP Conf. Series: Earth and Environmental Science **173** (2018) 012040 doi:10.1088/1755-1315/173/1/012040

[19] G. Calogero, J.-H. Yum, A. Sinopoli, G. Di Marco, M. Gratzel M. and Md K. Nazeeruddin, "Anthocyanins and betalains as light-harvesting pigments for dye-sensitized solar cells", *Sol. Energy,* vol. 86, pp. 1563-1575, 2012.

[20] O. Adedokun, Y.K. Sanusi and A.O. Awodugba. Pigment Extracts of Citrus Peels as Light Sensitizers for Dye-Sensitized Solar Cells. Journal of Materials Sciences and Applications, 3(1): 1-7, 2017.

[21] Chou, T. P., Zhang, Q., Fryxell, G. E., and Cao, G. Z. (2007). Hierarchically Structured ZnO Film for Dye-Sensitized Solar Cells with Enhanced Energy Conversion Efficiency" *WILEY, Advanced Material,* 19 (18): 2588- 2592

[22] S.M. Milenkovic, J.B. Zvezdanovic, T.D. Andelkovic and D.Z. Markovic, "The identification of chlorophyll and its derivatives in the pigment mixtures: HPLC-chromatography, visible and mass spectroscopy studies", *Advanced Technology,* vol. 1, pp. 16-24, 2012.

[23] Liu, T.M., Chu, S.W., Sun, C.K., Lin, B.L., Cheng, P.C., and Johnson, I. (2001). Multiphoton Confocal Microscopy Using a Femtosecond Cr:Forsterite Laser. *Scanning*, 23: 249-254.

[24] Lai, W. H., Hsun, Y., Teoh, L. G. and Hona, M. H. (2008). Commercial and natural dyes as photosensitizers for a water-based dye-sensitized solar cell loaded with gold nanoparticles. *Journal of Photochemistry and Photobiology A: Chemistry,* 195: 307-313.

[25] Furukawa, H., Oba T., Tamiaki, H., and Watanabe, T. (1999). Diastereoselective Self-Assemblies of Chlorophylls a and a′, *Journal Physical Chemistry B*, 103: 7398-7405.

[26] Longyue, Z., Songyuan, D., Weiwei, X., and Kongjia, W. (2006). Dye-Sensitized Solar Cells Based on ZnO Films", *IOPScience, Plasma Science & Technology,* 8 (2): 172-175.

[27] Rani, S., Suri, P., Shishodia, P.K., and Mehra, R.M. (2008). Synthesis of nanocrystalline ZnO powder via sol-gel route for dye-sensitized solar cells. *Solar Energy Materials and Solar Cells,* 92: 1639-1645.

[28] El-Agez, T.M., El Tayyan, A.A., Alkahlout, A., Taya, S.A., and Abdel-Latif, M.S. (2012). Dye sensitized solar cells based on ZnO films and natural dyes. *International Journal of Materials and Chemistry,* 2 (3): 105-110.

Synthesis of Alumina-Supported Chicken Eggshell Catalyst for Transesterification of Waste Cooking Oil

Baaki Monica Alueshima[1*], Elizabeth J Eterigho[2] and Omale David Friday[3]

[1, 2 & 3]Chemical Engineering Department, Federal University of Technology, Minna, Niger State, P.M.B. 65, Nigeria

* *Corresponding author*; baakimonica@gmail.com Telephone; 07038261982, 08072564141

Abstract

This research presents the synthesis of alumina-supported chicken eggshell for the transesterification of waste cooking oil. The major materials used in the research are from renewable sources because the research focuses on addressing issues associated with renewable energy and the environment. Biodiesel was produced from waste cooking oil using alumina-supported calcium oxide from chicken eggshell. The alumina was leached from kaolinite clay using hydrochloric acid while the eggshell was from chicken. The prepared catalyst was characterized using XRF, FT-IR, BET, and TGA. Using a 2^4 factorial design, the catalyst synthesis was optimized and the effect of the parameters used in catalyst synthesis such as the impregnation ratio and time, calcination temperature and time, on the yield of biodiesel from waste cooking oil were investigated. The optimization study showed that; impregnation ratio of 1.17:1, 1.32 hours of impregnation, calcination at 740.31°C for 2.62 hours gave the best results of 75.4% with 95% confidence level. The conditions for transesterification were methanol to oil molar ratio of 12:1, 6wt% catalyst, 60°C reaction time, and mixing speed of 250rpm for ninety minutes. The result of the characterization of the biodiesel produced using the CaO/Al_2O_3 produced proves that the biodiesel compares favourably with ASTM standards.

Keywords: Biodiesel, waste cooking oil, eggshell, kaolin, alumina, calcium oxide, transesterification, optimization.

1.0 Introduction

Rising world population has culminated higher demand for energy; unfortunately, this has not been met with higher supply of energy [1]. This increased demand has put a lot of pressure on fossil fuels which is the major source from which this global energy demand is met. Since these fuels are non-renewable, they are being depleted. In addition, global warming is another major environmental concern caused by the exploration and use of fossil fuels. Hence, the need for alternative renewable fuels that are environmentally friendly [2]. Among these alternatives are solar energy, wind energy, hydrogen (H_2) energy and most importantly biodiesel which is considered as a way out of the issues of energy sustainability and environmental pollution. Out of many methods for biodiesel synthesis from vegetable oils, transesterification happens to be widely used owing to its economy and simplicity. Transesterification is a reaction where the alkoxy group present in an ester compound is exchanged by another alcohol in the presence of either an acid or base catalyst [3]. Several alcohols can be employed for the process; however, methanol and ethanol are the most commonly used. The feedstock for biodiesel production includes edible oils such as canola, rapeseed, sunflower, corn and vegetable oil [4]. However, the use of these edible oils been discouraged because of it competition as food [5]. Therefore non-edible oils such as Jatropha, neem, rubber seed and castor oil can provide an alternative because of its incredible potential for biodiesel product ion with low acid and good oxidation stability [6]. But the cost of these oils is also not sustainable as it results in the very high cost of the biodiesel produced. This has drawn attention to the potential of waste cooking oil as a potential feedstock for biodiesel production. Using the waste oil for biodiesel means recycling the oil, hence curbing the problem of improper disposal and high cost of biodiesel feedstock since these waste cooking oil is readily available [7].

Hitherto, the catalysts used in the synthesis of biodiesel via transesterification have been majorly homogeneous acid or alkaline based catalyst such as sulphuric acid and sodium hydroxide [8]. The advantage of this catalyst is that the ester yield is high and the reaction occurs at a comparatively fast rate. But in most cases, the resulting cost of production using homogeneous catalyst is high because of the need to wash off the soap formed during the production process, thereby increasing the total production cost and limiting biodiesel commercialisation [9]. In recent times, heterogeneous catalysts have gotten so much attention in the biodiesel production process because of their reusability, environmental friendliness, low cost, high quality yield and efficiency in transesterification of triglycerides when compared to their homogeneous counterparts. Even of greater interest is the fact that some of these heterogeneous catalysts could be sourced from both domestic and industrial wastes, as well as biological resources, [10]. Hitherto, these materials were regarded as waste and discarded. But with several research and developments in biodiesel production, they have been reported to be of great value in the production of biodiesel. This is because of the great advantages inherent in these materials as catalysts. They are biodegradable, they can be reused, and they produce a high quality biodiesel with high yield [11]. Because of their environmental acceptability, they are been regarded as green catalysts. Their use is economically sustainable because they occur in abundance in our environment [12]. Their use in biodiesel synthesis as catalysts provides a solution to the problem of disposal. Their production process is also not expensive thereby ensuring a cost effective biodiesel production process, [10]. Research has presented these locally sourced heterogeneous catalysts as good alternatives to the synthetic heterogeneous catalysts, since the biodiesel produced should compete favourably with petroleum diesel as against that produced using homogenous catalysts [13].

Calcium oxide is an example of a heterogeneous catalyst that is abundantly and cheaply available all around our environment. It occurs naturally in animal bones, and shells such as eggshells, oyster shells, and snail shells [14]. According to [11], eggshells are a rich source for calcium oxide as heterogeneous catalyst for the production of biodiesel. In as much as heterogeneous catalysts have the advantage of high conversion and yield, researchers have encountered the problem of catalyst leaching with the use of calcium oxide [15] Several authors have reported that after a few cycles, the solid catalyst leach into the reacting medium thereby reducing the quantity and quality of the ester produced [16]. In order to curb this challenge, several supports have been developed upon which these solid catalysts are anchored inorder to

IOP Conf. Series: Earth and Environmental Science **173** (2018) 012041 doi:10.1088/1755-1315/173/1/012041

reduce leaching and increase the number of times they can be reused for a reaction. Anchoring the catalyst on a support enhances the value of the catalyst and reduces the cost of production. Catalyst supports consist of different materials. Silica, alumina, potassium hydroxide and chromium have been used by different researchers. A study by [4] and [17] presents gamma-alumina as a good support for eggshell derived calcium oxide. The alumina used in this work was sourced from kaolinite clay.

Calcium oxide supported with alumina is an example of solid catalyst that can be synthesized with characteristics that permit catalytic properties that will improve the reaction [18]. In view of all these, this research attempts to explore the use of calcium oxide from eggshell with alumina support as heterogeneous catalyst in the production of biodiesel from waste cooking oil. The research also attempts to emphasize the use of local content as the catalyst produced is entirely from local materials.

2.0 Experimental

2.1 Catalyst synthesis

Kaolinite clay was gotten from Kutigi in Niger state. Alumina was extracted from kaolinite clay using hydrochloric acid by leaching [19], [20]. The waste chicken eggshells were collected from a local restaurant in Minna, Niger state; washed thoroughly, dried at 100°C for 12 hours, grinded to 125μm mesh size and calcined at 900°C for 4 hours. The calcium oxide was dissolved in 50ml of distilled water containing 10g of alumina. The quantity of the calcium oxide used was in ratios of 1:1 and 2:1 with the alumina. The slurry was stirred for a period of 1 and 2 hours at room temperature using a magnetic stirrer. After which the slurry was oven dried at 100 °C for 12 hours and the dry sample was calcined over a temperature range of 650 to 750 °C over a time range of 2 to 3 hours. The impregnation ratio, time, calcination temperature and time were varied using 2^4 factorial design. The catalyst was characterized using FT-IR, XRF, BET, and TGA, [17].

2.2 Transesterification

100ml of the waste cooking oil was pre-treated before it was poured into a conical flask and heated to a temperature of 60^0C. A mixture of concentrated H_2SO_4 (1 % w /w) with methanol (30 % v/v) was separately heated at (60^0C) and added to the heated oil in the flask. The mixture was stirred for 1 hour and allowed to settle for 2 hours in a separating funnel. The clean oil was then withdrawn. 30 ml of pre-treated waste cooking oil was poured into a conical flask and heated in a water bath until the oil attained a temperature of 60°C. 50 ml of methanol and 6 wt. % catalyst was added to the oil and the mixture was returned back to the water bath at 60°C and mixture stirred as 250 rpm for 90 minutes. The methanol to oil ratio was 12:1, [17]. The conical flask was equipped with a stopper to prevent the escape of methanol by evaporation. After 90 minutes the mixture was removed and the catalyst was separated by filtration. The liquid was then poured into a separating funnel and left overnight for separation of the biodiesel from the glycerol by gravity. The glycerol which formed the lower layer was collected and the biodiesel yield was determined, [21]. The catalyst reusability was tested for 5 runs under the same conditions of reaction, [12]. The biodiesel was characterized to determine the physicochemical properties and the fatty acid profile was determined by gas chromatography.

3.0 Results and Discussion

3.1 Catalyst Analysis

The X-ray fluorescence carried out on the CaO/Al_2O_3 catalyst showed that CaO and alumina were present as major components in the catalyst in nearly equal amounts. While other oxides like silicon oxide, zinc oxide, titanium oxide, and iron oxide were present in very minute quantities as impurities, the alumina and calcium oxide have the largest percentage compositions of 46.163% and 45.472% respectively. This result reflects the 1:1 ratio of impregnation used in preparing the catalyst. For the FTIR, in the O–H stretching region, the sample shows prominent bands at 3672 - 3996cm^{-1} corresponds to Al–OH stretching. Inner hydroxyl groups, lying between the tetrahedral and octahedral sheets, give the absorption at 3672cm^{-1}. A strong band at 3620cm^{-1} is related to the in phase symmetric stretching and a weak absorption at 3561cm^{-1} is assigned to out-of-plane stretching vibrations. The band observed at 3063 - 3562cm^{-1}, assigned to the high amount of water physisorbed on the surface of the clay [22]. In the bending region mode, the catalyst show a series of IR bands with peak maxima at 1065, 1427 and 1512cm^{-1}. The peak at 1512cm^{-1} is quite intense and can be attributed to the bending vibration mode of physisorbed water on the surface of catalyst produced due to leaching [23]. The 1065 and 1427cm^{-1} bands is due to the stretching vibrations of the CO_3^{2-} which is as a result of the complete decomposition of the $CaCO_3$ in the chicken eggshell [24]. IR peaks at 2654 - 2809cm^{-1} and 2021 - 2577cm^{-1}can be assigned to the Al–Al–OH and Si–O–Al vibration of the catalyst sheet. Again well resolved strong bands in the 525 and 578cm^{-1} regions are due to Si–O stretching in the catalyst. Strong IR spectral lines characteristic for CaO are placed in the far-infrared range (~400cm-1 and 290cm-1) and one weak band is in the range from 500cm^{-1} to 560cm^{-1}. This peak is due to the formation of amorphous catalyst produced.

The summary of the BET analysis revealed a very high surface area of 340m^2/g for the alumina supported calcium oxide catalyst with a pore volume and mean pore radius of 0.1351cm^3g^{-1} and 7.208 **Å** respectively. The surface area is actually high compared to the previous studies reported in various literatures. A study by [25] reported a surface area of 83.77m^2g^{-1} for alumina supported calcium oxide catalyst, though the alumina loading was just 30% unlike the 100% loading used in this research. This result also varies with the report by [17] where the surface area for the calcium oxide supported with alumina in a 1:1 ratio was 82.74m^2g^{-1}. A study by [26] reported that alumina has very large specific area which enhances high dispersion of active sites. According to [27], a combination of high surface area, large pore volume and smaller pore sizes allows triglycerides of different sizes to enter the pores of the catalyst, and as well gives a large surface for proper transesterification. This combination therefore makes for a highly active catalyst.

The thermal stability of the catalyst was determined via thermo gravimetric analysis (TGA). A plot of the results obtained showed three different stages of weight loss with increasing temperature. At the first stage, the weight of the catalyst was seen to have reduced steadily between temperature ranges of 60°C -160°C. This was followed by a second stage between 160°C and 360°C which showed a drastic weight loss with increasing temperature. According to [24], this weight loss could be attributed to loss in organic matter. The last stage of weight loss was similar to the first stage, the weight loss occurred steadily between 360°C and 600°C. Above 600°C, the catalyst maintained a constant weight. This means that the catalyst is thermally stable above 600°C. Figure 1 is a TGA plot of the CaO/Al_2O_3 catalyst.

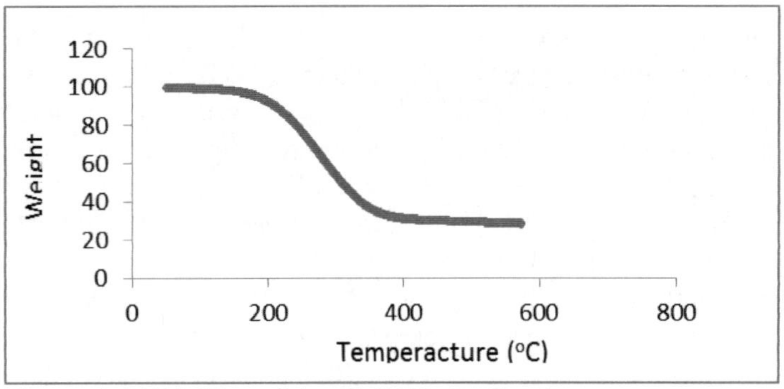

Figure 1: TGA Plot of the CaO/Al_2O_3 Catalyst.

3.2 Design of Experiment

The design of experiment was done using a 2^4 experimental design to investigate the effect of the process variables on the yield of the biodiesel. Table 1 shows the experimental design and the results obtained.

Table 1: Experimental Design Catalyst

Std	Run	F1	F2	F3	F4	Response 1	Response 2
1	7	1:1	1	650	2	82.59	40.03
2	11	2:1	1	650	2	87.63	23.2
3	3	1:1	2	650	2	85.04	48.51
4	12	2:1	2	650	2	81.80	28.62
5	2	1:1	1	750	2	74.30	76.83
6	10	2:1	1	750	2	77.63	50.02
7	15	1:1	2	750	2	73.01	85.89
8	14	2:1	2	750	2	79.92	57.02
9	9	1:1	1	650	3	84.09	68.07
10	5	2:1	1	650	3	88.72	37.23
11	8	1:1	2	650	3	78.27	71.06
12	1	2:1	2	650	3	83.54	34.05
13	16	1:1	1	750	3	79.20	82.40
14	6	2:1	1	750	3	81.71	48.23
15	13	1:1	2	750	3	76.67	90.10
16	4	2:1	2	750	3	78.05	55.71

**where; F1 = Catalyst impregnation ratio, F2 = Impregnation time (hours), F3 = Calcination temperature (^oC) and F4 = Calcination time (hours). Response 1 = Percentage catalyst yield; Response 2 = Percentage Biodiesel yield.*

3.3 Analysis of Variance (ANOVA) on Yield of Biodiesel

The adequacy and significance of the experimental data obtained from the investigations of the effect of process parameters used in catalyst production on the yield of biodiesel was emphasized using the analysis of variance (ANOVA). Tables 2 and 3 give the details of the analysis of variance for the 2^4 factorial design for the effect of process parameters used in calcium oxide-alumina as a catalyst on the yield of biodiesel.

Table 2: ANOVA for 2^4 Factorial Design on the Yield of Biodiesel

Source	Sum of Squares	Df	Mean Square	F Value	p-value Prob > F	
Model	6625.89	13	509.68	624.53	0.0016	Significant
A-Impregnation Ratio	3272.13	1	3272.13	4009.44	0.0002	
B-Impregnation Time	126.28	1	126.28	154.74	0.0064	
C-Calcination Temp.	2387.06	1	2387.06	2924.93	0.0003	
D-Calcination Time	367.97	1	367.97	450.88	0.0022	
AB	8.28	1	8.28	10.15	0.0860	
AC	24.18	1	24.18	29.63	0.0321	
AD	121.06	1	121.06	148.33	0.0067	
BC	19.21	1	19.21	23.53	0.0400	
BD	14.01	1	14.01	17.16	0.0536	
CD	250.98	1	250.98	307.54	0.0032	
ABC	3.02	1	3.02	3.70	0.1944	
ACD	20.82	1	20.82	25.51	0.0370	
BCD	10.91	1	10.91	13.36	0.0674	
Residual	1.63	2	0.82			
Cor Total	6627.52	15				

The Model F-value of 624.53 implies that the model is significant. There is only a 0.16% chance that a Model F-value this large could occur due to noise (induced variation under normal operating conditions by uncontrollable factors).

IOP Conf. Series: Earth and Environmental Science **173** (2018) 012041 doi:10.1088/1755-1315/173/1/012041

Prob > F values less than 0.0500 indicates model terms are significant. In this case A, B, C, D, AC, AD, BC, CD, ACD are significant model terms. Values greater than 0.1000 indicate the model terms are not significant. In this experiment, only one model term; ABC has a value above 0.1000 which is the only insignificant term in the developed 2^4 factorial model. The values of R^2, adjusted-R^2 and standard deviation give an understanding of the accuracy of the model that has been developed. R^2 reflects the extent to which the model perfectly estimated the experimental data points. In this case, there is a reasonable agreement between the predicted R-squared value of 0.9842 and the adjusted R-squared value of 0.9982. This close agreement between the predicted R-squared value and the adjusted R-squared value reflects the closeness of the experimental data to the predicted values and also the extent of accuracy of the developed model, its correctness and applicability. Adequate precision gives the ratio of signal to noise. Usually, a ratio above 4 is desirable, which means that the 78.98 ratio gotten from this model is an indication of an accurate signal. This model can therefore be used to navigate the design space. Determining the value of the coefficient of variance is of utmost importance because the value gives the ratio of the standard error of estimate to the mean value of the observed response as a percentage and also helps to measure the reproducibility of the model. The coefficient of variance for this model was found to be 1.61%. This value is less than 10% of the biodiesel yield which makes the model reproducible. Table 4 gives details of the statistical parameters obtained from the analysis of variance (ANOVA).

Table 3: Statistical Parameters Obtained from the Analysis of Variance (ANOVA) on Biodiesel Yield

Variable	**Biodiesel**
Standard deviation	0.9
Mean	56.06
Coefficient of variation (%)	1.61
PRESS	104.46
R- Squared (R^2)	0.9998
Adjusted-R^2	0.9982
Pred. R^2	0.9842
Adeq. Precision	78.98

The model equation in terms of coded factors for biodiesel yield is given as;

$$Y = +56.06 - 14.30 * A + 2.81 * B + 12.21 * C + 4.80 * D - 0.72 * A * B - 1.23 * A * C - 2.75 * A * D + 1.10 * B * C - 0.94 * B * D - 3.96 * C * D + 0.43 * A * B * C + 3.69 * A * C * D + 1.14 * A * C * D + 0.83 * B * C * D \quad (1)$$

But values greater than 0.1000 indicates that the values are insignificant and as such ABC is insignificant which tends to zero and the final model equation for biodiesel yield is given as;

$$Y = +56.06 - 14.30 * A + 2.81 * B + 12.21 * C + 4.80 * D - 0.72 * A * B - 1.23 * A * C - 2.75 * A * D + 1.10 * B * C - 0.94 * B * D - 3.96 * C * D + 3.69 * A * C * D + 1.14 * A * C * D + 0.83 * B * C * D \quad (2)$$

Equation 2 describes how the yield of biodiesel was affected by the parameters used in the production of the calcium oxide-alumina catalyst.

3.4 Influence of parameters used in catalyst production on yield of biodiesel.

The analysis of the experimental data obtained showed that all the four factors had a considerable effect on the biodiesel yield with the impregnation ratio and the calcination temperature being of greater effect because of their F-values. But in terms of interaction of the factors, the interaction between the calcination temperature and calcination time had the greatest effect on the yield of biodiesel as shown in Table 3 above. There should be a better response in the percentage yield of biodiesel when the calcium oxide-alumina catalyst is calcined at a higher temperature and time with lower impregnation ratio and time. This result agrees with the study done by [4] and [17] where calcium oxide from chicken eggshell impregnated with gamma alumina in a 1:1 ratio and calcined at 718°C gave the optimum biodiesel yield of 98%.

The three-dimensional response for the effect of the interactions of these parameters, according to the 2^4 factorial design is shown in Figures 2a to 2f which clearly represents the effect of the interactions of the process variables used in the production of the catalyst on the yield of the biodiesel.

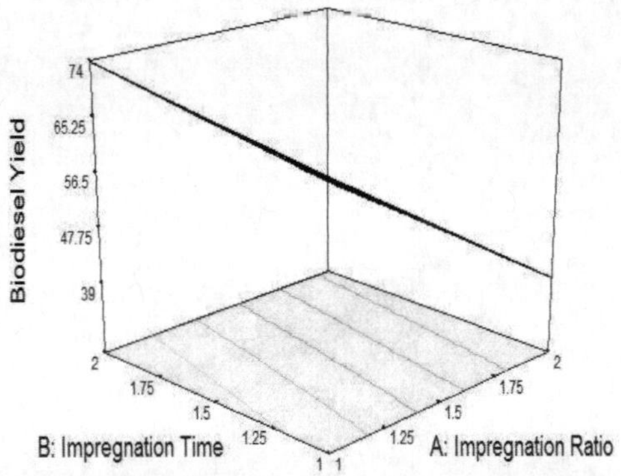

Fig. 2a: Effect of impregnation ratio and impregnation time on biodiesel yield.

High yields of biodiesel were gotten with a combination of lower impregnation ratio and higher impregnation time. From the model equation of biodiesel response, (Eqn. 1) impregnation ratio alone has a significant negative effect on the biodiesel response while the impregnation time has a significant positive effect. Since the impregnation ratio-time term of the biodiesel yield model equation is a linear term, the effect of the impregnation ratio will be more pronounced on the response. This means that loading a higher quantity of the calcium oxide on the alumina will require more time for impregnation in order to get a higher biodiesel yield. Lower biodiesel yields at lower impregnation time and high impregnation ratio could be due to improper mixing since the time of impregnation is shorter and catalyst loading higher. This result also agrees with the report by [17] that once the loading of calcium oxide on alumina increases beyond a ratio of 1:1, the yield of biodiesel begins to drop because of blockage of catalyst pore spaces by the precursor.

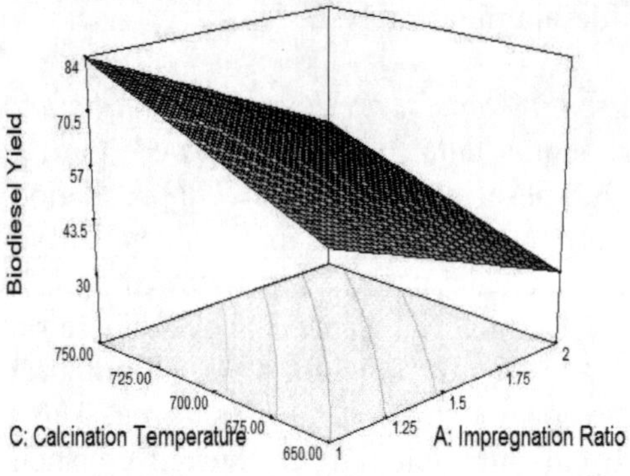

Fig. 2b: Effect of impregnation ratio and calcination temperature on biodiesel yield.

This represents the effect of the interaction of calcination temperature and impregnation ratio on the yield of biodiesel. The biodiesel yield can be seen to have increased with an increase in temperature and a decrease in impregnation ratio. This means that in order to get a higher yield, a combination of lower ratio of impregnation and a higher temperature of calcination is suitable.

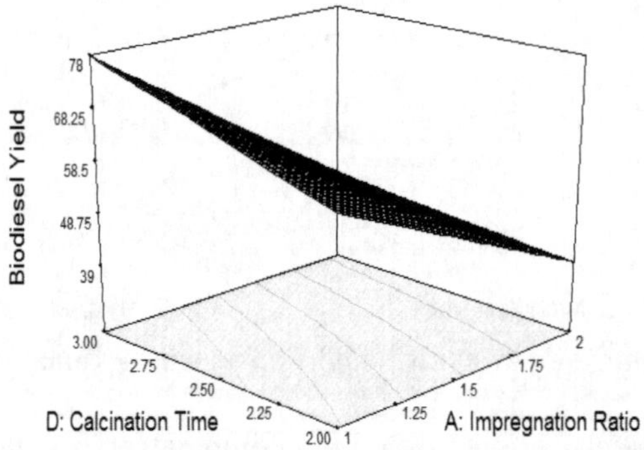

Fig. 2c: Effect of impregnation ratio and calcination time on biodiesel yield.

From the plot, a better biodiesel response is observed with a combination of longer hours of calcination and a lower impregnation ratio. It can also be seen that between two to three hours of calcination, the response is high with low impregnation ratio. But as the impregnation ratio is increased, the response begins to drop even with long hours of calcination. This can be explained from the model equation for biodiesel yield (Equation 1), while calcination time has a significant positive effect on the biodiesel response, the impregnation ratio has a significant negative effect on the response and a combination of these terms gives a significant negative effect on the yield of biodiesel. This means that the ratio used in impregnation is of utmost importance as it has a strong effect on the biodiesel response [4].

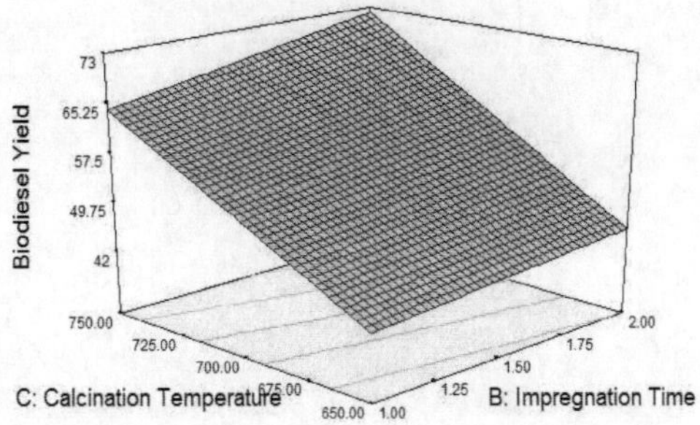

Fig. 2d: Effect of impregnation time and calcination temperature on biodiesel yield.

It is observed that the yield of biodiesel increased with increase in calcination temperature and impregnation time. Worthy of note is the fact that there seems to be a very slight difference in the yield at lower impregnation time and that at the higher. From the plot, there seems to be a greater effect on the yield from the calcination temperature than that from the hours used in impregnation. From Equation 1, the calcination temperature-impregnation time term of the equation is linear and has a positive term. This implies that the interaction of these factors has a significant positive effect on the biodiesel response. This means that when the catalyst is impregnated for longer hours and calcined at higher temperatures, the yield will be higher.

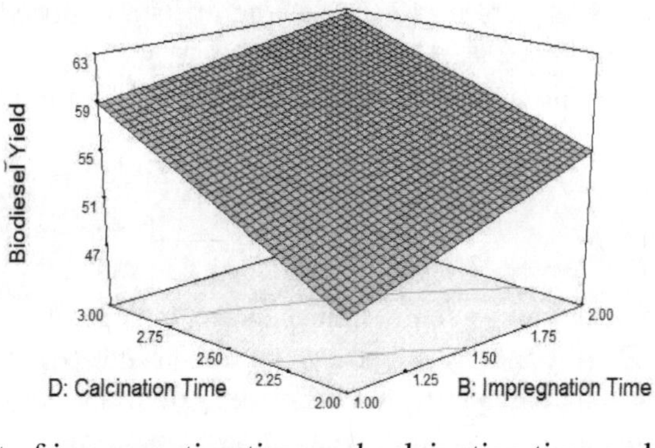

Fig. 2e: Effect of impregnation time and calcination time on biodiesel yield.

Figure 2e shows a positive effect on the response from the interaction of calcination time and impregnation time. This can be seen as the biodiesel yield increased linearly with increase in calcination time and impregnation time.

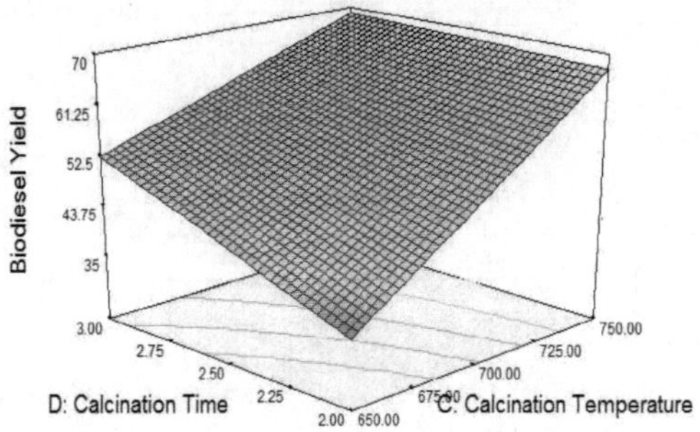

Fig. 2f: Effect of calcination temperature and calcination time on biodiesel yield.

There is a similar occurrence as observed in Figure 2e, the biodiesel yield increased from low to high with increase in calcination time and calcination temperature with calcination temperature giving a greater effect on the response.

Table 4.12: Results of the model evaluation

Run	Impreg. Ratio	Impreg. Time (hr)	Cal. T. (°C)	Cal. Time (hr)	Biodiesel Yield (%)	
					Experimental	Predicted
1	1.17:1	1.32	740.31	2.62	76.02	75.4

Cal=Calcination; T=Temperature; Impreg=Impregnation

Comparing the results obtained in table 4.12 above, the error margin between the experimental and predicted biodiesel yields is 0.62%, it can be conclusively said that the developed model is accurate and precise enough to predict the amount of biodiesel yield at any point in time using the produced catalyst.

4.0 Conclusion

The interaction between the calcination temperature and calcination time had the greatest effect on the yield of biodiesel. There could be a better response in the percentage yield of biodiesel when the calcium oxide-alumina catalyst from chicken eggshell and kaolinite clay is calcined at a higher temperature and time with lower impregnation ratio and time.

Optimum conditions of the reaction obtained were; impregnation ratio of 1.17:1, 1.32 hours of impregnation, calcination at 740.31°C for 2.62 hours. The biodiesel yield from these optimum conditions was predicted to be 75.4% with 95% confidence level. The conditions for transesterification were 12:1 methanol to oil molar ratio, 6wt% catalyst, 60°C reaction time, and mixing speed of 250rpm for ninety minutes. This result closely agrees with the reports from previous literature.

IOP Conf. Series: Earth and Environmental Science **173** (2018) 012041 doi:10.1088/1755-1315/173/1/012041

REFERENCES

[1] Jakeira, M. R., Fazal, M. A., and Haseeb, A. S. (2014). Influence ofdifferentfactorsonthestabilityofbiodiesel: Areview. *Renewable and Sustainable Energy Reviews*, 154-163.

[2] Tinwala, F., Mohanty, P., Parmar, S., Patel, A., and Pant, K. K. (2015). Intermediate pyrolysis of agro-industrial biomasses in bench-scale pyrolyser: Product yields and its characterization. *Bioresource Technology*, 258-264.

[3] Sajjadi, B., Raman, A. A., and Arandiyan, H. (2016). A comprehensive review on properties of edible and non-edible vegetable oil-basedbiodiesel: Composition,specifications and prediction models. *Renewable and Sustainable Energy Reviews*, 62-92.

[4] Zabeti, M., Wan Daud, A. W., and Aroua, M. K. (2009). Activity of solid catalysts for biodiesel production: A review. *Fuel Processing Technology*, 770-777. doi:doi:10.1016/j.fuproc.2009.03.0

[5] Verma, P., and Sharma, M. P. (2016). Review of process parameters for biodiesel production from different feedstocks. *Renewable and Sustainable Energy Reviews*, 1063-1071.

[6] Baroi, C., Yanful, E. K., and Bergougnou, M. A. (2009). Biodiesel Production from Jatropha curcas Oil Using Potassium Carbonate as an unsupported catalyst. *Internation Journal of Chemical Reactor Engineering, 7*, 1-20.

[7] Acquah, N., Obeng, G., and E., M. (2016). Repititive use of vegetable cooking oil and effects on physico-chemical properties - case of frying with red fish (lutjanus fulgens). *Journal of science and technology, 6*(1), 4-14.

[8] Sharma, Y. C., Singh, B., and Korstad, J. (2010). Application of an Efficient Nonconventional Heterogeneous Catalyst for Biodiesel Synthesis from Pongamia pinnata Oil. *Energy Fuels*, 1-9.

[9] Fadhil, A., Aziz, A. M., and Al-Tamer, M. H. (2016). Biodiesel production from Silybum marianum L. seed oil with high FFA content using sulfonated carbon catalyst for esterification and base catalyst for transesterification. *Energy Conversion and Management*, 255-265.

[10] Sanjay, B. (2013). Heterogeneous catalyst derived from natural resources for biodiesel production: A review. *Research journal of chemical sciences, 3*(6), 95-101.

[11] Jagadale, S., and Jugulkar, L. (2012). Review of various reaction parameters and other factors affecting on production of chicken fat based biodiesel. *international journal od modern engineering research, 2*(2), 407-411.

[12] Tan, Y., Abdullah, M., Nalasco-Hipolito, C., and Taufiq-Yap, Y. (2015). waste ostrich- and chicken eggshells as catalyst for biodiesel production from waste cooking oil: Catalyst charaterization and biodiesel yield performance. *Applied energy, 160*, 58-70.

[13] Semwal, S., Arora, A. K., Badoni, R. P., and Tuli, D. K. (2011). Biodiesel production using heterogeneous catalysts. *Bioresource Technology*, 2151-2161. doi:doi:10.1016/j.biortech.2010.10.080

[14] Pandit, P., and Fulekar, M. (2017). Eggshell waste as heterogeneous nanocatalyst for biodiesel production: optimized by response surface methodology. *journal of environmental management, 198*, 319-329.

[15] Elizabeth J. Eterigho, T. S. Farrow and S. E. Ejejigbe (2017),''Sulphated Zirconia Catalyst Prepared from Solid Sulphates by Non-aqueous Method'', Iranica Journal of Energy and Environment, Vol. 8, Issue 2, pp 142 – 146 www.IJEE.net

[16] Singh, V. y. (2017). Effect of co-solvent on biodiesel production using calcium aluminium oxide as a reusable catalyst and waste vegetable oil. *Fuel, 203*, 360-369.

[17] Zabeti, M., Daud, W. M., and Aroua, M. K. (2010). Biodiesel Production using alumina supported calcium oxide: An optimization study. *Fuel processing technology, 91*, 243-248.

[18] Muthu, K., and Viruthagiri, T. (2015). Study on solid base calcium oxide as a heterogeneous catalyst for the production of biodiesel. *Journal of advanced chemical sciences, 1*(4), 160-163.

[19] Ajemba, O. R., and Onukwuli, O. D. (2012). Kinetic model for Ukpor clay dissolution in Hydrochloric acid solution. *Journal of emerging trends in engineering and applied sciences, 3*(3), 448-454.

[20] Olaremu, A. (2015). sequencial leaching for the production of alumina from nigerian clay. *International journal of engineering technology, management and applied sciences, 3*(7), 103-109.

[21] Mahesh, S., Ramanathan, A., Begum, K., and Narayanan, A. (2015). Biodiesel production from waste cooking oil using KBr impregnated CaO as catalyst. *Energy conversion and management, 91*, 442-450.

[22] Madejova, J. (2003). FTIR Techniques in Clay Mineral Studies. *Vibration Spectroscopy, 31*, 1-10.

[23] Dudkin, B., Loukhina, I., Avvakumov, E., & Isupov, V. (2004). Application of Mechnochemical Treatment of Disintegration of Kaolinite with Sulphuric Acid. *Chem.Sust. Dev., 12*, 327-330.

[24] Rohazriny, R., Razi, A., Naimah, I., Nasrul, H., and Abidin, C. (2014). Characterization of Calcium oxide catalyst from eggshell waste. *Advances in Environmental Biology, 8*(22), 35-38.

[25] Asri, N., Diah, A., Puspita, S., Bambang, P., and Suprapto. (2015). Utilization of Waste Cooking Oil for Biodiesel Production Using Alumina Supported Base Catalyst. *3rd International Conference on Biological, Chemical & Environmental Sciences* (pp. 59-63). Kuala Lampur, Malaysia: BCES.

[26] Theodoro, P., Pedro, A., & Silva, E. A. (2017). Synthesis of bifunctional calcium oxide supported on gamma alumina for biofuel production from the transesterification of crambe oil . *Acid base catalysis*, 7-10.

[27] Sulaiman, N. B. (2017). Response surface methodology for the optimum production of biodiesel over Cr/Ca/r-Al2O3 catalyst: catalytic performance and physicochemical studies. *Renewable energy, 113*, 697-705.

Evolvement of Thermal Energy Storage Systems

I.E. Ogunrinola; M.L. Akinyemi; H.O. Boyo; O. Maxwell; A.Akinpelu; T.E. Arijaje

Department of Physics, Covenant University Ota, Nigeria

iyanuoluwa.ogunrinola@covenantuniversity.edu.ng

Abstract. A world with constant and unhindered supply of power from thermal sources is a feasible possibility. It is left to us researcher to find out the most optimal way to collect and store energy in this form. This review work seeks for thermal storage materials and methods to promote harnessing the thermal component of renewable and abundant solar energy at Covenant University, Ogun state, in Nigeria. The efficient implementation of thermal renewable energy sources is still being faced by a few challenges such as storage and efficiency. The review seeks methods of incorporating cost effective and efficient thermal energy storage techniques into the University campus energy system. Such techniques can be in form of thermal energy storage materials such as $LiNO_3$-NaCL containing 87 wt% of $LiNO_3$ and 13 wt% of NaCL with latent heat capacity of 300 kJ/kg and melting temperature of 220 °C. This paper reviews the evolution as well as trends of thermal energy storage systems considering the sensible and latent heat storage materials. Results of previous researches are discussed and these include the thermal properties of various proposed heat storage and heat transfer materials such as the heat capacity, latent heat capacity, melting point, thermal conductivity and maximum temperature of stability. These results were reviewed for the two categories of storage materials considered namely: latent and sensible heat materials.

Keywords: Thermal energy, thermal storage materials, sensible heat

1. Introduction

Over the years, production of energy from non-conventional energy sources have faced the following major challenges such as efficiency and storage. Industrial heat energy waste can be optimally utilized if proper energy storage technologies are employed. Exhaust heat energy from industries and power plants can be reused for domestic as well as the production of electricity at a small to medium scale. Another reason for energy storage is to compensate for the intermittent nature of renewable energy sources [1].

For the past few decades, researchers have come up with different technologies and methods of constructing a viable thermal storage system. The ideal thermal storage system consists of heat transfer fluid as well as heat storage material with both high thermal conductivity and thermal capacity and durability as well. In addition to these properties, this system will be cheap and portable [2,3]. According to [1], there are three (3) major key points to be considered as regards an efficient thermal storage system. These are: the environmental impact, technical properties of the system and cost. The technical properties are: storage capacity, thermal conductivity, thermal stability over time, compatibility and thermal loss [1].

Researchers have identified three main modes of thermal energy storage. These are sensible heat storage, latent heat storage and chemical heat storage. When a material stores or releases heat by the rise and fall of its temperature, the material can be said to be a sensible heat storage material. This material can either be in the solid or liquid phase. Examples of solid sensible heat materials are iron, steel and concrete. Their known working temperature is between 200 °C to 1200 °C. these materials have high thermal conductivity. Examples of liquid state sensible heat storage materials are nitrate salts, carbonate salts and synthetic oil. Molten salts are the most viable for use in thermal systems as a result of the following properties they possess: high thermal conductivity, low viscosity, low vapour pressure, thermal stability, non-toxic and non-flammable [4,5,6]. Sensible heat materials are relatively low cost compared to the other two class of materials. As cheap as these materials are, they have very low thermal capacity and this puts them at a disadvantage. Thermal systems built with these materials will be large compared to the amount of heat energy they store [1].

For the second category, latent heat storage materials store and release thermal energy when they change phase. They are also called phase change materials (PCM). Their thermal capacity is far greater than those of sensible heat storage materials [1]. Paraffin is an example of a phase change material. They absorb and release heat almost isothermally. There is no significant change in temperature during thermal absorption and dissipation. The major disadvantage of these materials are they have very low thermal conductivity. Their thermal conductivities in watts per meter-kelvin vary between 0.2 W/(mk) – 0.7 W/(mk) [5].

Solar energy is the most abundant source of energy today and every day the earth receives a huge dose of this energy. Most of it goes to waste. Another source of waste energy is industrial thermal waste. This wasted energy from the sources listed above can be put to good use through proper storage. This study is a review article on the evolution of thermal energy storage systems. The approaches as well as results of past and current researchers will be looked into.

2. Evolution of Thermal Storage Systems

Various improvements on the technology for thermal storage have been proposed over the years. From the time of sensible heat storage materials being the dominant storage media to the discovery of phase change materials and then to the use of highly reversible chemical reactions that absorb and discharge heat. This section is a review of some of the methods employed for optimal and cost effective thermal storage.

2.1. Sensible Heat Storage Materials

This section focuses on past work done on analyzing materials that fall under the first two categories which are the sensible heat storage and the latent heat storage materials. As regards sensible heat storage systems, [6] experimented on the most effective mixture of molten salts that will possess the following properties: low vapour pressure, low freezing point, good chemical stability at high temperatures and good thermal properties. They tested the melting temperatures of the following salts: 50-80 wt% of KNO_3, 0-25 wt% $LiNO_3$, and 10-45 wt% $Ca(NO_3)_2$. Their salt mixtures had the maximum melting temperature to be 450 °C. their viscosities were found to be lower than 10cP when within the range of operating temperature. The salts were stable and cost effective as well.

The study of sunflower oil as a viable sensible thermal storage material was the aim of [7]. Data was recorded on its performance under different conditions during charging, retention for 24 hours and discharging.

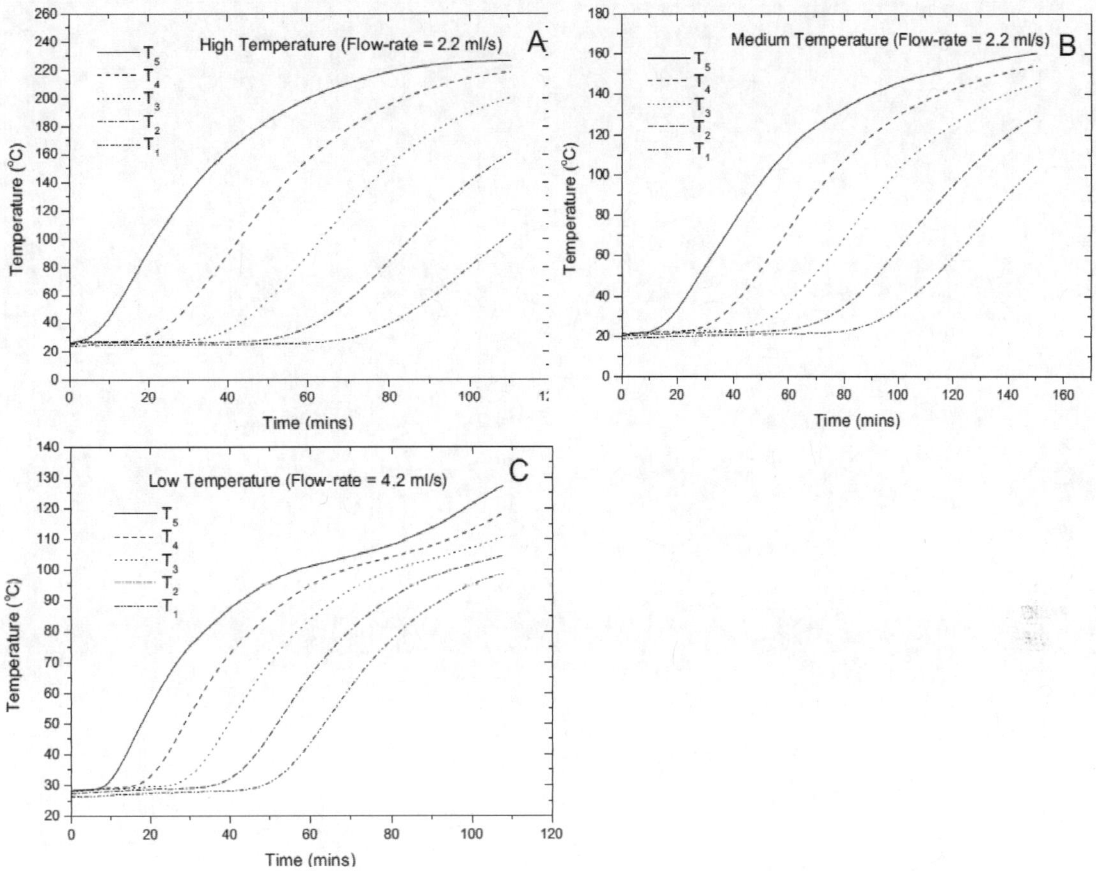

Fig. 1. Results of low, medium and high temperature charging conditions of sunflower oil [7]

From figure 1, as a result of the top of the tank having higher temperatures, thermal stratification was observed. Due to higher flow rates used in low temperature charging, the thermal stratification observed was significantly reduced. Higher flow rates resulted in faster heat transfer rate. In the case of the high temperature charging, the low flow rate enhanced thermal stratification. This served as a major factor that determined the performance of the system. For high, medium and low temperature charging, 230, 160 and 160 °C were recorded respectively at the top of the tank.

IOP Conf. Series: Earth and Environmental Science **173** (2018) 012042 doi:10.1088/1755-1315/173/1/012042

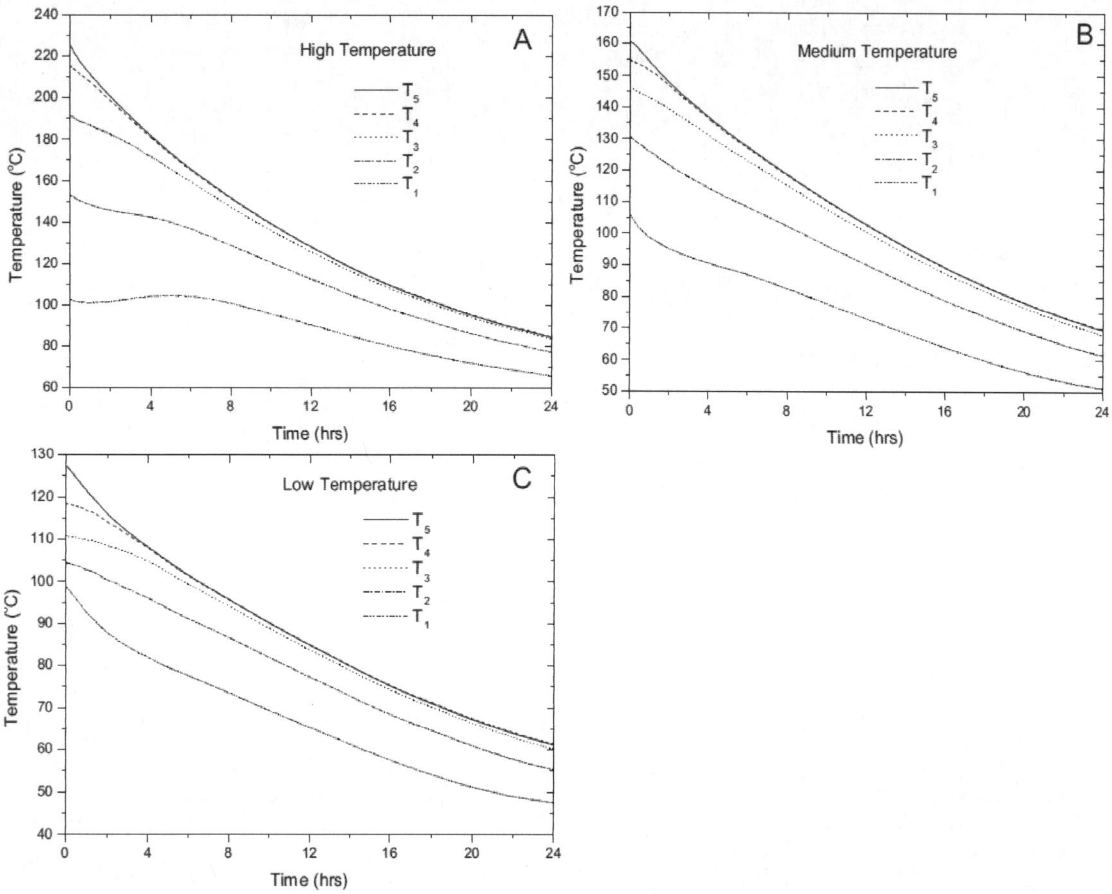

Fig. 2. Heat retention for a 24 hour period [7]

Figure 2 shows data from the system during a 24 hour heat retention for three categories studied. There was a higher rate of heat loss at the top 3 levels. This was attributed to heat loss due to convection. Another possibility for this was that the lower levels (level 1 and 2) had higher densities and therefore increasing their thermal density with time. Thermal loss was more at higher temperatures than at lower temperature during retention. This was as a result of different thermal masses along the height of the storage tank.

Concrete and sand are known to be low cost sensible thermal storage materials. However, the cracks associated with repeated expansion and contraction reduces the thermal conductivity of the system with time. This has led to the damage of the pipes that convey the heat transfer fluid as well. Sand is a good alternative but the pores present in it also limits thermal conductivity since they are basically filled with air [8]. They [8] made a comparison between concrete without cracks, concrete with cracks and sand saturated with heat transfer fluid (Hitec$^{(R)}$). Hitec$^{(R)}$ is a commercially available heat transfer fluid.

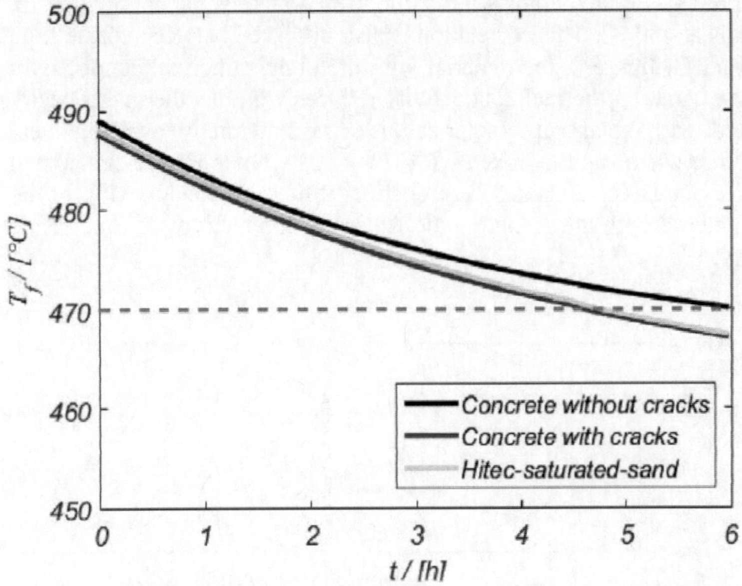

Fig. 3. Temperature at the exit point of heat transfer fluid within a 6 hour period [8]

In figure 3, the system containing concrete without crack maintained a heat transfer fluid temperature above the cut-off temperature indicated with the red dashed line throughout the six hour period. However, the system containing concrete with cracks had a 3 °C drop below the cut-off temperature in the last 1.5 hour of the six hour period. For sand saturated with Hitec®, there was a slight improvement over the performance of cracked concrete. They concluded that sand saturated with Hitec® was 4.5 % more efficient compared to cracked concrete.

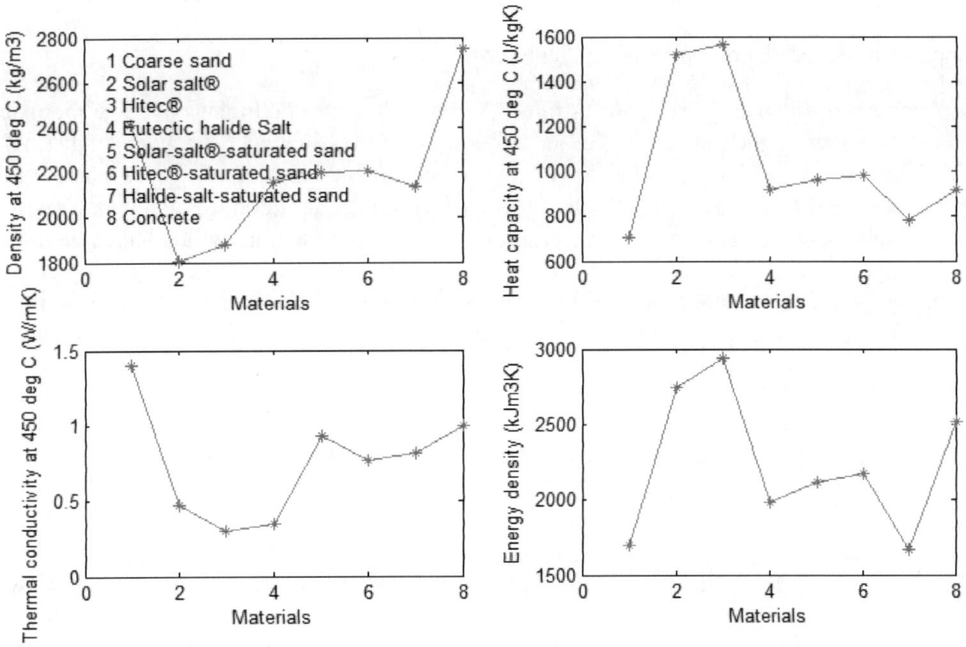

Fig. 4. Graphical illustration of properties of materials at 450 °C

Figure 4 was plotted from the data in [8]. Figure 4 shows a direct correlation between energy density and heat capacity for all the materials at 450 °C. This correlation is also observed between coarse sand and concrete. From the data presented in figure 4, the material with the highest thermal conductivity and lowest thermal capacity is coarse sand. Its thermal conductivity reduces while its thermal capacity increases significantly when mixed with solar-salt® (solar-salt®is a commercially available heat transfer fluid). This same effect is true when sand is mixed with Hitec® as well as Halide-salt. When mixing sand with any of these fluids, one has to consider the reduction of thermal conductivity for the increase in thermal capacity. The material with the second highest thermal conductivity is concrete. It has the third lowest thermal capacity.

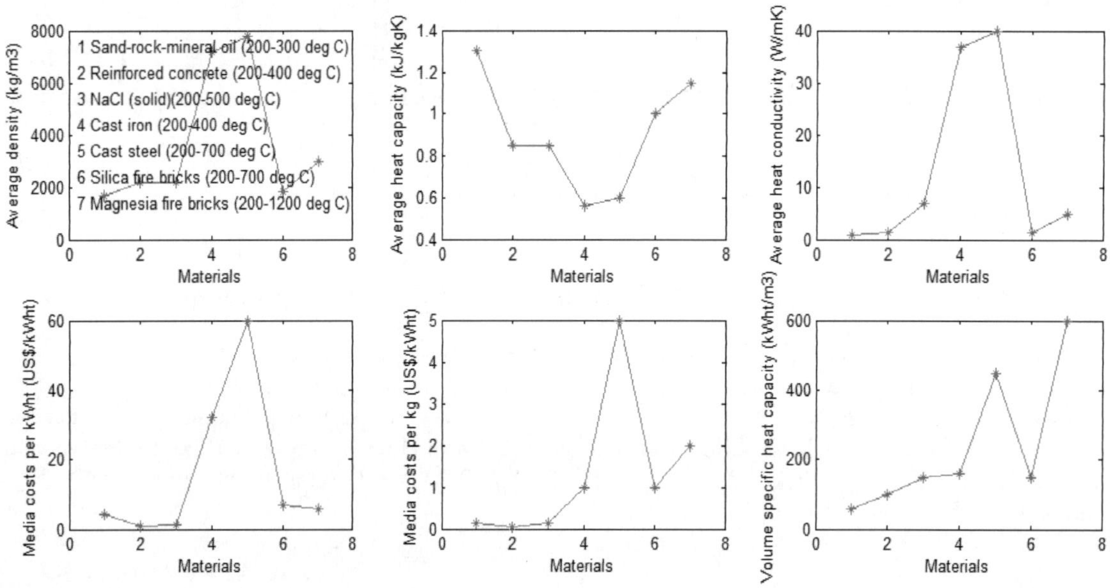

Fig. 5. Comparison of material properties and their costs

Figure 5 was plotted from the data in [13]. From figure 5, materials with the highest average thermal conductivity have the lowest thermal capacity and vice versa. Cast steel has the highest thermal conductivity and has the second largest value for volume specific heat capacity. The draw back with cast iron is that it is the costliest. Cast iron and cast steel have considerably good thermal conductivity and their thermal capacity can be enhanced by incorporating other materials that will enhance thermal capacity. The drawback is cost. Enhancement techniques employing the use of sand-rock-mineral oil which has the highest average thermal capacity in the list as well as its relatively low cost and cast iron having a very high average thermal conductivity with a good price compared to others will yield optimal results.

2.2. *Latent Heat Storage Materials*

Latent heat storage materials are more efficient when it comes to heat capacity. Within a very small temperature range, a very large amount of heat can be stored within the material. In latent heat storage, there are three (3) ranges of temperature: low temperature thermal storage (<120 °C), medium temperature thermal storage (120-300 °C) and high temperature thermal storage (>300 °C) [9].

Industrial heat energy waste is a good source for medium thermal energy storage for phase change materials with melting temperature in the range of 120 °C to 300 °C[10]. $LiNO_3$-NaCl with 87-13% mix proportion was worked on by [10]. A melting temperature of 220 °C was recorded. A latent heat capacity of 300 kJ/kg was recorded as well. The temperature at which the salt is stable is within the range of 400 °C and 450 °C.

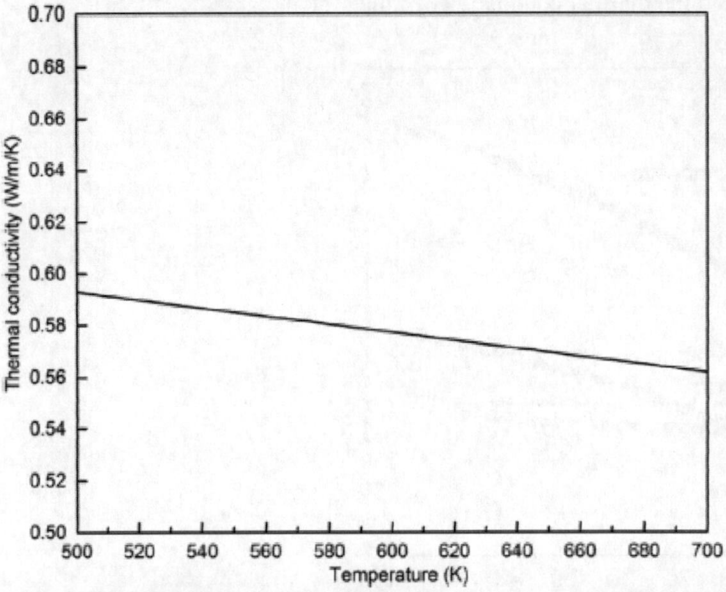

Fig. 6. Thermal conductivity of $LiNO_3$-NaCl (87-13%) as it varies with temperature [10]

From figure 6, they plotted the predicted thermal conductivity of $LiNO_3$-NaCl (87-13%). It's thermal conductivity as it melts reduces as temperature increases. Below 400 °C, the predicted thermal conductivity is between 0.57 W/m/K and 0.6 W/m/K. they reported the thermal conductivity to be higher than that of many other organic phase change materials as well as a few nitrate mixtures.

The use of metal foam as well as microencapsulation of phase change materials as a medium to increase thermal conductivity of a latent heat storage based system was made by [11]. They had three (3) sets of storage systems for comparison. Figure 7 shows details on the setup of the system before and after adding the microencapsulated and pure phase change material. The first was the phase change material (paraffin) without microencapsulation. The second was phase change material without encapsulation but with metal foam included. The last was microencapsulated phase change material with metal foam enhancement.

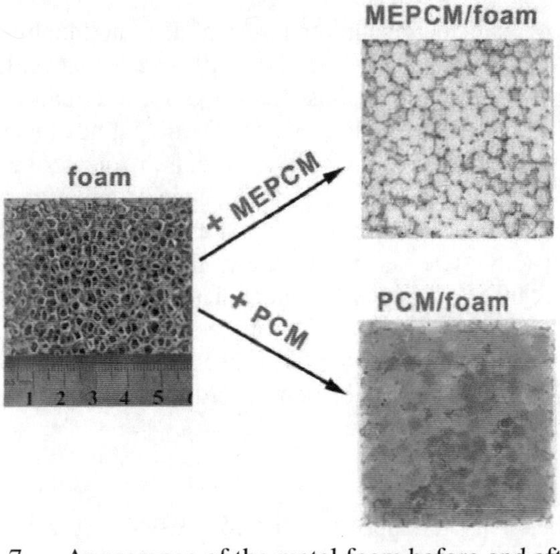

Fig. 7. Appearance of the metal foam before and after injecting the phase change material [11]

The porosity and thickness of the metal foams affects the thermal properties of the system. They also found that the metal foam enhanced the thermal conductivity of the system.

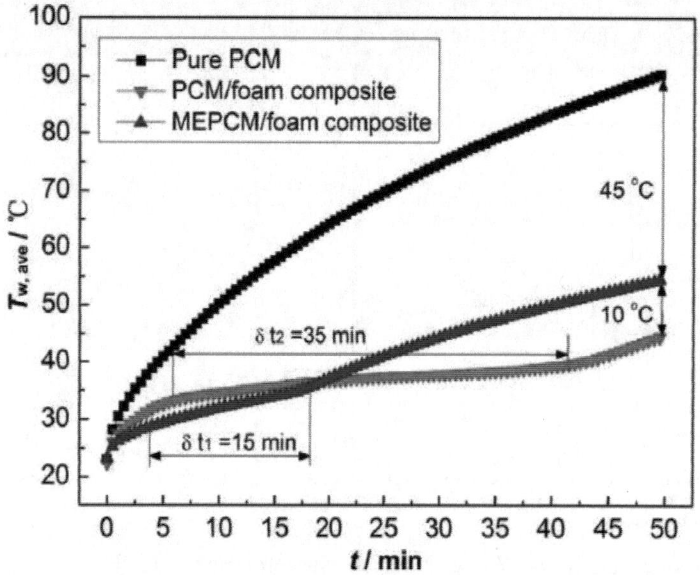

Fig. 8. Reaction of the surface temperature for the different phase change material setups [11]

The time-related arithmetic-averaged temperatures of the top side of each setup were measured from five (5) different points. Figure 8 shows the results for the three setups. In figure 8, a very large amount of temperature reduction was recorded for the setup containing microencapsulated phase change material/foam system as well as for that containing the phase change material/foam system when compared to the setup containing only phase change material. There is a temperature reduction of 45 °C for the microencapsulated phase change material/foam setup and a 55 °C reduction for the phase change material/foam setup as compared to the pure phase change material setup at the fiftieth minute. This is because the thermal conductivity of these two setups have been enhanced by the metal foam contained in it.

3. Conclusion

From the information provided in this review, we can merge the strengths of the thermophysical properties of both sensible and latent heat storage materials together utilizing enhancements [8,11,12]employed in these articles. Sensible heat storage materials are good for thermal fluids because they are cheap and alsohave considerably high thermal conductivity. Their cost advantage is also important because a significant volume of heat transfer media will be required for an average system. Thermal conductivity is important for a heat transfer fluid so as to enhance efficient thermal charging and discharging. On the other hand, latent heat storage materials are ideal as the storage material because of their high latent thermal capacity. The issue with these materials is their low thermal conductivity. This according to literature can be enhanced by microencapsulation and/or the use of metal foams [8,11,12]. This has enhanced the thermal conductivity of the systems they worked on. It is believed that further enhancement of thermal conductivity of latent heat storage materials as well as enhancement of heat capacity of sensible heat materials can be done in order to further enhance the efficiency of thermal storage systems.

Acknowledgement

The authors would like to appreciate Covenant University for support of the costs associated with this paper.

References

[1] Tian, Y., & Zhao, C. Y. (2013). A review of solar collectors and thermal energy storage in solar thermal applications. *Applied energy*, *104*, 538-553.J. Clerk Maxwell, A Treatise on Electricity and Magnetism, 3rd ed., vol. 2. Oxford: Clarendon, 1892, pp.68-73.

[2] Zalba, B., Marın, J. M., Cabeza, L. F., & Mehling, H. (2003). Review on thermal energy storage with phase change: materials, heat transfer analysis and applications. *Applied thermal engineering*, *23*(3), 251-283.

[3] Sharma, A., Tyagi, V. V., Chen, C. R., & Buddhi, D. (2009). Review on thermal energy storage with phase change materials and applications. *Renewable and Sustainable energy reviews*, *13*(2), 318-345.

[4] Hale, M. (2000). Survey of thermal storage for parabolic trough power plants. *NREL Report, No. NREL/SR-550-27925*, 1-28.

[5] Zhao, C. Y., Zhou, D., & Wu, Z. G. (2010). Heat transfer enhancement of phase change materials (PCMs) in low and high temperature thermal storage by using porous materials. In *2010 14th International Heat Transfer Conference* (pp. 435-441). American Society of Mechanical Engineers.

[6] Zhao, C. Y., & Wu, Z. G. (2011). Thermal property characterization of a low melting-temperature ternary nitrate salt mixture for thermal energy storage systems. *Solar Energy Materials and Solar Cells*, *95*(12), 3341-3346.

[7] Mawire, A. (2016). Performance of Sunflower Oil as a sensible heat storage medium for domestic applications. *Journal of Energy Storage*, *5*, 1-9.

[8] Xu, B., Han, J., Kumar, A., Li, P., & Yang, Y. (2017). Thermal storage using sand saturated by thermal-conductive fluid and comparison with the use of concrete. *Journal of Energy Storage*, *13*, 85-95.

[9] Zhou, D., & Eames, P. (2016). Thermal characterization of binary sodium/lithium nitrate salts for latent heat storage at medium temperatures. *Solar Energy Materials and Solar Cells*, *157*, 1019-1025.

[10] Zhou, D., & Eames, P. (2017). A study of a eutectic salt of lithium nitrate and sodium chloride (87–13%) for latent heat storage. *Solar Energy Materials and Solar Cells*, *167*, 157-161.

[11] Li, W., Hou, R., Wan, H., Liu, P., He, G., & Qin, F. (2017). A new strategy for enhanced latent heat energy storage with microencapsulated phase change material saturated in metal foam. *Solar Energy Materials and Solar Cells*, *171*, 197-204.

[12] Németh, B., Németh, Á. S., Tóth, J., Fodor-Kardos, A., Gyenis, J., & Feczkó, T. (2015). Consolidated microcapsules with double alginate shell containing paraffin for latent heat storage. *Solar Energy Materials and Solar Cells*, *143*, 397-405.

[13] Gil, A., Medrano, M., Martorell, I., Lázaro, A., Dolado, P., Zalba, B., & Cabeza, L. F. (2010). State of the art on high temperature thermal energy storage for power generation. Part 1—Concepts, materials and modellization. *Renewable and Sustainable Energy Reviews*, *14*(1), 31-55.

IOP Conf. Series: Earth and Environmental Science **173** (2018) 012043 doi:10.1088/1755-1315/173/1/012043

Wood fuel analysis of some selected wood species within Ibadan

O.E. Ogunsola[1], O. Adeleke[1] and A.T. Aruna[2]

[1]Department of Physics, University of Ibadan, Ibadan, Nigeria.
[2]Nigerian Institute of Science Laboratory Technology, Samonda, Ibadan, Nigeria.

seyiogunsola22@gmail.com

Abstract. Wood waste (saw dust) from sawmills if well harnessed is a potential alternative source of energy for electricity production. This is due to its prospects in the production of fuel such as biogas (methane, carbon dioxide and other trace gases) having high calorific value that can be used to drive a gas turbine in the production of electricity. However, since the electrical energy generated mainly through the hydro power generating stations along with the other power plants including the conventional coal and gas currently in existence in the country is not meeting the energy demand of Nigerians, it is therefore necessary to introduce additional sources of alternative energy in order to be able to augment these existing energy sources, and wood waste is the proposed alternative source of fuel energy used in this study. The fuel properties and characterization of six different wood waste samples (Iroko (Milicia excelsa), Masonia (Masonia altissima), Afara (Terminalia superba), Arere (Annona senegalensis), Ayin (Anogeissus leiocarpus) and Obeche (Triplochiton scleroxylon)) obtained from sawmills located within Bodija, Sango and Idi Ayunre areas of Ibadan were determined by following the procedure of the American Society Standard Test Method (ASTM E870-82(2013)) for the analysis of wood fuels. The result of characterization of the wood samples showed a range of values for the gross calorific values (20.14MJ/Kg to 22.30MJ/Kg), volatile matter (74.22% to 82.62%) and ash content (0.605% to 5.03%). The overall ratings of the fuel properties of the six wood samples in terms of their average ranking were obtained as 3.25 (Iroko), 3.13 (Masonia), 3.25 (Afara), 2.0 (Arere), 2.6 (Ayin) and 1.75 (Obeche) which suggests that wood waste from Obeche has the best fuel property because of its least rating while that of Afara and Iroko were the worst.
Keywords: Wood waste, fuel properties, alternative energy source, overall ratings.

1. Introduction

Nigeria is endowed with abundance of renewable and non-renewable energy resources including hydro, wind and solar energy. However, the present dependence of the Nigerian government on hydro and fossil fuel resources (petroleum, coal and gas) to generate electricity is not enough to meet the energy demands of the populace because of its unavailability to the majority of them. The increase in the demand for electricity in Nigeria had also resulted in the interest in other alternative sources of energy including wood wastes (biomass) by various researchers. Biomass is the primary source of energy for nearly 50% of the World's population [1] and wood biomass is a major renewable energy source in the developing world, representing a significant proportion of the rural energy supply. Biomass is defined as organic material, available on renewable basis, which are produced directly or indirectly from living organisms without contamination from other substances or effluents. It is a nonfossilized and biodegradable organic material that originated from plant, animals and micro-organisms which could be residues/waste from forestry and agricultural product including the waste from the municipalities. Biomass is available in varying forms throughout the developing world. It can come from densely forested areas of the world or sparsely vegetated arid regions as well as farmlands and garbage collection centers. Worldwide, biomass is the fourth largest energy resource after coal, oil, and natural gas estimated at about 10% of global primary energy [2]. Biomass broadly falls into four types based on primary source which are wood and agricultural products, municipal solid waste, landfill gas and biogas, and alcohol fuels and so on. Biomass has high input but variable moisture content and is made up of carbon, oxygen, nitrogen, sulphur and inorganic elements [3].

Plant biomass is the largest potential source of biomass when compared to others. Most of these potentials lie in wood processing by-products referred to as wood shavings such as sawdust [4]. UNFCC [5] reported in 2005 that biomass generally has high moisture content and is made up of carbon, hydrogen, oxygen, nitrogen, sulphur and inorganic elements. The uncontrollable release of global warming gases to the atmosphere mainly from fossil fuel resource with harmful effect to damaging the ozone layer has made it very necessary to search for a technology with various methods to developing an alternative energy production in which sawdust from agricultural and forestry biomass is examined as the alternative energy source [6]. The combustion of agricultural biomass from sawdust (and other biomass) ranges from the utilization of wood stove in domestic application to industrial power plant applications which also generate greenhouse gases [7]. However, this biogas which is harmful to the ozone layer can be converted into useful valuable gaseous and liquid resource through the process called gasification [8].

Wood pellets are actually manufactured from waste products from other wood processing industries and sawmills. Although man rely on sawmill waste and forest harvest residuals for the bulk of their fibre supply, this sawdust could also be used to generate electrical power through a power plant that uses wood pellets as fuel. This kind of power plant which produces steam that runs a turbine to produce electricity is known as the biomass power plant or biopower plant.

2. Materials and Methods

The quantity (i.e. volume) of the daily amount of saw dust (wood waste) generated from the various wood species: Iroko (Milicia excelsa), Masonia (Masonia altissima), Afara (Terminalia superba), Arere (Annona senegalensis), Ayin (Anogeissus leiocarpus) and Obeche (Triplochiton scleroxylon) [9] obtained at the plank markets in Bodija (BM), Sango (SM) and Idi Ayunre (IM) areas of Ibadan land (Fig. 1) were estimated. Also, wood pellets were produced from the wood wastes (biomass) obtained from the various wood species, through the procedures of drying, milling, pelletizing, and cooling. The analyses of these wood species were determined by following the procedure of the American Society Standard Test Method [10] for the analysis of wood fuels. The thermal properties of the wood pellets obtained from the various wood species were carried out in order to determine their moisture content, ash content, volatile matter content, and fixed carbon content. The elementary analysis of these wood pellets was carried out in order to determine which of the elemental composition of Carbon, Hydrogen, Oxygen, Sulphur and Nitrogen is available in them. The fuel characteristics of these various wood species including their gross calorific values were determined from their wood pellets and compared with the values obtained using the Dulong-Berthelot model.

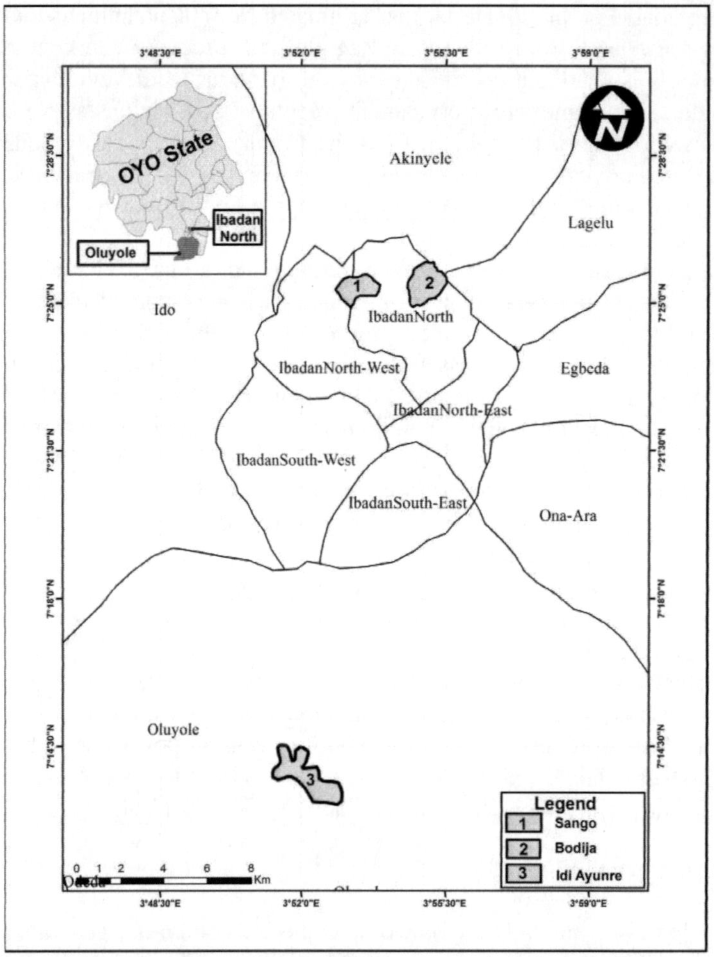

Fig. 1: Map of Oyo state showing plank market locations in Ibadan land

2.1 *Estimation of Volume Waste*

The process of milling a lumber (log) into planks generates a lot of wood waste which is a biomass that can be used for electricity generation. Similarly, the process of milling planks also generates waste on the average from a lumber (log of wood) and approximately 45-55% of the log will not be part of the final wood product [11]. It was estimated that the total volume of planks per truck (V_{PT}) in Bodija, Sango and Idi Ayunre markets where the survey was conducted is obtained as:

$$V_{PT} = 400V_P \tag{1}$$

This was based on the fact that each truck delivers 400 logs per truck.

The estimated volume of plank (V_P) from a piece of plank is (12x12x1) ($inches^3$) = (144 $inches^3$), and this is equivalent to $3.658m^3$. Also the total daily plank volume (V_{DT}) offloaded was estimated as:

$$V_{DT} = N_{LD}V_P \tag{2}$$

where,

N_{LD}= Number of lumber per day

V_P = Volume of Plank.

Therefore,

$$V_{DT} = N_{LD} \times 3.658m^3 \tag{3}$$

Also, from the survey carried out, it was estimated that the volume of wood milled daily was 55% of the total 100% wood species supplied to millers at the various sawmills considered in this work. Out of this 55% volume of the daily milled wood, 34% of it accounted for the daily milled wood waste (V_{DW}) produced i.e.

$$V_{DW} = 34\% \times 55\% V_{DT} \tag{4}$$

∴ Daily milled wood waste,

$$V_{DW} = 0.187 V_{DT} \tag{5}$$

Thus,

$$V_{DW} = 0.187 V_{DT} = 0.187 \times N_{LD} \times 3.658 \tag{6}$$

$$\Rightarrow V_{DW} = 0.684 N_{LD} \tag{7}$$

Thus, the volume of weekly milled wood waste (V_{WW}) based on the six (6) working days the millers usually work in a week is:

$$V_{WW} = 6 \times V_{DW} \tag{8}$$

$$\Rightarrow V_{WW} = 6 \times 0.684 N_{LD} \tag{9}$$

$$\therefore V_{WW} = 2.74 N_{LD} \tag{10}$$

Then the total Volume of weekly milled wood waste (V_{WW}) is modeled as:

$$V_{WW} = 2.74 N_{LD} \tag{11}$$

2.2 *Biomass Analysis*

Biomass is a key resource for renewable energy generation [12]. This analysis was carried out by making use of the following techniques:

2.2.1 *Proximate Analysis* The proximate analysis of the biomass materials was determined by heating approximately 2g of the sun-dried mass of each wood species with particle size of 420μm using an electric furnace at a temperature of 500 ˚C for four (4) hours. It was cooled thereafter in a desiccator and weighed in order to determine the ash content of the samples. The percentage of volatile matter of the biomass was also determined.

2.2.1.1 *Estimation of Moisture Content* The moisture content of a biomass material (wood waste) which is directly related to a fuel's calorific value is obtained by first determining its initial weight before sun-drying it. After sun-drying, the samples were continually reweighed until a constant weight was attained (i.e. the final dry weight). The moisture content (MC) was obtained as:

$$MC = \frac{Initialweight - Finalweight \times 100\%}{Finalweight} \tag{12}$$

2.2.1.2 *Ash Content Determination* The ash content which is a measure of the mineral content and other inorganic matter in a biomass is used in conjunction with other procedures to determine the total composite of the biomass sample [13]. The ash content (AC) is calculated as:

$$AC = \frac{M}{M_s} \times 100\% \tag{13}$$

where, M is the mass of the ash and M_s is the mass of the sun-dried sample

IOP Conf. Series: Earth and Environmental Science **173** (2018) 012043 doi:10.1088/1755-1315/173/1/012043

2.2.1.3 *Volatile Matter Content* The percentage volatile matter of the biomass materials (wood shavings) was calculated by using 2 g each of the sundried biomass materials (wood shavings) with particle size of 425 μm placed in a porcelain crucible. Each sample was first sun-dried and then kept in a furnace at a temperature of 500˚C for 10 min and weighed after cooling in a desiccator. The volatile matter (VM) was then calculated as follows:

$$VM = \frac{X - Y}{X} \times 100\% \tag{14}$$

where,
X = weight of the sun-dried sample
Y = weight of the sample after 10 min in the furnace at 500˚C.

2.2.1.4 *Fixed Carbon* The fixed carbon is the solid combustible residue that remains after a coal particle is heated and the volatile matter is expelled. The fixed carbon content of a coal is determined by subtracting the percentages of moisture, volatile matter, and ash from a sample [14]. It is also the solid residue, other than ash, obtained by destructive distillation of a coal, determined by definite prescribed methods [15].
Fixed carbon (FC) can be expressed as:

$$FC(\%) = 100 - \%MC - \%AC - \%VM \tag{15}$$

2.2.2 *Ultimate Analysis* The ultimate analysis is the elementary analysis performed on the biomass in order to determine the elemental composition of carbon, hydrogen, oxygen, sulfur and nitrogen present in it. It also includes the amount of carbon monoxide, methane and carbon dioxide present in the synthesis gas obtained from the gasification of the biomass. It was this synthesis gas that when cleaned could be used to generate electricity. The energy output actually depends on the types of wood species used as fuel.

2.3 *Fuel Characteristics*

Fuel characteristics include the estimation of the calorific values which indicates the amount of heat obtained from the complete combustion with oxygen in a standard calorimeter

2.3.1 *Estimation of Gross Calorific Value* The gross calorific value of the samples of biomass materials was determined using an adiabatic bomb calorimeter. About 0.2 g each of these samples was burnt using the bomb calorimeter until complete combustion was obtained. The differences between the maximum and the minimum temperatures obtained for each of these samples were used to compute the gross calorific values of the biomass materials (wood waste) given as:

$$Q = \frac{(C_{Water} + C_{Cal})(T_2 - T_1)}{W_t} \times 100 \tag{16}$$

where,
Q = Calorific value of species (kJ/kg)
C_{water} = Heat capacity of water
C_{cal} = Heat capacity of the bomb calorimeter
$T_2 - T_1$ = Rise in temperature
W_t = Weight of the biomass material sample (kg)

2.3.2 *Dulong-Berthelot Model* The Dulong-Berthelot model which was originally developed for coal but modified by [16] was utilized in determining the heating value of biomass. This modified form is expressed as:

$$GCV = 349.1C + 1178.3H + 100.5S - 103.4O - 15.1N - 21.1ASH(kJ/kg) \quad (17)$$

where, C, H, S, O, N and ASH were the percentages of carbon, hydrogen, sulfur, oxygen, nitrogen and ash obtained from ultimate analysis.

2.4 *Bulk Density of the various Wood Species*

The density of biomass (wood shavings) plays a very important role in the determination of its fuel value and the denser a wood is, the more heat per volume it contained. The bulk densities of the various wood species are determined from their sun-dried masses using the expression:

Bulk density of wood (kg/m^3) = Mass of Specimen / Volume of Specimen (18)

3. Results and Discussion

The analysis of the six different types of wood species (Iroko (Milicia excelsa), Masonia (Masonia altissima), Afara (Terminalia superba), Arere (Annona senegalensis), Ayin (Anogeissus leiocarpus) and Obeche (Triplochiton scleroxylon)) obtained from sawmills located within Bodija, Sango and Idi Ayunre areas of Ibadan were carried out using laboratory investigations

3.1 *Volume of Wood Waste generated at the plank markets*

Table 1 is the result of the summary of the average volume of wood wastes generated on daily basis for the three plank markets considered in Ibadan for this study. Bodija being the largest of these plank markets has the largest average number of truck delivering lumber to the market and the highest average volume of waste generated on daily basis 957.6m^3. The total volume of wood waste generated by these three plank markets on daily basis is 1607.4m^3.

Table 1: Summary of the average volume of wood waste generated daily at plank markets in Bodija, Sango and Idi Ayunre areas of Ibadan

AVERAGE TRUCK NUMBER	IBADAN NORTH LG		OLUYOLE LG	TOTAL
	Bodija	Sango	Idi Ayunre	
AVERAGE TRUCK NUMBER PER WEEK (NT)	21	3.25	11	35.25
AVERAGE LUMBER PER WEEK (N_L)	8,400	1,300	4,400	14,100
AVERAGE LUMBER PER DAY (N_{LD})	1400	217	733	2350

AVERAGE WASTE VOLUME GENERATED DAILY (m^3)	957.6	148.4	501.4	1607.4

3.2 *Proximate Analysis*

The result of the proximate analysis of the various wood species obtained for this study (Table 2) showed that the percentage moisture content of Ayin was the highest, while that of Obeche was the least (Fig. 2). This indicates that Obeche will give the highest fuel calorific value (i.e. best fuel property) amongst the other wood species considered in this work. Also, it will be the one with the highest durability value in terms of their being stored. The ash content of the various wood species varied from 0.61% for Ayin to 5.03% for Afara (Fig. 3). The high ash content of Afara indicates that it has high mineral matter. The ash content of all the various wood species used in this study was lower than 6%, with Afara and Iroko having values of 4.6% and 5.0% respectively. All the other species had an ash content of less than 4%, a value at which slagging of ash (i.e. an incombustible by product) usually occurs for biomass fuels (wood shavings). Hence, it is likely that when sawdust of species like Masonia, Arere, Ayin and Obeche are used as fuel, slagging would not occur. On the contrary, slagging might occur when wood species like Afara and Iroko are used as biomass fuel. The volatile matter of the various six wood species used in this study varied from 74.22% Afara to 82.60% Obeche, with Obeche having the highest value and Afara the least (Fig. 4). The volatile matter of all other wood species was greater than 78% with the exception of Afara. The volatile matter content of a biomass material when high indicates that during the process of combustion, most of it will volatize and burn as gas. The fixed mass for these wood species varied from 3.97% (Iroko) to 17.75% (Afara) (Fig. 5). The high fixed carbon content indicates that the biomass will require a long combustion time. The fixed carbon which actually gives a rough estimate of the heating value of a fuel acts as the main heat generator during the burning of a biomass. Hence, the higher the content of carbon of the biomass fuel the more likely that the species would have higher heating value.

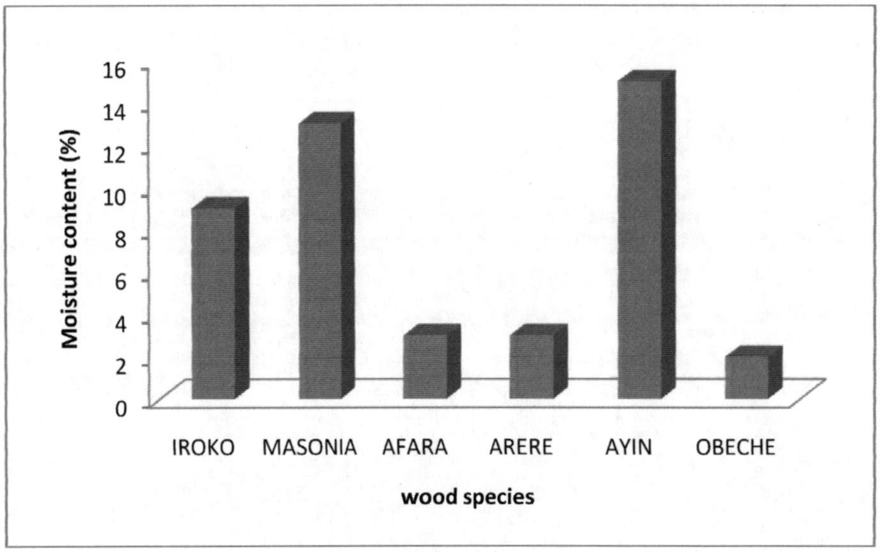

Fig.2: Percentage moisture content for the various wood species

Table 2: Proximate Analysis of data for selected wood waste biomass materials

Local Government	Saw Mill Location	Types of wood species	Moisture Content (%)	Ash Content (%)	Volatile Matter (%)	Fixed Carbon (%)
Ibadan North	Bodija SM1	Iroko	9.0	4.62	82.41	3.97
Ibadan North	Bodija SM2	Masonia	13.0	2.01	80.93	4.06
Ibadan North	Sango SM3	Afara	3.0	5.03	74.22	17.75
Ibadan North	Sango SM4	Arere	3.0	2.75	78.54	15.71
Oluyole	Idi Ayunre SM5	Ayin	15.0	0.61	80.50	3.89
Oluyole	Idi Ayunre SM6	Obeche	2.0	3.51	82.60	11.89

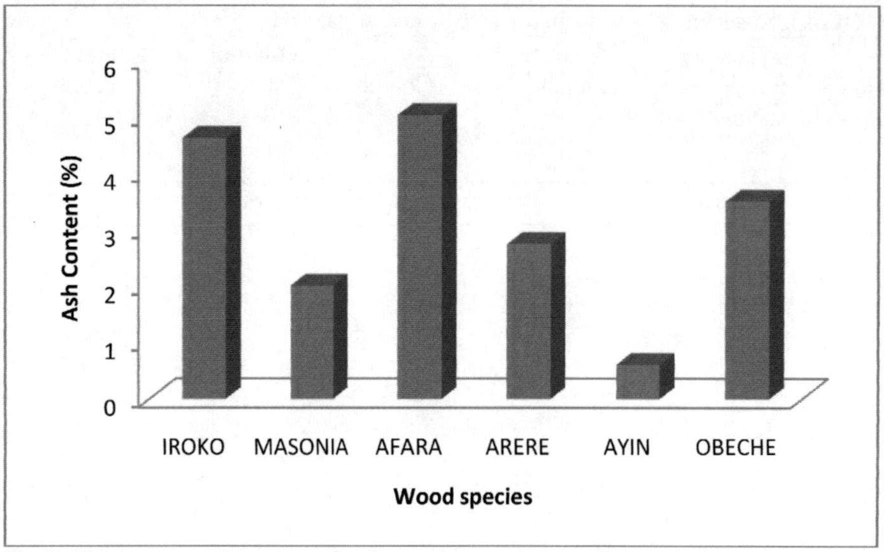

Fig.3: Percentage ash content for the various wood species

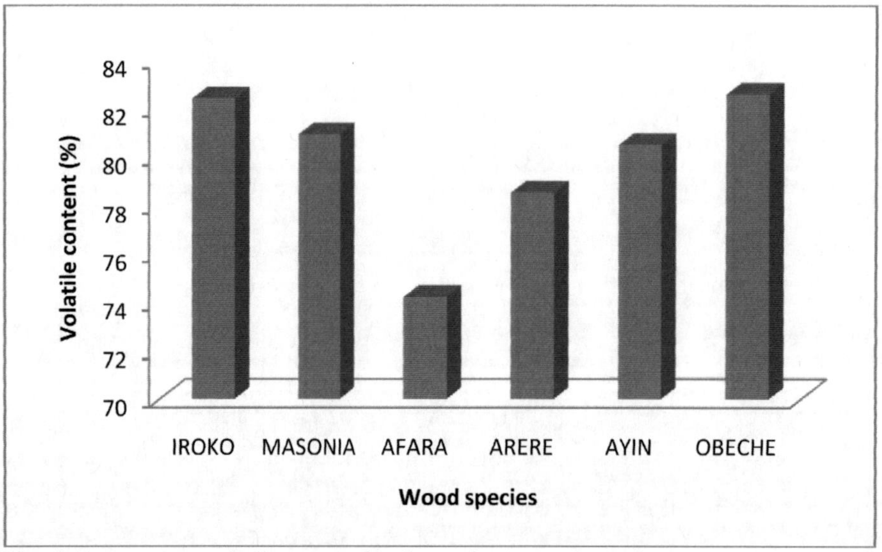

Fig. 4: Percentage volatile matter for the various wood species

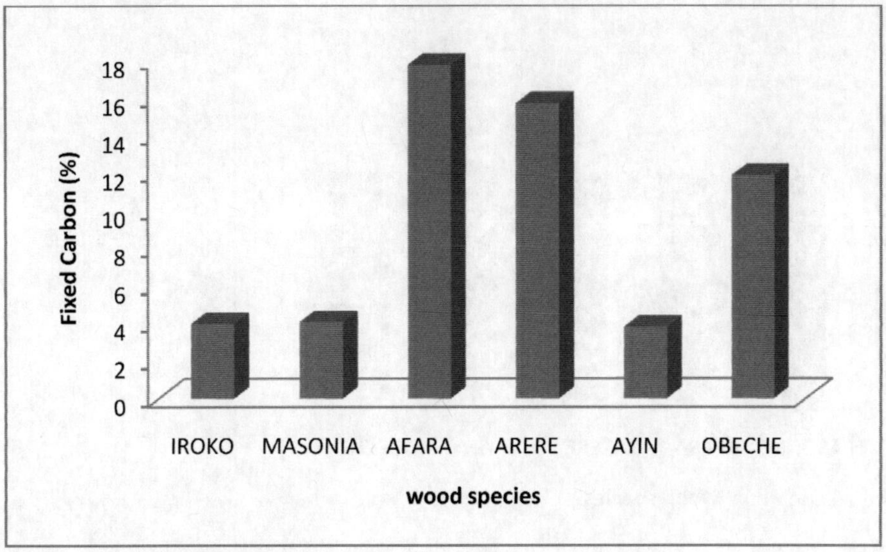

Fig.5: Percentage fixed carbon for the various wood species

3.3 *Ultimate Analysis*

The result of the ultimate analysis (Table 3) for the various wood species considered in this study shows that the values of Hydrogen content in the biomass materials ranged from 3.82% for Afara to 5.24% for Iroko (Fig. 6). With the exception of Iroko having 5.24%, the other wood species used in the study were below the expected range of 5% to 6% for a dry biomass matter with a good heating value. Methane content of the wood specie used ranged from 7.50% for Afara to 8.90% for Arere (Fig. 7), while the carbon dioxide content of the wood species ranged from 29.44% for Arere to 38.21% for Iroko (Fig. 8). The carbon monoxide content also ranged from 22.42% for Arere to 26.44% for Iroko (Fig. 9).

Table 3: Ultimate Analysis of data for selected wood waste biomass materials

Local Government	Saw Mill Location	Types of wood species	H_2 (% vol)	CH_4 (% vol)	CO_2 (% vol)	CO (% vol)
Ibadan North	Bodija SM1	Iroko	5.24	8.02	38.21	26.44
Ibadan North	Bodija SM2	Masonia	4.03	7.59	37.22	25.32
Ibadan North	Sango SM3	Afara	3.82	7.50	36.41	23.12
Ibadan North	Sango SM4	Arere	3.85	8.90	29.44	22.42
Oluyole	Idi Ayunre SM5	Ayin	4.08	8.63	31.43	23.42
Oluyole	Idi Ayunre SM6	Obeche	3.85	8.90	29.44	22.42

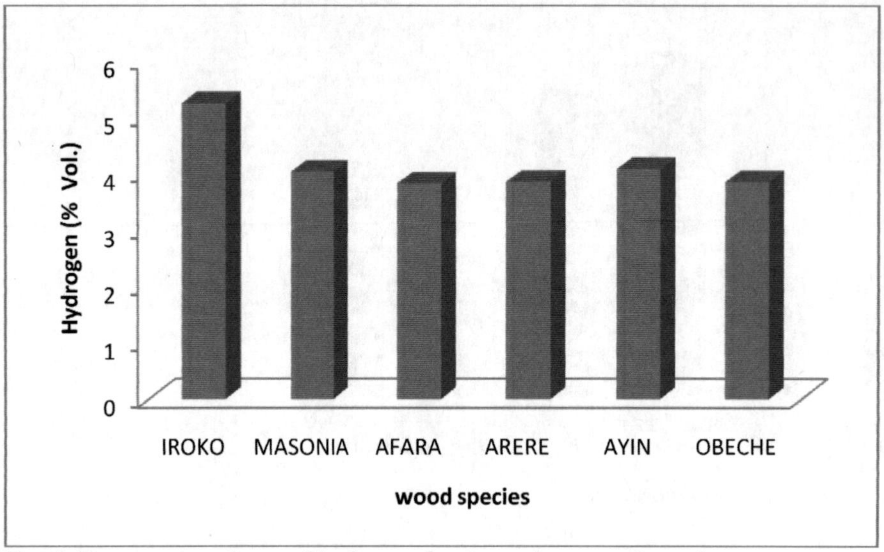

Fig. 6: Percentage volume of hydrogen for the various wood species

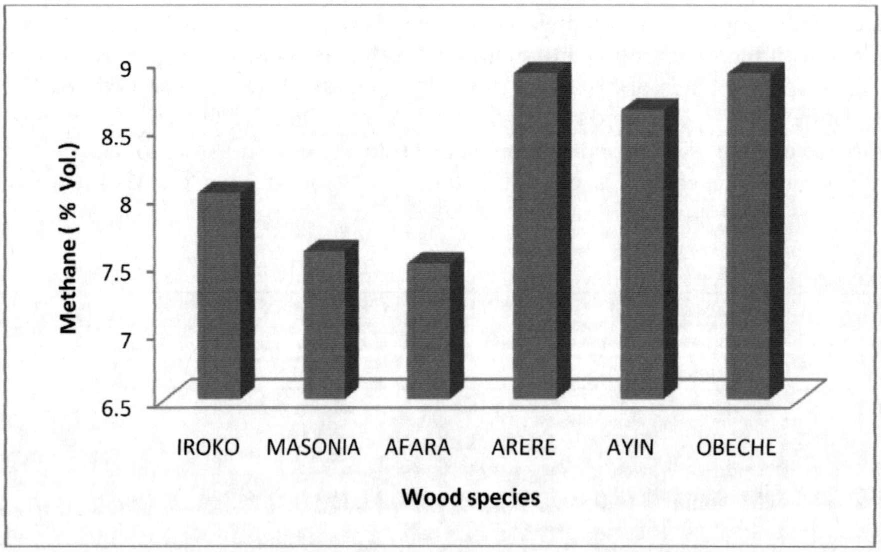

Fig. 7: Percentage volume of methane for the various wood species

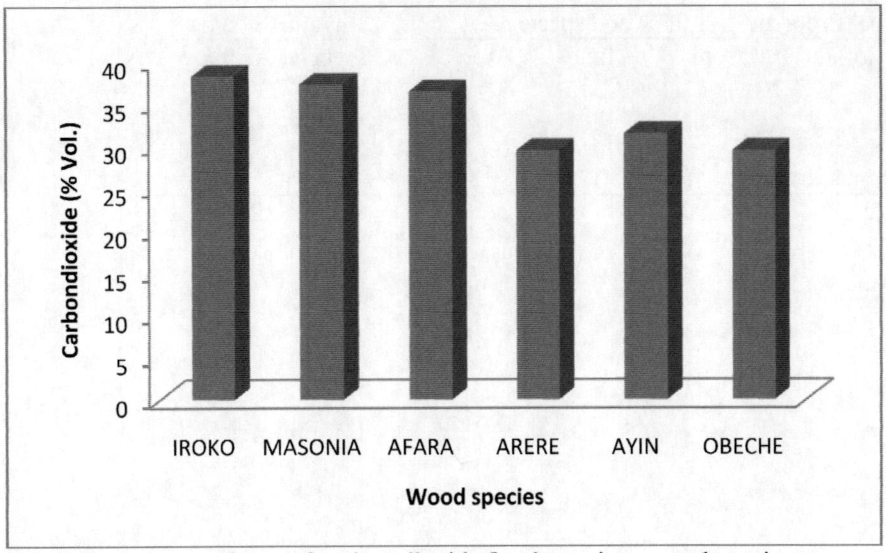

Fig. 8: Percentage volume of carbon dioxide for the various wood species

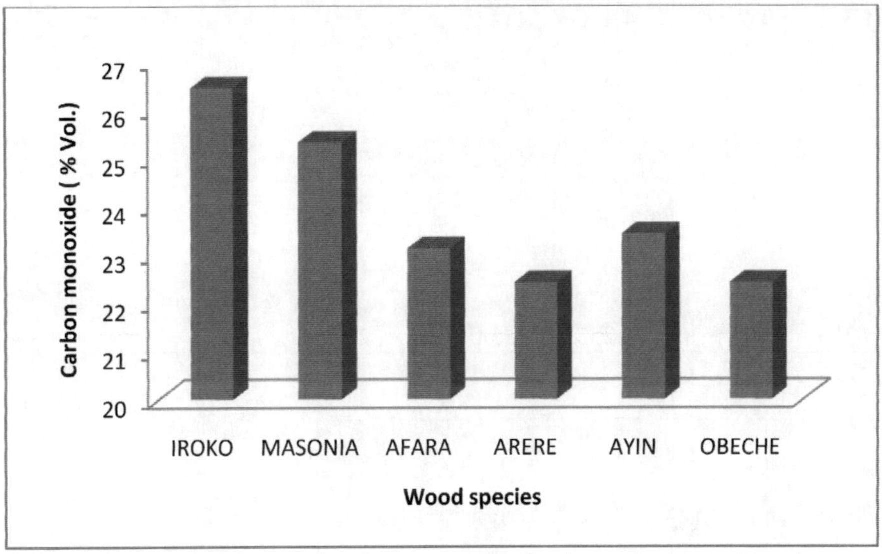

Fig. 9: Percentage volume of carbon monoxide for the various wood species

3.4 Fuel Properties

Table 4 shows the ranking of the fuel properties for the six wood species considered. The fuel properties were assigned values between 0-1, 1-2, 2-3, 3-4, 4-5, with 1 as the best and 5 as the worst. Significant variation was noticed in the fuel properties of these six different wood species with Obeche (1.75) having the least value, while Afara (3.25) had the highest value. This suggests that wood waste from Obeche has the best fuel properties, while the wood waste form Afara has the worst wood fuel properties. Thus, it must be noted that the lower rating values has the best wood species in terms of the fuel properties while the higher ratings has the worst wood species in terms of fuel properties (Fig. 10).

Table 4: Ranking of fuel Properties for the wood species

Properties	Iroko	Masonia	Afara	Arere	Ayin	Obeche
Wood density (Kg/m^3)	5	4	3	2	0	2
Gross Calorific Value (MJ/Kg)	5	5	3	1	2	1
Volatile Matter (%)	4	3	5	4	1	4
Ash Content (%)	5	2	3	1	4	1
Hydrogen (% Vol.)	0	3	5	2	3	2
Methane (% Vol.)	4	3	4	1	2	1
Carbon Dioxide (% Vol.)	2	0	1	2	5	0
Carbon Monoxide (% Vol.)	1	5	2	3	4	3
Fuel rating	3.25	3.13	3.25	2.00	2.63	1.75

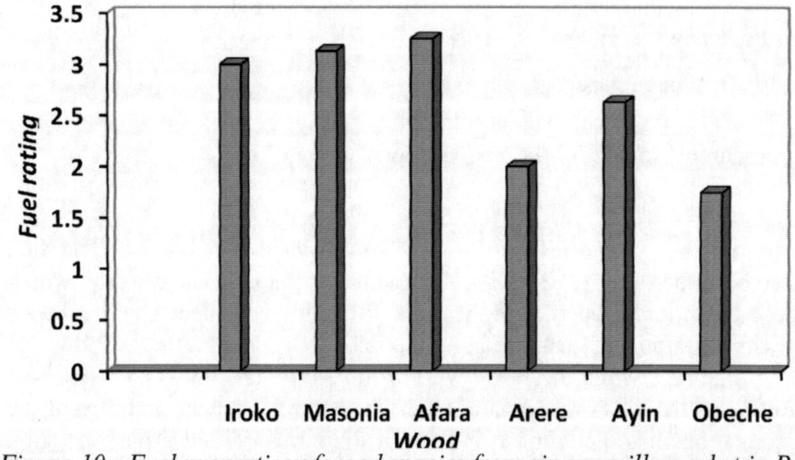

Figure 10: Fuel properties of wood species from six sawmills market in Bodija, Sango and Idi Ayunre areas of Ibadan with its rating

IOP Conf. Series: Earth and Environmental Science **173** (2018) 012043 doi:10.1088/1755-1315/173/1/012043

3.5 Gross Caloric Value of the Samples

The result of the gross calorific value of the six various wood waste species used (Table 5) shows that the least gross calorific value of 20.14 MJ/kg was obtained for Obeche, while the highest value of 22.30 MJ/kg was obtained for Ayin (Fig. 11). Comparison of the experimental value of GCV with the estimated Dulong-Berthelot model is as shown in Table 6. These values are consistent with those obtained from experiment only that they are lower due to the absent of components like sulphur and nitrogen in saw dust which reduces the calorific value (Fig. 12). This is also important because it shows that the saw dust obtained from the biomass from these wood species produces less ozone depletion greenhouse gases. However, it must be noted that despite that Arere was ranked fifth in the rating of the calorific value of the wood species but the overall rating value indicated that it has the second best fuel properties. Hence, it shows that calorific value should not be the only factor to be considered in determining the suitability of a biomass material for fuel briquettes.

Table 5: Gross Calorific Values for selected wood waste biomass materials

Types of wood species	Iroko	Masonia	Afara	Arere	Ayin	Obeche
Gross calorific value (MJ/kg)	20.20	21.50	20.68	22.15	22.30	20.14

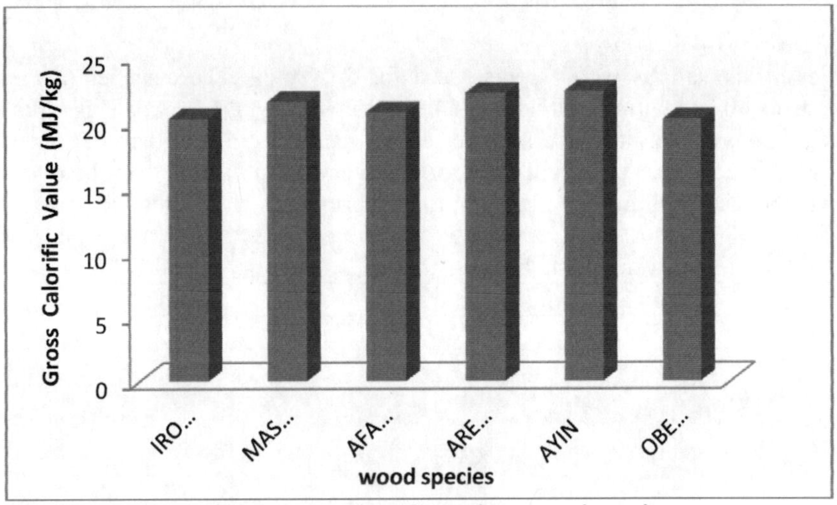

Fig. 11: Gross calorific value against the various wood species

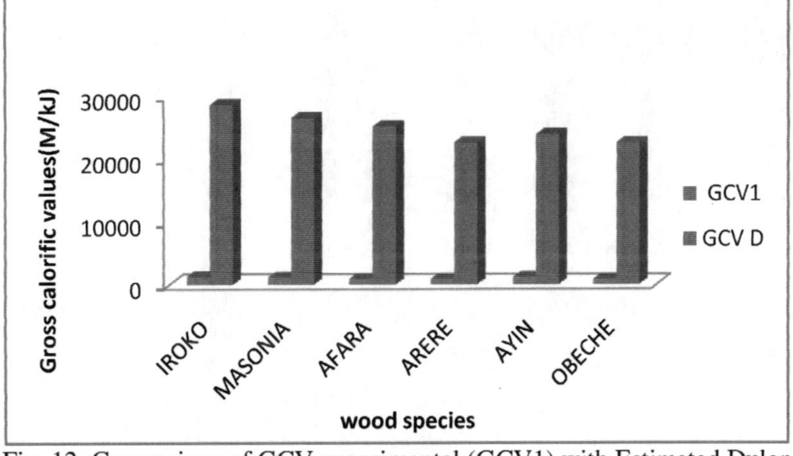

Fig. 12: Comparison of GCV experimental (GCV1) with Estimated Dulong-Berthelot model (GCVD)

Table 6: Comparison of GCV experimental with Estimated Dulong-Berthelot model

Wood waste species	Gross Calorific value Experimental	Gross Calorific Value Dulong-Berthelot model
IROKO	1289.42	28646.04
MASONIA	1135.00	26528.84
AFARA	865.47	25176.88
ARERE	874.04	22582.83
AYIN	1191.17	23942.82
OBECHE	848.43	22566.81

3.6 Bulk Density of the various Wood Species

Table 7 shows the densities of six various wood species used for this study. The densities of the various wood species range from 403.12 kg/m^3 for Iroko to 754.11 kg/m^3 for Ayin (Fig. 13). This result therefore suggests that Ayin is more likely to have a higher energy per unit volume than the other wood species considered in this study, because it will burn for a longer period of time than the other wood species. Woods with higher densities are also likely to produce briquettes with higher densities.

Table 7: Wood density value for selected wood waste biomass materials

Types of wood species	Iroko	Masonia	Afara	Arere	Ayin	Obeche
Wood density (kg/m^3)	403.12	440.52	562.74	560.05	754.11	734.69

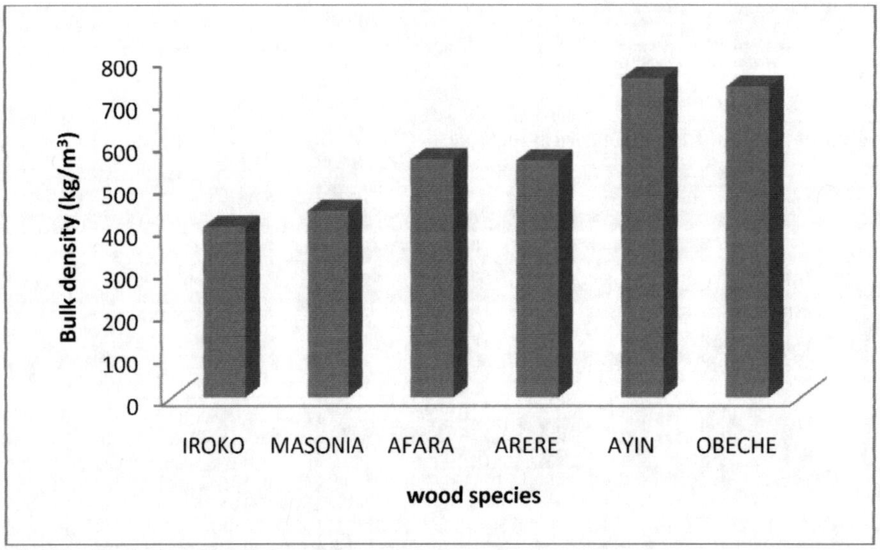

Fig.13: Bulk density of the various wood species

4. Conclusion

Wood biomass which is a major renewable energy source of energy in the developing world represents a significant proportion of the rural energy supply, and its potential lies in the residual waste usually referred to as wood shavings (sawdust). It is from these wastes that electrical power could be generated in a power plant that uses wood pellets as fuel. The estimated volume of wood milled daily from the various sawmills considered in this work was 55% of the total 100% wood species supplied to millers. Out of this 55% volume of the daily milled wood, 34% of it accounted for the daily milled wood waste. It was determined that the total volume of wood waste generated by the three plank markets in Ibadan on a daily basis amounts to 1607.4 m^3. The analysis of the six different types of wood species (Iroko, Masonia, Afara , Arere, Ayin and Obeche) obtained from the sawmills located within Bodija, Sango and Idi Ayunre areas of Ibadan were through laboratory investigations. This includes the utilization of both the proximate and ultimate analysis, and the estimation of the calorific values in the determination of their fuel characteristics.

The result of the proximate analysis of the various wood species obtained for this study showed that the percentage moisture content of Ayin was the highest, while that of Obeche was the least. This is an indication that Obeche will give the highest fuel calorific value (i.e. best fuel property) amongst the other wood species of wood considered in this work. Also, amongst these various wood species; Afara has the highest ash content which is an indication of high mineral matter. The volatile matter of Obeche being the highest and that of Afara the least indicates that during combustion process most of Obeche would not volatize, but would completely be burnt when utilized in cooking, while that of Afara would be otherwise. However, Afara has the highest fixed mass amongst these wood species with Iroko being the least. This high fixed carbon content indicates that the biomass will require a long combustion time. The result of the ultimate analysis shows that Iroko has the highest Hydrogen content, while Afara was the least. The methane content of Arere was the highest, while that of Afara was the least. Also, the carbon dioxide content of Iroko was the highest, while that of Arere was the least. Similarly, just as in the carbon dioxide content for the biomass, the carbon monoxide content of Iroko was the highest, while Arere was the least.

The fuel characteristics of these wood species showed that Afara has the highest value, while Obeche was the least. This suggests that wood waste from Obeche has the best fuel properties, while that of Afara was the worst. From the result of the gross calorific value (GCV), Ayin has the highest value, while Obeche had the least value. The comparison of the experimental value of the GCV with the estimated Dulong-Berthelot model showed consistency with those obtained from experiment, but of lower values due to the absent of components like sulphur and nitrogen in the saw dust obtained from these various wood species. This is an indication that saw dust obtained from these wood species produces less ozone depletion greenhouse gases.

The result of the bulk densities shows that Ayin has the highest value, while Iroko had the least value. This suggests that Ayin is more likely to have a higher energy per unit volume than the other species of wood used in this study, because it will burn for a longer period of time than the other species of wood. In summary, it can be concluded that amongst the six wood species studied in this work Obeche has the best fuel properties with both Arere and Afara being the worst.

References

[1] Karekezi S. and Kithyom, W. (2006). Bioenergy and Agriculture: Promises and Challenges. Bioenergy and the Poor. In: 2020 Vision for Food, Agriculture, and the Environment. International Food Policy Research Institute, Washington DC, USA

[2] Ladanai, S. and Vinterback, J. (2009). Global Potential of Sustainable Biomass for Energy. SLU, Institutionen for energy och teknik, Swedish University of Agricultural Sciences, Department of Energy and Technology, Report 013, ISSN 1654-9406, Uppsala

IOP Conf. Series: Earth and Environmental Science **173** (2018) 012043 doi:10.1088/1755-1315/173/1/012043

[3] Muhammad, M.B. (2016). Effect of Gasification operating Parameters on Quality of Syngas produced using sawdust feedstock. A dissertation submitted to the school of postgraduate studies Ahmadu Bello University, Zaria. In partial fulfillment of the requirement for the award of Master of Science Degree in Chemical Engineering. Accessed April 18, 2018 from kubanni.abu.edu.ng/jspui/bitstream/123456789/8068/…/

[4] Probsten, H.V; and Kicks, C.A. (1982). Briquetting Agricultural Waste as an Energy Source. Journal of Environmental Science, Computer Research & Technology (IJERT), Vol. 2 Issue 2.

[5] UNFCC (2005). Clarifications of Definition of Biomass and consideration of changes in Carbon Pools due to a CDM Project Activity. Accessed 18 April, 2018 from https://cdm.unfcc.int/Reference/Guidclarif/ar/methAR_guid07_v01.pdf

[6] Bergerson, J. and Keith, D. (2006). Life Cycle Assessment of oil sands Technologies Alberta Energy Futures Project Institute for Sustainable Energy, Environment and Economy (ISEE), University of Calgary.

[7] Christus, J.S., Joseph, S.S. and Thyagarajan, K. (2014). Performance Studies on Downdraft Gasifier with Biomass Energy Sources Available in Villages. Amercan Journal of Applied Science, 11(4), 611-622. DOI:10.3844/ajassp.2014.611.622.

[8] Reed, T.B., and Das, A.I. (1988). Hand book of Downdraft Gasifier Engine System. Solar Technical Information Program, Solar Energy Rresearch Institute (SERI) Colarado.
Basu, P. (2010). Biomass Gasification and Pyrolysis: Practical Design and Theory. Academic Press/ Elsevier, USA, 376pp.

[9] Hickey, M. and King, C. (2001). *The Cambridge Illustrated Glossary of Botanical Terms*. Cambridge University Press, UK. Accessed April 18, 2018 from http://en.wikipedia.org/wiki/Municipal_solid_waste

[10] ASTM E 870-82(2013), Standard Test Methods for Analysis of Wood Fuels, ASTM International, West Conshohocken, PA, 2013, www.astm.org

[11] Berberet, A., Ory, T., Poulos, V. and H. Powell. (2010). Milling the Future Sustainable Forestry and Sawmill Operations in Northern Winsconsin, Geography 565 Undergraduate Major Colloquim, 59pp.

[12] Biomass Analytical / Solid fuel Analysis. (2018). Assessed March 1, 2018 from www.laborec.be/ENG/Services/biomass-analysis/

[13] Sluiter, A., Hames, B. Ruiz, R., Scarlata, C., Sluiter, J. and D. Templeton. (2008). Determination of Ash in Biomass. Laboratory Analytical Procedure (LAP) National Renewable Energy Laboratory Technical Report NREL/FP-510-42622

[14] Encyclopedia Brittanica (2018) Fixed Carbon- Coal. Accessed April 25, 2018 from http://www.brittanica.com/topic/fixed-carbon

[15] ASTM (1981). Coal Resource Classification System of the U.S. Geological Survey. Accessed 26 March, 2018 from https://pubs.gov/circ/1983/0891/report.pdf
Wood, G.H., Kehn, T.M., Carter, M.D., Culberston, M.D (1983) Coal Resource Classification System of the U.S. Geological Survey-Geological Suvey Circular 891. Accessed 26 March, 2018 from https://pubs.usgs.gov/circ/1983/0891/report.pdf

[16] Channiwala, S.A. and Parikh, P.P. (2002). A unified correlation for estimating HHV of solid, liquid and gaseous fuels. Fuel, vol 81(8), pp. 1051-1063

IOP Conf. Series: Earth and Environmental Science **173** (2018) 012044 doi:10.1088/1755-1315/173/1/012044

Investigating the effect of solar trackers on solar energy harnessing in the Tropics

A. O. Ndubuisi, E. Imolorhe, M. L. Akinyemi

Department of Physics, College of Science and Technology, Covenant University, Ota, Ogun State, Nigeria

amanda.ndubuisi@covenantuniversity.edu.ng

Abstract. Solar panels produce higher output power when the solarradiation reaching it is perpendicular to its surface. Trackers are used to orient the solar panels perpendicular to the incident irradiation for maximum power output.In the temperate region, there is low irradiation compared to the Tropical region, thereby necessitating the use of solar trackers in order to harness as much energy as possible. This research seeks to investigate the effect of using trackers in the Tropical region by making a comparative analysis of the power and voltage output between a static solar panel and a panel mounted on a solar tracker. A cost-benefit analysis is also carried out. The results showed that there was an increase in output power with the use of a solar tracker, relative to a static photovoltaic module by 23.87%. It was concluded that use of trackers in the tropical region could be beneficial if it is used for a large generating solar plant or during the raining season, when there is more cloud cover.

1. Introduction

Photovoltaic solar energy is being explored as an alternative and renewable energy source due to depletion of fossil fuels and for reasons of environmental concern. However, generation of electricity using this technology is still considered undeveloped in comparison with fossil fuels due to the low conversion efficiency and production costs of photovoltaic modules. Research has shown that solar trackers have the ability to cause an increase in the efficiency of Solar Photovoltaic Systems [1, 2, 3]. The benefit of using a solar tracking system is that the output power and energy yield can be increased as the photovoltaic module is always oriented perpendicular to the incoming solar radiation even as the sun's location changes throughout the day.

There are two major factors which affect the increase in efficiency using solar trackers; the mode of operation of the tracking system and the geographical location of the installation of the photovoltaic system. Based on the geographical location, varying temperature, changes in climate and the geographical coordinates (longitude and latitude) affect the energy yield of the photovoltaic modules. For instance, using trackers in places like Europe which is in the temperate region can cause a 50% increase in energy gain when compared a fixed system [4]. In tropical regions, the increase in output power from the photovoltaic system is not as high as that of the temperate region, although it has been reported that the energy gain could be as high as 30% in some places like Tunisia [5], Taiwan [6], and Malaysia [7].

Despite these reports, there is a bit of uncertainty regarding the need for the trackers since the irradiation per day is quite high when compared to temperate regions. Also, a tracking system might

IOP Conf. Series: Earth and Environmental Science **173** (2018) 012044 doi:10.1088/1755-1315/173/1/012044

be unnecessary for tropical regions due to the overheating of the solar panels, from excessive exposure to solar irradiance in a hot climate, which consequently affects the performance of the modules. The aim of this research is to ascertain whether or not solar tracking is needed in the tropical region by showing the effect of solar tracking on the output of a PV cell in the tropics. We make a comparative study between a static photovoltaic solar panel and a one-axis mobility panel installed in Covenant University, Ogun state, Nigeria.

1.1. Studyarea
The study was carried out in Covenant University, Ogun state, the southwestern part of Nigeria (6.6718° N, 3.1581° E). This area is immensely influenced by the tropical savannah climate with distinct wet and dry seasons. The dry period lasts for about 130 days and runs through November till February, while the rainy season lasts for seven and eight months, between April and October with a short dry season in August. The rain in this area is defined by trade winds and its seasonal shift. The mean annual rainfall and temperatures are about 1,270 mm and 28 °C respectively, while the estimated mean annual potential evaporation is 1,100 mm. The daytime lasts from 10 to 13.5 hours, because the zones on both sides of the equator that are shaped by higher daily temporal (time of day climate) and smaller seasonal variations in temperature (25 °C annual averages) and due to that, there is a high irradiation all year around.

2. Methodology

2.1. Description of panels
Technical data at standard test conditions of the photovoltaic modules used are listed below:

Output Tolerance= $\pm 5\%$
Current at $P_{max}(I_{mp})$ = 0.57A
Voltage at $P_{max}(V_{mp})$ =17.5V
Short-Circuit Current (I_{SC}) = 0.63A
Open-Circuit Voltage (V_{OC}) = 22.05V
Mass = 1.2kg
Dimension = 360mm x 245mm x 23mm

2.2. Position of Photovoltaic (PV) modules
The PV modules for the control (stationary) and variable (tracking) systems were placed on the Terrace of the College of Science and Technology (CST) building, Covenant University, Ogun state, Nigeria. The location is suitable as it has a very wide-open air space and allows maximum exposure of the PV module with no shading. The solar panels were oriented towards 22.5° SW, horizontally. While the stationary panel was inclined at an angle of 45° vertically, the tracking panel was left to vary over the x-axis (horizontally).

2.3. Data acquisition
The photovoltaic module was mounted on the solar tracker in order to follow the sun's radiation over the angles of -4o to 53o on the x-axis each day for a period of 10 days, corresponding to time range of about 10.00-18:00 Hrs. Two pairs of an Ammeter and a Voltmeter were connected to the output of the panels (a pair for each panel). The solar tracker passes through each angle and stops at the point where the Ammeter and Voltmeter display the highest values of current and voltage, respectively, at a particular time. The angle at the point of maximum values of current and voltage is the angle where the incident irradiation of the sun is perpendicular to the panel. The values at this point for the both panels were recorded.

3. Results

Recorded in Table 1, are the average values of the measurements for short-circuit current (Isc), open-circuit voltage (Voc) and the power output (P) for the period of 10 days, between 10:00 to 18:00 Hrs, daily.

Table 1: Average Values of Isc, Voc and P.

	NO TRACKER			TRACKER			
Time (Hrs)	Voc(V)	Isc(A)	P(W)	Voc(V)	Isc(A)	P(W)	Angle(o)
10:00	18.40	0.33	4.36	19.32	0.55	7.66	-4
12:00	18.55	0.36	4.84	19.04	0.54	7.45	-3
14:00	18.95	0.49	6.68	19.13	0.55	7.51	20.8
16:00	19.03	0.53	7.29	19.02	0.54	7.42	38.5
18:00	18.23	0.51	6.62	18.26	0.52	6.84	52.1

It is seen that the output power for the mobile systems is significantly higher than that of the static system, although the power output of both systems increases as the sun rises and decreases as the sun sets. There is an increase in efficiency by 23.87% which denotes an improvement in power yield. The performance of open circuit voltage, short circuit current and power for both systems can be seen in figures 3, 4 and 5 respectively. In the three figures, the photovoltaic module which is mounted on the tracker performs better than the static panel. However, in the evening when the incoming solar radiation is minimal, they both exhibit a low performance which is typical at low radiation.

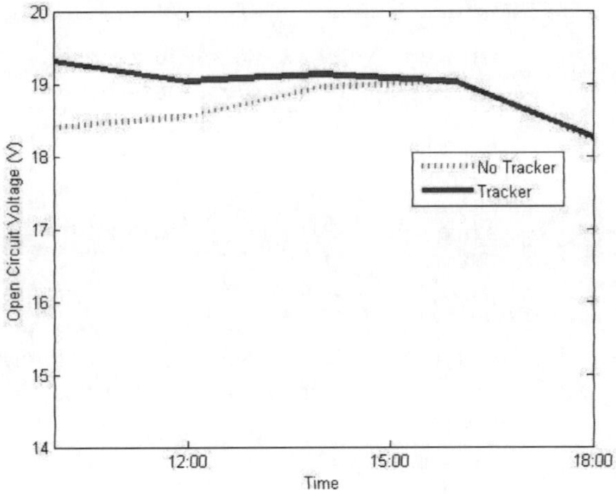

Figure 1: V_{OC} (Non-Tracked [NT] and Tracked [T])

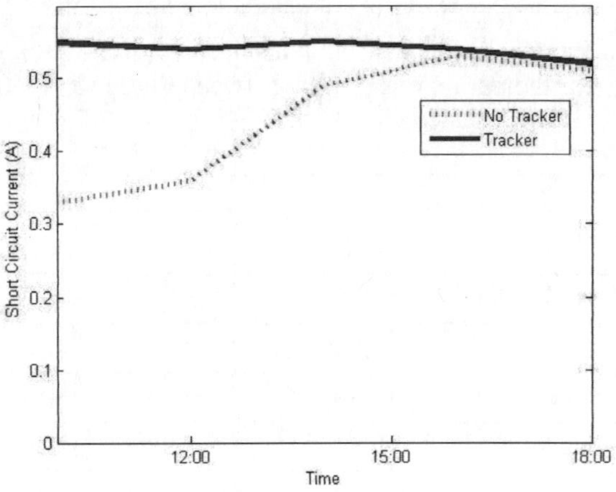

Figure 2: I_{SC} (Non-Tracked [NT] and Tracked [T])

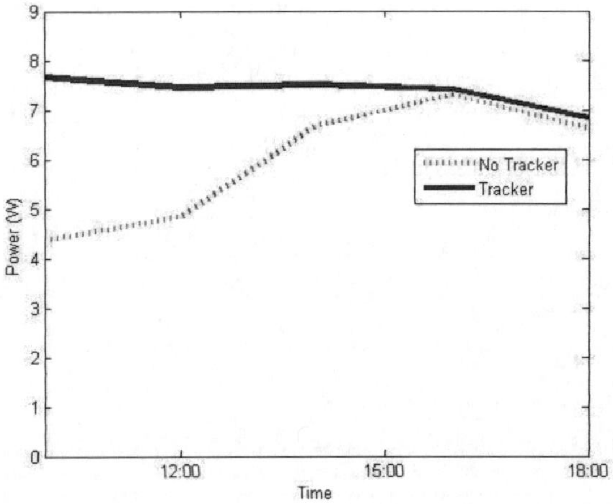

Figure 3: Power (Non-Tracked [NT] and Tracked [T])

It is shown in Table 2, the average results of Isc, Voc and P on a daily basis for both systems. From the analysis of the averaged values, we see again that the mobile system performs better than the static system. The variation in the daily power output is due to atmospheric conditions and other factors which affect the efficiency of the module such as cell temperature.

Table 2: Average Results for each day

	No Tracker			Tracker		
	Voc(V)	Isc(A)	P(W)	Voc(V)	Isc(A)	P(W)
Day 1	18.65	0.45	6.09	19.08	0.54	7.47
Day 2	18.87	0.41	5.63	19.22	0.55	7.59
Day 3	18.77	0.40	5.33	19.13	0.55	7.51
Day 4	18.95	0.50	6.79	19.25	0.55	7.61
Day 5	18.61	0.44	5.90	18.89	0.54	7.32
Day 6	19.07	0.50	6.88	19.31	0.55	7.66
Day 7	18.51	0.47	6.21	18.71	0.53	7.19
Day 8	18.49	0.46	6.16	18.75	0.53	7.22
Day 9	18.23	0.41	5.30	18.51	0.53	7.04
Day 10	18.17	0.40	5.25	18.69	0.53	7.18

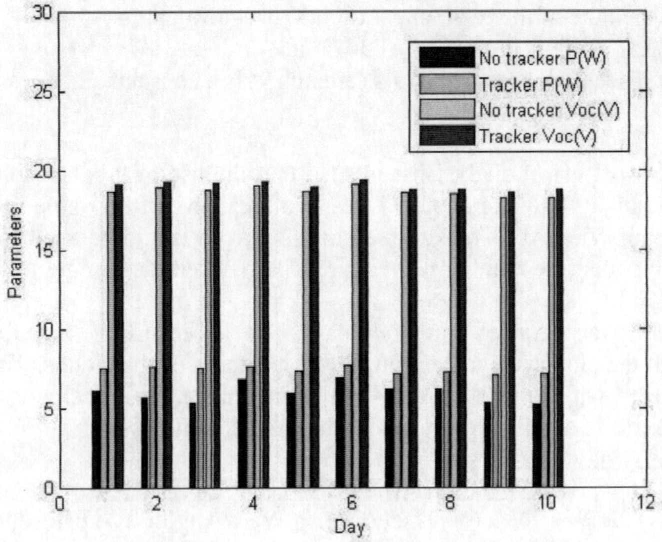

Figure 4: Average value of each parameter over 10 days

3.1. Cost-benefit analysis

3.1.1. Production cost/ Capital expenditure. The list of parts or components of the system needed for the implementation of the tracker is shown in Table 3.

Table 3: Capital expenditure

S/N	Parts/Components	Quantity	Cost
1	Frame/Stand	1	₦7,000
2	LDR	2	₦20
3	Resistor	7	₦70
4	AT Mega 16 MC	1	₦1000
5	LM 328	1	₦100
6	L293D	1	₦200
7	Variable Resistors	2	₦20
8	Stepper Motor	1	₦2000
9	Battery	1	₦1300
	TOTAL		₦11,710

3.1.2. Operational Cost/ Maintenance Expenditure. The average cost of maintenance is so small that it is negligible.

3.1.3. Cost Effect. Average increase in power= 23.87%

Average power generated= 5.96W (Non tracked)/7.38W(Tracked)

Average electricity cost in Nigeria= ₦29.04/KWhr

Total net time for the experiment = 8hrs each day x 10 days => 80hrs
The Gain in power generation over the time of the experiment (80hrs) is equivalent to:
₦14.72/KWhr x 5.96W x 80hrs x 23.87% => ₦3.31 => ~ ₦0.0413/hr.
Note: The cost of the panel was not discussed, because in this research, it is a constant and does not change anything if included.

3.1.4. Cost versus benefit. From the Cost effect analysis, it is seen that the use of a solar tracker saves about ₦0.0413 per hour. This means it will take about 33 years to break even (this value is for this research and its specifications alone). The exact tracker used in this project could be used for a 20W solar panel and higher; this implies that the number of years can be reduced, depending on the power efficiency of the Tracker.

The highest energy increase (Table 4) was recorded on cloudy days. This is because the rays from the sun are refracted randomly through the clouds dependent on the arrangement of the clouds. Thus, the incident angle of irradiation is altered with respect to the cloud arrangement. This ascertains the fact that the implementation solar trackers are significant especially during the rainy season.

Table 4: Highest energy increase recorded

	Time (Hrs)	Voc(V)	Isc(A)	P(W)	Voc(V)	Isc(A)	P(W)	Angle(^{o})	Difference
Day 2	10:00	18.80	0.09	1.21	20.10	0.57	8.29	-36	7.08
Day 3	10:00	18.60	0.09	1.18	19.80	0.57	8.04	-36	6.86
Day 10	10:00	17.85	0.10	1.26	19.85	0.57	8.08	-35	6.82
Day 3	12:00	18.60	0.31	4.16	19.00	0.54	7.41	-10	3.25
Day 9	12:00	18.44	0.31	4.10	18.88	0.54	7.31	-10	3.22

4. Conclusion and Recommendation

The power generated in the photovoltaic module mounted on the tracker yields a higher power output of 23.87% in comparison to the static PV module. The use of a solar tracker also saves ₦0.0413 per hour. Thus, we have in view, a plausibility of relevance of trackers in the tropical region, but it must be noted that the relevance increases directly with the size of the generating plant and power efficiency of the Trackers.

With respect to the results of the analyses carried out, we commend that the use of solar trackers in the tropical region would be very utilitarian in the following cases:

- Large Generating Plants: Use of trackers in the tropical region could have greater relevance when it is used in a large generating plant. This is because the bigger the tracking system (in singular size or quantity), the more cost effective it would be. The amount of time taken to break even is also taken into consideration. Thus, a small scale generating plant would not efficiently express the potentials of Trackers in the tropical region.
- Rainy season: When the atmosphere is cloudy, the irradiation of the sun is randomly reflected and refracted through the clouds. This makes the angle of highest irradiation intensity to vary randomly; thus, a tracker is needed.

Acknowledgment
The authors thank Covenant University for conference support

References

[1] Eke, R.and Sentruk, A. (2012). Performance comparison of a double-axis sun tracking versus fixed PV system. *Solar Energy,* **86**: 2665-2672

[2] Dakkak, M. and Babeli, A. (2012). Design and performance study of a PV tracking system (100W- 24Vdc/220Vac). *Energy Procedia,* **19**:91-95

[3] Senpinar, A. and Cebeci, M. (2012). Evaluation of output power for fixed and two-axis tracking PV arrays. *Applied Energy,* **92**: 677-685

[4] Huld, T., Beyer, H. G., Gottschalg, R. and Topic, M. (2010). Mapping the performance of PV modules, effects of module type and data averaging. *Solar Energy*, **84**(2): 324-338

[5] Maatallah, T., Alimi, S. E. and Nassrallah, S. B. (2011). Performance modeling and investigation of fixed, single and dual-axis tracking photovoltaic panel in Monastir city, Tunisia. *Renewable and Sustainable Energy Reviews,* **15**(8): 4053-4066

[6] Chang, T. P. (2009). Output energy of a photovoltaic module mounted on a single-axis tracking system. *Applied Energy*, **86**(10): 2071-2078

[7] Samsudin, K., Noor, A. M. and Rahman, A. R. A. (2008). Evaluation of genetic algorithm based solar tracking system for photovoltaic panels. International Conference on Sustainable Energy Technologies ICSET, Singapore. IEEE

Projected Range and Sputter Yield of Ne^+ and Ar^+ in the Sputtering of Lead and Tin Perovskites

Oluwole E. Oyewande[1, 2] **and Akinwumi Akinpelu**[1]

[1]Department of Physics, College of Science and Technology, Covenant University, Ota, Ogun State, Nigeria
[2]Department of Physics, Faculty of Sciences, University of Ibadan, Ibadan, Nigeria

oluwole.oyewande@covenantuniversity.edu.ng
akinwumi.akinpelu@covenantuniversity.edu.ng

Abstract. Metal halide perovskites have been the subject of intense theoretical and experimental research in recent years due to their huge potential over their silicon based counterparts for tunable optoelectronic applications in high-tech device innovation. The current best perovskite for solar cell applications, with a power conversion efficiency of 22%, methyl ammonium lead iodide ($CH_3NH_3PbI_3$), is toxic due to the presence of lead and is therefore harmful in solar cell applications despite its low concentration in solar cells. Hence, research exploits are geared towards perovskites without lead. Unfortunately, this has taken back the gains in PCEs by about 15%, and a lot is being done for improvement. In this paper, we performed molecular dynamics and Monte Carlo simulations of ion-beam sputtering of lead and tin perovskites to determine differences between the sputtering characteristics of lead perovskite and a lead-substituted perovskite (tin perovskite). Our results show that they both exhibit similar sputtering characteristics of linear projected ion range, and maximum yield around 78° ion incidence.

Keywords: Solar cells; Sputtering; Perovskites; Sputter yield; Range of ions.

1. Introduction

Lead halide perovskites have huge potential for applications in solution-based photovoltaics (e.g. solar cells) with greater power conversion efficiencies (PCEs) than silicon-based photovoltaics [1]. However, they suffer a large setback in these applications due to the toxicity of lead and its solubility in water. There has therefore been an intense research interest with promising results in recent years on non-toxic metal replacements [2 - 6]. One advantage of perovskite based photovoltaics over their silicon based counterparts is their low-cost fabrication due to the possibility of producing them in a variety of ways, including the use of vacuum techniques [7].

A vacuum processing and fabrication technique is ion-beam sputtering of materials, which is an area of current intensive research as a cost-effective method of surface analysis, processing and fabrication of self-organized nanostructures for optoelectronic applications [8 – 14]. A new radio-frequency-sputtering method of production of lead-iodide perovskite was proposed in Ref. 15. This comprises of a deposition of thin films of lead sulphide, their conversion to perovskites by placement in an iodine atmosphere, and subsequent immersion into a solution of methyl ammonium (i.e. solution based). Whereas in Ref. 16, a non-solution based method of perovskite film fabrication by sputtering was proposed. They confirmed that sputter processed perovskite films showed similar characteristics as their solution-processed counterparts and produced perovskite solar cells with higher PCEs.

In a sputtering process, an incoming ion collides with an atom of the target material and sets off secondary collision cascades leading to kinetic and thermal agitations of the targets atoms. Noble gas ions are commonly used in sputtering experiments due to their inert nature and low reactivity with the sputtered targets. By virtue of the nature of the sputtering process and the need for kinetic details within the earliest impact times, molecular dynamics simulation can be exploited. A number of enquiries have taken this approach, focusing on specific sputtering conditions (e.g. [17- 20]). However, statistical collisional data is also required and this is obtained by statistical physics Monte Carlo simulations.

A molecular dynamics simulation suite developed by Ziegler and Biersack [21 – 23], Stopping and Range of Ions in Matter (SRIM), and its more extended counterpart, TRansport of Ions in Matter (TRIM), which includes SRIM and statistical simulation and calculation of some sputtering parameters with Monte Carlo techniques, are versatile tools in this regard [24]. They are reference and very popular simulation software packages with a large number of scientific citations to them every year [23]. Signatures of the ion-target interaction are the collision cascade parameters and sputter yield. These can be drastically changed by minute changes in the target composition. It is therefore of interest to know the effect of lead replacement on the sputter characteristics of metal halide perovskites. This has not yet been studied, to the knowledge of the authors of this article.

The most suitable material for this comparison is $CH_3NH_3SnI_3$ which is a metal halide perovskite with similar characteristics as $CH_3NH_3PbI_3$ but with a reportedly much lower PCE of 6.4% [2]. It, nonetheless, has the highest PCE among all the lead-substituted perovskites [7].However, Sn or Pb are good occupants of the B-site of a perovskite ABX_3 structure, where the monovalent organic cation (e.g. methyl ammonium, $CH_3NH_3^+$) occupying the position A, or the halide counter-ion (e.g. iodine, I^-) occupying the X-site of the perovskite structure, can be changed to tune the properties of the perovskite.

In this paper, we studied the range and sputter yield of ions in $CH_3NH_3PbI_3$ (lead) and $CH_3NH_3SnI_3$ (tin) perovskites for different ion energies and incidence angles by molecular dynamics and Monte Carlo simulations using SRIM and TRIM. We used ions of inert gases Ne and Ar in the sputtering of these perovskites. The rest of the paper is organized as follows. In the next section we elaborate on the methods we have used to obtain our results. For reproducibility, we provide details of the simulation set-up on the SRIM and TRIM graphical user interface and the values used for all other parameters. We present and discuss our results in section 3, and provide our conclusions in section 4.

2. Methodology

In this section we provide the specific details of our simulation. Details of the theoretical background for the calculations and simulation algorithms embedded in the TRIM and SRIM packages are discussed in the papers by Ziegler and Biersack [21, 25]. SRIM was used to perform molecular dynamics simulations of the range of inert gas ions Ne^+ and Ar^+, with energies varied from 1 keV to 10 keV, at normal incidence on the two targets, namely, lead perovskite $CH_3NH_3Pb`I_3$ and tin perovskite $CH_3NH_3SnI_3$. While TRIM was used to perform molecular dynamics and Monte Carlo simulations for the number of each component of the perovskites yielded as a result of bombardment of the perovskite by an incident ion, for varied incidence angles from 0° to 89°, and for ion energies 1 keV and 5 keV.

In both SRIM and TRIM set-ups we used perovskite wafer thickness of 35 nm, and built up the lead perovskite from its composites in the stoichiometric ratio 1:3:1:3:1:3 for C, H, N, H, Pb/Sn, I, respectively, and densities 4.16g/cm^3 [26] and 3.51g/cm^3 [27] for the lead and tin perovskites, respectively. We also used 1000 ions for each simulation. A detailed calculation with full damage cascade was performed for each simulation (i.e. each complete run of TRIM or SRIM for a particular set-up).

3. Results and discussion

The results of our simulations are presented and discussed in this section. We observed the same trends for the two ions (Ne^+ and Ar^+), though with higher values for Ar^+, and therefore present the results for only Ne ion here. Figure 1 shows the results of the projected range of the ions in the perovskites for different ion energies varied from 1 keV to 10 keV. These exhibitions of almost the same values of the linear range of the ions for the different perovskites indicate that the two materials are remarkably similar as regards their stopping power to energetic particle irradiation.

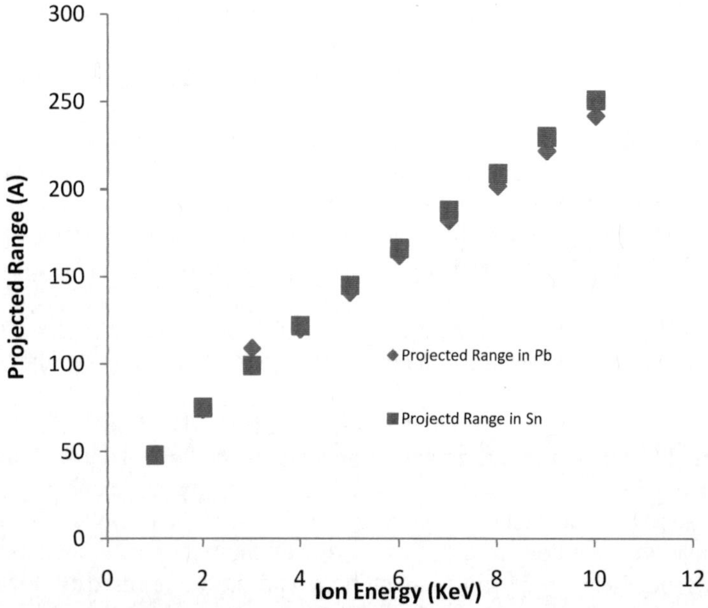

Figure 1: Projected range of Ne^+ ion lead (diamond symbol) and tin (square) perovskite targets for differention energies from 1 keV to 10 keV.

Sputter yield results are presented in the figures that follow, one for each element in the perovskite, namely, C, H[1] (H of the methyl molecule), N, H[2] (H of the ammonium molecule), Pb/Sn, I. Figure 2 shows the result for the number of C atoms ejected per incident Ne^+ ion on the lead and tin perovskites at different angles of incidence varied from 0° to 89°, and for Ne^+ energy of 1 keV and 5 keV. The same variables apply to Figs. 3 – 7, but for the other elements, respectively.For all results the yield is expectedly higher for higher ion energy, and a maximum yield occurs at the same $\theta \sim 78°$ for both energies. Also, for 5 keV the yield obtained from the sputtering of tin perovskite is lower than that of the sputtering of lead perovskite for normal incidence, and then increases with angle of incidence, relative to the yield on lead perovskite, until it becomes higher around the angle of incidence at which maximum yield occurs.

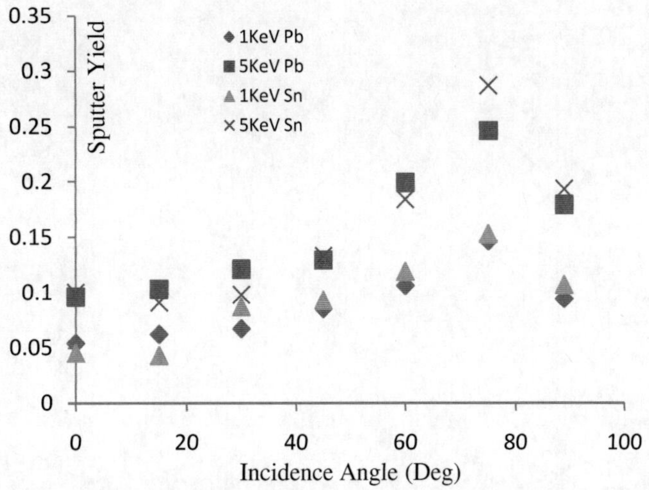

Figure 2: Sputter yield (C atoms/ion) for the ejection of C atoms from Ne^+ bombardment of the perovskites at different angles of incidence and ion energy 1 keV and 5 keV.

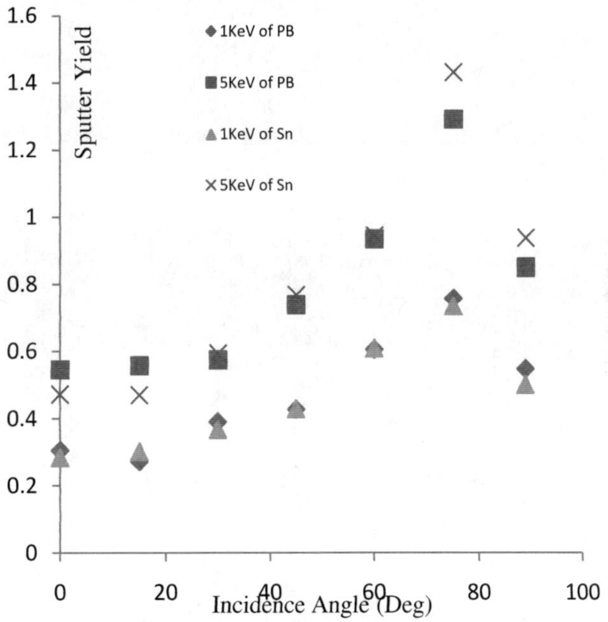

Figure 3: Sputter yield (H^1 atoms/ion) for the sputter erosion of H atoms of the methyl molecule from Ne^+ bombardment of the perovskites at different angles of incidence and ion energy 1 keV and 5 keV.

However, for N erosion of Fig. 4, a first instance of the higher value of the sputter yield of the ion on Sn perovskite occurs at a lower angle of about 45° and then a second instance at the angle for maximum yield. Though, this could be a mere statistical fluctuation, it is noteworthy as it is exhibited in the H^1 result of Fig. 3.Note that in Fig. 6 only two sets of data are available for each (Pb or Sn) since Pb is not available in tin perovskite or vice versa. The values of I atom yields are the highest Fig. 7, followed by H atom, Pb/Sn atom, N atom, and the lowest yields are those of the C atom.

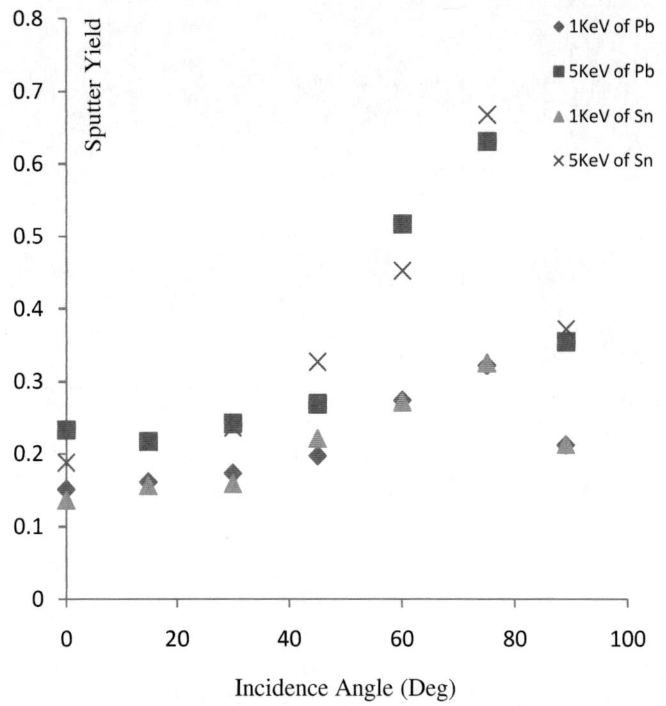

Figure 4: Sputter yield (N atoms/ion) for the erosion of N atoms from Ne^+ bombardment of the perovskites at different angles of incidence and ion energy 1 keV and 5 keV.

The two H atom yields are similar, with the yield of H^1 being slightly higher (about 1.4) than that of H^2 (about 1.3).

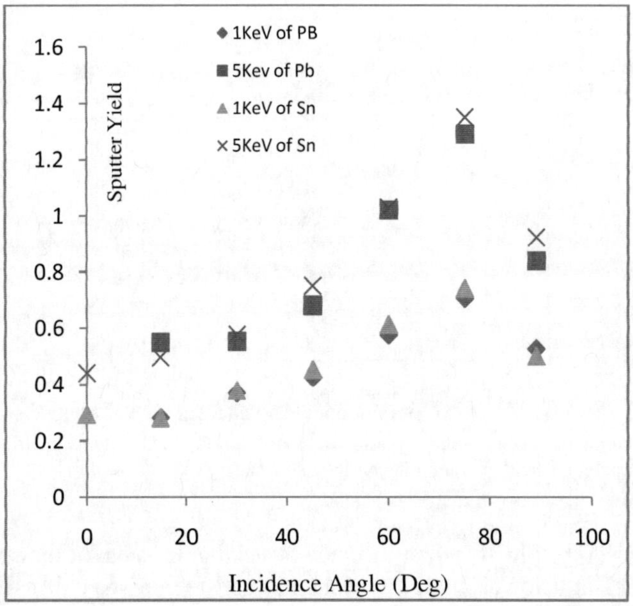

Figure 5: Sputter yield (H^2 atoms/ion) for the ejection of H atoms of the ammonium molecule from Ne^+ bombardment of the perovskites at different angles of incidence and ion energy 1 keV and 5 keV.

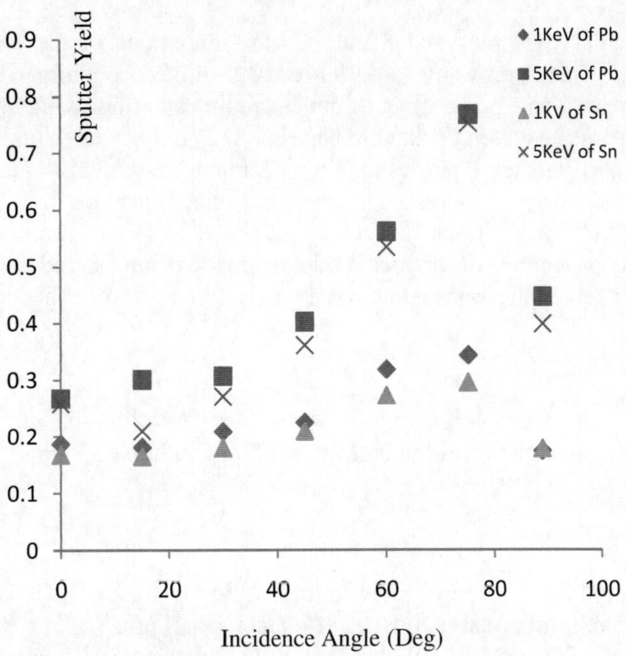

Figure 6: Sputter yield (Pb/Sn atoms/ion) for the erosion of Pb/Sn atoms from Ne^+ bombardment of the perovskites at different angles of incidence and ion energy 1 keV and 5 keV.

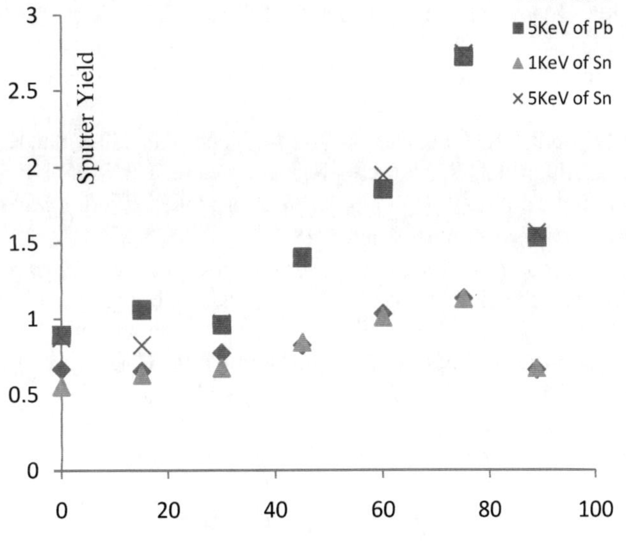

Figure 7: Sputter yield (atoms/ion) for the erosion of I atoms from Ne^+ bombardment of the perovskites at different angles of incidence and ion energy 1 keV and 5 keV.

4. Conclusion

We performed molecular dynamics and Monte Carlo simulations of the ion sputtering of methyl ammonium lead iodide (lead perovskite) and its lead-substituted counterpart, methyl ammonium tin iodide (tin perovskite), as a new perspective of understanding and improving upon the desired physical properties of safer perovskite materials that are based on the possible substitutes for lead.

Currently, these substitutes give perovskites with a much lower PCE than lead perovskite. In our studies, reported in this paper, we found the sputtering results of lead perovskite and its best substitute so far, tin perovskite, to be very similar. In fact, so similar as to serve as another reason why the quest to improve upon the properties of tin perovskites, towards the desired better PCE in tin halide perovskite based solar cell applications, should not be abandoned despite its current very poor PCE.

Acknowledgement

The authors will like to appreciate Covenant University for conference support.

References

[1] National Renewable Energy Laboratory. Best Research-Cell Efficiencies Chart. http://www.nrel.gov/pv/assets/images/efficiency_chart.jpg.

[2] Noel NK, Stranks SD, Abate A, Wehrenfennig C, Guamera S, Haghighirad AA, Sadhanala A, Eperon GE, Pathak SK, Johnston MB, Petrozza A, Herz LM, Snaith HJ,*Energy Environ Sci* 7: 3061(2014).

[3] Krishnamoorthy T, Ding H, Yan C, Leong WL, Baikie T, Zhang Z, Sherburne M, Li S, Asta M, Mathews N, Mhaisalkar SG, J Mater Chem A 3: 23829(2015).

[4] Cui XP, Jiang KJ, Huang JH, Zhang QQ, Su MJ, Yang LM, Song Y-L, Zhou X-Q, Synth Met 209: 247(2015).

[5] Park BW, Philippe B, Zhang X, Rensmo H, Boschloo G, Johansson EMJ, Adv Mater 27: 6806(2015).

[6] Harikesh PC, Mulmudi HK, Ghosh B, Goh TW, Teng YT, Thirumal K, Lockrey M, Weber K, Koh TM,Li S, Mhaisalkar S, Mathews N, Chem Mater 28: 7496(2016).

[7] Hoefler SF, Trimmel G, Rath T,*Progress on lead-free metal halide perovskites for photovoltaic applications: a review*, MonatshChem 148: 795 – 826(2017).

[8] Oyewande EO, *Modelling and simulation of surface morphology driven by ion bombardment,*Institute for Theoretical Physics. Goettingen : University of Goettingen. (2006). 1 - 106, Ph. D. Thesis. Publication URL: https://ediss.uni-goettingen.de/handle/11858/00-1735-0000-0006-B596-C.

[9] Munoz-Garcia J,Vazquez L, Castro M, Gago R, Redondo-Cubero A, Moreno-Barrado A and Cuerno R, *Self-organized nanopatterning of silicon surfaces by ion beam sputtering,*Materials Science and Engineering B, Vol. 86, pp. 1 – 44(2014). And references therein..

[10] Oyewande OE, Hartmann AK, Kree R, Phys. Rev. B, Vol. 71, 195405 (2005). And references therein..

[11] Oyewande EO, Kree R and Hartmann AK, *Morphological regions and oblique-incidence dot formation in a model of surface sputtering*, Phys. Rev. B, Vol. 73, 115434 (2006).

[12] Oyewande EO, Kree R and Hartmann AK,*Numerical analysis of quantum dots on off-normal incidence ion sputtered surfaces*, Phys. Rev. B, Vol. 75, 155325(2007).

[13] Kree R, Hartmann AK, Geyer U and Kölbel M,*Long-time effects in a simulation model of sputter erosion,* Phys. Rev. B, **65**, 193403(2002).

[14] Oyewande OE, *A unified spatio-temporal framework of the Cuerno-Barabasi stochastic continuum model of surface sputtering*, Commun. Theor. Phys. **58**, 165 – 170(2012).

[15] José Maria Clemente da Silva Filho, Viktor A. Ermakov and Francisco Chagas Marques,*Perovskite thin film synthesised from sputtered lead sulphide*, Scientific Reports 8, 1563(2018).

[16] Raifuku I, Ishikawa Y, Bourgeteau T, Bonnassieux Y, Cabarrocas PRI and Uraoka Y, *Fabrication of perovskite solar cells using sputter-processed $CH_3NH_3PbI_3$ films*, Applied Physics Express 10, 094101(2017).

[17] Feix M, Hartmann AK, Kree R, Munoz-Garcia J, and Cuerno R, Phys. Rev. B **71**, 125407(2005).

[18] Moeller W, *TRI3DYN - Collisional computer simulation of the dynamic evolution of 3-dimensional nanostructures under ion irradiation*, Nucl. Instrum. Meth. Phys. Res. B, Vol. 322, pp. 23 – 33(2014).

[19] Hofsäss H, Zhang K, Mutzke A, *Simulation of ion beam sputtering with SDTrimSP, TRIDYN and SRIM*, Appl. Surf. Sci. **310**, 134 – 141(2014).

[20] Shulga VI,*Sputter yield of rippled surfaces: A simulation study*,Nucl. Instrum. Meth. Phys. Res. B **412**, 207 – 213(2017).

[21] Ziegler JF, Biersack JP and Littmark U, *The Stopping and Range of Ions in Solids,*New York :Pergamon(1985).

[22] http://www.srim.org

[23] Ziegler JF, Ziegler MD and Biersack JP,*The stopping and range of ions in matter*, Nucl. Instrum. Meth. Phys. Res. B **268**, 1818 – 1823(2010).

[24] Martinie S, Saad-Saoud T, Moindjie S, Munteanu D and Autran JL,*Behavioral modeling of SRIM tables for numerical simulation*, Nucl. Instrum. Meth. Phys. Res. B **322**, 2 – 6(2014).

[25] Wilson WD, Haggmark LG and Biersack JP, *Calculations of nuclear stopping, ranges, and straggling in the low-energy region*, Phys. Rev. B **15**, 2458 (1997).

[26] Huang J, Jiang K, Cui X, Zhang Q, Gao M, Su M and Yang L, *Direct conversion of of $CH_3NH_3PbI_3$ from electrodeposited PbO for highly efficient planar perovskite solar cells*, Sci. Rep. **5**, 15889(2015).

[27] Kim C, Huan TD,Krishnan S, and Ramprasad A, *A hybrid organic inorganic perovskite dataset*, Sci. Data **4**, 170057(2017).

Comparative studies on the kinetics of biogas purification using activated Carbon and Zeolite

Abdulsalam Abdullahi[1], Mohammed Alhassan[1], Abubakar Garba Isah[1], Kovo Abdulsalami Sani[1], Olanrewaju A. Olalekan[2]

[1]Chemical Engineering Department, Federal University of Technology, Minna, Nigeria
[2]O.A.O. Technologies, Kaduna.

abdulsalamabdul81@gmail.com, moh.alhass@futminna.edu.ng, abubaka.isah@futminna.edu.ng, kovo@futminna.edu.ng, oaotechnologies2016@gmail.com

Abstract. The study focus on CO_2 adsorption using activated carbon and zeolite for the purification of the produced biogas and developed kinetic study for the AC and zeolite CBV 8014 adsorption. Furthermore, model equations were developed for the upgraded biogas purity as a function of adsorption time and adsorbent dosage to predict biogas purity. The adsorbents and kinetic study was evaluated using a cylindrical glass column. The biogas produced from chicken droppings and cow dung were characterized and found to contains CH_4 gas (53.26%), CO_2 (33.44%) and other gases in small quantity. The highest purity of 84.34% and 91.01% was obtained for AC and zeolite CBV 8014 respectively at 20 min adsorption time and the model equations developed well represent the experimental data. The adsorption capacity of the adsorbents was evaluated using a cylindrical glass column and was found to be in the order of zeolite CBV 8014 (91.01%) > AC (84.34%) at 0 – 20 min adsorption time with adsorption increasing over time. The kinetic study of the biogas purification shows that the CO_2 adsorption onto AC and zeolite CBV 8014 obeys the pseudo-second-order model, due to its regression coefficient (R^2) value closer to unity. The equilibrium and kinetic study of CO_2 adsorption onto AC and zeolite CBV 8014 shows that the magnitude of activation energy for both adsorbent are lesser than 40 kJ/mol, indicating the occurrence of physisorption mechanism. The intra-particle diffusion model also indicates that intra-particle diffusion is not the sole rate-limiting step in the adsorption process for both adsorbent (AC and zeolite CBV 804). The zeolite CBV 8014 has higher CO_2 adsorption capacity than AC.

Keywords: *Activated Carbon, Adsorption, Biogas, Carbon dioxide, Zeolite CBV 8014.*

1. INTRODUCTION

Globally, energy is known to be an integral part of any developed economic and society. It has been a fundamental factor in poverty alleviation of any society [1]. The availability and access to energy ingredient of national development and in improving the standard of living of any nation. However, global energy demand has continued to rise, while availability and access to energy are dwindling. The dependence on petroleum based fossil fuel and petroleum products as the only chief source of energy has led to global climatic changes, environmental degradation and pollution, increase in sea level as a result of the melting of the ice caps resulting from temperature increase. This problems therefore, lead to environmental and health problems. Petroleum based fossil fuel, though finite, it damages the ecosystem in many and diverse ways, acid rain, destruction of the ozone layer, and exerting greater influences upon biogeochemical cycles [2]. These has led credence to the need to turn energy systems towards sustainability.

Africa as a content with very weak law against environmental pollution is facing greater environmental challenges such as; flooding, air and water pollution caused by degradation of the ecosystem due to pollution from petroleum fossil fuel. Reference [3] argued that less than 10% of the total population of 21 sub-Saharan African countries have access to electricity. Thus, the need for alternative renewable sources of energy from renewable and sustainable locally available resources cannot be over emphasized. In developing countries like Nigeria, lack of access to adequate energy has remained a major hindrance to improving key indicators of human development [4]. Currently over 60% of the population in Nigeria depends almost entirely on wood as sources of energy for cooking, heating and agro-processing activities. Furthermore, electricity which is a major requirement for modern economies is not readily available and or better still unreliable and in short supply. Thus, there is an urgent need to sources alternative renewable forms of energy before the petroleum based oil supplies run dry. Hence, in the midterm 2008 – 2015 and long term, 2016 – 2025, Nigeria envisions an energy transition from crude oil to renewable energy [5]. However, an alternative source of energy must be technically feasible, economically competitive, environmentally acceptable and readily available and accessible. One key and possible alternative to petroleum based fuel is the use of biofuel resources.

Biofuel resources are key renewable energy sources and relatively distributed. The production of biofuels has become a global agenda. Biofuel includes, biogas, bioethanol and biodiesel. Biogas is can be produce from many different environments to include sewage sludge, landfills and during anaerobic degradation of biomass material [6]. Biogas comprises mainly of CH_4 (methane: 45 – 75% vol.), CO_2 (carbon dioxide: 25-55% vol.), and other compounds in trace amount to include H_2S (hydrogen sulfide: a couple of thousand parts per million), H_2O and other trace gas compounds (N_2, CO, O_2 etc.) [7]. Methane is a potent greenhouse gas if emitted into the atmosphere, however it also represent a valuable renewable source of energy, with the potential to reduce Green House Gas (GHG) emissions when it is collected and substituted for petroleum based fuels.

In Nigeria, biogas utilization is on the low side despite huge reserves of biogas feedstock within the country. For example, animal and agricultural waste and residues generated in Nigeria stood at a million tons per day [2]. This implies that from animal and agricultural waste alone, Nigeria can generate over 4.75×10^9 MJ per annum indicating that that a large proportion of Nigeria's energy need can be sourced from biogas. In order to upgrade raw biogas to a high purity of over 95% methane to ensure environmentally friendly energy source and reduce emission of large GHG (CO_2), CO_2 generated from the gas need to be captured. Amine based and physical adsorption are the most commonly used for CO_2 capture.

Studies have shown that amine based adsorbents of CO_2 are associated with several problems to includes; high regeneration energy requirement, corrosive tendency and degradation of amines [9]. Selecting a suitable separation process is determined by adsorption and diffusion activity which characterizes most activated carbon (AC) materials also limits their applications [10]. However, many studies have documented successful use of AC and zeolite for CO_2 capture. Reference [10] reported the comparative studies of CO_2 capture using acid and base modified activated carbon from sugarcane bagasse and observed that acid and base modified sugarcane bagasse AC produced have enhanced capacity for CO_2 adsorption. Several studies on the use of AC for CO_2 adsorption have shown that AC are with a highly developed porous structure with good adsorption properties. Therefore, the relative abundance, low cost, minimal sensitivity to moisture and high thermal stability make AC to have distinctive characteristics and also placed it in a vantage position as an exceptionally viable and cheap adsorbent for CO_2 capture.

Biogas is said to comprise of methane (45-75% by volume CH_4), carbon dioxide (25-55% vol. CO_2) and trace quantities of hydrogen sulfide (H_2S), ammonia (NH_3), volatile organic compounds (VOCs), siloxanes, carbon monoxide (CO), oxygen (O_2), and water (H_2O) [8]. The concentration of CO_2 in the atmosphere was projected to increase from pre-industrial concentration of 270 ppm to a high concentration of 407 ppm as of mid-2017 [11]. There is therefore the need to prevent the release of large quantities of CO_2 into the atmosphere. CO_2 emissions to the atmosphere can be effectively reduced through several methods which include extraction and sequestration from GH gases emission with chemical solvents and physical adsorption method commonly used methods of GHG emission. This necessitate the need to investigate and optimized the purification (CO_2 capture) of biogas produced from animal waste. The aim of this research work is to investigate the production of biogas and kinetic study of CO_2 capture from biogas produced from animal waste.

2. METHODOLOGY

2.1. *Materials*

The materials and major equipment used in the course of the research are Cow dung and Chicken Droppings obtained from FUT Minna and Minna respectively in Niger State. The Activated Carbon was obtained from Kaduna Polytechnic and Zeolite CBV 8014 from Chemical Engineering Department, FUT Minna. Analytical grade ammonium hydroxide and distilled water were obtained from Chemical Engineering Laboratory, Kaduna Polytechnic and Chemical Engineering Laboratory, FUT Minna.

2.2. *Sample Collection*

Cow dung was collected from the Federal University of Technology Farm, Gidan Kwano, Minna, Niger State while chicken droppings were collected from the poultry farm in Minna. The cow dung and chicken droppings collected were weighed according to the calculated ratio and mixed together to ensure homogeneity of the substrate.

2.3. *Substrate Preparation and Biogas Production*

The experiments were carried using an improvise batch laboratory scale bio-digester (10 liters sea way gallon). The digesters was closed with a cork and a puncture hole was bored at the top of the cork. A rubber tube of 10 mm in diameter was inserted through the hole and then sealed with gum which made the digester air tight completely (anaerobic).

The feed stocks used in this experiment are the prepared cow dung and chicken droppings above. From the experiment carried out by [12] shows that at 20% (1:4) and 25% (1:3) concentration of cow dung and chicken dropping respectively gives the highest biogas yield at ambient temperature. Therefore, 1kg (1000 g) of chicken dropping were mixed together with 4 liters (4000 mL) of distilled water to give a ratio of 1:4 and substrate concentration of 20%. Also 1 kg (1000 g) of cow dung with 3 liters (3000 mL) of water which give a ratio of 1:3 and substrate concentration of 25% was also prepared and mixed together. The mixing of both biomass were thoroughly carried out to ensure homogeneity and the substrate were charged into the 10 liters batch type digester. The digester was connected to 16 rim tire tube to collect the biogas generated for about 22 days. The anaerobic digestion take place at ambient temperature of 27 – 32 °C.

2.4. *Test for the Presence of Methane in Biogas Produced*

The presence of methane was tested by lighting flame on a Bunsen burner connected to the biogas reservoir. The gas that came out from the reservoir burns with blue flame which indicate the presence of methane (CH_4).

2.5. *Carbon (IV) Oxide Adsorption*

The glass adsorption column used for this study was fabricated in SEDI, Minna. The adsorption of CO_2 was carried out in the fabricated glass column while activated carbon (AC) and Zeolite CBV

8014 were used as adsorbents. The adsorption column was held firmly tight to a designed stand and biogas delivery pipe was connected from biogas storage (tire tube) to the top of the adsorption column that contains the adsorbent. Then the purified biogas pass through the delivery tube to the blood bag (purified biogas storage) for analysis. In order to optimize the CO_2 capture, adsorbent dosage and contact time were varied from 1g – 4g and 0 – 20 min respectively.

1 g AC of particle size 0.8 – 1 mm was poured into the adsorption column and firmly corked. The apparatus was connected to the tire tube containing the biogas produced, then the biogas flow rate control valve was opened and the gas was released in to the adsorption column at the rate of about 60 mL/min and held at atmospheric pressure and ambient temperature, over the time intervals of 2.5, 5, 7.5, 10, 15 and 20 min. The treated biogas was then collected in the treated biogas collector (blood bag). The weights of the AC samples were measured with an electronic weighing balance at every time intervals to note the difference in weight of the AC samples until a constant weight was reached, which indicates its maximum adsorption capacity. The same procedure was repeated for the other AC samples of 2g, 3g and 4g. After the completion of the AC sample, the same procedure was also repeated using Zeolite CBV 8014 as adsorbent for 1 – 4 g Zeolite sample and time intervals of 2.5, 5, 7.5, 10, 15 and 20 min. Prior to usage, the zeolite was activated with ammonium hydroxide (NH_4OH) in an oven at a temperature of 550 ^oC for 2 hrs. The overall experiment for each adsorbent sample comprise of 16 runs each obtained from a full factorial experimental design. The amount of CO_2 adsorbed from each experimental run were calculated from equation 1 given by [12].

$$\text{Amount of CO2 Adsorbed} = \frac{Q_t(mg) - Q_o(mg)}{Q_o} \quad (1)$$

where Q_o = initial weight of adsorbent before adsorption, Q_t = final weight of adsorbent after adsorption at time, t

2.6. *Adsorption Kinetics*

Adsorption kinetics studies gives the capacity of the adsorbent and the equilibrium relationships between adsorbent and adsorbate described by adsorption isotherms which are usually the ratio between the quantity adsorbed and the remaining in solution at fixed temperature at equilibrium [13]. A sorption isotherm (also adsorption isotherm) describes the equilibrium of the sorption of a material at a surface (more general at a surface boundary) at constant temperature. It represents the amount of material bound at the surface (the sorbate) as a function of the material present in the gas phase and/or in the solution. The adsorption kinetics for AC and zeolite CBV 8014 were further studied using the model discuss below.

2.6.1. Lagergren's pseudo-first-order and pseudo-second order models. Lagergren's pseudo-first-order and pseudo-second order models were used to test the kinetic behavior of the two adsorbent and the conformity of the experimental work and kinetic model were analyzed by the regression coefficient (R^2) values. The linearized form of the Lagergren pseudo-first-order model is generally expressed by equation 2 given by [14].

$$\log(q_e - q_t) = \log q_e - \frac{k_1(t)}{2.0303} \quad (2)$$

where qe and qt = the amount of CO_2 adsorbed in mg/g at equilibrium and at time t, respectively, k1 = the rate constant of pseudo-first-order model (min-1).

According to [14], the pseudo-second-order model and initial rate of adsorption, h (mg/g.min) is given by equation 3.

$$\frac{t}{q_t} = \frac{1}{k_2 q_e^2} + \frac{1(t)}{q_t} = \frac{t}{q_t} = \frac{1}{h} + \frac{1(t)}{q_t} \quad (3)$$

where

$$h = k_2 qe^2 \quad (4)$$

According to [14], applying the pseudo-second-order kinetic model in the adsorption process result in the plot of 1/qt against time (t) which gives a straight line with 1/qe and 1/h as the slope and y-intercept respectively.

2.6.2. Activation energy. According to [14], the Arrhenius equation shown in equation 5 below can be used to estimate the activation energy of the adsorption process.

$$\ln k = -\frac{E_a}{RT} + \ln k_o \quad (5)$$

where k = the pseudo-second-order kinetic model rate constant (g/mg.min), E_a = the activation energy of the adsorption process (J/mol), R = the gas constant (8.314 J/mol.K), T = the adsorption temperature in Kelvin, k_o = the temperature independent factor (g/mg.min).

2.6.3. Intra-particle diffusion model. Ho *et al.* [15] gives the intra-particle diffusion model as expressed by equation 6.

$$q_t = k_{id} t^{1/2} + C_i \quad (6)$$

where qt = the amount of CO2 adsorbed at any time t (mg/g), kid = the intra-particle rate constant (mg/g.min1/2), t = the contact time (min1/2). According to [14], the adsorption process obeys the intra-particle diffusion model if a straight linear plot that passes through the origin is observed.

3. RESULTS AND DISCUSSION

3.1. *Biogas Production*

Table 1 present the composition of the biogas produced from chicken droppings and cow dung substrate (unpurified). The CH_4 present in the raw biogas is observed to be 53.26%, which is in conformity with the compositions of raw biogas reported by [14] for a commercial biogas plant. From Table 1, the values obtained for all component were in conformity with the standard value except for H_2 and O_2 which are higher than the standard value. The slightly higher value of H_2S as observe from Table 1 may partially be attributed to the sulfur in the substrate. In pure biogas, O_2 and H_2S content should be negligible as possible.

Table 1: Compositions of produced biogas (unpurified).

Component	Value (%)	Standard Value
Methane (CH_4)	53.26	45 – 70
Carbon (IV) Oxide (CO_2)	33.44	25 – 55
Carbon monoxide (CO)	1.26	0 – 2
Hydrogen (H_2)	3.53	0 – 2.5
Oxygen (O_2)	8.04	0 – 3
Hydrogen Sulphide (H_2S) (ppm)	4700	10 – 20000

3.2. *Biogas Production*

Raw biogas produced from digestion is roughly 33.44% CO_2 and other gases (CO, O_2, H_2 and H_2S), it is not high quality enough to be used as fuel gas for machinery. The corrosive nature of H2S alone is enough to destroy the internals of the machinery. Also, the need to upgrade raw biogas to a high purity of over 95% methane to ensure environmentally friendly energy source and reduce emission of large GHG generated from the gas necessitate the purification of raw biogas gas. The composition of CH_4 is taken to imply the biogas purity.

It was observed that the biogas purity increases slightly from 53.26% – 55.96 % for an adsorption time of 0 – 20 min using 1g of AC adsorbent while biogas purity increases from 53.26% – 61.15% for an adsorption time of 0 – 20 min using 2g of AC adsorbent. For the same adsorption time, the biogas purity increases from 53.26% – 68.39% using 3g of AC adsorbent, however, highest purity of 84.34% was observed at 20 min adsorption time using 4g of AC adsorbent. This indicate that purity of CH4 gas increases with time and adsorbent dosage with 4g AC adsorbent at 20 min having the highest purity of 84.34% using AC. This implies that more impurity gases (CO_2, CO, O_2 H_2 and H_2S) are remove with more adsorbent dosage due to availability of more adsorption site [16].

It was also observed that at adsorption time of 0 – 20 min, the purity of the biogas produced increases from 53.26% – 56.75% using 1g CBV 8014 zeolite while the biogas purity increases from 53.26% – 61.61 % at 0 – 20 min adsorption time using 2g CBV 8014 zeolite adsorbent. At the same adsorption time, biogas purity increases from 53.26% – 72.42% using 3g of CBV 8014 zeolite adsorbent while biogas purity of 91.01% was observed at adsorption time of 20 min using 4g of CBV 8014 zeolite adsorbent. This indicate that the biogas purity increases with time and increase in CBV 8014 zeolite dosage [16]. For 1 g – 2 g of both AC and CBV 8014 zeolite it was observed that there was no significant different in the biogas purity at adsorption time of 0 – 20 min. This further indicates that there is significant removal of more impurity gases (CO_2, CO, O_2, H_2 and H_2S) from 3 g – 4 g CBV 8014 zeolite dosage compare to highest of 61.61 % and 61.15% CH_4 for 1 g – 2 g CBV 8014 zeolite and AC dosage respectively. Comparing the two adsorbent (AC and zeolite CBV 8014), it was observed that the purification of raw biogas using zeolite CBV 8014 adsorbent gave a better purity compared to AC adsorbent as the adsorption time increases from 0 – 20 min. Thus indicating that the longer the contact time, the higher the removal efficiency of the adsorbents.

3.3. *Statistical Analysis*

A second-order model is useful in approximating a portion of the true response with parabolic curvature. A second-order model includes all the terms in the first-order model, plus all quadratic terms like $\beta_{11}x_1^2$ and all cross product terms like $\beta_{13}x_{1i}x_{3j}$. It is usually expressed as equation 7;

$$\sum_{j=1}^{q}\beta_j x_j + \sum_{j=1}^{q}\beta_{jj}x_j^2 + \sum\sum_{i<j}\beta_{ij}x_i x_j + \varepsilon \qquad (7)$$

Minitab 17 was used to carry out a factorial design analysis for the purity of the upgraded gas. Appropriate second order polynomial equation was used to represent the relationship between the input parameters (Adsorption time and Adsorbent dosage) and the output response (upgraded biogas purity in term of CH_4 composition). Analysis of regression, prediction equation, factorial plots (interactive and main effect) was carried out. The model to represent the relationship between the input parameter and the corresponding response output parameter for both AC and zeolite CBV 8014 was obtained using the experimental data for regression coefficients of a second order polynomial equation. The significance of the regression coefficients was tested using a t-test, VIF and the p-values were used to test the significance of the effect of each variables in the polynomial equation. The result shows that the individual term and the interaction between the terms as well as the quadratic term are very significance ($p<0.05$) [17]. The responses generated were inputted into the Minitab software to generate results presented in Table 2 to Table 3.

Table 2: Model Coefficient for biogas upgrading using AC

Term	Coefficient Estimate	T-Value	P-Value	VIF
Constant	56.15	25.57	0.000	
A	0.521	2.09	0.041	26.04
B	-6.24	-4.05	0.001	35.07
A^2	-0.04111	-4.26	0.000	21.04
B^2	1.648	5.67	0.000	32.25
AB	0.4095	9.37	0.000	8.82

From Table 2, the model equation describing the response from the interaction of time and adsorbent dosage during biogas purification using A.C was generated and the model equation for biogas purity using AC in term of actual factor is given in equation 7.

$$\text{Response} = 56.15 + 0.521A - 6.24B - 0.04111A^2 + 1.648B^2 + 0.4095AB \qquad (8)$$

where A = Adsorption Time (min), B = Adsorbent Dosage (g)

The regression coefficient of the model equation for biogas purity using AC and zeolite CBV 8014 are shown in Table 2 and 3 from the table it can be observed that for biogas purity (CH_4) the R-sq, R-sq(adj) and R-sq(pred) are 97.91%, 97.33% and 94.86% respectively for AC, which implies that about 97.91% of the variable in the response (biogas purity) could be capture and explained by the model equation. It also indicate the high significance of the model. The R-sq(pred) value of 94.86% shows reasonable conformity with the R-sq(adj) value of 97.33% which implies a good agreement between the observed and predicted value [18]. It can be observed from Figure 1 the responses from both experimental and simulated results correlate very well with $R^2 = 0.9791$. Suggesting that the model equation generated can be used to predict purification efficiency of the AC when used to purify biogas.

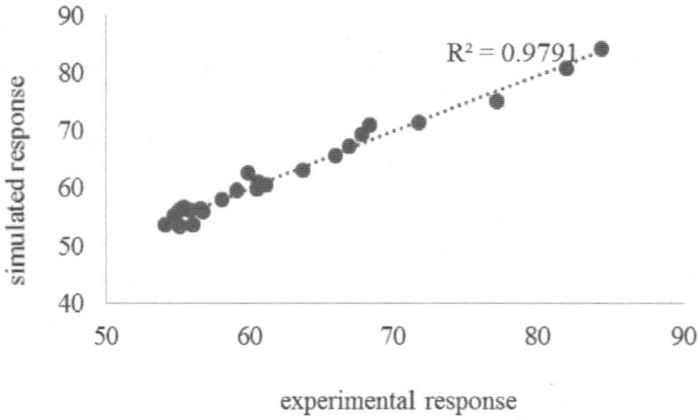

Figure 1: The plot of the experimental responses against simulated responses for activated carbon.

Table 3: Model Coefficient for biogas purity using zeolite CBV 8014

Term	Coefficient Estimate	T-Value	P-Value	VIF
Constant	55.67	15.11	0.000	
A	1.086	2.60	0.018	26.04
B	-8.62	-3.34	0.004	35.07
A^2	-0.0697	-4.31	0.000	21.04
B^2	2.428	4.98	0.000	32.25
AB	0.4781	6.52	0.000	8.82

The model equation for upgraded biogas purity using zeolite CBV 8014in term of actual factor is given in equation 7.

$$\text{Response (Purity)} = 55.67 + 1.086A - 8.62B - 0.0697A^2 + 2.428B^2 + 0.4781AB \quad (9)$$

where A = Adsorption Time (min), B = Adsorbent Dosage (g)

For biogas purity using zeolite CBV 8014, the R-sq, R-sq(adj) and R-sq(pred) are 96.67%, 95.74% and 91.86% respectively. This implies that about 95.74% of the variable in the response (biogas purity) could be capture and explained by the model and also indicate the high significance of the model. The R-sq(pred) value of 91.86% shows reasonable conformity with the R-sq(adj) value of 95.74%. This implies a good agreement between the observed and predicted value [18]. However, the R-sq, R-sq(adj) and R-sq(pred) for AC are higher than those of zeolite CBV 8014 which implies that the obtained experimental data for AC are highly correlated than those of zeolite CBV 8014. Figure 2 show that model equation generated for the purification of biogas using zeolite CBV 8014 can be used to predict purity of the biogas as the prediction gave a correlation coefficient (R^2) = 0.9667.

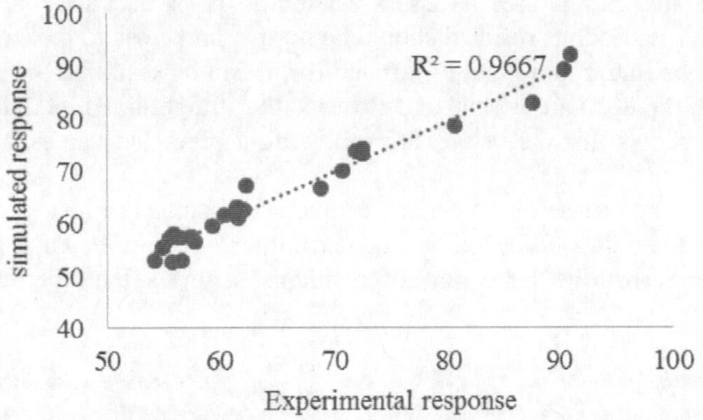

Figure 2: The plot of the experimental responses against simulated responses of Zeolite.

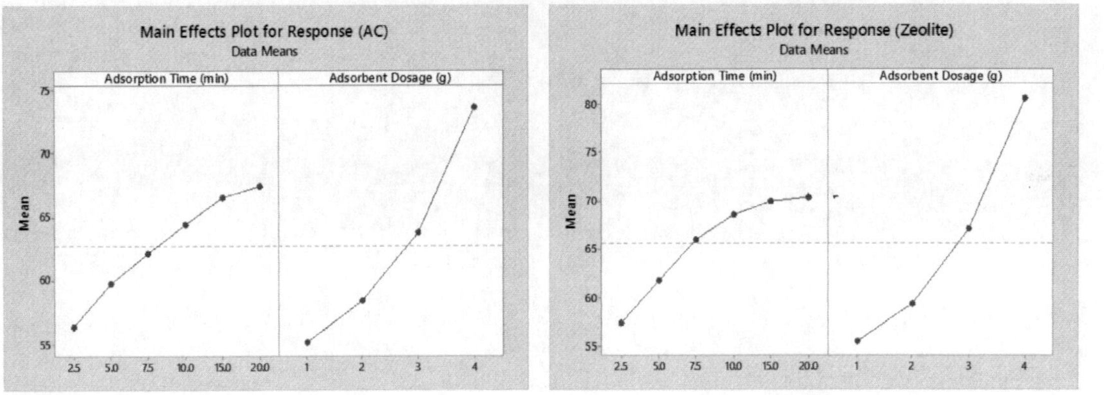

(a) (b)

Figure 3: Main effect plot for biogas purity using (a) AC and (b) zeolite CBV 8014.

Figure 3 (a) and (b) are the main effect plot for the biogas purity. The main effect plot is used to examine the difference between level mean for one or more factors. It is believed that there is a main effect when different levels of a factor affect the response in different manner (Montgomery, 2004). From Figure 3 (a) and (b) it can be observed that the mean responses increases slightly from 15 – 20 min for both AC and zeolite CBV 8014, which implies that further increase in time will not have significant on the mean response which is in conformity with the findings of [10] who shows that impurity adsorption flattens as time increase beyond 20 min which could be due to the adsorbent reaching its maximum adsorption capacity. For the adsorbent dosage, it has the highest effect on purity. The mean response increases almost linearly with increase in adsorbent dosage from 2 – 4 g which could be attributed to availability of more adsorption site [10].

From Table 2 and 3, it can be seen that the linear, quadratic and interaction term of the model equation for biogas purity using AC and zeolite CBV 8014 have significant effect ($p<0.05$). Also the VIF values is an indication of the correlation of the predictors. The VIFs are not close to 1, but higher which indicates that the predictors are correlated. The VIFs are high because of the interaction terms which are correlated with main effect terms because they include the main effects terms. This also indicates that the biogas purity model developed in equation 8 and 9 well represent the experimental data.

Residual plot was also use to analyze the observed and predicted responses. Residual plot present the differences between the observed and predicted or fitted values. From Figure 3 and 4, it can be seen that the point in the residual plot are randomly dispersed around the horizontal axis, this implies that a linear regression is appropriate for the data which indicate a good fit for linear model [12].

The normal probability plot is used to assess whether or not a data set is approximately normally distributed. If a model is good, its residual should be normal and close to the straight line plot with no large deviation from the line. From Figure 4 (a) and (b), it can be seen that the residual of the plot for both AC and zeolite CBV 8014 are normal and closer to the straight line plot, indicating a good model. Figure 4 (a) and (b) shows that the values are following a predicted values which indicate that the model are correct.

Histogram gives a rough sense of the density of numerical data. The histogram of residuals is used to determine whether the data are skewed or whether outliers exist in the data. From Figure 4 (a) and (b), it can be observe that the histogram of residuals suggests that the residuals are normally distributed.

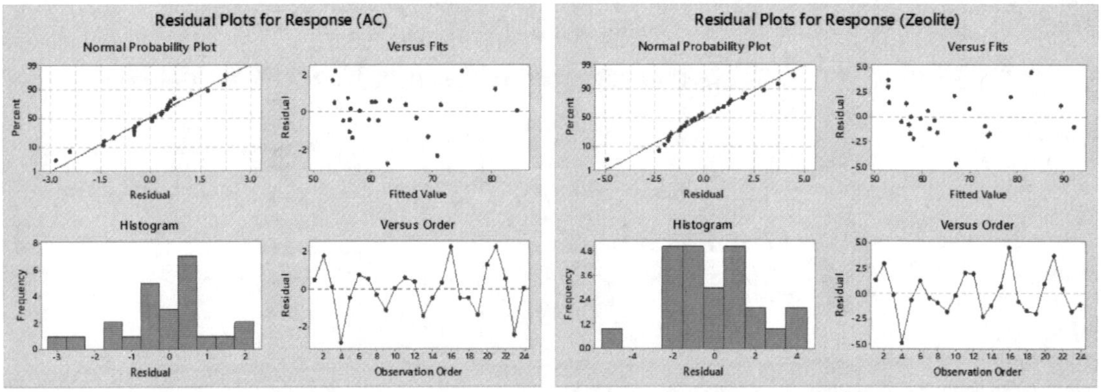

(a) (b)

Figure 4: Four-in-One Residual plot using (a) AC Adsorbent and (b) zeolite CBV 8014 Adsorbent

Residuals versus fits plot are used to verify the assumption that the residuals have a constant variance. Residual versus fits plot is a scatter plot with residual on the y-axis and fit on the x-axis. The data for the biogas purity (Figure 4 a and b) shows that the errors are randomly scattered about zero which shows that the residual are fitted and unbiased. The findings of this studies are in agreement with the null hypothesis that the residuals have a constant variance.

3.4. *Physiochemical Properties of Adsorbent Sample*

Microporous materials have become important in the separation or capture of gases such as CO_2, CO, O_2, N_2, H_2, H_2S etc. The surface area and pore size of any adsorbent is a fundamental property to effectiveness of adsorbent materials [16]. Activated carbon and zeolite over the years have become an important adsorbent materials that have been used for several purposes. The adsorbent used for the CO_2 and other gas adsorption are zeolite CBV 8014 and KOH treated activated carbon (AC). It was found that AC prepared by KOH activation is highly micro porous when compared to that produced from $ZnCl_2$ or H_3PO_4 activation or other form of activation [19]. Besides, KOH also enhances the specific surface area and the formation of –OH functional groups on the carbon surface [20]. The –OH functional group makes the produced AC to be more susceptible to chemical reaction with KOH, which consequently leads to the formation of a higher degree of mesopores and micropores. Table 4 present the surface area and pore volume of the AC and CBV 8014 zeolite used for the kinetic studies of the gas adsorption. From Table 4, it was noted that the KOH-activated carbons used possess microporosity with a surface area of 784 m^2/g which implies high adsorption capacity. The AC used has low bulk density of 0.42 g/cm^3 as observed from Table 4. However, it has been reported that low density AC enhances adsorption in gas and liquid phase systems [21].

Table 4: Physiochemical properties of adsorbents

Parameter	AC Sample	Zeolite CBV 8014
Appearance	Black	White
pH	8.4	-
Density, g/cm^3	0.42	0.64
Surface area, $m^2/g/$	784	700
Pore volume, g/cm^3 / Pore diameter (Å)	0.93 (6.72 Å)	7.0

The zeolite CBV 8014 use in the study was activated using NH_4OH at a temperature of 550 °C in an oven for 2 hr. The surface structure of zeolites are usually heterogeneous which means that it consists of a finite number of different kinds of adsorption sites and the heterogeneity of the surface of zeolites led to its consideration for use in CO_2 capture and adsorption of other gases [22]. The properties of the zeolite CBV 8014 used are presented in Table 4. From Table 4, the pore diameter of the zeolite CBV 8014 is 7.10 Å, this implies that the pore diameter of the zeolite CBV 8014 is sufficient for the CO_2 (3.614 Å) to enter into the zeolites CBV 8014 channels. From the properties of the AC and zeolite CBV 8014 presented in Table 4, it can be observed that the CO_2 adsorption capacity of AC is lower than that of zeolite CBV 8014. This is due to the fact that zeolite CBV 8014 has a highly developed porous structure with good adsorption properties and high affinity for CO_2 adsorption than AC. [23]. From literature (Table 2.2) the pore sizes of the various gases are available which are all smaller than that of AC (6.72 Å) and CBV 8014 (7.0 Å) adsorbent which is an indication that the adsorbent have sufficient pore size to adsorbed the impurity gases (CO_2, CO, O_2, H_2 and H_2S) present in the biogas with 2.89 Å – 3.623 Å (Table 2.2).

3.5. *Carbon Dioxide Adsorption Capacity*

Figure 5 and 6 illustrates the adsorption profiles at different adsorbent dosage (1 g – 4 g) for both the AC and CBV 8014 zeolite respectively. From Figure 5 and 6, it can be observed that as time increases from 0 – 20 min, the adsorption of CO_2 and other gases increases for both adsorbent up to a time of 10 min. A sharp and linear increase was noticed at 0 – 10 min for all adsorbent dosage which then tends to flattens as the time increases from 10 – 20 min. This is in conformity with the report of [10] that CO_2 adsorption increases sharp at 5 – 22 min using AC.

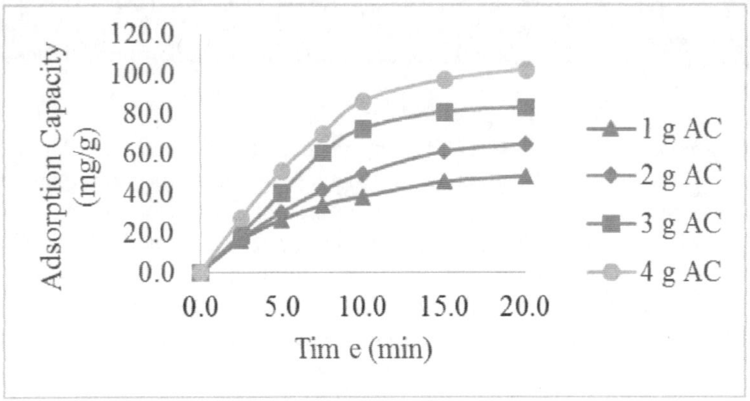

Figure 5: CO_2 adsorption capacity vs time at various AC dosage

The rate of CO_2 adsorption gradually decreased from 10 – 20 min, this might be due to the adsorbents approaching its saturation or maximum adsorption capacity. However, the rapid adsorption rate observed between 0 – 10 min while slow adsorption rate was observed from 10 – 20 min which could be attributed to the high sorbent surface area (Table 4) which is accountable for interaction with the CO_2 and other trace molecules. Over time, the adsorption rate is observed to be on the decreased due to reduction of active sites that may slow down the adsorption process from taking place. In addition, [10] suggest that the rapid adsorption at the beginning of the process is due to external surface of the adsorbent, and is followed by slower internal diffusion process.

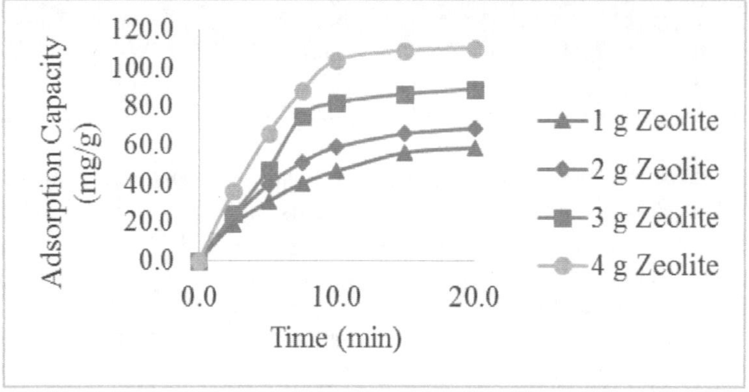

Figure 6: CO_2 adsorption capacity vs time at various zeolite CBV 8014 dosage.

From Figure 5 and 6, it can also be observed that adsorption of CO_2 was in the order zeolite CBV 8014 > AC, which implies that at all dosage, CBV 8014 zeolite has the best adsorption capacity than the AC investigated. This high adsorption capacity observed from CBV 8014 zeolite could be attributed to its high surface area due to the facts that the pore diameter of the CBV 8014 zeolite is 7.0 Å compared to 2.89 Å – 3.623 Å (Table 2.2) for the impurity gases (CO_2, CO, O_2 H_2 and H_2S) present in the biogas, which makes it sufficient for the CO_2 to enter into the zeolites and AC channels. Therefore, from Figure 5 and 6, it can be observed that the rate of adsorption in CBV 8014 zeolite are higher than in AC.

3.6. *Adsorption Kinetic Study*

Adsorption kinetic analysis establishes an adsorption uptake rate and examine the residence time for the adsorption process to be completed. It is also a baseline on determining the adsorption dynamics as well as its performance in real application. Lagergren pseudo first-order model and pseudo-second-order model was examine in this study. The conformity of the experimental work and kinetic model were analyzed by the R2 values. Furthermore, Arrhenius equation was use to estimate the activation energy of the adsorption process and also to determine if the adsorption process obeys the intra-particle diffusion model.

3.6.1. Lagergren pseudo first-order model. Lagergren pseudo-first-order model was used to test the adsorption kinetics of the system using Equation 2 to evaluate the adsorption dynamics. A straight line plot of log (qe-qt) versus time (t) at various adsorbent height of 3.35 cm – 13 cm (1g – 4 g) for both AC and zeolite CBV 8014 was made, and the parameters were tabulated in Table 5. It was observed that the pseudo-first-order kinetic model for both AC and zeolite fit well with the experimental data, with R^2 value found to be within the range of 0.913 to 0.9935 for AC and 0.868 – 0.9704 for zeolite. This implies that AC well obeyed the pseudo-first-order model when compared to zeolite.

Table 5: Lagergren's pseudo-first-order model kinetic parameters

Adsorbent	Parameter	1 g	2 g	3 g	4 g
AC	K1 (min^{-1})	0.030	0.049	0.085	0.189
	R^2	0.913	0.962	0.936	0.994
Zeolite	K1	0.036	0.047	0.082	0.218
	R^2	0.927	0.899	0.868	0.970

The first order adsorption rate constants K1 for both adsorbent are presented in Table 5. From Table 5, the can be observed that the first order adsorption rate constants K for zeolite CBV 8014 are higher than those of the AC. However, first order adsorption rate constants K for both AC and zeolite compared favorably with K values of 0.097, 0.1252 and 0.1140 reported for CO_2 adsorption using based modified activated carbon, acid modified activated carbon and unmodified activated carbon respectively by [10].

3.6.2. Lagergren's pseudo-second-order model. Lagergren pseudo-second-order model was also used to study the adsorption kinetics of the system using Equation 3 to further evaluate the adsorption dynamics. A straight line plot of 1/qt against time (t) was made at various adsorbent height of 3.35 cm – 13 cm (1g – 4 g) for both of the AC and zeolite CBV 8014, and the kinetic parameters obtained were tabulated in Table 6. It was observed that the pseudo-second-order kinetic model for AC at 1 g (3.25 cm bed height), 2 g (6.5 cm bed height) and 4 g (13 cm bed height) only fit well with the experimental data, with R^2 value found to be within the range of 0.8668 to 0.9982. Likewise for zeolite, it was observed that the pseudo-second-order kinetic model at 1g, 2 g, 3 g and 4g adsorbent dosage fit well with the experimental data, with R^2 value of 0.9017 – 9951.

Table 6: Lagergren's pseudo-second-order model kinetic parameters

Adsorbent	Parameter	1 g	2 g	3 g	4 g
AC	K2 (min^{-1})	0.002	0.0007	0.0005	0.0006
	R^2	0.998	0.9827	0.8668	0.9621
Zeolite	K2	0.002	0.0015	0.0009	0.0010
	R^2	0.995	0.9882	0.9017	0.9621

The second order adsorption rate constants K2 for both adsorbent are presented in Table 6. It can be observed from Table 6, that the second order adsorption rate constants K for zeolite CBV 8014 are higher than those of the AC. However, second order adsorption rate constants K for both AC and zeolite compared favorably with K values of 0.0105, 0.0052 and 0.0042 for an adsorption time of 0 – 25 min reported for CO_2 adsorption using based modified activated carbon, acid modified activated carbon and unmodified activated carbon respectively by [10]. The study revealed that the pseudo second order kinetics model for AC and zeolite well fitted the CO_2 adsorption kinetics data than pseudo first order kinetics model. This is because pseudo-first-order model R^2 value for both AC and zeolite CBV 8014 is within the range of 0.95 – 0.998 while pseudo-second-order model R^2 value for both AC and zeolite is 0.8668 – 0.9982.

3.6.3. Activation energy. In adsorption process, the magnitude of the activation energy (Ea) is very significant, as it provides information on the adsorption mechanism. The value of activation energy for physical adsorption is usually in the range of 5 – 40 kJ/mol. However, chemisorption process involves formation of strong bonding between the adsorbate (CO_2) and adsorbent with activation energy in the range of 40 – 800 kJ/mol [24]. Arrhenius equation that is shown in equation 5 is applied to determine the activation energy of the adsorption process. Table 7 shows the activation energy (Ea) for AC and zeolite CBV 8014 adsorbent used in the study for CO_2 adsorption.

Table 7: Activation energy for CO_2 adsorption in AC and zeolite

Adsorbent	Parameter	1 g	2 g	3 g	4 g
AC	Ea (kJ/mol)	14.37	17.12	19.23	17.84
Zeolite	Ea kJ/mol)	15.71	16.18	18.30	17.14

From Table 7, it can be observed that the activation energy value for the study fall within the range of 5 – 40 kJ/mol for physical adsorption [24]. The lower activation energy found in this work implies that the CO_2 adsorption onto both AC and zeolite CBV 8014 is a diffusion-controlled process (phisiosorption) rather than a chemically-controlled process (chemisorption). The results obtained in this study were slightly higher than 5.77, 13.02 and 13.55 kJ/mol reported for based modified activated carbon, acid modified activated carbon and unmodified activated carbon respectively [10] but compare favorably with 17.51 and 14.27 kJ/mol reported for optimized and commercially available activated carbons respectively [13].

3.6.4. Intra-particle diffusion model. The intra-particle diffusion model rate processes are generally presented in respect to square root of time. In contrast to Lagergren pseudo-first-order and pseudo-second-order model, intra-particle diffusion model provides an insight of the mechanism in adsorption process. Adsorption consists of few steps involving the transport of adsorbate from the bulk phase to the surface of solid sorbent, and is followed by the molecules diffusion into the interior of the pores. Intra-particle diffusion, normally is a rate-limiting step in many of adsorption processes described as a slow process. A plot of qt against the square root of time gives the intra-particle rate constant.

Table 8: Intra-particle diffusion rate constant for CO_2 adsorption in AC and zeolite.

Adsorbent	Parameter	1 g	2 g	3 g	4 g
AC	k_{id} (mg/g.min$^{1/2}$)	11.203	17.048	22.68	26.35
Zeolite	k_{id} (mg/g.min$^{1/2}$)	14.072	18.892	26.914	25.677

From Table 8, it can be observed that the intra-particle diffusion rate constant for the adsorption of CO_2 in zeolite are higher at all dosage than those of AC. This implies that the rate of intra-particle diffusion in zeolite is higher than that of AC.

4. CONCLUSION

The biogas produced contains high CH_4 gas (53.26%), CO_2 (33.44%) and other gases in small quantity. The study shows the practicality of both AC and zeolite CBV 8014 to be used as adsorbent for CO_2 capture in industrial applications. The pore diameter of the CBV 8014 zeolite (10 Å) enable it to be most effective in CO_2 adsorption compared to AC.

Comparability in the CO_2 adsorptions capacity indicates the practicality of the AC and zeolite CBV 8014 to be used in industrial applications. The adsorption capacity of the adsorbents was found to be in the order zeolite CBV 8014 > AC at 0 – 20 min and 1-4g adsorbent. The kinetic study shows that the CO_2 adsorption into AC and zeolite CBV 8014 obeys the pseudo-second-order model, due to its R2 value closer to unity. The kinetic study of CO_2 adsorption onto AC and zeolite CBV 8014 shows that the magnitude of activation energy for both adsorbent are less than 40 kJ/mol, indicating the occurrence of physiosorption mechanism. The intra-particle diffusion model study indicates that intra-particle diffusion is the sole rate-limiting step in the adsorption process for both adsorbent (AC and zeolite CBV 8014). In addition, lower activation energy found in this work confirm that the CO_2 adsorption onto AC and zeolite CBV 8014 is a diffusion-controlled process, rather than the chemically-controlled.

REFERENCES

[1] A. O. Aderemi, M. O. Ilori, H. O. Aderemi, and J. F. K. Akinbami, "Assessment off Electrical Energy use Efficiency in Nigeria food Industry," *African Journal of food Science*. Vol. 3, No. 8, pp. 206-216, 2009.

[2] J. I. Eze and Agbo, K. E., "Maximizing the potentials of biogas through upgrading," *American Journal of Scientific and Industrial Research*. 1(3), pp. 604-609, 2010.

[3] A. M. Mshandete and W. Parawira, "Biogas Technology Research in Selected SubSaharan Africa," *African Journal of Biotechnology*. Vol. 8. No. 2, pp. 116-125, 2009. Available online at http://www.academicjournals.org/AJB, Accessed: 02/08/2017.

[4] O. D. Onafeso, "The Need for Government Ownership of Solar Energy Production in the Achievement of the Millennium Development Goals in Nigeria," *Nigeria Journal of Solar Energy*, Vol. 16, pp. 143-154, 2006.

[5] B. K. Highina, I. M. Bugaje and B. Umar, "Biodiesel Production from Jatropha Caucus Oil in a Batch Reactor Using Zinc Oxide as Catalyst", *Journal of Petroleum Technology and Alternative Fuels*. Vol. 2, No. 9, pp. 146-149, 2011. Available at http://www.academicjournals.org/JPTAF. Accessed: 02/10/2017.

[6] D. Keeney and M. Muller, "Ethanol Production: Environmental Effects", Institute for Agriculture and Trade Policy, Minneapolis, 2007. Available at http://www.iatp.org./iatp.publications, Accessed: 15/10/2017.

[7] Q. Zhao, E. Leonhardt, C. MacConnell, C. Frear and S. Chen, *Purification Technologies for Biogas Generated by Anaerobic Digestion*, Climate Friendly Farming, CSANR Research Report, Vol. 001, pp. 1 – 20, 2010.

[8] G. Murshid, A. M. Shariff and L. K. Keong, "Thermo physical analysis of 2-amino-2-methyl-1-propanol solvent for carbon dioxide removal," *Chem Eng Trans*., Vol. 25, pp. 45–50, 2011. DOI:10.3303/CET1125008.

[9] M. Alhassan, I. Andrew, M. Auta, M. Umaru, U. G. Garba, A. G. Isah and B. Alhassan, "Comparative studies of CO2 capture using acid and base modified activated carbon from sugarcane bagasse," *Biofuels*, 2017. DOI:10.1080/17597269.2017.1306680.

[10] E. Dlugokencky, "Annual Mean Carbon Dioxide Data," Earth System Research Laboratory. *National Oceanic and Atmospheric Administration*. Retrieved 12 February 2016.

[11] A. Mohammed, G. I. Abubakar, M. J. Jibril and U. G. Mohammed, "Modelling and Optimization of Biogas Production from Anaerobic Digestion of Agricultural Waste: A Factorial Design Analysis," *Nigeria Journal of Engineering and Applied Science (NJEAS)*, Vol. 3, No. 1, pp. 24 – 25, 2016.

[12] A. Dabrowski, "Adsorption: from theory to practice," *Adv. Colloid Interface Sci*, Vol. 93, No. 3, pp. 362-371, 2001.

[13] N. A. Rashidi, S. Yusup and H. L. Lam, Kinetic studies on carbon dioxide capture using activated carbon. Chem Eng Trans. Vol. 35, pp. 361–366, 2013.

[14] Y. S. Ho, J. C. Ng and G. McKay, "Kinetics of pollutant sorption by sorbents: Review," *Separ Purif Rev*., Vol. 29, pp. 180–232, 2000.

[15] Y. G. Gor, M. Thommes, K. A. Cychosz and A. V. Neimark, "Quenched solid density functional theory method for characterization of mesopore carbons by nitrogen adsorption," *Carbon*, Vol. 50, pp. 1583, 2012.

[16] P. R. Krishna and S. N. Srivastava, "Sorption of Distillery Spent Wash onto Fly Ash: Kinetics and Mass Transfer Studies". *Chem. Eng. J.*, Vol. 146, No. 1, pp. 90-97, 2009. https://doi.org/10.1016/j.cej.2008.05.021.

[17] B. Sajeena, P. P. Jose and G. Madhu, Optimization of Process Parameter Affecting Biogas Production from Organic Fraction of Municipal Solid Waste via Anaerobic Digestion. *International Journal of Environmental, Ecologocal Geological and Geographysical Engineering*, Vol. 8, No. 1, pp. 43 – 48, 2014.

[18] S. E. Abechi, C. E. Gimba, A. Uzairu. and Y. A. Dallatu, "Preparation and characterization of

activated carbon from palm kernel shell by chemical activation," *Research Journal of Chemical Science*, Vol. 3, pp. 54, 2013.

[19] B. Viswanathan, B. Indra, P. Neel and T. K. Varadarajan, "Methods of Activation and Specific Applications of Carbon Materials, *National Centre for Catalysis Research*, Chennai, Vol. 1, 2009.

[20] F. T. Ademiluyi and E. O. David-West, "Effect of Chemical Activation on the Adsorption of Heavymetals Using Activated Carbons Fromwastematerials", *International Scholarly Research Network*, pp. 1-5, 2012.

[21] S. Pakseresth, M. Kazemeini and M. M. Akbarnejad, "Equilibrium isotherms for CO, CO_2, CH_4 and C_2H_4 on the 5A molecular sieve by a simple volumetric apparatus," *Separation and Purification Technology*, Vol. 28, pp. 53-60, 2002.

[22] L. Qiuli and A. S. George, The role of adsorbent pore size distribution in multicomponent adsorption on activated carbon, *Carbon*, Vol. 42, Issue 15, pp. 3133-3142, 2004.

[23] A. Özcan, E. M. Öncü and A. S. Özcan, Kinetics, isotherm and thermodynamic studies of adsorption of acid blue 193 from aqueous solutions onto natural sepiolite, Colloids and Surfaces A: *Physicochemical and Engineering Aspects*, Vol. 277, pp. 90-97, 2006.

Spectral responses of *B.vulgaris* dye-sensitized solar cells to change in electrolyte

T.J. Abodunrin[1,*], A.O. Boyo[2], M.R. Usikalu[1], O. Kesinro[1]

[1]Department of Physics, Covenant University, Ota, Ogun State, Nigeria
[2]Department of Physics, Lagos State University, Ojo, Lagos State, Nigeria

temitope.abodunrin@covenantuniversity.edu.ng

Abstract. Dye sensitized solar cells still offer an encouraging option in the photovoltaic family. It is endowed with several attributes such as ability to generating green energy at low cost, ease of fabrication and yearlong availability of raw materials. Betalain an important constituent of *B.vulgaris* shows diverse changes to temperature and solvents as they alter the fluidity of its membrane. This experiment seeks to explore the latter with a keen interest on the IPCE and efficiency of beetroot dye-sensitized solar cells (DSCs). The Doctor blade method of application of TiO_2 photoanode using ethanol as precursor and high temperature sintering reveals diverse spectral responses in the efficiency of the DSCs with the most efficient *B.vulgaris* dye recording 0.27% and the highest IPCE of the dye revealed as 28.34%.

1. Introduction

Till date, efficiency of solar cells is an utmost limiting factor in dye-sensitized solar cells. In order to be effective, the monolayer of organic dye must be able to harvest a wide crop of insolation. Diverse photo sensitizers differ in their characteristic absorption of available light, output efficiency and cost. Our choice of *B.vulgaris* is based on its efficiency and low cost from previous research work [1].

A lot of organic, inorganic and hybrid dyes have been used as photo sensitizers [2-3]. This is as a result of the potential high cost and scarcity of ruthenium, N719 and N3 dyes, which are also not ecologically friendly. Thus, several metal-free organic dyes have been applied in DSCs [4]. In recent studies, natural organic dye extracts examples include anthocyanin, chlorophyll, tannin, and carotene from various parts of plants such as seeds, fruits, flowers and leaves have been productively used as sensitizers in DSCs [5]. Hao and some researchers successfully extracted dye sensitizer from the leaves of capsicum, black rice, Rosa xanthina, kelp and Erythrina variegata flower for application in DSCs. The result was an IPCE of 0.327 % from Black rice [6]. Another researcher Wongcha-ree and some others used a dye cocktail from natural extracts of blue pea and rosella in ratio 1:1 as sensitizer in DSCs, the result was a slight improvement over the last, the IPCE recorded was 0.37% [7].

A crucial factor that affects the performance of DSCs is the choice of photoanode material. TiO_2 is a preferred photoanode because of its wide band gap of 3.2 eV and biocompatibility with other organic products [8]. We report here the results of a series of experiments carried out on raw extracts of the *B.vulgaris* specie. Beta vulgaris belongs to phyla Chenopodiaceae and comes in an assortment of bulb with colours alternating from yellow to red. The most popular variety is deep red *B.vulgaris* which is usually either cooked or served raw in salad or juice. There is more awareness in the choice of natural

IOP Conf. Series: Earth and Environmental Science **173** (2018) 012047 doi:10.1088/1755-1315/173/1/012047

food colours due to the prevalent danger of ingesting indigestible synthetic dye compounds. This research seeks to explore *B.vulgaris* because it is less used in food processing relative to anthocyanins and carotenoids even though, it water-soluble and is stable between pH 3 and 7 [9].

2. Materials and Method

63.7 g of *B.vulgaris* was blended without the peel. It was mixed with 1000 ml of water to extract the dye. The mixture was fed into a Stuart RE 300 B series rotary evaporator to extract the dye. The dye extract was dissolved in 1:10 distilled water (DW) to obtain the phytochemicals. The same ratio was used to obtain the UV/VIS spectrograph with Thermoscientific Evolution 60S spectrophotometer. The *B.vulgaris* DSCs were fabricated using an active area of 3.14 cm^2. The TiO_2 paste was uniformly blended and applied via doctor blade method of standard laboratory procedure and high temperature sintering at 450 ^{O}C [10]. The counter electrode was evenly coated with soot over a naked Bunsen flame and allowed to cool. The two electrodes were coupled together with binder clips. Four different electrolytes dissolved in ratio 1: 100 DW was injected in-between the conducting slides. A Voltcraft M-3850 digital multimeter was used to obtain the value of potential differences as the load was varied.

3. Results and Discussion

3.1 *Phytochemical Analysis*

The phytochemical screening result revealed the presence of a compound rich in chromophores as shown in Table 1.

Table 1: Phytochemical Screening Result of *B.vulgaris*

Flavonoid	Saponin	Steroid	Phenol	Carotenoid	Tannin	Alkaloid	Triterpenoid	Cardiac Glycoside	Quinones
+	+	+	+	+	+	+	+	+	-

Key: + means presence while – means absent

3.2 *UV/VIS Spectroscopy*

The UV/VIS spectrograph revealed that *B.vulgaris* records maximum absorbance at 316 nm wavelength. This is a near ruthenium absorption characteristic behaviour [11]. According to Beer Lambert's law,

$$A_\lambda = -\log T \quad (1)$$

$$A = \varepsilon c l \quad (2)$$

where A is absorbance in a.u, T is the transmittance in %, ε is the molar absorption coefficient in mol/cm^3, c is the concentration of the solution in mol/cm^3 and l is the path length in cm. A *(B.vulgaris)* from Figure 1=3.252 a.u.

IOP Conf. Series: Earth and Environmental Science **173** (2018) 012047 doi:10.1088/1755-1315/173/1/012047

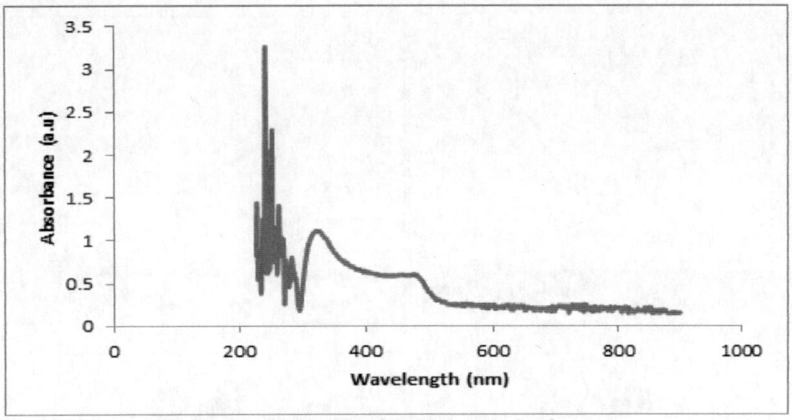

Figure 1: UV/VIS of *B.Vulgaris* leaf extract

3.3 Photovoltaic Result

Photovoltaic Result for different electrolyte sensitized DSCs are illustrated in Figure 2 (a) to (d). The interboundary kinematics and redox reactions produce the differences in the I-V curve which provides the parameters for determining efficiency and IPCE from Equations (3) and (4).

$$\eta = \frac{I_{SC} V_{OC} ff}{A \times P_{in}} \tag{3}$$

$$IPCE = \frac{1240 \times I_{SC}}{P \times \lambda} \tag{4}$$

where η is efficiency in %, I_{sc} is the short circuit current in mA, V_{oc} is the open circuit voltage in volts, *ff* is the fill factor, A is the area of exposure in cm^2, P_{in} is the power input in Watts and λ is the wavelength in nm. The results are shown in Table 2. The best *ff*, I_{sc}, V_{oc}, P_{max} and η was produced by *B.vulgaris* with KI although it is a non-uniform curve attributed to redox reactions occurring spontaneously all through charge transport. The best incident photon to conversion efficiency was recorded in KBr although the efficiency was poor. This is probably due less favourable redox reactions and interboundary kinematics such as trap sites and recombination losses.

Table 2: Photovoltaic Spectra of *B.vulgaris* with Four Electrolytes

Parameters	**I_{sc} (mA)**	**V_{oc} (V)**	**P_{max} (W)**	***ff***	**η (%)**	**IPCE (%)**
KI	0.090	320	8.5	2.95	0.27	4.15
KCl	0.047	165	0.99	0.128	0.003	18.62
KBr	0.052	190	0.72	0.073	0.002	28.34
HgCl2	0.018	120	1.97	0.91	0.06	3.59

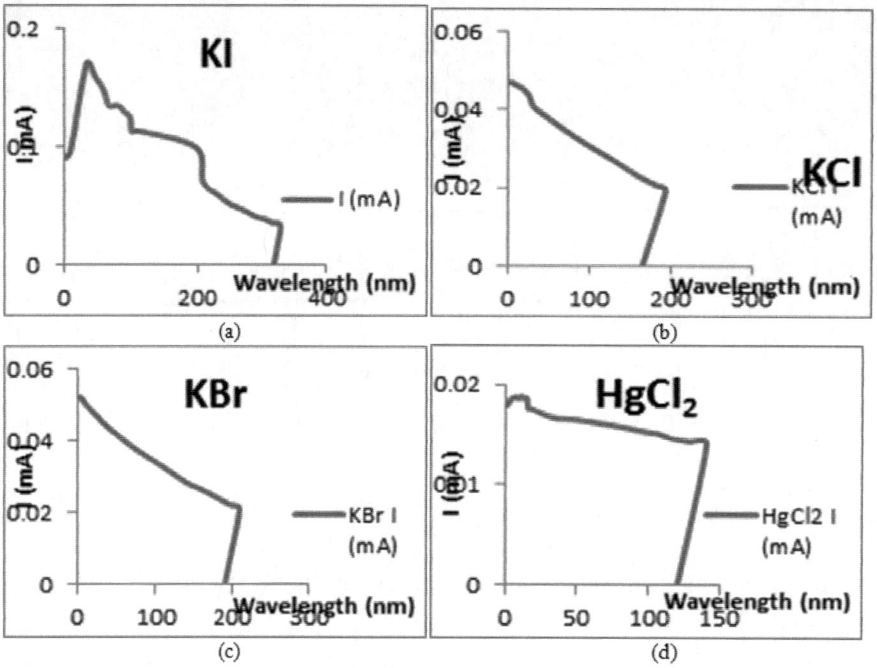

Figure 2: Illustrates *B.vulgaris* I-V curve with (a) KI electrolyte, (b) KCl electrolyte, (c) KBr electrolyte and (d) HgCl2 electrolyte respectively.

4. Conclusion and Recommendations

B.vulgaris leaf extract records a relatively high V_{oc} with the different electrolyte used as sensitizer. The best electrolyte recommended for future use either in dye cocktails or for a different photoanode is KI. It produces the fastest charge transport which promotes better interfacial boundary relationship with the molecules of the dye, TiO_2 and electrolyte.

Acknowledgment

The authors are grateful to the Center for Research, Innovation and Discovery (CUCRID), Covenant University Ota, Nigeria for sponsoring this research.

References

[1] Grätzel, M., 2009. Recent advances in sensitizedmesoscopic solar cells. Acc. Chem. Res., 42: 1788- 1798. DOI: 10.1021/ar900141y PMID: 19715294.
[2] Wang, H., P.G. Nicholson, L. Peter, S.M. Zafeeruddin and M. Grätzel, 2010. Transport and interfacial transfer of electrons in dye-sensitized solar cells utilizing a co(dbbip)2 redox shuttle. J. Phys. Chem.C., 114: 14300-14306. DOI: 10.1021/jp105753k
[3] MichaelGrätzel, Journal of photochemistry and photobiology, 2003
[4] Ciprian Iliescu, Bangtao Chen, Jianmin Miao, “On the wet etching of Pyrex glass”, Sensors and Actuators A: Physical, 143 (2008) 154–161
[5] Siva R, Status of natural dyes and dye-yielding plants in India, *CurrSci*, 2007, **92**(7), 916-925.
[6] Pandey A.K., Tyagi V.V., Jeyraj A., Selvaraj L., Rahim N.A. and Tyagi S.K. (2016). Recent advances in solar photovoltaic systems for emerging trends and advanced applications. *Renewable Sustainable Energy Rev.*, 53, 859-884.
[7] Federico B. and Claudio G. (2016). Natural Polymers for Dye-Sensitised Solar Cells: Electrolytes and Electrodes Redox electrolyte Cathode. *Encyclopaedia of Polymer science and*

IOP Conf. Series: Earth and Environmental Science **173** (2018) 012047 doi:10.1088/1755-1315/173/1/012047

Technology.
[8] Brian O'regan and Grfitzeli M. (1991). A Low-Cost, High-Efficiency Solar Cell Based on Dyesensitized Colloidal TiO2 Films. *Nature*, 353(6346), 737-740.
[9] Senthil T.S., Muthukumarasamy N., Velauthapillai D.,Agilan S., Thambidurai M. and Balasundaraprabhu R.(2011). Natural dye (cyanidin 3-O-glucoside) sensitized nanocrystalline TiO2 solar cell fabricated using liquid electrolyte/quasi-solid-state polymer electrolyte. *Renew.Energy*, 36(9), 2484-2488.
[10] Mohammed A.A., Ahmad A.S.S. and Azeez W.A. (2015).Fabrication of Dye Sensitised Solar Cell Based on Titanium Dioxide (TiO2). *Advances in Materials Physics and Chemistry*, 5(9), 361-367.
[11] Hemmatzadeh R. and Jamali A. (2015). Enhancing the optical absorption of anthocyanins for dye sensitised solar cells. *Journal of Renewable and Sustainable Energy,* 7(1),013120.

Evaluation of biogas production and pressure from composite of poultry droppings and lemon grass using strain gage rosette

Adekoya Olaoluwa[1], Akan Williams[1], Ndubuisi Amanda[2], Adebayo Timileyin[3], Henry Boyo[2]

[1]Department of Chemistry, Covenant University, Ota, Ogun State, Nigeria
[2]Department of Physics, Covenant University, Ota, Ogun State, Nigeria
[3]Department of Physics, University of Lagos, Lagos State, Nigeria

olaoluwaadekoya2@gmail.com

Abstract. The study explored the production of biogas from a composite of Lemon grass and Poultry droppings by studying the total gas pressure exerted on a sealed near-cylindrical plastic container used as a digester. The composite pre-fermented substrates were mixed with water and the formed slurry was digested for a month. The temperature was kept relatively constant by lagging the digester with fiberglass wool. The exerted stress on the digester by the produced biogas was determined using a tri-axial quarter-bridge strain gage rectangular rosette, carefully fixed to the external surface of the plastic digester. Subsequently, the total pressure exerted on the wall of the digester container was determined. The daily gas production potential in form of computed pressure is presented. Results showed a maximum gas pressure of 31,200 pascal above atmosphere produced from the composite over the period of four weeks. The research demonstrated that pressure changes at relatively constant volume can be used to monitor gas production.

1. Introduction

Biogas energy production is a clean energy technology. Pollution caused by animal manure especially in developing countries can be recycled and reduced. The digestion of organic materials by micro-organisms produces biogas and production of the biogas can be very small in laboratory scale experiments [1]. The level of biogas measured may be used to predict the state of degradation in anaerobic digestion. Unfortunately, several researchers working on biogas in laboratories in most developing countries lack access to sophisticated gas measuring equipment. Accurate measurement of the produced gas pressure could be very difficult task. Advanced research and accuracies of data may be limited especially where gas volumes are to be measured conventionally [5]. Small perturbation factors such as micro changes in temperature and pressure may affect measurements of biogas yield. Most calibrated volume measuring devices may not be sensitive enough to detect micro changes in biogas volume.

1.1. Bubbles and pressure in biogas

Bubbles are formed from within the slurry when microbes digest it. Whenever we see bubbles rising in the slurry, we literally say gas is being formed. The formation of bubbles require very high internal pressure. From the fundamental principles of bubbles formation in biogas production, more bubbles implies more gas accumulation and hence more pressure in the digester. That may suggest to us that direct measurement of the pressure activity inside the digester tells more about biogas formation rather than volume displacement measurement [6]. Gas will always occupy any available space or volume. For an ideal gas it is specified as 22.4 liters at STP and serves as a reference to convert gas pressure to volume.The formation of the bubbles is due to excess pressure 'p_{excess}' inside it as described by the Laplace's law.

$$p_{excess} = {2S}/{r} \quad (1)$$

S = surface tension, r = bubble radius. Let 'ρ' be the density of the medium and 'g' gravity, then (Bubble external pressure) = (static pressure of liquid at bubble height h) + (pressure in the digester)

$$p_{ext} = h\rho g + p_{digester} \quad (2)$$

Bubble is formed within the slurry when, the Internal Pressure of bubble 'p_{int}' equals Excess pressure 'p_{excess}' plus External pressure of bubble 'p_{ext}'

$$p_{int} = {2S}/{r} + p_{ext} \quad (3)$$

New bubbles are formed with extremely high internal pressure because they will have a smaller radius 'r' making 2S/r very large.

1.2. Reasons to measure pressure
Pressure measurements and controls is important in digesters because it can influence key processes such as liquid-vapour equilibrium, fluid dynamics and the rate of chemical reaction.A well designed pressure measuring method can be used for comparison of biodegradability caused by microbial activities in biogas production using composite materials [3]. The mechanical Bourdon tube pressure gauge is simple with excellent sensitivity for higher pressure applications but its slow response to change in pressure and hysteresis makes it unsuitable for low pressure application.

1.3. Research aims
The study here is intended to, develop the method for measuring total biogas in a plastic digester; by using strain gage rosette to detect changes in biogas production pressure. A set of three (3) strain gauges (rosette) were attached to the surface of the digester, to measure strains in different directions, providing precise evaluation of the surface stresses, from which pressure is deduced.
The study described helps to overcome volumetric measurement insensitivity problem at low volume or pressure. It may not be as simple as volumetric measurement but it could be more informative regarding to micro changes in biogas production per second per minute per hour on a daily bases.The study also aims at measuring the production of biogas, using a mixture of chicken droppings and *Cymbopogoncitratus* (lemon grass) as feedstock in a plastic bio digester, since co-digestion of both digestible material produced high gas quantity from the digester.Calibration of the designed automated meter is easy but not discussed here and can give more precise measurement of micro changes in biogas production.

2. Materials
The biomass used were lemon grass and chicken droppings. The lemon grass was obtained from Covenant University, while the chicken droppings were collected from a poultry farm around Lagos. They were mixed homogenously.Other materials used for the study were the following:
- A twenty (20) litre size water dispenser plastic bottle as anaerobic digester.
- Strain gauge sensorBF350 (3 pieces) (Figure 1) and temperature sensor LM 35 (2 pieces).
- Bridge amplifiers HX711 (3 modules), as shown in Figure 2.
- Arduino microcontroller for data logging to PC, enabled through WiFi (Figure 2).
- Tyre tubes with valve used to cap the digester and Fibre glass wool for insulation (Figure 3).

IOP Conf. Series: Earth and Environmental Science **173** (2018) 012048 doi:10.1088/1755-1315/173/1/012048

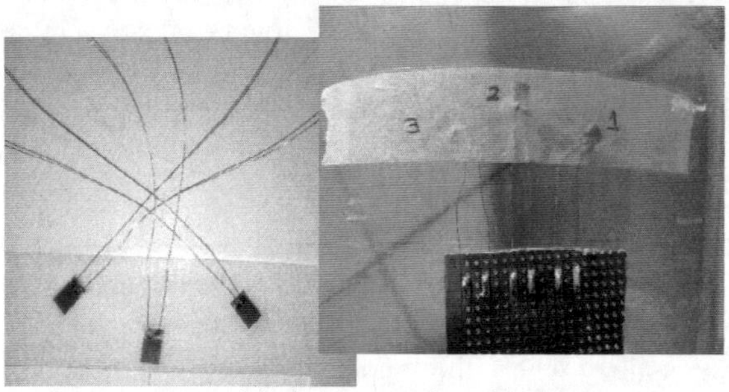

Figure 1: A set of strain gage sensors assembled at an angle of 45 degrees apart (LHS) and glued to the plastic digester (RHS)

.

Figure 2: Strain gage bridge amplifiers on board with ADC, microcontroller and Wifi customised for datalogging.

Figure 3: (LHS) Digester plastic before loading it with feedstocks. Attached to the lower side are the strain gages and temperature sensors which are connected to the data acquisition board. (RHS) Insulated digesters capped with tyre tube during data acquisition.

The digester structure primarily is a plastic almost cylindrical bottle strong enough to withstand weight and gas pressure of the slurry. The tank was loaded with the feedstocks (70% chicken droppings and 30% lemon grass). The slurry was allowed to fill about half of the digester space leaving half height of the digester as space for the producedbiogas. The digester was sealed air tight, wrapped with fibre glass wool as thermal insulator and left for thirty (30) days to digest while measurements were taken (Figure 3).The experiment was carried out at the laboratory of Covenant University Centre for Research and Innovative Development.

3. Method

Affixed with glue to the digester are a set of three strain gage sensors whose resistance varies with applied force and also converts pressure into a change in electrical resistance which can then be measured. Each strain gage signal from the BF350 sensor is connected in a Wheatstone bridge configuration, amplified (x60) and the bridge voltage is finally converted from analog to digital signal using a 24-bit ADC converter on a BF711 moduleon-board [7], [8]. The digital signals are collated in an Arduino microcontroller and sent to PC using a Wifi module. Two (2) temperature sensors type LM35 data signals were also processed in the same microcontroller and sent to PC. The data retrieved from PC is shown in a spreadsheet format (Table 1).

3.1. ADC_count to Bridge output voltage to Strain value

The bridge output voltage 'e' was scaled from the ADC_count using the fact that, a fixed Bridge excitation voltage 'Vin' (= 3.9V) will result to an ADC count of 2^{24} after it has been amplified with a Voltage gain (=60).

The Gage factor 'Ks' of the strain gage is Ks=2.0, and Strain 'ε', can be determined from the expression

$$\varepsilon = \frac{4}{K_s} * \frac{e}{V_{in}} \tag{4}$$

3.2. Measured Strain to Principal Strain Conversion

The signal from each strain gage is given as ε1, ε2, and ε3 respectively [2], [4].

The maximum principal strain is given as,

$$\varepsilon_{max} = \frac{1}{2}\left\{\varepsilon_1 + \varepsilon_3 + \sqrt{2|(\varepsilon_1 - \varepsilon_2)^2 + (\varepsilon_2 - \varepsilon_3)^2|}\right\} \tag{5}$$

Minimum principal strain is,

$$\varepsilon_{min} = \frac{1}{2}\left\{\varepsilon_1 + \varepsilon_3 - \sqrt{2|(\varepsilon_1 - \varepsilon_2)^2 + (\varepsilon_2 - \varepsilon_3)^2|}\right\} \tag{6}$$

Direction of principal strain from the ε1-axis is given as,

$$\theta = \frac{1}{2}\tan^{-1}\left[\frac{2\varepsilon_2 - \varepsilon_1 - \varepsilon_3}{\varepsilon_1 - \varepsilon_3}\right] \tag{7}$$

3.3. Principal Strain to Principal Stress Conversion

Let 'v' be the Poisson's ratio (=0.3) and 'E' the Young's modulus (=30*10^6psi),

Maximum Principal Stress is,

$$\sigma_{max} = \frac{E}{2(1-v^2)}\left\{(1+v)(\varepsilon_1 + \varepsilon_3) + (1-v) * \sqrt{2|(\varepsilon_1 - \varepsilon_2)^2 + (\varepsilon_2 - \varepsilon_3)^2|}\right\} \tag{8}$$

Minimum Principal Stress is,

$$\sigma_{min} = \frac{E}{2(1-v^2)}\left\{(1+v)(\varepsilon_1 + \varepsilon_3) - (1-v) * \sqrt{2|(\varepsilon_1 - \varepsilon_2)^2 + (\varepsilon_2 - \varepsilon_3)^2|}\right\} \tag{9}$$

3.4. Stress to Pressure Conversion

For the thin-walled cylindrical digester of thickness't'(=9.466E-4meter) and radius 'R' (=12.32E-3 meter), the forces exerted by biogas in the hoop or circumferential direction will result to a pressure 'p', given as

$$p = \frac{t}{R} * \sigma_{max} \tag{10}$$

And the forces exerted by biogas in the longitudinal direction will result to a pressure of

$$p = \frac{2t}{R} * \sigma_{min} \tag{11}$$

The average of both pressures in equation 10 and equation 11 was used. Although if the bottle is isotropic and have uniform thickness all over, both pressure values will be equal.

4. Results and Discussion

4.1. Results

There was a progressive increase in the production of biogas within the period of thirty (30) days(Figure 4). A third degree cubic curve fits the data set with a correlation coefficient of 0.9885 and an equation $p = 1.2932t^3 - 90.748t^2 + 2655.3t - 16.864$

(12)

Where 'p' denotes the pressure in Pascal and 'd' is the day number starting from the first day of digestion as day 1 onward.

Physical observation of the graph in figure 4 reveals; a steep rise in gas production from day 1 to 11, followed with a slower steady rise in gas production between days 19 to 30. There was a slowdown in gas production between 12 and day 17.

Mathematically, equation 12 can be shown to be 'non-stationary point of inflection and rising'.

The first derivative of equation 12 gives 'p''

$$p' = 3.8796t^2 - 181.496t + 2655.3 \tag{13}$$

Itssecond derivative gives 'p'''

$$p'' = 7.3592t - 1811.496 \tag{14}$$

The 2[nd]derivative is zero (p''=0)at t=24.66 days which indicates the day of minimum gas production. The 1st derivative (Figure 5)gives the rate of gas productionthe day (t=24.66)as p'=538.88 Pascal/day. The gas pressure on day t=24.66 was p=29,669 Pascal.

From the temperature profile graph (Figure 6), the digester temperature consistently stayed above its ambient temperature.Cubic polynomialswere fitted to show the relationship between the ambient temperature '$T_{\_ambient}$' and that of the digester '$T_{\_digester}$'. The fitted trend lines showed a smooth divergence in temperatures overthe period of digestion. It signifies that bacterial activities are increasingly active during the period of gas production and most likely uncompleted. Equation 15 and equation 16 are the derived from the fitted function. The correlation of fit for the digester function is $R_{\_digester}{}^2$ (= 0.7827) and that of the ambient temperature function is $R_{\_ambient}{}^2$ (= 0.4405).

$$R_{\_digester}{}^2 = 0.7827 \text{ and } T_{\_digester} = 0.0017t^3 - 0.0713t^2 + 0.8743t + 26.131 \tag{15}$$

$$R_{\_ambient}{}^2 = 0.4405 \text{ and } T_{\_ambient} = 0.0013t^3 - 0.0534t^2 + 0.6429t + 26.015 \tag{16}$$

$R_{digester}{}^2$ (=0.7827) being greater than $R_{ambient}{}^2$ (=0.4405) indicates less temperature perturbations in the digester which is the effect of insulating the digester with Fibre glass wool.Both temperatures attained their peak on the eight day (30.0°C and 29.3 °C).

Both the digester temperatures and ambient temperatures fluctuate in a correlated pattern except that the amplitude of fluctuations of the digester is less than that of the ambient.

Table 1: A sample spreadsheet format of the data retrieved from the data-logger showing the tables of strain gage and temperature measurements every second.

Day	Hr	Min	Sec	ADC_1	ADC_2	ADC_3	T_digester	T_ambient
24/04/2018	19	21	30	1583114	1974904	1928112	26.43	25.22
24/04/2018	19	22	30	1580761	1975889	1927975	26.43	25.22
24/04/2018	19	23	30	1577929	1975203	1926529	26.43	25.22
24/04/2018	19	24	30	1576531	1975770	1926375	26.43	25.22
24/04/2018	19	25	31	1573728	1974779	1925197	26.43	25.22
24/04/2018	19	26	31	1573409	1975276	1925430	26.43	25.22
24/04/2018	19	27	31	1569824	1974253	1924486	26.43	25.22
24/04/2018	19	28	31	1568387	1974584	1924726	26.43	25.22
24/04/2018	19	29	31	1567205	1973470	1923760	26.43	25.22
24/04/2018	19	30	32	1566577	1973756	1924140	26.43	25.22
24/04/2018	19	31	32	1565398	1972676	1923145	26.43	25.22
24/04/2018	19	32	32	1564705	1973110	1923605	26.43	25.22

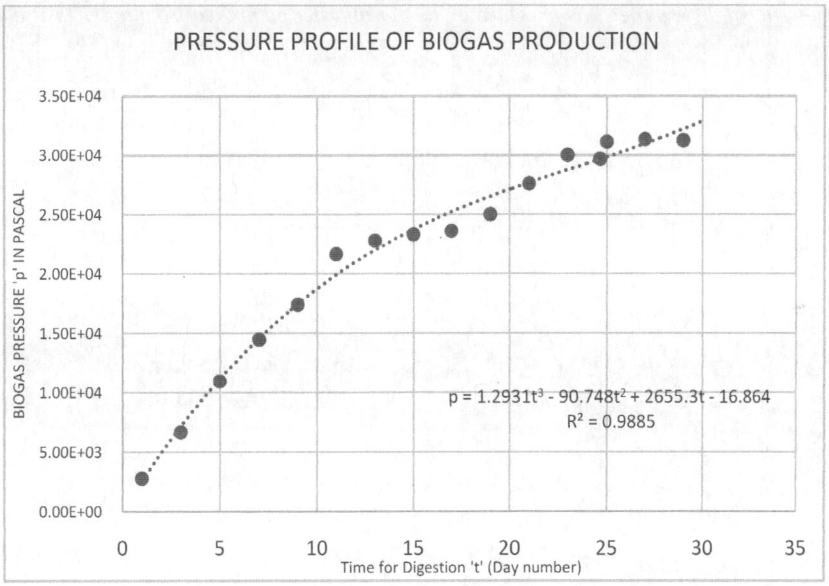

Figure 4: Shows the daily average gas pressure in the bio-digester computed from strain gage measurements.

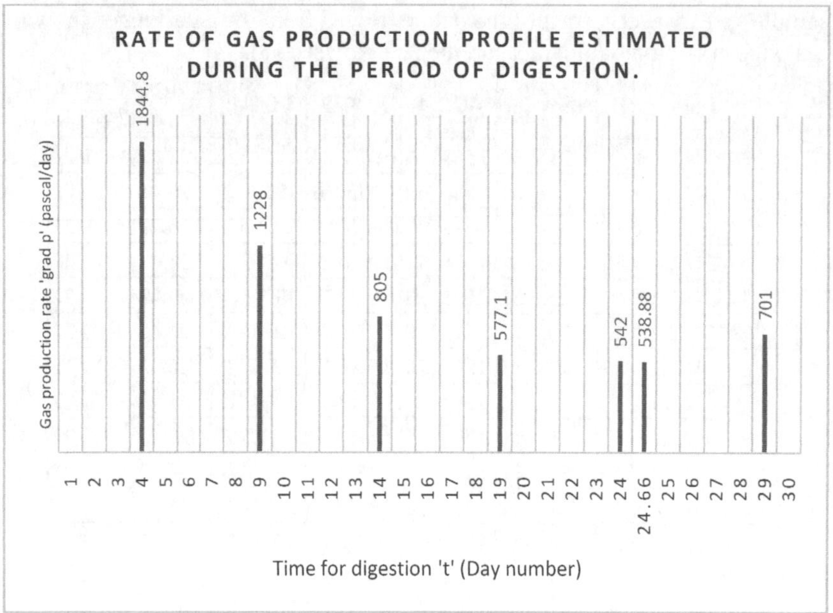

Figure 5: Shows the daily average gas pressure in the bio-digester computed from the pressure derivatives.

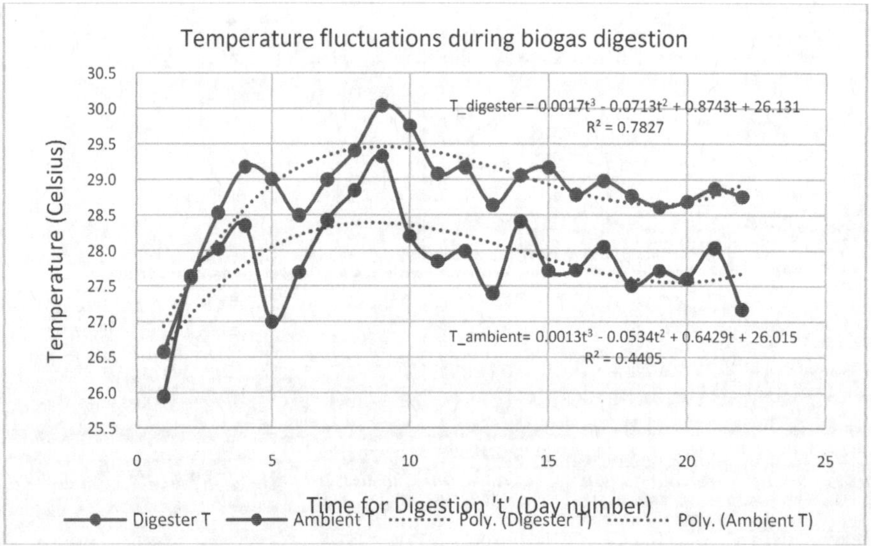

Figure 6: Temperature variations of the digester and that of its ambient showing the effect of insulating the digester with Fibre glass wool.

4.2. Discussion

Production of biogas from the composite of Lemon grass and Poultry droppings was studied using pressure computed from strain gage measurements of stress on the walls of the digester. The digestion for a period of one month inside a twenty litre plastic digester produced gas up to a pressure of 31.2 kP.

The sensitivity of using strain gage with instrumentation for measurement of biogas production especially at low gas volume or pressure was established by the perfect correlation of the data set within day 0 to day 10 of Figure 5. The biogas production pressure-profile of Figure 5 also showed three (3) distinct phases of biogas production corresponding to; aerobic digestion (days 1-11), end of aerobic to start of anaerobic digestion (days 12-17), and thirdly the anaerobic digestion (days 19 onward). As shown in Table 1, frequent changes in strain (hence pressure changes) are accurately captured every second because of the sensitivity of this method rather than conventional volume measurement. The temperature profile also could be used to predict the progress of the digestion if well insulated and compared with that of the ambient environment.

References

.

[1] "A simple device to measure biogas production in laboratory scale digesters," Biotechnology Letters,Vol. 8, pp. 719–720, 1986.

[2] Beer, F. P. and Johnston Jr., E. R. (1981) Mechanics of Materials. McGraw-Hill, New York, pp. 1-70, pp. 325-330.

[3] F. Tagliapietra, M. Cattani, L. Bailoni, S. Schiavon, "In vitro rumen fermentation:Effect of headspace pressure on the gas production kinetics of corn meal and meadowhay," Animal Feed Science and Technology,Vol. 158, p.p197–201, 2010.

[4] Figliola, R. S. and Beasley, D. E. (2000) Theory and Design for Mechanical Measurements. 3rd edition, Wiley, New York, pp. 432-438, pp. 292-293.

[5] M. Walker, Y. Zhang, S. Heaven, C. Banks, "Potential errors in the quantitative evaluation of biogas production in anaerobic digestion processes," Bioresource Technology, Vol. 100, p.p 6339-6346, 2009.

[6] M. K. Theodorou, B. A. Williams, M. S. Dhanoa, A. B. McAllan, J. France, "A simple gas production method using a pressure transducer to determine the fermentation kinetics of ruminant feeds," Animal Feed Science and Technology, Vol. 48,p.p 185-197, 1994.

[7] Perry, C.C., "Data-Reduction Algorithms forStrain Gage Rosette Measurements", Experimental Techniques. May, 1989, pp. 13-18.

[8] Users Manual for DAQ 6023E/6024E/6025E, National Instruments, Inc., Austin, Texas.